Seismic Interpretation:
The Physical Aspects

Nigel A. Anstey

being a record of the short course
The New Seismic Interpreter

Springer-Science+Business Media, B.V.

Copyright © 1977 by Springer Science+Business Media Dordrecht
Originally published by International Human Resources Development Corporation. in 1977
Softcover reprint of the hardcover 1st edition 1977

All rights reserved. No part of this book may be used or reproduced in any manner whatsoever without written permission of the publisher.

Library of Congress Catalog Card Number: 77-86312

ISBN 978-0-934634-18-2   ISBN 978-94-015-3924-1 (eBook)
DOI 10.1007/978-94-015-3924-1

ACKNOWLEDGEMENTS

Thanks are expressed to the authors and companies who have allowed the use of their material in the course.

TABLE OF CONTENTS

PART 1 (THE PRE-COURSE NOTES): INTRODUCTORY REVIEWS

Introduction

1.1 Review of elementary signal theory

1. The decibel scale
2. Time and frequency domains
3. The Fourier series
4. The Fourier integral
5. The phase spectrum
6. Linear operators
7. The spike and the impulse response
8. Superposition and convolution
9. The complete Fourier scheme
10. Low-pass and high-pass filtering
11. Minimum phase
12. The power spectrum
13. The autocorrelation function
14. The cross correlation function

1.2 Review of traditional seismic interpretation

1. The geophysical interpretation
2. The geological interpretation
3. The new era

PART 2: SEISMIC WAVE PROPAGATION

2.1 Elementary acoustic wave propagation

2.2 The nature of seismic waves

1. Seismic waves
2. The physical picture
3. What to measure?
4. The household electrical analogy
5. Acoustic impedance
6. The equations

7. Particle velocity and seismic velocity
8. Amplitude and energy
9. A fun problem

2.3 Properties of earth materials relevant to seismic propagation, and their dependence on geological composition and history

1. Density
2. Normally and abnormally pressured sections
3. Factors affecting velocity in a solid earth
4. Factors affecting velocity in a porous earth
5. The overall picture of velocity and density variation with depth
6. The problem of velocity in porous gas-saturated materials

2.4 The partition of plane seismic waves at plane interfaces

1. Reflection of plane waves at normal incidence
2. Transmission of plane waves at normal incidence
3. Measurements with a borehole geophone
4. Measurements with a surface geophone
5. Some important examples
6. Reflection and transmission of plane waves at inclined incidence

2.5 The loss mechanisms

1. Absorption
2. Transmission coefficients and short-path multiples
3. Scattering from inhomogeneities
4. The overall picture

2.6 The determination of reflection coefficients

1. The top-and-bottom method

        2. Methods employing multiple reflections
        3. The method of comparison with known values
        4. Removing the losses

2.7 Interference between reflections

        1. Interference in the time domain
        2. Interference in the frequency domain
        3. The consequences of interference for our measurements

2.8 Spherical waves, curved reflectors, and diffraction

        1. Spherical divergence
        2. Absorption and scattering for a spherical wave
        3. Reflection and transmission at interfaces
        4. Curved reflectors and velocity lenses
        5. Diffraction

2.9 Post-depositional geological processes as modifiers of seismic propagation conditions

PART 3: SEISMIC SIGNAL MEASUREMENTS IN DETAIL

3.1 The measurements which may be made

        1. Measurements on a single trace
        2. Additional measurements on individual records
        3. Additional measurements on many single traces
        4. Additional measurements on a gather
        5. Additional measurements on a common-offset set
        6. The use of stacked traces

3.2 Corrections required before these measurements have geological significance

        1. Spherical divergence
        2. Amplitude correction for partial failure of identical source units
        3   Amplitude correction for partial failure of geophones

        4. Correction problems involving both amplitude and frequency content (stationary)
        5. Correction problems involving both amplitude and frequency content (time-variant)
        6. Amplitude correction for sea-floor transmission coefficient
        7. Corrections for time
        8. General matters of concern

3.3 Detailed studies of the attributes of a borehole-geophone signal

        1. Amplitude of direct arrival
        2. Pulse shape of direct arrival
        3. Amplitude of discrete reflection

3.4 Detailed studies of the attributes of a two-way reflected signal

        1. Consequences of the two-way nature of the path
        2. Style of displays
        3. Illustration of factors affecting reflection strength
        4. Illustrations of polarity estimation
        5. Illustrations of frequency content
        6. Change of frequency content

3.5 Velocity measurement

        1. For stacking
        2. Continuous interval velocities
        3. Detail studies

PART 4: DIRECT HYDROCARBON DETECTION

4.1 Introduction

4.2 The criteria for direct recognition of a gas accumulation

        1. The gas-liquid contact
        2. Anomalous reflection coefficients
        3. Anomalous low velocities
        4. Inversions of polarity

       5. "Shadows"
       6. Diffractions
       7. The inter-relation of these criteria

4.3 Inferences from published illustrations

4.4 Porosity estimation in gas reservoirs

       1. Using the top-and-bottom method
       2. From velocity, in the presence of both liquid and gas
       3. Using the top-only method

4.5 The limitations and validation of direct detection

       1. The limitations
       2. Methods of internal validation

4.6 Direct hydrocarbon detection in older rocks

PART 5: INDIRECT HYDROCARBON DETECTION

5.1 Indirect applications of direct techniques

       1. Shallow gas as an indicator of deep hydrocarbons
       2. Sea-floor seeps
       3. Recognition of a permeable path to the basin margin
       4. Differential mineralization

5.2 Seismic stratigraphy

       1. Review of the objectives
       2. Reflection amplitude and continuity
       3. Angular relationships
       4. Sequences
       5. Eustatic cycles and basin dating
       6. The consequences for indirect hydrocarbon location

PART 6: MODELLING

Introduction

6.1 Structural modelling
6.2 Bright-spot and stratigraphic modelling
6.3 Combined structural and stratigraphic modelling
6.4 Modelling the cdp gather

PART 7: MIGRATION

7.1 Introduction
7.2 The classical technique for time migration
7.3 The choice of migration velocity
7.4 How much sophistication is worth while?
7.5 Approaches to three-dimensional migration

7.6 Depth conversion

    1. The importance of selecting the data
    2. The need for smoothing
    3. The smoothing of stacking velocities
    4. The smoothing of interval velocities
    5. The derivation of depth-conversion velocities

PART 8: SPECIFIC INTERPRETATION PROBLEMS

8.1 The incorporation of borehole and other data

    1. Discordance between stacking velocities and check-shot velocities
    2. The synthetic seismogram
    3. The incorporation of sparker data

8.2 Misties between surveys

    1. Survey and feathering problems
    2. Polarity conventions
    3. Processing problems
    4. Other problems of reflection shape

8.3 The special problems of land data

    1. Just differences
    2. Statics
    3. The correction of amplitudes

8.4 Enhancing the section for particular objectives

        1. The folly of compartmentalizing the seismic chain
        2. Problems of poor signal-to-noise ratio
        3. The importance of display
        4. Multiples, reverberations and ringing

8.5 Delineation with fewer wells

        1. The geophysicist remains involved after the discovery well
        2. The seismic expression of the target zone
        3. The resolution of faults
        4. The general search for anomalies associated with the target
        5. The delineation possibilities of a mobile source and a borehole geophone

8.6 The prediction of drilling hazards

        1. Hydrocarbon seeps
        2. Buried river channels
        3. Gas pockets
        4. Unconsolidated shale

8.7 The application of seismic work after a field is in production

        1. Estimating recovery factors conventionally
        2. The seismic contribution to the type of drive
        3. The gas reservoir
        4. The oil reservoir
        5. Limitations to be expected
        6. Summary

PART 9: ASPECTS RELATED TO EXPLORATION MANAGEMENT

9.1 Methods of conveying the essence of an interpretation to management

        1. Contour maps and their problems
        2. The conventional fence diagram
        3. The sculpted fence diagram

9.2 Cost-effective exploration

    1. Cost-effectiveness in the total exploration scheme
    2. Cost-effectiveness in seismic exploration specifically

9.3 Questions that management should properly ask

    1. General questions
    2. Questions before a seismic survey is approved
    3. Questions before a well is drilled
    4. Questions after a discovery

PART 1

INTRODUCTORY REVIEWS

(THE PRE-COURSE NOTES)

INTRODUCTION

In this course we shall assume that all participants are familiar with the essentials of seismic prospecting. Thus the rudiments of the field work — spreads, sources, arrays and digital recording — are assumed known. So also are the rudiments of processing — such processes as gain recovery, filtering, deconvolution, velocity analysis, and display.

Just as important, we shall assume that all participants have some feeling for the realities of seismic work — in the field, under real conditions.

Elementary signal theory and the basic techniques of interpretation are also assumed known. However, for certainty, the following pre-course notes include sections reviewing basic signal theory, geophysical aspects of interpretation, and geological aspects of interpretation. These reviews are not intended to be comprehensive. Their function is solely to cover, with the minimum possible discussion, the essential features which will be assumed to be known in the course. None of the course time will be spent on the material of these pre-course notes. Participants are advised that they will not derive full benefit from the course if this background is not known.

Most course participants will be already familiar with this material, and will need to do little more than read it through.

If, before the course, any participant requires further discussion of signal theory in the same non-rigorous style, he will find it in other writings of the present author, particularly:

"Wiggles", Journal of the CSEG, December 1965, pp.13-43.
"Correlation Techniques-A Review"' Journal of the CSEG, December 1966, pp.55-82.
"The Sectional Auto-Correlogram", Geophysical Prospecting, December 1966, pp.389-411.
"Seismic Prospecting Instruments"' Gebrüder Borntraeger, vol.1, part 2, pp.14-49.

The type of treatment adopted in the course will not go beyond this graphical non-rigorous style. However, the reader may choose to be familiar with other treatments, and possibly with the more-rigorous mathematical approaches. Complementary treatments are given by:

Lindseth, R., "The Nature of Digital Seismic Processing"'
  Journal of the CSEG, December 1967;
Cruz, R.B., "Fundamentals of Predictive Filtering",
  Journal of the CSEG, December 1973, pp.12-26.

More mathematical treatments are given by:

Finetti, I., Nicolich, R. and Sansin, S., "Review of
  the Basic Theoretical Assumptions in Seismic Digital
  Filtering", Geophysical Prospecting, September 1971, pp.292-320.
Kanasewich, E.R., "Time Sequence Analysis in Geophysics",
  University of Alberta Press, 1973.

If, before the course, any participant requires further discussion of traditional seismic interpretation, he is fortunate in that two new books have just appeared to fill the void on this subject. They are "Applied Geophysics" by Telford, Geldart, Sheriff and Keys (Cambridge University Press, 1976) and "Seismic Reflection Interpretation" by Fitch (Gebrüder Borntraeger, 1976). An entertaining account of specific interpretation hazards is given by Tucker and Yorston in the SEG Monograph "Pitfalls in Seismic Interpretation".

Aspects of petroleum geology relevant to the interpretation problem are covered in readable form by Chapman in "Petroleum Geology" (Elsevier, 1973); course participants requiring a three-page summary of the main points are referred to Chapman's pages 170-172.

Unpublished citations will be available for reference on the course.

Numerals in the left margin of this course manual refer to slide numbers. The suffix L signifies the left projector; the suffix B signifies a blank.

Reference to tapes apply to session numbers on the videotapes.

1-2

1.1 REVIEW OF ELEMENTARY SIGNAL THEORY

1.1.1 The decibel scale

The dB scale is used for expressing the ratio of two quantities (for example, amplitudes A_1 and A_2).

$$\text{Ratio in dB} = 20 \log A_2/A_1.$$

A conversion graph is given on page 1-4.

Much-used values (some approximate) which are worth committing to memory:

Ratio -	dB	+ Ratio
1	0	1
0.9	1	1.1
$1/\sqrt{2}$	3	$\sqrt{2}$
0.5	6	2
0.3	10	3
0.1	20	10
0.01	40	100
0.001	60	1000

If the ratio is greater than 1 the decibel value is positive; if the ratio is less than 1 it is negative. When ratios are multiplied the decibel equivalents are added.

1.1.2 Time and frequency domains

A normal seismic trace, representing the seismic signal as a function of time, is in the time domain. The same information may be represented in the frequency domain; the bridge between the two is given by Fourier.

1.1.3 The Fourier Series

Any repetitive waveform, however complicated, may be viewed as the addition of sine (or cosine) waves whose frequencies are integral multiples of that of the basic repetition. We call the basic repitition the "fundamental", and the frequencies which are 2, 3, 4 .... times that of the fundamental we call "harmonics".

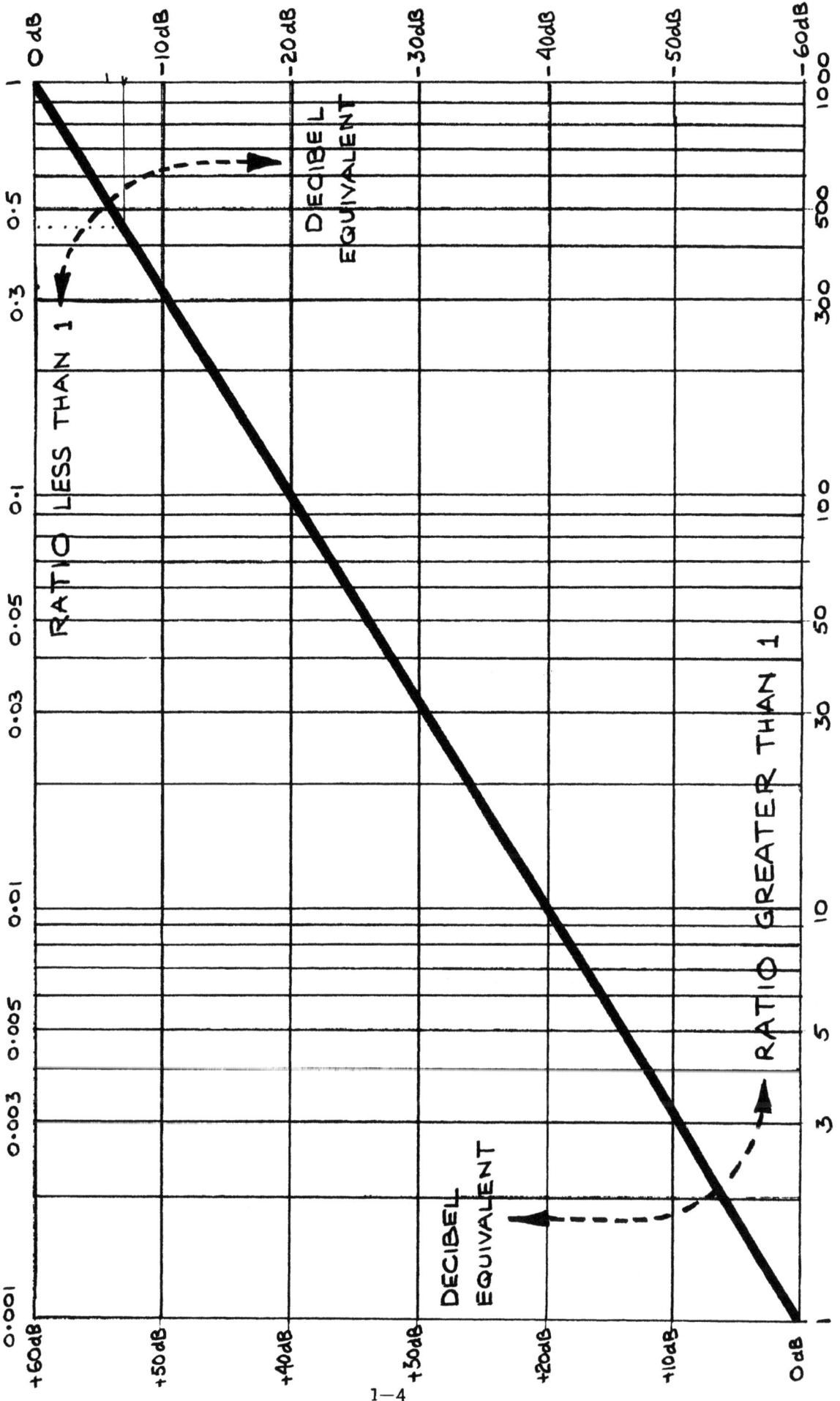

Figure 1.1.3A illustrates the effects which may be obtained by various combinations of two cosine waves having different amplitude, frequency and phase relationships.

The unique description of a waveform requires knowledge of the frequencies, amplitudes and phases of all components.

The summed or composite waveform is exactly equivalent to its components, and the components to their sum. Any linear operation performed on a waveform may equally well be regarded as performed on its components.

The frequency, amplitude and phase of all components can be represented by the amplitude and phase spectra. Figure 1.1.3B shows the line spectra corresponding to Figure 1.1.3A.

The amplitude and phase spectra, taken together, uniquely define their corresponding composite waveform. Conversely any repetitive waveform specifies a pair of spectra.

Thus there is an interchangeability between a waveform as a function of time and the two spectra showing amplitude and phase as functions of frequency. All effects concerned with repetitive waveforms can be considered in terms of the waveform itself, as a function of time (that is, in the time domain), or in terms of the amplitude and phase spectra (that is, in the frequency domain).

1.1.4 The Fourier Integral

If the components occur at infinitesimally small frequency spacing, the pattern repeats only after infinite time. Therefore a transient waveform, occurring only once, may be viewed as the addition of components which are harmonics of an infinitesimally small fundamental frequency.

The spectra of repetitive waveforms are line spectra; those of transient waveforms are continuous spectra.

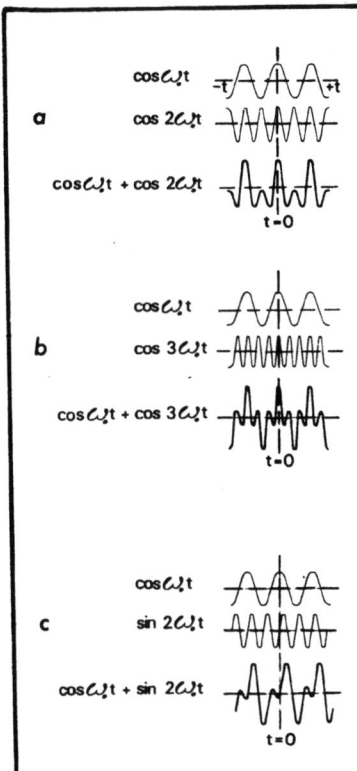
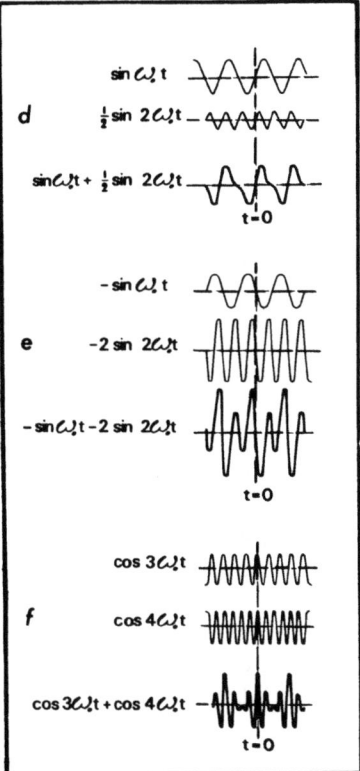

Figure 1.1.3A

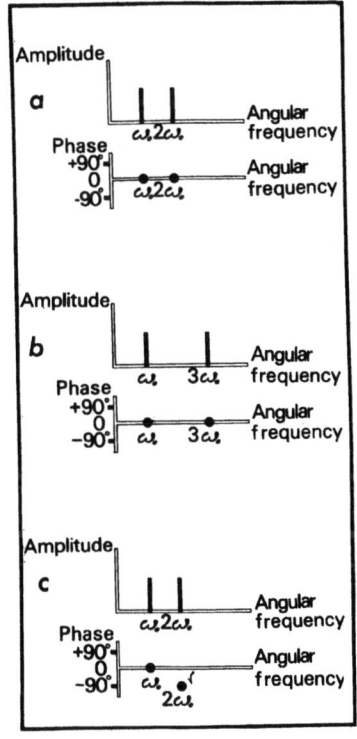
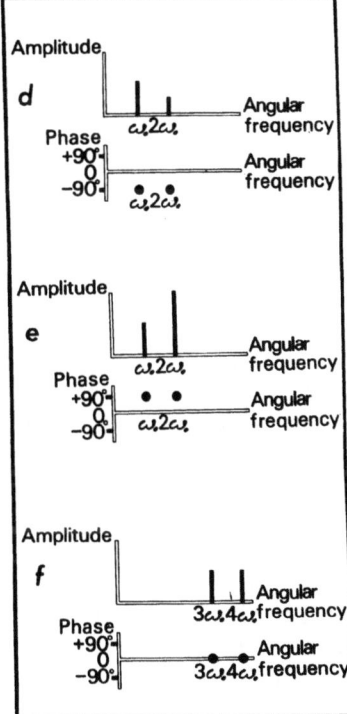

Figure 1.1.3B

The spectra of repetitive waveforms are line spectra; those of transient waveforms are continuous spectra.

Figures 1.1.4A and B illustrate the transition from line spectra to continuous spectra. They also illustrate the following statements.

The fine structure of a continuous spectrum defines the waveform at distant times.

The effective duration of a pulse cannot be less than the reciprocal of the bandwidth; short pulses must have large bandwidth.

Sharp shoulders on the amplitude spectrum prevent a pulse from dying quickly to zero.

Repetitive pulses have undulating amplitude spectra.

1.1.5 The phase spectrum

For cosine components, a zero phase spectrum produces a symmetrical pulse having the maximum amplitude and minimum duration allowed by the amplitude spectrum.

A phase spectrum which is a straight line passing through the phase axis at the origin or an integral multiple of 2π maintains the pulse shape given by zero phase, but imposes on the pulse a delay proportional to the slope of the line.

Any other phase spectrum, however, changes the shape of the waveform, reducing its peak amplitude, making it unsymmetrical, and dispersing it to occupy a greater time.

Figure 1.1.5A illustrates a typical amplitude and phase spectrum. Figure 1.1.5B illustrates the Fourier synthesis of a pulse from this amplitude spectrum, with zero phase and with the given phase spectrum.

The position of any individual component along the time axis is given by the phase angle, evident from the phase spectrum. However, the time at which such a component (and its neighbours) make their most obvious contribution to the corresponding transient waveform is dictated not by the actual value of the phase spectrum but by the slope of the phase spectrum.

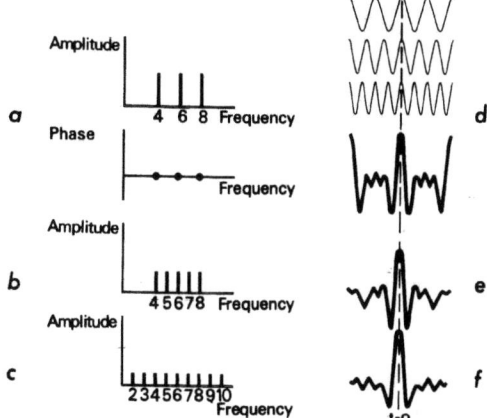

Figure 1.1.4A

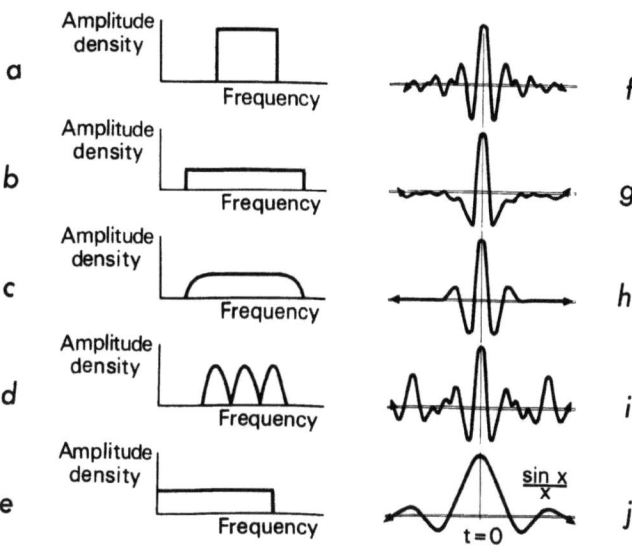

Figure 1.1.4B

1.1.6 <u>Linear operators</u>

An operator acts on an input waveform to produce an output waveform.

A linear operation is one whose action is the same whatever the magnitude or polarity of the input waveform.

The order in which a train of linear operations is performed may be changed without affecting the final result.

Figure 1.1.6A illustrates the concept of amplitude-frequency responses, for a weight hanging on a spring.

The behaviour of any linear system is uniquely specified by the combination of the amplitude-frequency response and the phase- frequency response.

To obtain the amplitude spectrum of an output we multiply the value of the amplitude spectrum of the input, for each frequency, by the value of the amplitude-frequency response, noting that if either term is zero the product is zero.

To obtain the phase spectrum of the output we add the value of the phase spectrum of the input, for each frequency, to the value of the phase-frequency response. In short, we multiply the amplitude curves and add the phase curves.

This is illustrated in Figure 1.1.6B.

The overall effect of a cascaded series of operators is obtained by multiplying all their amplitude-frequency responses and adding their phase-frequency responses.

1.1.7 <u>The spike and the impulse-response</u>

A unit impulse waveform (or spike) contains all frequencies equally (has a "white" amplitude spectrum). Figure 1.1.7A illustrates its properties.

The output of any linear system when a spike is used as the input is called the impulse-response of the system;

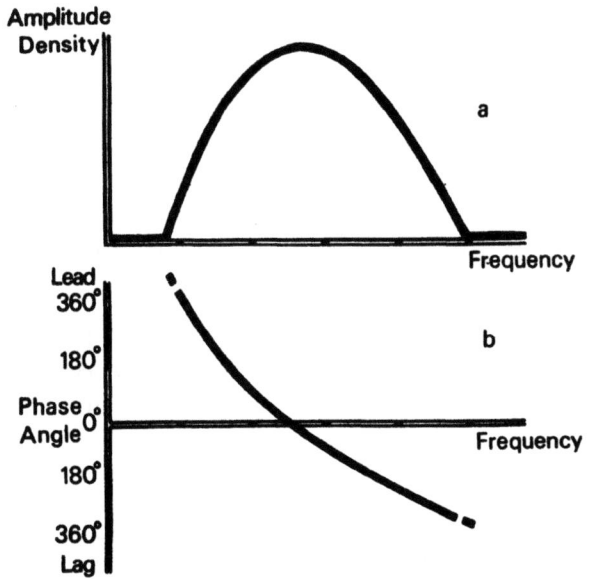

Figure 1.1.5A

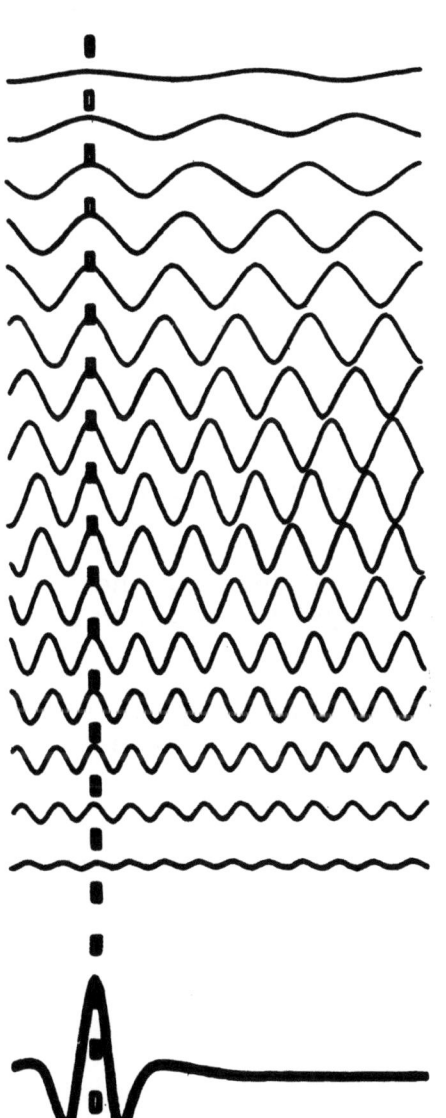 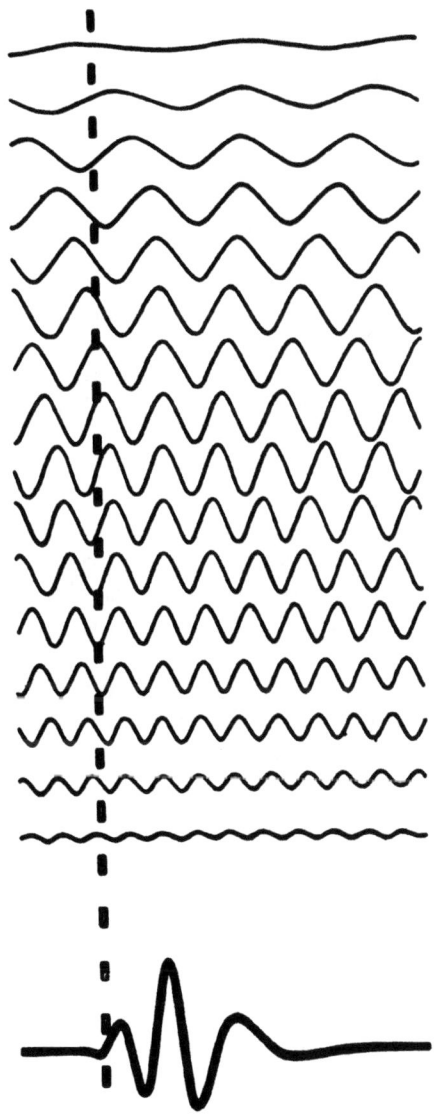

Figure 1.1.5B

1-10

it specifies the behaviour of the system uniquely. Figure 1.1.7B illustrates this.

The amplitude spectrum of the impulse-response of the system has the same form as the amplitude-frequency response of the system, and the phase spectrum as the phase-frequency response.

The two frequency-response curves and the impulse-response therefore represent alternative methods of specifying the behaviour of the linear system. The two methods are rigorously equivalent.

Broad-band (high-fidelity) systems have impulse-responses which are short and sharp; narrow-band systems have impulse-responses consisting of many cycles.

1.1.8 Superposition and convolution

For any linear system and for any input waveform, the output waveform is the superposition of all the impulse-responses obtained by regarding the input waveform as a succession of spikes.

In Figure 1.1.8A, b is the impulse response to input a. Curve d is the response to the two spikes c.

Figure 1.1.8B shows how the response b to a complicated input a is formed by superposition.

Physically, convolution is the same process as superposition.

For any linear system and for any input waveform, the output wave-form is the convolution of the system impulse-response with the input waveform.

The duration of the output waveform is always equal to the sum of the durations of the input waveform and the impulse-response.

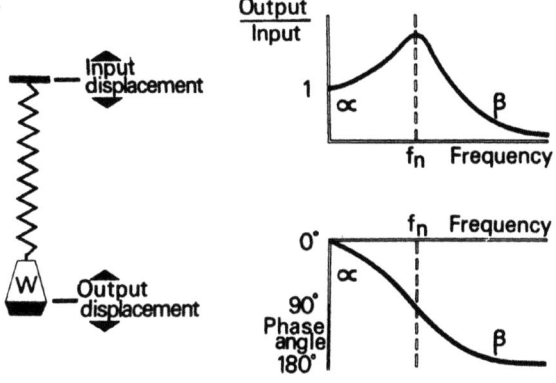

Figure 1.1.6A

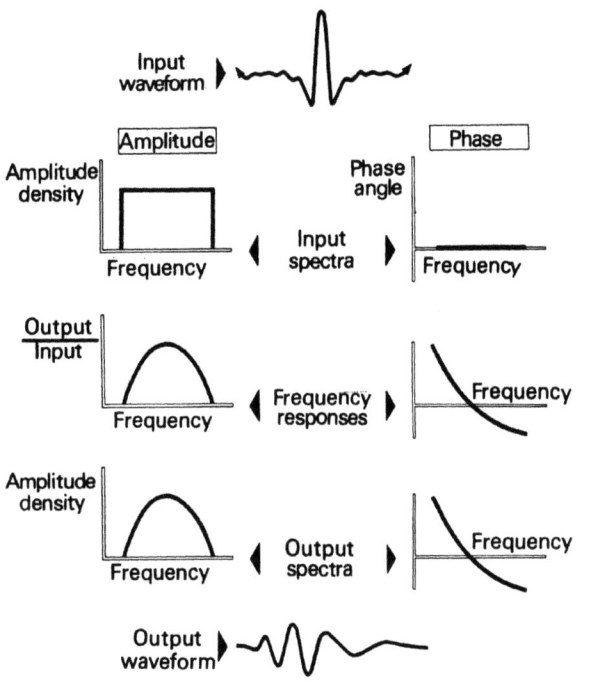

Figure 1.1.6B

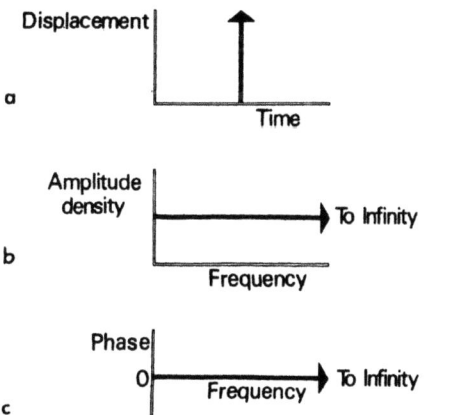

Figure 1.1.7A

1.1.9 The complete Fourier scheme

This is illustrated in Figure 1.1.9.

1.1.10 Low-pass and high-pass filtering

Low-pass (high-cut) filtering makes a pulse longer, and delays its maximum amplitude.

High-pass filtering adds to the number of oscillations in the pulse, decreases its amplitude stand-out, and reduces or removes any bias from the zero-line.

Integration and differentiation are examples of low-pass and high-pass filtering respectively.

1.1.11 Minimum phase

This is the phase characteristic of systems which produce a certain attenuation (a certain amplitude-frequency response) in the minimum time required by nature to effect such an attenuation.

The phase-frequency response of a minimum-phase system can be uniquely calculated from its amplitude-frequency response.

The convolution of two minimum-phase operators is also minimum phase.

The energy in a minimum-phase pulse is concentrated as far towards the front as the amplitude spectrum allows.

1.1.12 The power spectrum

For the present purposes, this is obtained by squaring the ordinates of the amplitude spectrum.

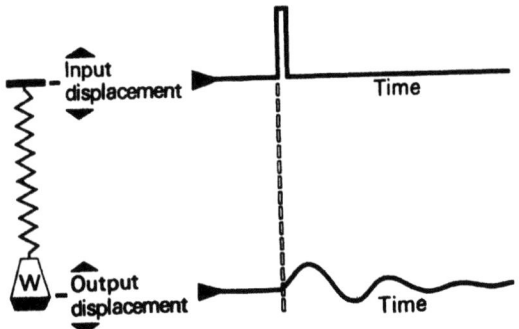

Figure 1.1.7B

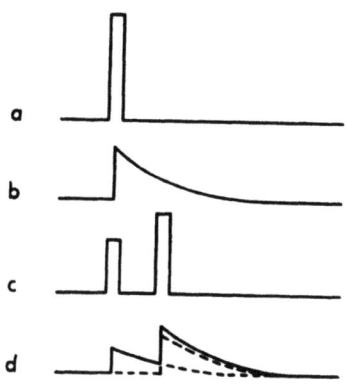

Figure 1.1.8A

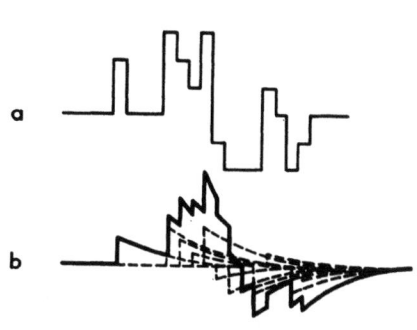

Figure 1.1.8B

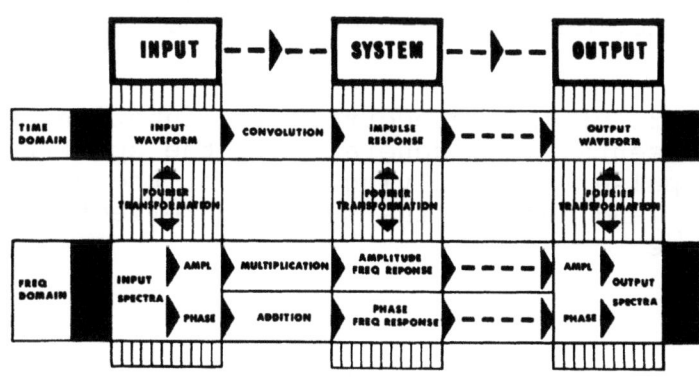

Figure 1.1.9

1-14

1.1.13 The autocorrelation function

The autocorrelation function is the zero-phase waveform obtained by adding all the components represented by the power spectrum. It is therefore symmetrical about its time origin, and does not have values higher than that at the time origin.

The autocorrelation function of a waveform is also a graph of the similarity between the waveform and a time-shifted version of itself, as a function of this time-shift.

Figure 1.1.13A illustrates the autocorrelation functions of a sine wave and of broad-band noise.

Figure 1.1.13B illustrates the construction of an auto-correlation function of a seismic pulse; the pulse a is kept fixed while a replica of itself slides past (as suggested by b-f), and for every shift position such as c all ordinates are cross-multiplied, and the products added, to give a single autocorrelation value appropriate to that shift.

For any portion of a seismic trace which may be regarded as the random superposition of many elementary pulses reasonably close to some average shape, the autocorrelation function of the complete portion may be regarded as the autocorrelation function of this average elementary shape, superimposed on a random low-amplitude background. This is suggested in Figure 1.1.13C, in which the upper trace has many seismic pulses at random time spacing, and the lower trace is the autocorrelation function of the upper trace.

Figure 1.1.13D shows at a and b the physical systems producing "ghost" or free-surface reflections, the corresponding seismic pulse doublet at c, and the corresponding form of the auto-correlation function (of the individual doublet or of the whole record) at d.

The central part of the autocorrelation function of a seismic trace represents, in a general way, the spectrum of the basic seismic pulse; the outlying parts of the autocorrelation function represent (and define) the multiple reflection activity present in the trace.

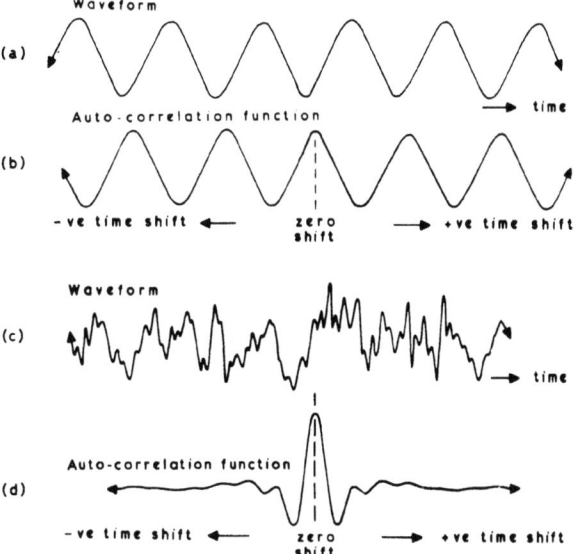

Figure 1.1.13A

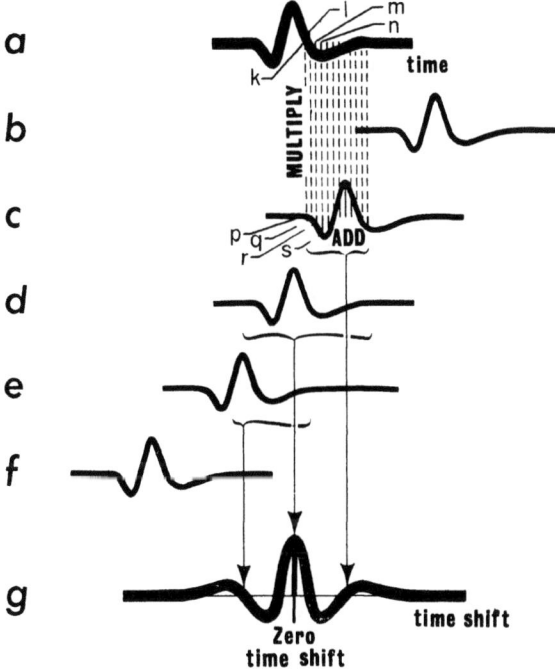

Figure 1.1.13B

Figure 1.1.13E shows the physical systems producing marine reverberations, with the corresponding seismic pulse trains and autocorrelation function. It illustrates that the autocorrelation of a reverberant pulse train may be viewed as the superposition of the autocorrelation of the basic pulse shape on the autocorrelation of the reverberant system. Train c and autocorrelation d represent the effect of a single reverberant system (in practice, particularly, the water-trapped part of a); train e and autocorrelation f represent the effect of both reverberant systems together (a plus b). Both cases are important; in the course they are termed trapped and reflected reverberations.

Figure 1.1.13F relates, for the two cases, the ratio A_1/A_0 with the product R of the upper (sea surface) and lower (sea floor) reflection coefficients. It is valid only when the reverberation time is long enough to allow substantial separation of the pulse-shape autocorrelations (that is, when the reverberation time is at least twice as long as the basic seismic pulse). The curve for reflected reverberations may be approximated (within 2% up to $|R|=0.6$) by $A_1/A_0 = 2R/(1 + R^2)$.

If non-reverberant noise is added to the pulse train, then (as suggested by Figure 1.1.13A, c and d) the amplitude A_0 is increased while the amplitude A_1 is unchanged. The presence of noise therefore decreases the estimate of reflection coefficient obtained from an autocorrelation function.

1.1.14 The cross-correlation function

The cross-correlation function of two waveforms is a graph of the similarity between the two waveforms as a function of the time shift between them.

This is illustrated in Figure 1.1.14. The function is computed by multiplying and summing, just as for the autocorrelation function. The figure illustrates the value of the cross-correlation function for the determination of static time shifts.

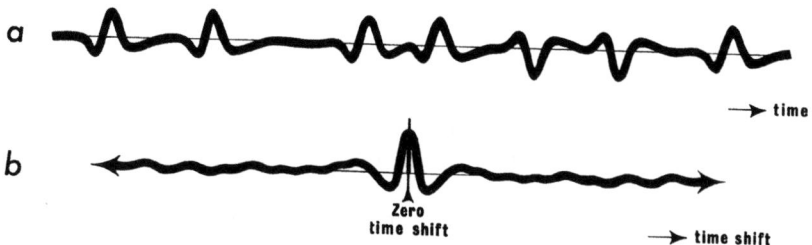

Figure 1.1.13C

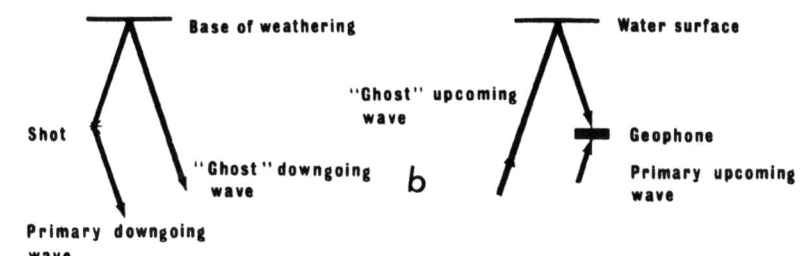

Figure 1.1.13D

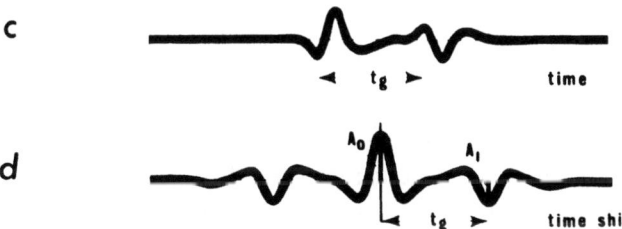

1-18

The crosscorrelation function of two waveforms contains only those frequencies common to both waveforms.

Correlation techniques are most powerful when the waveforms being correlated consist of many cycles.

1.2 REVIEW OF TRADITIONAL SEISMIC INTERPRETATION

1.2.1 The geophysical interpretation

This normally starts from stacked sections, according to the following sequence:

1. Inspection of the sections and test data for proper processing (for example, mute patterns, deconvolution operator lengths and windows, filters); in particular, a careful check on the aptness of the picking, manipulation and interpretation of <u>velocities</u> for stacking and migration.

2. Inspection of the sections for geological plausibility and harmonious inter-relation.

3. Picking of the sections, typically by colouring a trough, for each significant geological marker; checking line ties and loop closures. Picking usually starts on dip lines, in the deepest part of the basin.

4. Interpretation and marking of the fault patterns on the sections.

5. Digitization of time values for each picked horizon, at a suitable horizontal interval.

6. Posting of the time values and fault positions on a map.

7. Contouring of the time values (by machine, by hand, or by both). In this the geophysicist is careful not to view the contouring as a mere exercise in joining together points of like time, but as an attempt to represent his mental three-dimensional view of the <u>surface of a real geological solid</u>. To aid him in this he adopts several rules, of which the first is paramount:

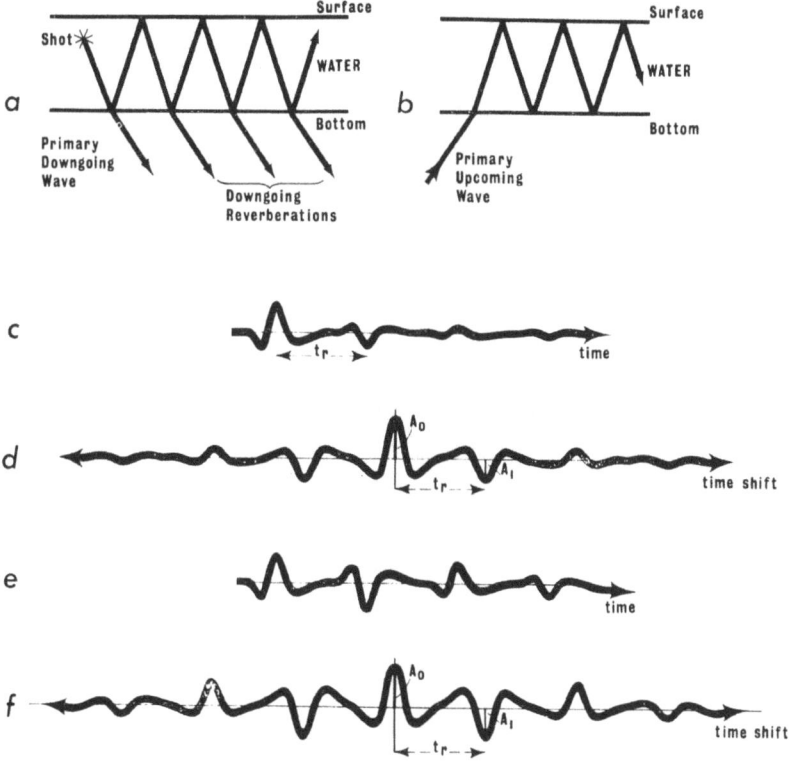

Figure 1.1.13E

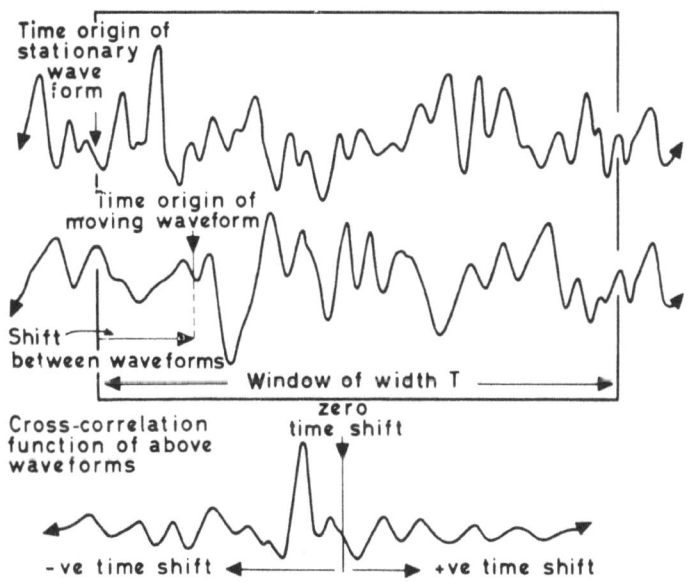

Figure 1.1.14

- He recognizes the trends. He establishes regional dip, searches for reversals, and then seeks the geological rationale for trends in these anomalies (whether folds, faults or reefs).

- He contours from dense data and simple geology toward sparse data and the complications.

- He is suspicious of a closed high within a low (except for peripheral sinks around piercements).

- He is suspicious of closed lows on top of a high.

- He looks twice at a low trend running toward a high.

- He is wary of like contours which run parallel over considerable distances.

8. Producing the contour map for each selected horizon, indicating highs and lows.

9. Repicking of the velocity analyses at the levels of the contoured horizons; conversion of the stacking velocity (usually assumed to be rms) to average velocity.

10. Smoothing, contouring and re-smoothing of the average velocity values, for each horizon, until they are judged to be geologically plausible.

11. Construction of depth maps from time and velocity maps.

12. Construction of time and depth isopachs, and construction of profiles and maps illustrating situations at different geological periods.

13. If the dips are appreciable but the time-picking is clear, migration of the contours; if the sections are too confused to allow positive picking, migration of the time sections first.

Throughout these procedures of traditional interpretation the prime concern is structure. Some lithologic studies may be attempted (through interval velocities), but the preoccupation is with the mapping of the structure.

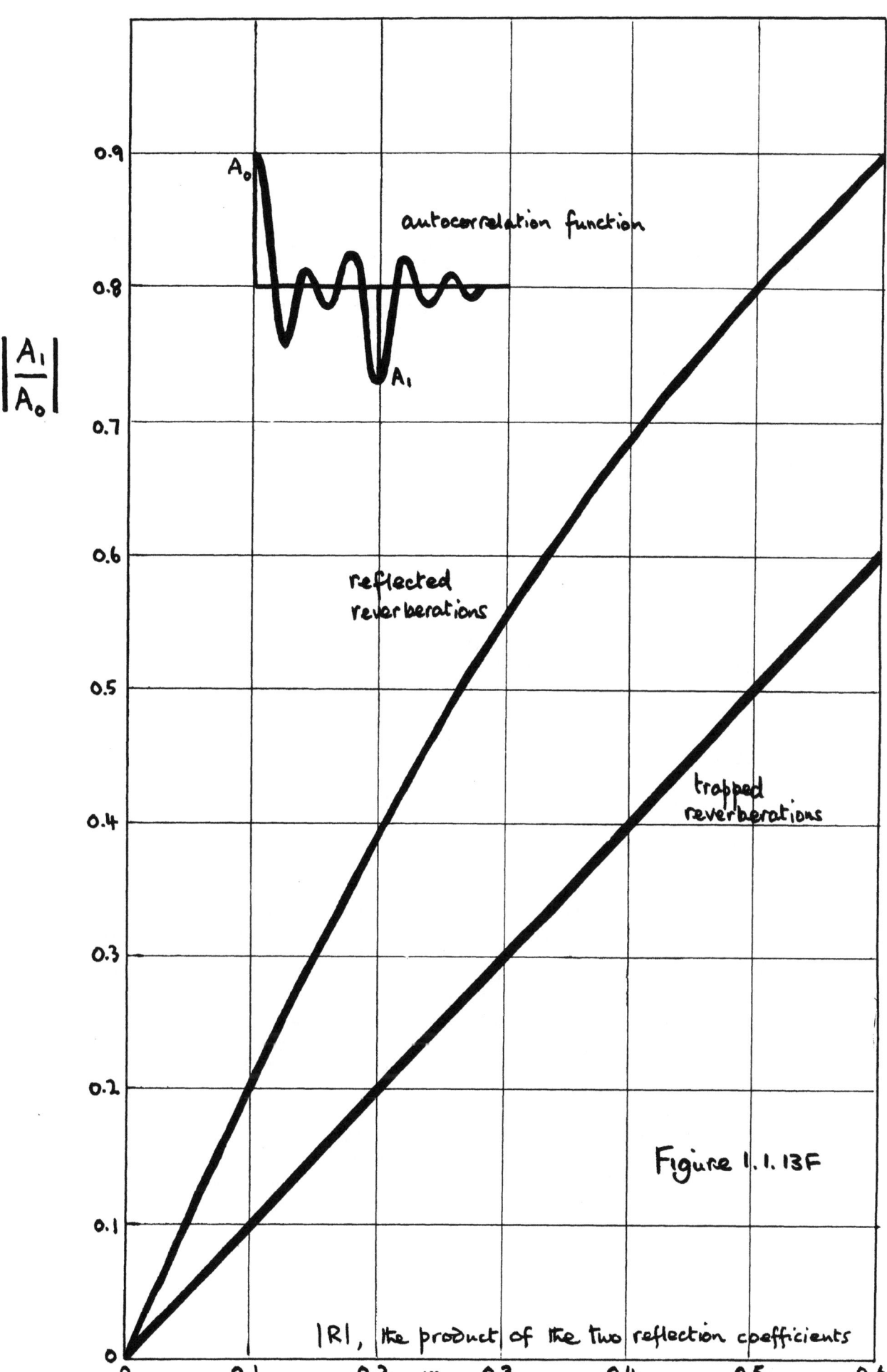

Figure 1.1.13F

1.2.2 The geological interpretation

This is concerned with the additional steps of assessing the geological chances of hydrocarbon accumulation in the structures indicated by the geophysical interpretation. In part, this is a matter of reconstructing the geological history, and of using the reconstruction to identify the probable lithology. The interpreter is particularly concerned with the following matters:

1. Identification of transgressive and regressive sequences, and of basin margins. The classical transgressive and regressive situations (land area decreasing and increasing, respectively) are illustrated in Figure 1.2.2A, and representative seismic expressions of such situations in Figure 1.2.2B. In general, the interpreter will be expecting structural accumulations associated with regressive sequences, and stratigraphic with transgressive.

2. Identification of growth structures. Ideally, the interpreter seeks to reconstruct the pattern of forces acting in past geological time — for example, the vertical forces due to density differences of upper and lower materials and the lateral forces due to crustal movements — and to reconcile this pattern with the evidence of diapirs, normal faults and reversed faults.

3. Identification of source rocks and reservoir rocks. Source rocks are likely to be fine-grained, containing organic debris, and deposited at considerable distance from the basin margin. One circumstance suggesting this would be a blanket sheet of even-layered nearly-parallel continuous reflections, conforming to the depositional topography with little evidence of thinning or onlap on the highs (Figure 1.2.2C). Reservoir rocks are likely to be coarse-grained sands deposited from a high-energy environment near a basin margin, or carbonates laid down (or built-up) in seas of suitable depth, temperature and biological condition.
The effort is therefore to identify from the seismic sections the basin conditions giving rise to each of these situations.

4. An understanding of the fluid potential field in past geological time, as a guide to likely primary and secondary migration. Although the details are not yet understood, the interpreter accepts the likelihood of primary migration

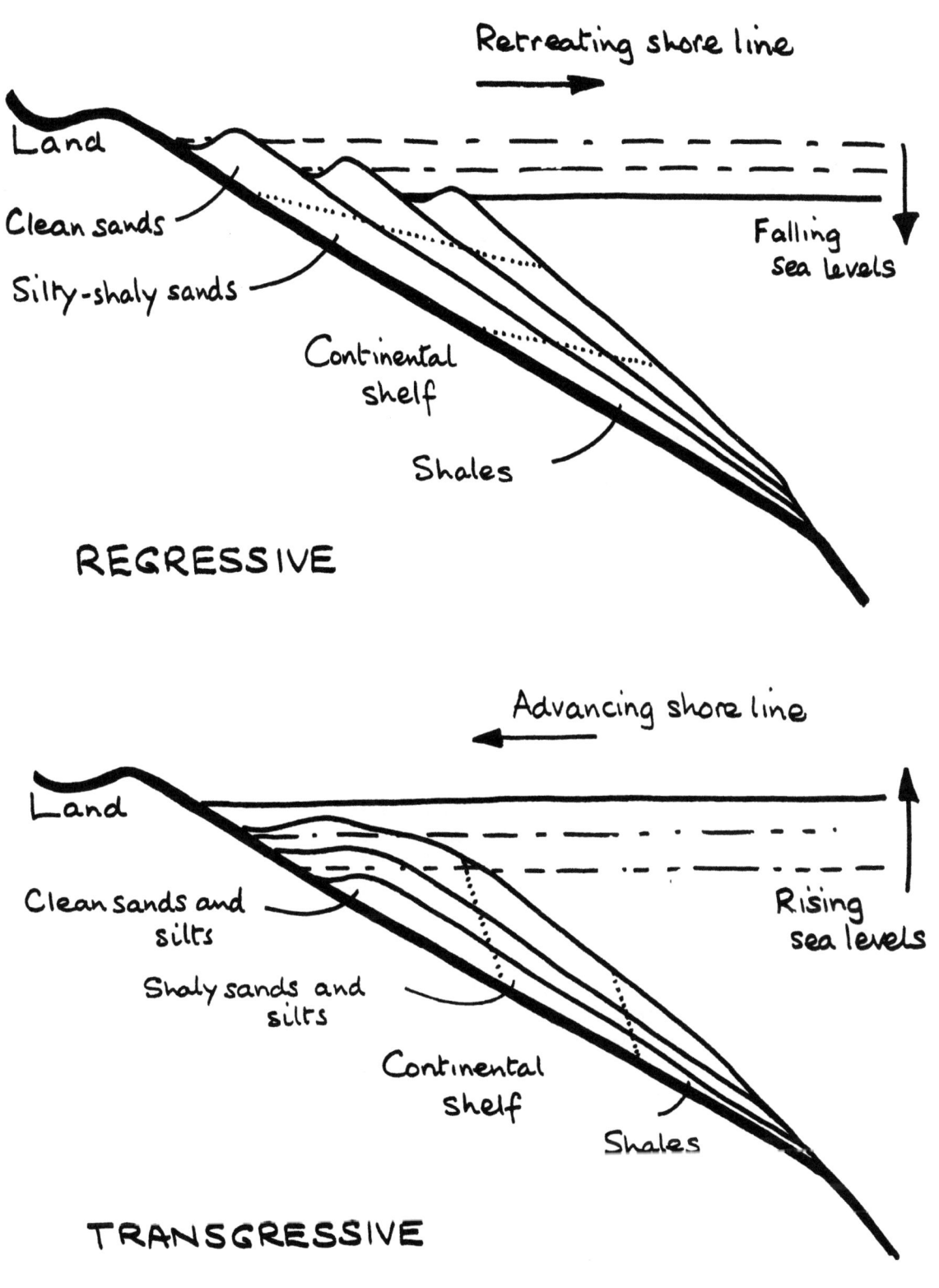

Figure 1.2.2A after Pirson

of petroleum by the agency of water squeezed out of buried
rocks under compaction, and so is concerned to establish
the likely direction of flow out of rocks believed to be
source rocks. For secondary migration, he is concerned
to identify a reservoir rock, a structural or stratigraphic
trapping situation, and a potential seal. He particularly
wishes to find a situation where the migration, trapping
and sealing processes occurred simultaneously, in growth
structures of one kind or other.

Following these interpretation stages, a prognosis is
prepared. This gives the geologist's best guess as to the
rock types likely to be encountered, and the geophysicist's
best guess (passed on to the driller with tongue in cheek)
as to the dips, the layer thickness, and the depth of the
target.

The result of all this, of course, is usually a dry
hole.

1.2.3 The new era

For more than forty years geophysicists and geologists
have been saying that they really should get together.
However, each has had plenty of problems of his own to
occupy him. What actually happened was that the geophysicist
continued to formalize his craft — until it could be
represented by unambiguous field specifications, instrument
specifications, computer programs and processing specifications.
In so doing he reduced himself to the status of
technician. Only in the picking of the sections and the
hand contouring was any of the old magic left — and that,
of course, was because they involve geological judgements.
The critical hydrocarbon judgements were made by the
geologist, and the task of the geophysicist was to feed him
structural information in a substantially hands-off manner.

Today there are some prospects where the geologist is
unnecessary; the geophysicist can identify the hydrocarbons
positively, and can estimate reserves, without help from

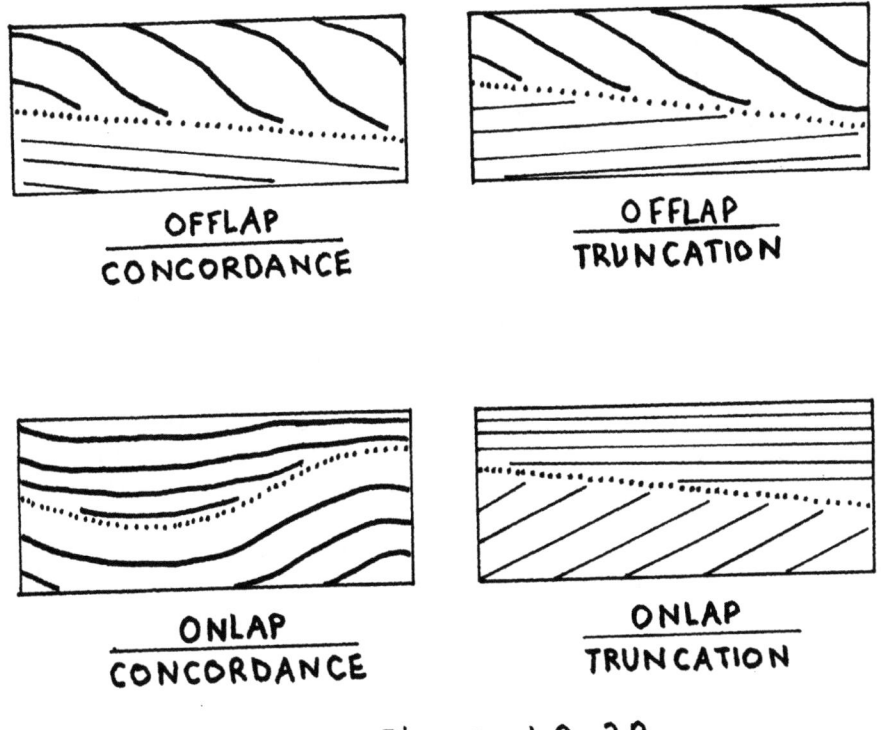

Figure 1.2.2B

(Figures after Sangree et al., 1974
Vail et al., 1975)

Figure 1.2.2C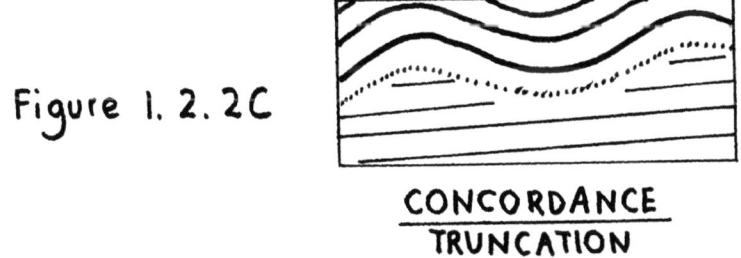

either geologist or driller. On the other hand, there are some prospects where the geophysicist can do little more than he did traditionally, and must still feel apologetic about the quality of his data. The general case, however, is now squarely between the two; the decision about hydrocarbon prospects is a joint decision between geophysicist and geologist.

The change lies in the geophysicist's new-found ability to transcend his preoccupation with <u>structure</u>; he begins to measure rock properties. The geophysicist who does this in isolation from geology will make himself a laughing-stock. The geologist who ignores the geophysicist's new power will make himself out-moded. And the company who does not compel the two to work together will make itself uncompetitive.

All of which constitutes the reason for the course. We look forward to an enjoyable and stimulating time together.

PART 2

SEISMIC WAVE PROPAGATION

2.1 ELEMENTARY ACOUSTIC WAVE PROPAGATION

Let us start with a little physics.

Before the geologists among us groan and prepare to close their ears, we should add very quickly that when the physicists talk physics is the time for the geologists to be monitoring very carefully. Left to themselves, the physicists will set up a physical model, will solve for its seismic response, and will say, "Look — here is the response of the earth." This is dangerous; what the physicists really have is the response of a model of the earth — which model may not be geologically realistic or even geologically possible. The distinction is not an idle one (as we shall see illustrated later in the course), and the geologists among us will perform a useful function if they will monitor our physics for its geological plausibility.

And this monitoring is equally important whether the physicist is considering the geometry of the layers or the physical properties of the layer materials.

Our discussion starts, then, with the generalities of acoustic propagation (of which seismic propagation constitutes a subdivision).

Acoustic waves transfer energy from one place to another by the agency of particle movement. A pressure is applied, the particles move, and the moving particles in their turn exert a pressure. Thus we may visualize a room with two doors, one of which is open and the other almost closed; the operation of slamming the first door causes the second to open — an acoustic wave has travelled across the room. The applied force on the first door causes the air particles to move, and to be locally compressed; the compression propagates across the room, and is capable of exerting a force on the second door.

The basic scientific principles which apply to accoustical situations are easy to accept. Important among them are these:

- Conservation of energy: in an acoustical system, the input energy is either transmitted, or reflected, or scattered, or converted into another wave type, or absorbed into heat; energy is neither created nor destroyed.

- The equations of motion (Newton's laws): these are the usual equations relating the force acting and the movement produced.

- Maintained contact at an interface: two materials in contact remain in contact as an acoustic wave passes across the interface — a vacuum does not appear between them.

- Reciprocity (White, 1960): in general, acoustic wave paths are reversible, so that source and receiver can be interchanged without changing the effects of the path.

- Linearity (section 1.1.6): in a linear acoustical system, waves can cross without affecting each other (like water waves on a pond, or speech heard against background music); as the waves cross, the resultant motion at any time is the simple sum of the component motions.

These basic foundations, representing the underlying principles for the solution of acoustical problems, are therefore quite simple and satisfying.

The details of the solutions, of course, are sometimes mathematically formidable. However, in this course we shall not be concerned with deriving mathematical solutions. In general, we shall first identify the basic physical principles (as we have just done), then establish a physical picture of the system, and finally just state and accept the mathematical solutions which are known to apply. We shall take the view that it is important for us to have a physical picture of the system in our minds, important for us to know the corresponding equation (which alone can enable us to make our approach quantitative), important for us to know the assumptions which limit the applicability of the equation, but not important (in the context of this course) for us to be able to derive the equation.

In general, the transfer of energy by an acoustic wave involves millions and millions of particles; the dimensions of any practically-realizable acoustic wave are large relative to those of a particle. In the case of an acoustic wave in air (for example, music) the "particle" is obviously a gas molecule; in the case of an acoustic wave in a heterogenous solid (for example, a seismic wave in a rock) it is less easy to be specific about the nature of a particle, but it is still clear that the relative dimensions are such that individual particles are not seen by the wave. However, large numbers of individually small inhomogeneities can cause a gross, smeared effect on the propagation; we know that we shall not be able to see the individual inhomogenities, but we are much interested to know whether their gross effect can be measured (and used to indicate their general presence).

Many phenomena exhibited by acoustic waves are most familiar to us by analogy; we think of acoustic "rays" in the sense of geometrical optics, and acoustic diffraction in terms of waves on the surface of water. The analogies are helpful, provided that we remember their limitations: an acoustic measurement is unlikely to be of _intensity_ (as in optics), the ratio of acoustic path length to wave length is unlikely to be of the order usual in optics, and any dispersion present in acoustic waves is of a nature quite different from that associated with water waves.

General reference: Kinsler and Frey

TAPE 2   2.2   THE NATURE OF SEISMIC WAVES

A seismic wave is an acoustic wave in a solid material — normally a rock. As with other wave types, it represents a propagating interchange between kinetic and potential forms of energy; the kinetic form represents the motion of the particles and the potential form represents the effect of the inter-particle (or "elastic") forces.

We should develop the physical picture a little. Let us take a big block of some representative earth material — in which the particles are in an undisturbed, evenly-spaced condition — and let us give a simple in-and-out motion to the left-hand edge.

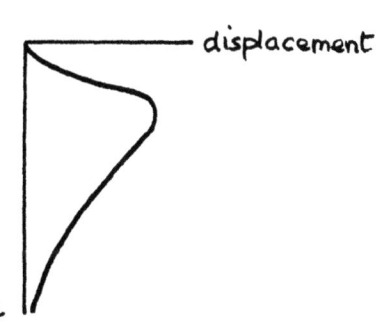

Let us say that we move the edge briskly to the right, and then, more slowly, we return to the original position; the displacement as a function of time is as sketched here:

At the moment that the left-hand edge comes back to its original position we <u>freeze</u> the motion — we imagine a photograph showing where all the particles are (p.2-6, top). The original compression has moved away to the right, and a rarefaction has formed behind it as the left-hand edge is pulled back.

We can plot the <u>displacement of the particles</u> from their original rest positions, as a function of the distance x (p.2-6, second sketch).

We can also plot the degree of compression of the particles (called excess pressure, or acoustic pressure, or sometimes just pressure) as a function of the distance x (p.2-6, third sketch).

2-4

On the basis of this plot, why should the disturbance propagate? Why does the region of compression not just cancel the region of rarefaction?

It is true that if the pressure measurement specified the whole system, the positive and negative pressures would equalize. But the particles are _moving_; they have kinetic as well as potential (compression) energy. This is evident if we plot the velocity of the particles (_particle velocity_) as a function of the distance x (p.2-6, bottom sketch).

In the case we are considering (a plane compressional wave), the particle-velocity plot and the acoustic-pressure plot have the same form. Where the particles are most compressed they have a large positive (forward) velocity. Just in front of this the particles have a smaller displacement from their rest position, and also a smaller velocity. Therefore the particles behind are coming on faster than those in front, so that the particles become more bunched to the front. That is, the region of positive pressure — the compression — moves forward.

Behind the pressure peak the particle velocity is still positive, but less so. Therefore this region tends to be replaced by a rarefaction; that is, the rarefaction — the negative pressure — moves forward too.

Then if we repeat the exercise for the tail of the disturbance, we find that in the same manner the whole thing moves to the right. The region of high pressure propagates, and the region of low pressure propagates, without any cancellation between them; the _pulse_ propagates, without loss and without change of form, to the right.

What should we measure in a seismic wave? Particle displacement? Acoustic pressure? Particle velocity? Something else?

For complete specification of the acoustic system described, we need a _pair_ of quantities, and it is convenient to take _acoustic pressure_ and _particle velocity_.

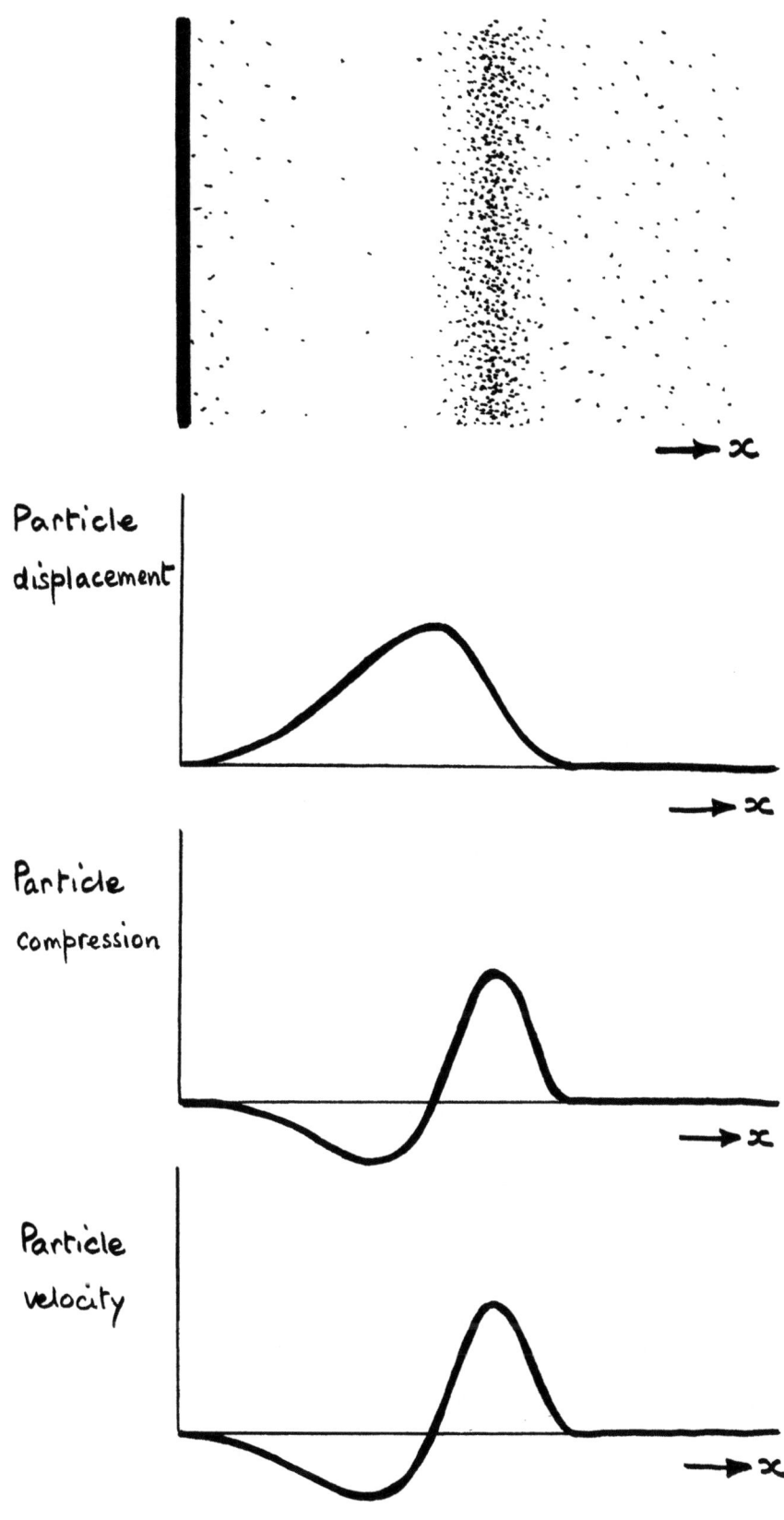

To see why we should need a pair of quantities, let us first seek help from a well-known analogy. For example, let us take the electrical analogy — perhaps the illustration of an electric coffee-pot — and let us review the variables which define the behaviour.

In the electrical case the basic variables are the _voltage_ of the supply and the _current_ drawn from the supply, the latter being decided by the resistance (or to be general the impedance) of our coffee-pot. The voltage represents an electrical _pressure_ (or potential); the current represents an electrical _flow_ (charge/second). The analogy is therefore easy to accept:

- The variable which acts like a _pressure_ is _voltage_ in the electrical case, _acoustic pressure_ in the acoustical case.

- The variable which acts like a _flow_ is _current_ in the electrical case, _particle velocity_ in the acoustical case.

- The variable which restricts the flow is _electrical impedance_ in the one case, _acoustic impedance_ in the other.

In the electrical situation, we also know that:

- If the household supply voltage is fixed (for example, 115 or 240 volts), small impedance produces large current flow (Ohm's law).

- The _power_ (the rate of generation of heat in the coffee-pot, or the ability to boil the water fast) is voltage times current (in watts).

- The _energy_ used (the total heat generated, or the work done) is power times time (in watt-hours).

- The _cost_ at the end of the month is for _energy_ (in kilowatt hours).

So two households — one on 240 volts and one on 115 — making the same amount of coffee in the same time must have coffee-pots of the same _power_. Further, the _energy_ used is the same, and if the price is the same the cost is the same. But the _impedance_ of the 240-volt coffee-pot must be higher than that of the 115-volt coffee-pot, in order for this to be true;

2-7

if the voltage (or electrical pressure) is higher, the current (or electrical velocity) must be lower, if their product is to be the same.

And so it is with acoustic impedance. Just as a material which conducts electrical energy has an electrical impedance, so a material which propagates acoustical energy has an acoustic impedance. Just as a fixed electrical voltage produces a large current in a material of low electrical impedance and a small current in a material of high impedance, so a fixed acoustic pressure produces a large particle velocity in a material of low acoustic impedance (such as a Quaternary clay) and a small particle velocity in a material of high acoustic impedance (such as a granite).

Now this is important, so we should be sure that we are completely comfortable with it. In a Quaternary clay a little pressure produces a lot of particle motion, and so the acoustic impedance is low. On the other hand, a granite can withstand a great deal of pressure with only a little particle motion, and so the acoustic impedance is high.

By the same token, if we visualize two acoustic wave packets of the same _energy_, one propagating in clay and one in granite, then we know that the acoustic pressure must be small in the clay and large in the granite; the particle velocity, on the other hand, must be large in the clay and small in the granite. This is very important.

How could we convey a feeling for acoustic impedance to someone without a physical background? Probably we would call it acoustic _hardness_. It is not quite what the geologist infers from the ring of the rock under his hammer — not quite that sort of hardness; neither is it quite the sort of hardness which we estimate in looking at the erosional behaviour at outcrop. But it is of this general type of _hardness_ property. As we shall see later, it is not easy for a geophysicist to take a sample of rock in his hand and to estimate its _velocity_ from the "feel" of the rock; the property which he is better able to estimate by this means is the _acoustic impedance_.

2-8

In terms of the other seismic properties of a rock, the acoustic impedance is the product of seismic velocity and bulk density.

The equations which formalize this physical view are as follows:

If seismic velocity	=	V
and rock density	=	ρ,
then acoustic impedance r	=	ρV (for plane waves).
If acoustic pressure	=	p
and particle velocity	=	v,
then pressure p	=	rv
	=	ρVv.
If the acoustic intensity I	=	the average flow of energy over unit area in unit time,
then intensity I	=	$½p^2/r$.

(The acoustic intensity I is equivalent to power in the electrical analogy; the electrical engineers ordinarily eliminate the $½$ in the expression $½V^2/R$ by considering rms values.)

It is most important that we have clear in our minds the distinction between <u>seismic</u> velocity and <u>particle</u> velocity. Seismic velocities, <u>of course</u>, are expressed in thousands of metres per second, and represent the <u>speed</u> at which energy is transported. Particle velocities, on the other hand, are expressed in millionths of a metre per second, and represent the <u>size</u> (not the speed) of the seismic disturbance.

Thus many of us will have stood by a Vibroseis vibrator and watched the baseplate — at the low frequencies one can actually follow the baseplate motion by eye, and assess the magnitude of its vibration. At 10 Hz this peak-to-peak displacement might be as much as 1 or 2 cm, from which we may compute that the particle velocity of the baseplate (and hence of the soil in direct contact with it) is of the order of 1 m/s. This is quite different (both in magnitude and in meaning) from the seismic velocity of the wave leaving the baseplate on its way into the earth; this seismic velocity, of course, is likely to be of the order of 1500-2500 m/s.

The signals received by our geophones, obviously, have even smaller particle velocities. The first break on a near geophone, for land work using dynamite, might well be 10^{-3} m/s, while a deep reflection might have a particle velocity of only 10^{-8} m/s. (While we have this example before us, let us note that the <u>displacement</u> of the geophone corresponding to this particle velocity is only about 10^{-10} m, which is less than one thousandth of the wavelength of visible light.)

Only very close to a charge of dynamite does the particle velocity compare to the seismic velocity. If it exceeds the seismic velocity, we have the situation described by the aircraft engineers as "supersonic"; this is a very lossy and non-linear situation which we always seek to minimize.

A normal land geophone measures particle velocity. A normal marine streamer hydrophone measures acoustic pressure. Provided the plane-wave assumption is justified (which it substantially is, in this context and at normal reflection times), the particle velocity measurement and the acoustic-pressure measurement have the same form, as indicated on p.2-6; therefore normally-shot land surveys should tie with normally-shot marine surveys. (In practice they may not, as we all know; however, the reasons for this are concerned with factors other than the distinction beween the two types of measurement.)

At this stage let us inject an aside about terminology — specifically about the terms "amplitude" and "energy". When the interpreter uses the term "amplitude" he signifies the magnitude of the acoustic pressure if he is working a marine survey, or the magnitude of the particle velocity if he is working a land survey. These are precise and satisfactory terms. However, when the English-speaking interpreter uses the term "energy", he often does so loosely, meaning nothing more than "event", or "arrival", or some unspecified measure of size. This loose practice really should be stopped. Energy is a precisely-defined scientific term (power x time); if the term is misused we find that when we want to measure energy we have no word for it.

Some of the concepts of this section can be illustrated by considering the problem sketched on the opposite page. In the upper sketch, a bar is struck at the two ends simultaneously with equal blows from two hammers. Equal disturbances propagate along the bar towards each other. Provided that the blows are not so intense as to make the propagation non-linear, the two disturbances cross each other and continue on their way, and, at the instant at which they cross, the resultant disturbance is the simple sum of the two equal individual disturbances. All is well.

In the lower sketch, the arrangement at the right-hand end is as before. However, the left-hand end is modified so that the disturbance generated is a rarefaction, rather than a compression. This negative disturbance propagates along the bar as before. At the instant when the two disturbances cross, they must clearly cancel. <u>At this instant, what has happened to the energy in the bar?</u>

We must define our measurement more precisely — "disturbance" is too vague. The pulses illustrated are clearly of acoustic pressure; it is true that the pressure at the instant of crossing is double in the upper sketch, zero in the lower. But we need a pair of measurements to specify an acoustical system. Let us draw the particle-velocity measurements below the pressure pulses illustrated. Let us say that at the right-hand end the particle-velocity pulse has the same form and polarity as the pressure pulse. Then at the left-hand end of the upper sketch we must draw an inverted particle-velocity pulse, since the pulse is travelling in the opposite direction; a compression is always a compression, whichever way it is travelling, but a compression which moves particles to the "front" of the disturbance must involve particle velocities which are to the left in the left-going wave and to the right in the right-going wave. The same argument means that the particle-velocity measurement on the right-going pulse of the lower sketch must be positive. It follows that at the instant of crossing the particles in the upper sketch are intensely compressed, but not moving; in the lower sketch they are moving fast, to the left. In one case the energy is concentrated in potential form, in the other in kinetic form; the total energy in the bar is, of course, constant at all times.

2-11

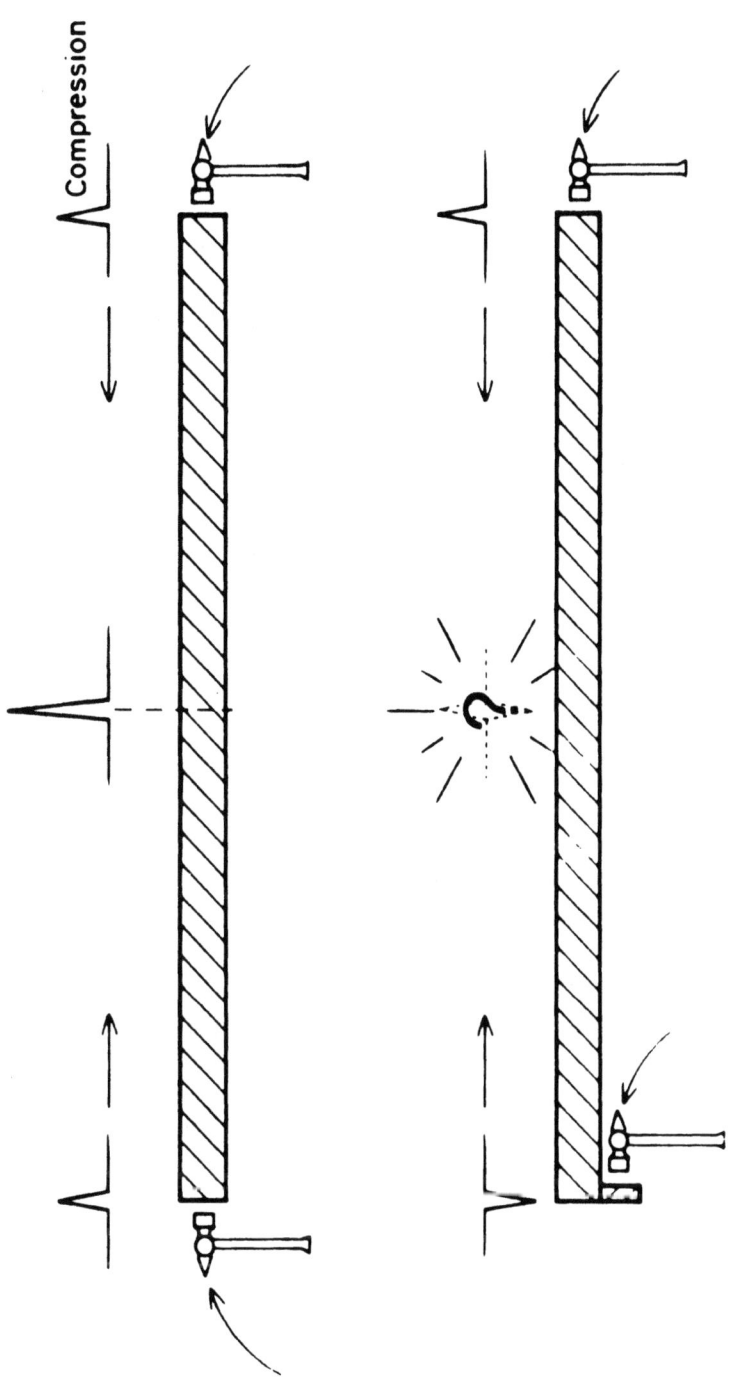

2-12

2.3 **PROPERTIES OF EARTH MATERIALS RELEVANT TO SEISMIC PROPAGATION, AND THEIR DEPENDENCE ON GEOLOGICAL COMPOSITION AND HISTORY**

As we all know, we can often determine seismic interval velocities (increasingly accurately, with modern techniques and well-placed care). In this course we shall learn how to determine acoustic impedances, under favourable conditions. It follows that we can determine the densities, as the ratio between acoustic impedance and velocity. Consequently we need to study how these three properties — velocity, acoustic impedance and density — relate to the type of rock and the condition of the rock.

Again we search first for a satisfying physical picture.

2.3.1 Density (Rieke and Chilingarian, 1974; Dapples, 1972)

What relation can we expect between density and depth?

We go back to the very simplest depositional model, wherein the products of land erosion are brought down to the sea by rivers; the sediments are deposited (directly or indirectly) with the coarse-grained sediments tending to be close to land and the finer-grained clays, silts and oozes tending to be further from land.

We consider the fine-grained sediments first, since in terms of volume they far exceed any other sedimentation; 75-80% of the earth's sediments are clays and shales. At the time of deposition, the water content by volume of a clay may be 60% or more. However, the density of this mix exceeds that of the simple water it replaces, and so there is a compacting effect on the clay deposits below the most recent sediments. This compaction squeezes out the water from the deeper deposits, upwards back into the sea. As more clay is deposited on top,

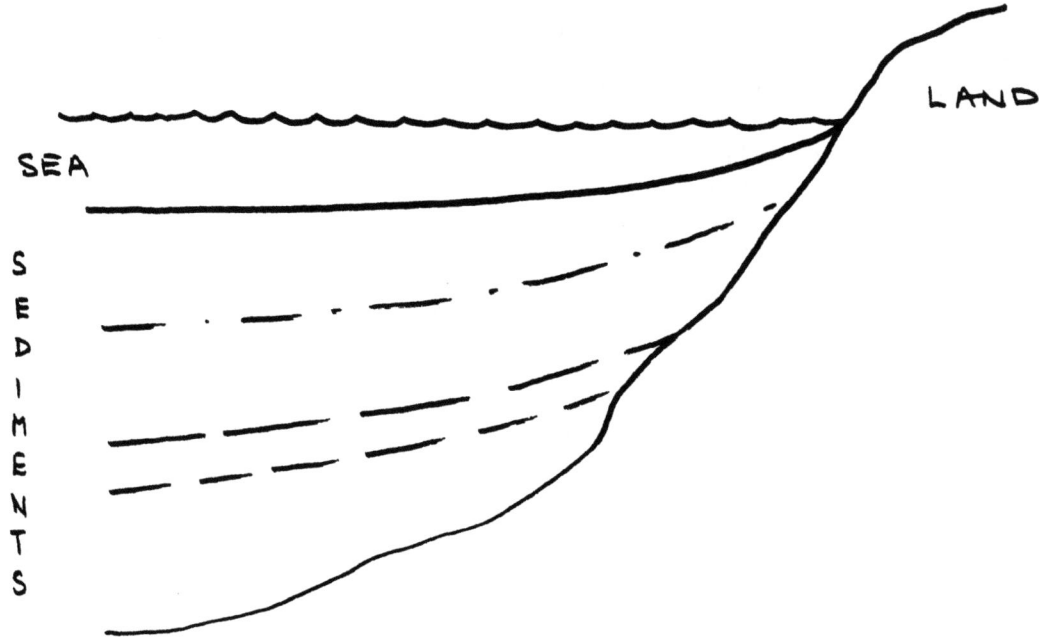

2-14

more compaction occurs below, and more water is squeezed out. The fine-grained deposits become clay rocks and shales under the influence of this compaction.

12(B)

Isostatic adjustment then causes the uplift of the land mass and the subsidence of the basin, allowing further deposition, more compaction of the deeper sediments, and more squeezing out of water. In massive shale deposition, therefore, we expect a progressive — even smooth — reduction of water content with depth, until in the limit there is substantially no water left in the rock. This reduction of water content brings a concomitant compaction and increase of density.

The curve on p.2-16 (given by Baldwin) represents a reasonable composite of the published data on compaction.

It is evident that, for clays, very considerable compaction occurs, particularly at shallow depth. At these shallow depths, the mechanism for increase of density is almost entirely the squeezing-out of pore water. At shallow-to-medium depths there is some rearrangement of the clay particles, and some exclusion of ionically-bound water not released at shallower depths. At great depths, when nearly all the water has gone, the remaining small compaction is primarily simple elastic compression, involving a reduction in size of the particles themselves.

For coarse-grained sediments (particularly the important sands) there is an initial adjustment into an efficient packing of the particles — which may occur during actual deposition — and then any further compaction is achieved primarily by truly elastic compression. Porosity is reduced (and water is squeezed out) as this compression occurs, but the effect is comparatively small.

Up to this first stage of complexity, therefore, we expect a simple and unique relationship between porosity and density, for each rock material. Let us quantify it.

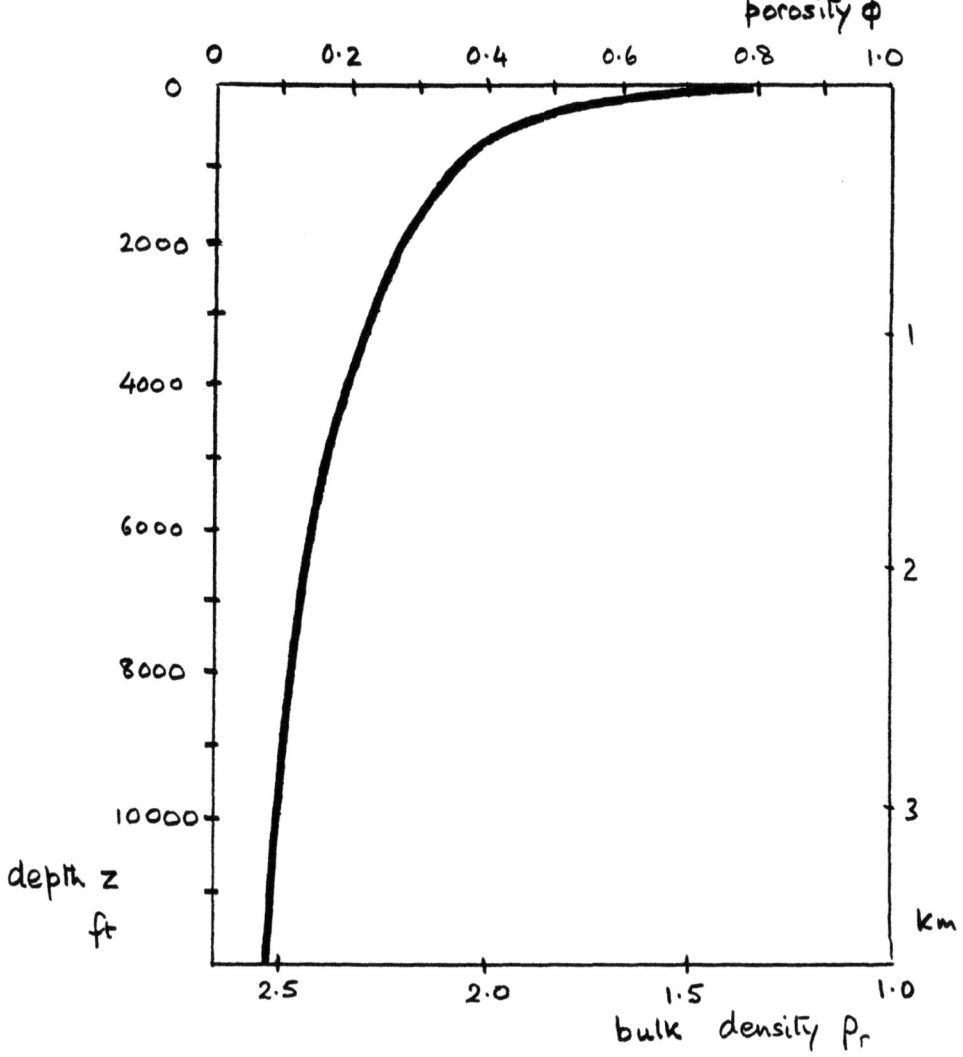

after Baldwin, 1971

(see also Rieke and Chilingarian, fig. 55)

Terms:
- ϕ = porosity,
- ρ_r (r for rock) = bulk density of the rock in its natural state — with pores filled,
- ρ_m (m for matrix) = density of the solid material — the sand grains themselves, or the clay particles,
- ρ_f (f for filling) = density of the material filling the pores,

then
$$\rho_r = \rho_m(1-\phi) + \rho_f \phi$$

This is a very important relation (not least because it is definite). Although we have said that we will not derive equations in this course, it is worth noting that the relation follows from considering a unit cube of the material, from moving all the filling material to one side of the cube, and from writing down the expressions for the mass of the cube in its two conditions. This concept is useful in connection with a later discussion.

If ρ_f approaches zero (gas saturation at shallow depth), then as an approximation:
$$\rho_r = \rho_m(1-\phi).$$

Sometimes we need the equation rearranged:
$$\phi = (\rho_m - \rho_r)/(\rho_m - \rho_f).$$

Usual values for ρ_m, the matrix density, are:

Sand (pure quartz grains)	2.65	Gypsum	2.32
Limestone, pure	2.71	Anhydrite	2.96
Dolomite, pure	2.85	Salt	2.05 - 2.16
Average for carbonates	2.8	Shale	2.6 - 2.65

Usual values for ρ_f, the density of fluid filling, are:

Brine (depending on salinity) 1.04-1.07   Crude oil 0.8 - 0.88
Methane   0.0007   at 0°C, 760 mm Hg

All values are in tonnes/m^3 (or g/cm^3).

The density of gas is very dependent, of course, on the pressure and temperature. To a first approximation, the density can be calculated from the ideal gas law

$$P_1 / \rho_1 T_1 = P_o / \rho_o T_o$$

where the suffix 1 relates to the reservoir conditions of pressure and temperature, and the suffix o relates to 0°C and 760mm Hg. A better approximation can be obtained by including a Z-factor:

$$P_1 / Z_1 \rho_1 T_1 = P_o / Z_o \rho_o T_o.$$

The locally-applicable Z-factor is available from the reservoir engineers (Burcik).

As shore lines advance and recede it is inevitable that some coarse-grained sediments acquire fine-grained sediments within their pores. Further, as a rock becomes deeply buried, in the presence of mineral-rich circulating waters, it is inevitable that some or all of the pore space should become cemented. The porosity-density formula may be extended very easily, to take into account as many components in the pore-filling as we wish. A good illustration is the case of water-saturation in a gas or oil reservoir:

$$\rho_r = \rho_m (1-\emptyset) + \rho_f \emptyset + (\rho_w - \rho_f) S_w \emptyset,$$

where ρ_w is the density of water and S_w is the water-saturation factor (the proportion of pore space filled with water). In general,

$$\rho_r = a\rho_a + b\rho_b + c\rho_c + d\rho_d,$$

where ρ_a is the density of the component whose fractional volume content is a.

13
14
The density of a multi-component rock therefore constitutes no problem for us; the equations are simple, physically meaningful, and <u>certain</u>. We should note in particular that the equations do not <u>depend</u> on the shape or nature of the pores.

15(B) So much for the compaction associated with the overburden. How about tectonically-applied pressures after sedimentation? Clearly, additional pressure applied by tectonic processes must lead to additional compaction, in generally the same way as overburden pressure.

2-18

What happens when the overburden and its pressure are eroded, or the tectonic pressure removed? Shales retain their compacted density; the only (minor) expansion is that small part of the final compaction which is truly elastic. Clean sandstones regain the original density and porosity which they had when they first became efficiently packed. Cemented sandstones, of course, are likely to expand less. Limestones are unpredictable. Simple relaxation is usually complicated by chemical change and/or the development of fracture porosity. This topic is resumed later

TAPE 3 2.3.2 <u>Normally and abnormally pressured sections</u> (Chapman, 1972)

Before we go on to velocity, we must digress for a moment to consider abnormal pressure, and its effect on the density-porosity-depth relations we have just discussed.

First, from our school physics, we recall that the overburden pressure (the "head" of water) in the deep ocean is given by $\rho g z$ where ρ is the density of water, g is the acceleration due to gravity, and z is the depth. So the pressure <u>gradient</u> is just ρg. Doing the arithmetic, we find that, for the salt ocean, the pressure increases at 10.5 kN/m^2/m. Here we will adopt the increasingly general usage of P (pascal) for the SI unit of pressure (N/m^2); the pressure gradient is thus 10.5 kP/m. In old-fashioned British units, it is 0.465 psi/ft. If the water were fresh, it would be 9.7 kP/m (0.43 psi/ft). This pressure we call <u>hydrostatic</u>.

By analogy, we can think of an overburden pressure existing in totally dry earth. If, for example, the effective overburden density is 2.3 times larger than for fresh water (as is typical), the overburden pressure increases at 22.6 kP/m (1 psi/ft). This pressure we call <u>geostatic</u> (or lithostatic).

A normally pressured section is most easily visualized as one in which all the pore spaces in all the rocks are freely interconnected — there are no seals, and no impermeable beds. This is unrealistic, but helpful. Then if a load cell is buried at A (p.2-20, lower sketch) in such a way that it records the

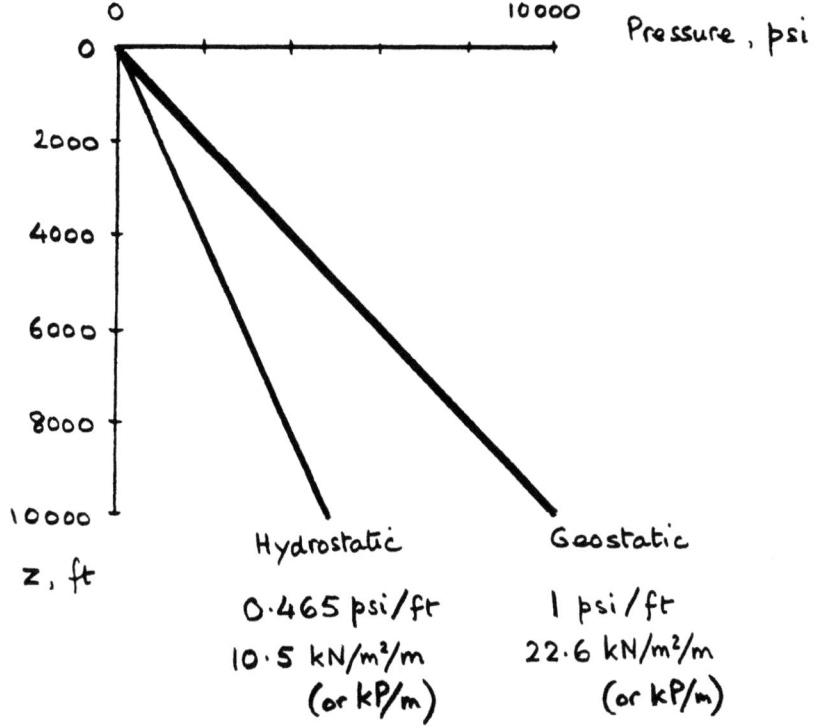

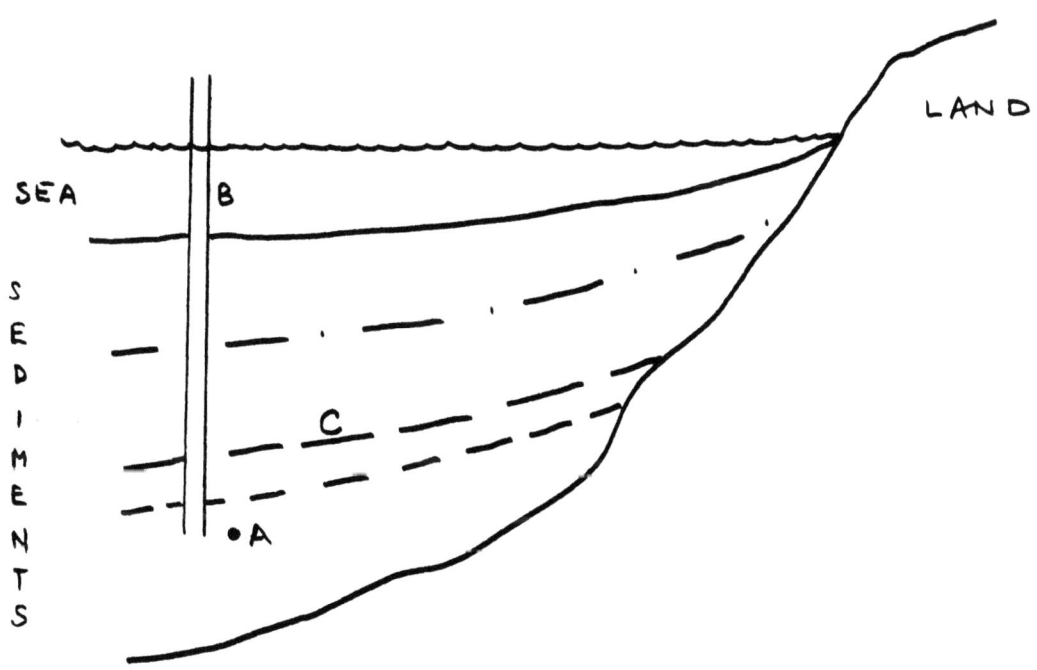

2-20

pressure acting on a horizontal plane in the rock matrix itself, it registers a geostatic pressure of 22.6 MP/m if it is at 1000m (or 10,000 psi if it is at 10,000 ft). Further, if we drill a hole to the same depth, and let the water find its own level in the casing B, it settles exactly at sea level — telling us that the fluid pressure in the rock pores at the bottom of the hole is simple hydrostatic.

In fact, to have a normally-pressured section, it is not necessary to have all the pores open and freely connected. It is adequate that the sediments should have been deposited SUFFICIENTLY SLOWLY for all the water squeezed out of the compacting formations below to have been able to escape to the surface.

If we imagine a totally impermeable sheet C laid down on the sea floor during deposition, and then sedimentation continuing normally thereafter, it is still possible for the section below the sheet to be normally pressured — provided that the water squeezed out of the deep compacting sediments can make its way to the basin margin and escape there.

But if that path is cut off by some tectonic complication, then the expected compaction of the deep sediments, as the overburden thickens, cannot occur. If the water cannot escape, it begins to suffer increasing compression — it begins to contribute to the support of the overburden. This means that the pressure in the water in the pores rises from its normal hydrostatic value towards the geostatic value. In principle it can become equal to the geostatic pressure; more usually, abnormal pore pressures rising to perhaps 70% of this are encountered (16 kP/m, or 0.7 psi/ft).

Formations in which the fluid pressure rises above hydrostatic are said to be overpressured (or undercompacted).

Sometimes impermeable sheets, such as that visualized above, do occur; salt would be a fairly frequent example. However, oddly enough, a highly permeable layer, such as a coarse-grained sand, can have an effect equivalent to that of an impermeable sheet. For the provision of an easy path for water to leave the top of the shale underlying the sand means that the shale starts to compact from the top down (in addition to the usual compaction from the bottom up). The compacted region at the top of the shale becomes less permeable by reason of the compaction, and the shale may therefore seal itself into an overpressured condition.

The same thing can happen when a local sand body overlies a shale. Because the sand is more dense than the uncompacted shale it loads the top surface of the shale; because it is permeable it allows water to flow easily out of the top surface of the shale; and so the shale seals itself into a locally overpressured condition. This situation can be important, in that it brings together a likely source rock (the thick — and possibly organic — shale), a likely reservoir rock (the coarse-grained permeable sand), and an anomaly of the fluid pressure field.

In a normally-pressured section, we have said, the pressure on the rock matrix (holding the rock grains in contact) increases at 22.6 kP/m (1 psi/ft), while the pressure in the pore fluid (trying to force the rock grains apart) increases at 10.5 kP/m (0.465 psi/ft). The difference between the two (the resultant forcing the rock to compact) therefore increases at 12.1 kP/m (0.535 psi/ft. In an overpressured zone, however, this differential pressure may decrease drastically. We shall see later how important this is to the seismic measurements in total; already we see that in a highly overpressured zone, where the differential pressure holding the grains in contact is small, the increase of density with depth is much reduced relative to the normal compaction curve.

Our later studies will show us the value of establishing a density compaction curve — density against depth — for our own area. We shall find that this is particularly important wherever (as is often true) the majority of the section is clay or shale. Establishing this curve becomes one of the interpreter's concerns.

20(B) Where no boreholes exist, the interpreter must either accept p.2-16 (the Baldwin curve) or one of the individual curves of which it is a composite. We shall see later this yields non-specific working techniques helpful in reconnaissance. However, where boreholes exist in his area, the interpreter seeks to make the compaction curve more specific. This he is best advised to do in cooperation with a log analyst, who may indeed be producing the curve wanted by the seismic interpreter as a by-product of his own studies (Pirson, 1970; Matthews, 1972; Sarmiento, 1961).

The objective is (i) to identify the shales traversed by the borehole (from cores, or cuttings, or comparison between logs),

(ii) to exclude from the first analysis those indicated by the density log to be overpressured,

(iii) to plot the accepted values of density as a function of depth (in the manner of page 2-16), and to reassess values clearly not conforming to a reasonably smooth curve, and

(iv) in the case of the overpressured zones, to project back the observed densities to the compaction curve just obtained, to make conclusions about the depth at which those zones were buried when they became overpressured, and to check the geologic reasonableness of the picture emerging.

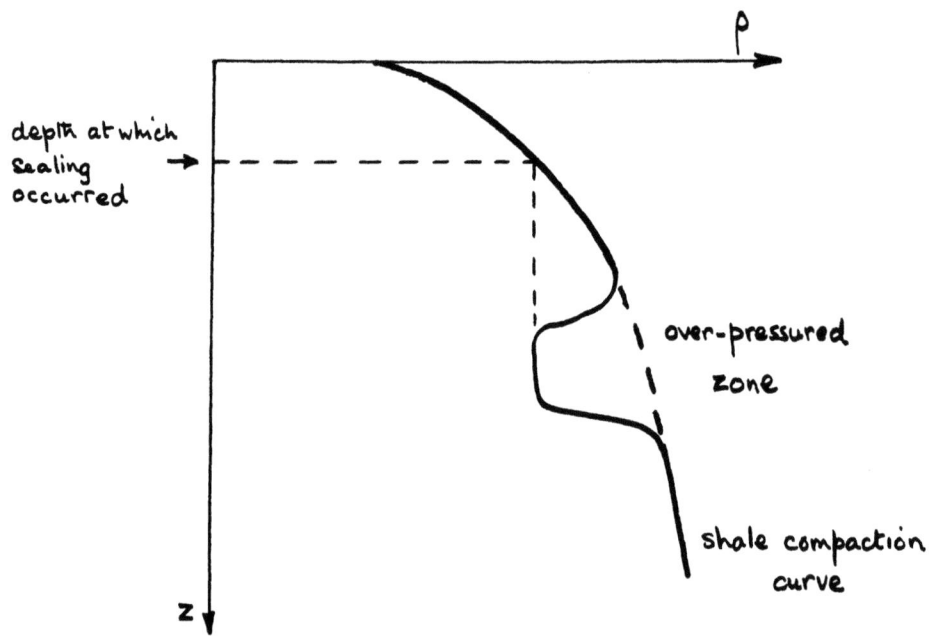

2-23

As we shall see in section 8.6.4, the records of drilling mud weight are also relevant to these determinations.

Summarizing the density discussion thus far, we see that the density of marine sediments, in their unchanged water-saturated condition, is decided by just two factors — the density of the grains or particles, and the porosity. Geologic age, as such, does not have a uniquely definable effect. In the sense that older may mean deeper, it is true that age may imply more compaction — less porosity, and therefore more density. It is also true that age implies more chance, and more time, for cementation (or other replacement of the pore water); this must generally increase the density. However, it is geologic <u>circumstances</u> during geologic time, rather than the time itself, <u>which defines</u> the resultant density.

2.3.3 Factors affecting velocity in a solid earth

We turn now to the second important seismic variable — velocity. It is much more complicated than density.

We probably still remember the sense of confusion we felt when, as novice geophysicists, we looked up the velocities of particular rocks in the text-books. There we might have found shales listed as having velocities all the way from 1600 to 5000 m/s (5500 to 16000 ft/s), and sandstones from 1700 to 6000 m/s (6000 to 19000ft/s). These ranges are so large, and they overlap to such a degree, that we might well think it impossible to use velocity as a tool for lithologic identification. One of our concerns in studying velocity, therefore, is to establish the circumstances which govern the velocity, and so to <u>narrow the range</u> for a particular set of conditions.

In the acoustic sense, the properties which define the seismic velocity V are the elasticity E and density ρ. The basic equation is

$$V = \sqrt{\frac{E}{\rho}}.$$

Since we know from elementary observation that dense rocks tend to have high velocities, this tells us immediately that the effect of <u>elasticity</u> on velocity must be much greater than the effect <u>of density</u>. (It also reminds us that large elasticity means very stiff -- not very elastic in the sense of the man-in-the street.) If a velocity shows a range of 2:1, it is certain that the elasticity shows a range of 4:1 (because of the square root), and likely that the elasticity shows a range of 5:1 (to offset the effect of density).

The physicists among us will remember that there are several moduli of elasticity. To obtain the seismic velocity we must select the <u>appropriate</u> modulus of elasticity, the appropriate type of E. The three which concern us here are Young's modulus e, the bulk modulus k, and the rigidity modulus u. For normal compressional waves in solid rock, the appropriate value of E is $k + 4\mu/3$. For the same in fluids (having no rigidity), the appropriate value of E is just k.

But for compressional waves is a solid rod, the appropriate value of E is Young's modulus e. The reason why compressing a semi-infinite block is not the same as compressing a rod is that the rod allows <u>sideways</u> release of the compression; we shall see later how important this is.

For the moment, however, we look first at a <u>dry solid earth</u> of a single composition. For this there will be a definite value of bulk modulus, a definite value of rigidity modulus, and a definite value of density. The seismic velocity is therefore definite:

$$V = \sqrt{\frac{1}{\rho}\left(k + \frac{4}{3}\mu\right)}.$$

Next we look at a more realistic case: a dry solid earth of <u>mixed</u> composition (for example, a totally cemented sand).

Then we remember from p.2-3 that the seismic wave, which involves the vibration of millions of grains and millions of samples of cement, does not see these inhomogeneities individually; the rock grains are very small compared with the dimensions of the seismic disturbance, and the wave behaves as if the material were not of two separate compositions, but of one average composition.

The problem, then, is to define the appropriate average.

A satisfactory solution for the solid case is the time average, given by the time-average equation:

$$\frac{1}{V} = \frac{1 - \emptyset}{V_1} + \frac{\emptyset}{V_2},$$

where V_1 and V_2 are the velocities of the two components (matrix and cement) and $\emptyset$ is now a cement-filled "porosity". Although the equation has a similar (reciprocal) form to that for density (p.2-17), it must be stressed that the time-average equation is an empirical approximation — a matter of good fortune, in contradistinction to the inevitability of the density equation.

If we recall the unit cube we used to demonstrate the density equation, the physical implication of the time-average equation becomes clear. As before, we move all the filling material to one side; its left-to-right dimension is $\emptyset$. Then instead of writing expressions for the mass, we write expressions for the time for a seismic wave to traverse the cube from left to right. 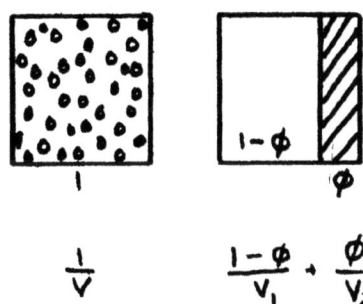 The two expressions are the left and right sides of the time-average equation. This equation therefore says that the velocity encountered in the natural rock is the same as would be obtained by passing the wave first through the matrix material and then separately through the filling material. Physical intuition would certainly not have required this to be so, and we can see that where the time-average equation applies it is indeed a matter of good fortune.

TAPE 4

The density of our dry solid rock is necessarily fairly large, since there is no void porosity. The elasticity is also large; the rock is difficult to compress by the seismic wave, since any movement of the particles can only be by compression of the particles themselves, against strong intermolecular forces. Again the effect of the elasticity dominates over the effect of the density, so that the velocity $(E/\rho)^{1/2}$ is also high.

Now let us apply a geostatic overburden pressure. The increased compaction is small, for the same reason that the elasticity is high. Nevertheless some increase of density does occur. However, a rather greater increase occurs in the elasticity, as the rock becomes stiffer. The velocity therefore increases.

But as we apply more and more geostatic pressure, we find that less and less increase of velocity occurs. Thus we see the basis for the well-known nature of the general velocity-depth relation shown in the sketch. This general form of relation would apply even if, miraculously, the whole section had the same geologic age — if it were all deposited overnight.

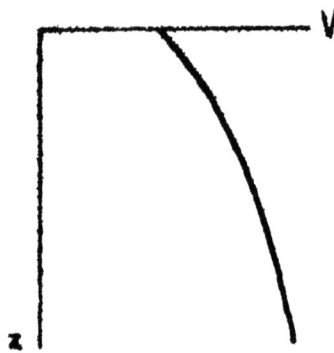

In practice, of course, the solid parts of the earth do not interest us very much. Now we must ask what happens when we introduce fluid-filled pores.

2.3.4 Factors affecting velocity in a porous earth

The pores make an enormous difference. To give us some physical feeling for why this should be so, let us consider a very-much-simplified model.

We start with an infinite layer of solid rock, sandwiched between two infinite plates, and we consider a seismic wave passing through it vertically over the area A (p.2-30). Then the last section has told us that the velocity is fairly high, and that it increases somewhat if we apply geostatic pressure to the top plate. We obtain the right-hand curve in the top sketch of p.2-32.

Nothing would be changed if the layer of solid rock was actually composed of cubes, all in intimate contact, as suggested in the top sketch of p.2-30. The cubes could be real cubic crystals, or they could be notional cubes consisting of a sand grain embedded in cement.

As a first step in introducing porosity, we now move apart, very slightly, the columns of cubes (middle sketch, p.2-30). The voids contain, let us say, nothing.

Then three effects are consequent:

1. The density is reduced, strictly according to the porosity.

2. The elasticity is reduced. In an elastic sense, each column is just a spring; the stiffness is reduced, therefore, in that the same area of concern (A) now includes fewer springs. This effect is also strictly according to the porosity; since $V = \sqrt{E/\rho}$, effects 1 and 2 tend to cancel.

3. The individual cubes can now <u>deform sideways</u> — vertical compression can occur without necessitating so much deformation of the molecules themselves. This means that a <u>different modulus of elasticity</u> is now appropriate in the equation for velocity.

As a result of this third effect, the stiffness of the rock decreases dramatically, and so does its velocity.

We can become more realistic, and still preserve this physical conclusion, by turning some of the cubes on edge rather than by opening up vertical spaces. This gives us something beginning to look like a rock (bottom sketch, p.2-30). The stiffness of this arrangement (and hence its velocity) is still low, by reason of the fact that the grains can deform readily into the pore spaces.

Now we apply the geostatic overburden pressure (left curve, upper sketch, p.2-32). The initial velocity at zero depth is much lower than the previous (solid) curve, but the final velocity at very great depth — obtained when all the deformation possible has occurred, and there are no more voids in the rock — is much the same. The rate of increase of velocity with depth is therefore considerably greater (dotted).

There is very little change if the pores contain gas instead of vacuum; the gas is very easily compressed, and therefore it has negligible capacity to resist the deformation of the solid material into the pore spaces.

Now we fill the voids with water. The immediate effect is that the deformation of the cubes into the pore spaces is <u>resisted</u> by the stiffness of the water. Again the effect of <u>elasticity</u> dominates over the increase of density. Although water is more compressible than most rocks, it is still fairly incompressible; therefore the velocity rises considerably from the gas-saturated state (though not to that of the solid state).

Now let us provide a little hole in the top plate, and explore the effect of applying overburden pressure. Since the hole allows all the water to be squeezed out under great pressure, the water, and its effect, are removed at great depths. Thus we have the dashed curve (top sketch, p.2-32) for the water-saturated case.

Here we should take note of an important fact: <u>the effect of fluid-saturation becomes less and less with depth</u>, in a normally-pressured section.

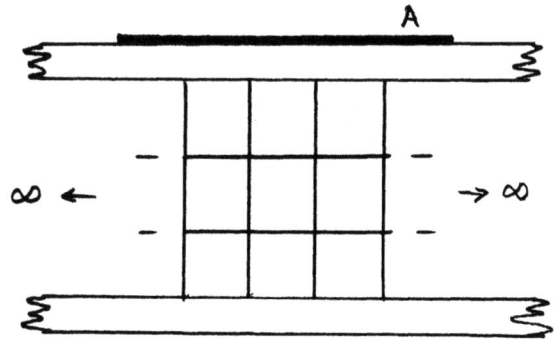

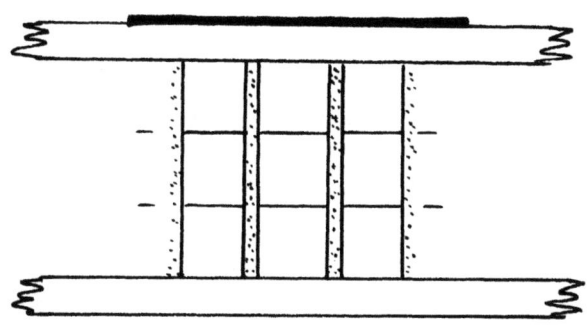

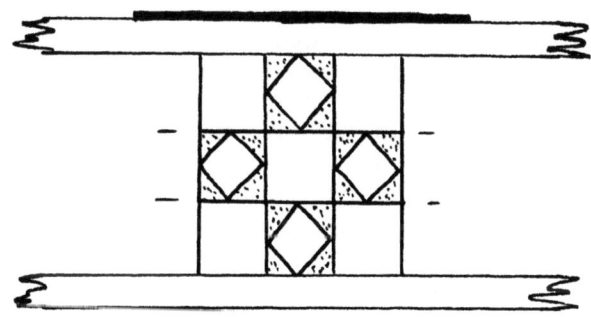

2-30

To explore what happens in an overpressured section, we can seal the little hole. Now the water cannot be expelled; it must always be present in the pore, whatever the overburden pressure. It is always there, allowing a measure of deformation of the solid grains into the water-filled pore. As a result, the velocity changes very little with depth (dotted line, lower sketch, p.2-32).

If we apply a certain geostatic pressure first (corresponding to some depth z_1) then the velocity (dashed line) remains constant at the value corresponding to this depth. In such a simple case, this velocity value defines the equivalent depth of the overpressured rock — the depth to which it was buried normally before overpressurization. This is similar to the density situation (p.2-23).

Perhaps it is difficult to visualize in the real earth, but if the water were pumped up until the grains were no longer held in contact, the rigidity of the material falls to zero, and the velocity of the water-saturated rock falls to a value closer to that of water.

A further conclusion from our simple normally-pressured model is that if the water became frozen the deformation of the grains would then be resisted by the solid ice; our colleagues working in the Arctic know well the velocity problems introduced by the permafrost. By the same token, the progressive replacement of water by solid cement must move the velocity up toward the totally-solid line.

In anticipation of our future discussion of fracturing, let us draw one more conclusion from the lower two diagrams of p.2-30. The bottom diagram corresponds very grossly to inter-granular porosity; let us suppose that it does indeed yield the gas-saturated and water-saturated curves of the top diagram of p.2-32. The middle diagram of p.2-30 (the one where we introduced the porosity by moving apart our columns of cubes) then corresponds very grossly to fracture porosity, with the dominant fracture direction aligned with the direction of the seismic path. In our

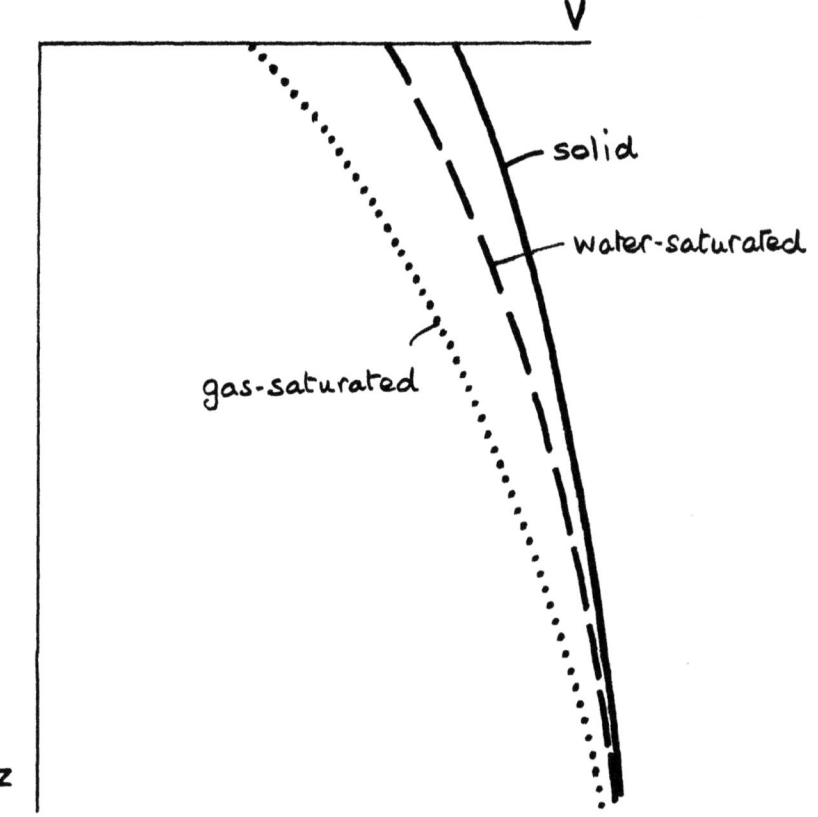

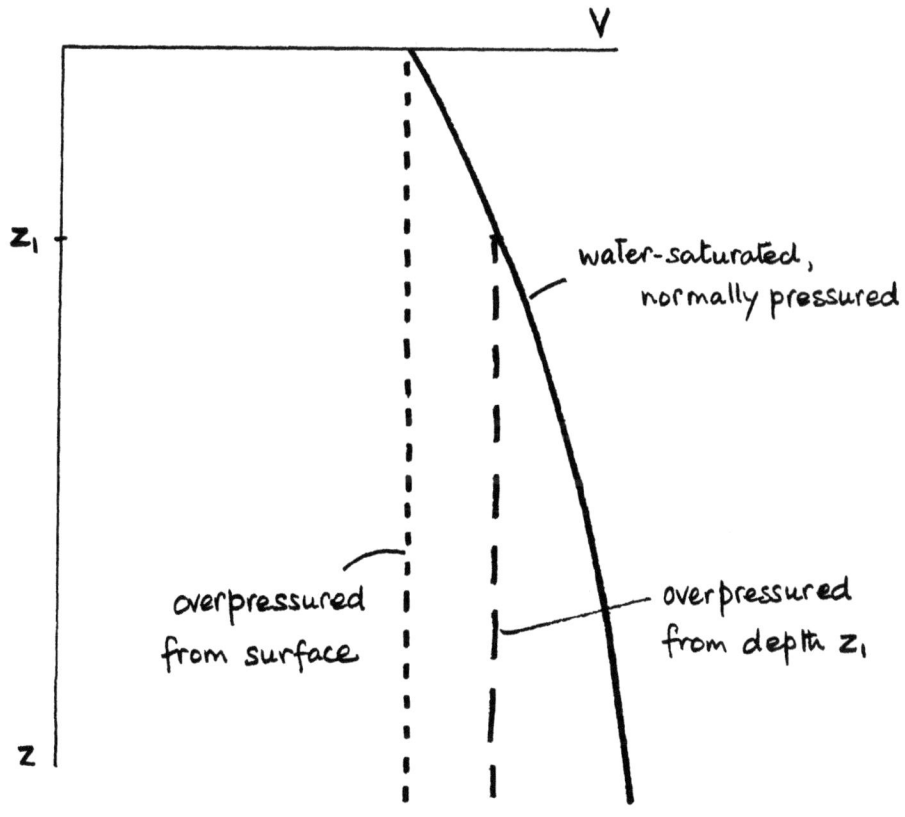

2-32

simple model, a major change of velocity follows the very first separation of the columns of cubes, and further separation (that is, increase of the porosity) makes no further difference. We should not expect, therefore, a simple velocity-porosity relationship in the case of fractured rocks; the velocity might be more closely related to the surface area of the void than to its volume.

However, a difference in the behaviour does become apparent when we include the effect of the overburden pressure; if the fractures are all narrow they close at small depth (so that the fluid-saturated curves of p.2-32 join the solid curve at a depth shallower than that suggested for intergranular porosity), while if they are wide this may not occur until great depth.

The last two paragraphs prepare us for the unwelcome but necessary conclusion that the velocity-porosity behaviour of a rock depends on the shape of the pores. This, we remember, is quite different from the simple and certain density-porosity behaviour. Needless to say, the dependence of velocity on the shape of the pores is a major complication.

This discussion of our simplified model has been shamelessly naive; it is justified because the literature contains many contradictions, and because a little physical insight — even if over-simplified — protects us from some of the errors, and helps us to predict what is likely to happen in a given set of circumstances.

Now let us look at some observed results — from the lab and from the field — to see to what extent our expectations are borne out in the real world. We restrict ourselves here to liquid pore-filling; gas is tackled later. We look at shale first, to ask whether the major compaction effects visible in density are observed also in velocity.

In the case of shale, we brace ourselves for two difficulties. First, some of the water content in a clay material is ionically bound; there is a possibility that this water will affect the elasticity of the material in a chemical way which is not covered

by our simple physical model. This first difficulty means
that although our density-depth behaviour has led us to expect
a simple relationship between water-content and depth, the
corresponding relation between velocity and depth need not be
so simple. This fact is usually expressed by the (otherwise
physically unsatisfying) statement: "The time-average equation
cannot be used for liquid saturation of shales because shale
has no true matrix velocity." Second, shales are so often at
least slightly overpressured that we must expect some difficulty
in defining "normal" conditions in real sections — and particularly in young sections.

This second difficulty becomes apparent when we consider
published versions of the velocity-compaction law for shales.
Immediately we are into dissent. Among the log analysts (who
are best qualified to know) there are two clear camps — whom
we shall call the log-loggers and the semi-loggers.

According to the semi-loggers, the velocity-compaction
relationship is best shown by plotting the sonic travel-time
(in μs/ft or μs/m) on a logarithmic scale against depth on a
linear scale; they then expect the plot to be a straight line
(for normally-pressured shales) of the form

$$z = A + B \ln \Delta t,$$

where Δt is the sonic travel-time and A and B are constants.

Plotted in the form favoured by geophysicists, with the
velocity increasing to the right, the semi-log velocity-
compaction relationship appears as shown on p. 2-36, for several
areas.

Since nothing hereafter will invalidate the application of
this approach over limited depth ranges, it is worth developing
a little. By reading off two pairs of coordinates on a velocity-
depth line, we can set up two simultaneous equations and solve
for A and B. For the Miocene-Oligocene line, for example, and
working in feet, we obtain A = 82,780 and B = -15,700. (B is
always negative.) For Herring's line, A = 73,500 and B = -14,000.

2-34

Then by manipulation of the equation $z = A + B \ln \Delta t$, we obtain

$$T = B e^{-A/B} (e^{z/B} - 1)$$
$$V_a = z/T$$
$$V_z = e^{(A-z)/B}$$

where T is the one-way seismic travel-time to depth z, in µs,

V_a is the average velocity from surface to depth z (or time T), in m/µs or ft/us,

and V_z is the local velocity at depth z, in m or ft/s.

By use of these equations we can bring the information into the form more familiar to the seismic interpreter — graphs of local and average velocity against depth and time. These are given on p.2-38 for the Miocene-Oligocene example.

It is then that we realize that the semi-loggers cannot be correct in general, because the velocity-depth graphs curve the wrong way. They are concave to the right, whereas physical argument indicates the opposite (p.2-27). We shall leave to section 8.6.4 our attempt to resolve this disagreement; for the moment our main concern is not with establishing the exact form of the velocity compaction law, but with noting the benefits to be derived from any law (even if it applies only over a limited depth range). One such benefit is in narrowing the range of velocities expected at a certain depth; we recall that this is a prerequisite to any attempt to identify lithology from velocity. Another benefit, obviously recognizable at this stage, is that if we know the average velocity to the top and bottom of a massive shale layer, the law allows us to compute the local velocity anywhere within the layer.

Another type of law (and one more likely to be correct in general) is the power law supported by the log-loggers; they say that the graphs of p.2-36 are linear if the depth scale also is plotted logarithmically. This is equivalent to saying that the velocity is proportional to some power of the depth. Popular

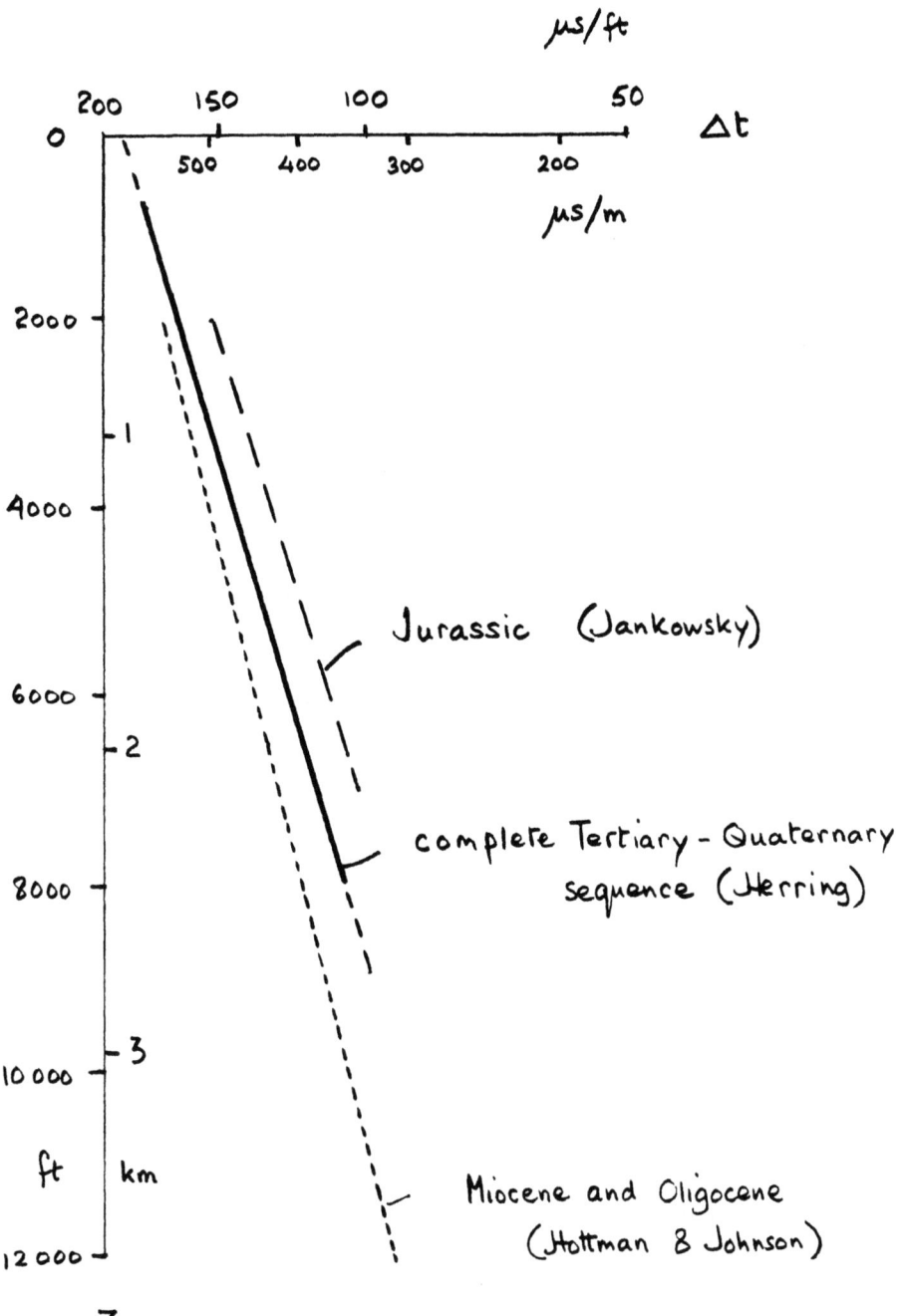

2-36

values for the power range from 1/4 to 1/6. If n is this power, we have that:

$$V_z = kz^n$$
$$T = z^{1-n}/k(1-n)$$
$$V_a = k(1-n)z^n = (1-n)V_z$$

(using the same symbols as on p.2-35).

A sample of the type of velocity-depth relationship resulting from a one-fifth power law is shown in the shale curve on p.2-60.

Over restricted depth intervals, either of these laws is satisfactorily approximated by that old trusted friend — the linear increase of velocity with depth, for which:

$$V_z = V_o + cz$$
$$T = 1/c \ \ln(V_o/c + z)$$
$$V_a = zc/\ln(V_o/c + z).$$

All these equations, of course, are well known (Kaufman, 1953); they are reproduced here merely for completeness.

We have said that, at this stage, we are less concerned with choosing between these three laws than we are with noting the general benefit we obtain from them — a way out of the apparently-hopeless situation where shales are quoted a velocity range too large to be useful for purposes of lithologic identification. Thus the knowledge which restores hope for the meaning of velocity is the nature of the shale compaction curve. As soon as we know the appropriate law then we know that a normally-pressured shale at a given depth cannot have less than a certain velocity. If it has less, then it is over-pressured. If it has more, then it is either contaminated with other rock materials, or modified by chemical processes or tectonic pressures or abnormal temperatures, or it is uplifted from its greatest historical depth.

TAPE 5   We now ask what restores hope for the meaning of velocity in the case of liquid-saturated <u>sandstones</u>. The answer is the time-average equation.

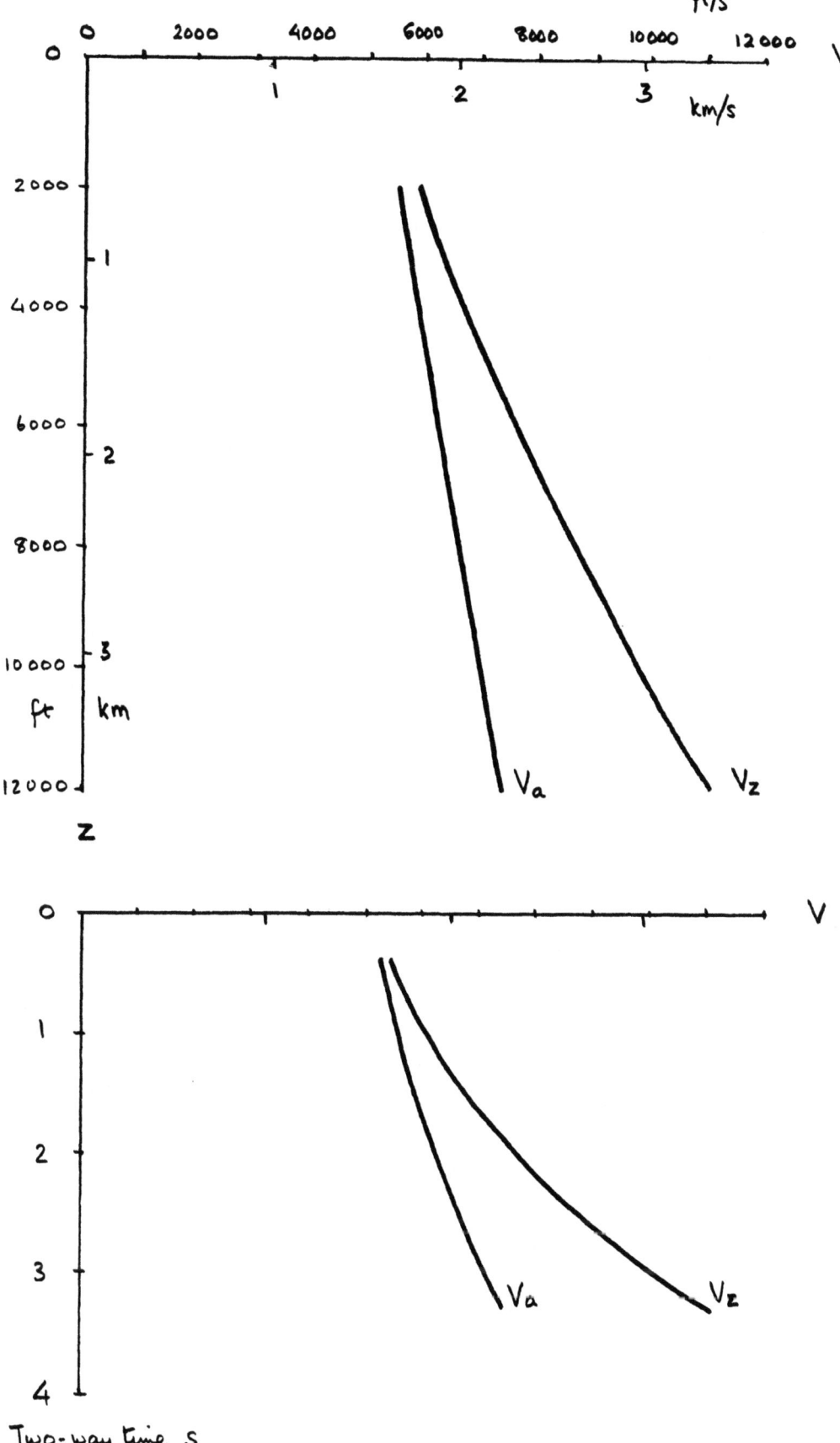

2-38

For liquid saturation, the time-average equation becomes:

$$\frac{1}{V_r} = \frac{1-\phi}{V_m} + \frac{\phi}{V_l}$$

(subscripts: r = rock, m = matrix, l = liquid).

For deep clean sands, the values usually adopted are

V_m = 5700 m/s   (18500 - 19000 ft/s)
V_l = 1600 m/s   (5250   ft/s) for water
V_l = 1300 m/s   (4270   ft/s for oil

We have said that the time-average equation is not generally applicable to shales; now we note that its application to sands is subject to two important provisos. The first is that the sand must be <u>normally pressured</u>; as we have seen, only if a permeable sand is totally sealed in impermeable material can the sand itself be overpressured. The second proviso concerns a break often observed in sand compaction curves. The dotted curve on p.2-40 represents the effect of <u>compaction alone</u> on loose sands. The full curve represents real observations on sands in place. Clearly, something significant happens to cause the velocity to rise above that expected by reason of compaction alone. The adjustment to an efficient packing almost certainly occurs at shallower depths, and so the accepted explanation is that of increasing cementation with age. The dashed line represents the velocity computed from the time-average equation for measured variation of porosity with depth. Therefore it is usually accepted that the time-average equation can be used for liquid-saturated porous sands, as the relation between velocity and porosity, <u>provided that</u> the rock is old enough to be at least slightly cemented and (as mentioned previously) the pore pressure is normal.

Another probable condition is that the porosity in the sand should be of <u>intergranular</u> type, without significant fracturing after consolidation; this is usually satisfied in the important practical cases, because the high-porosity sandstones do tend to have most of their porosity in the intergranular voids, and little of it in the thin voids or cracks.

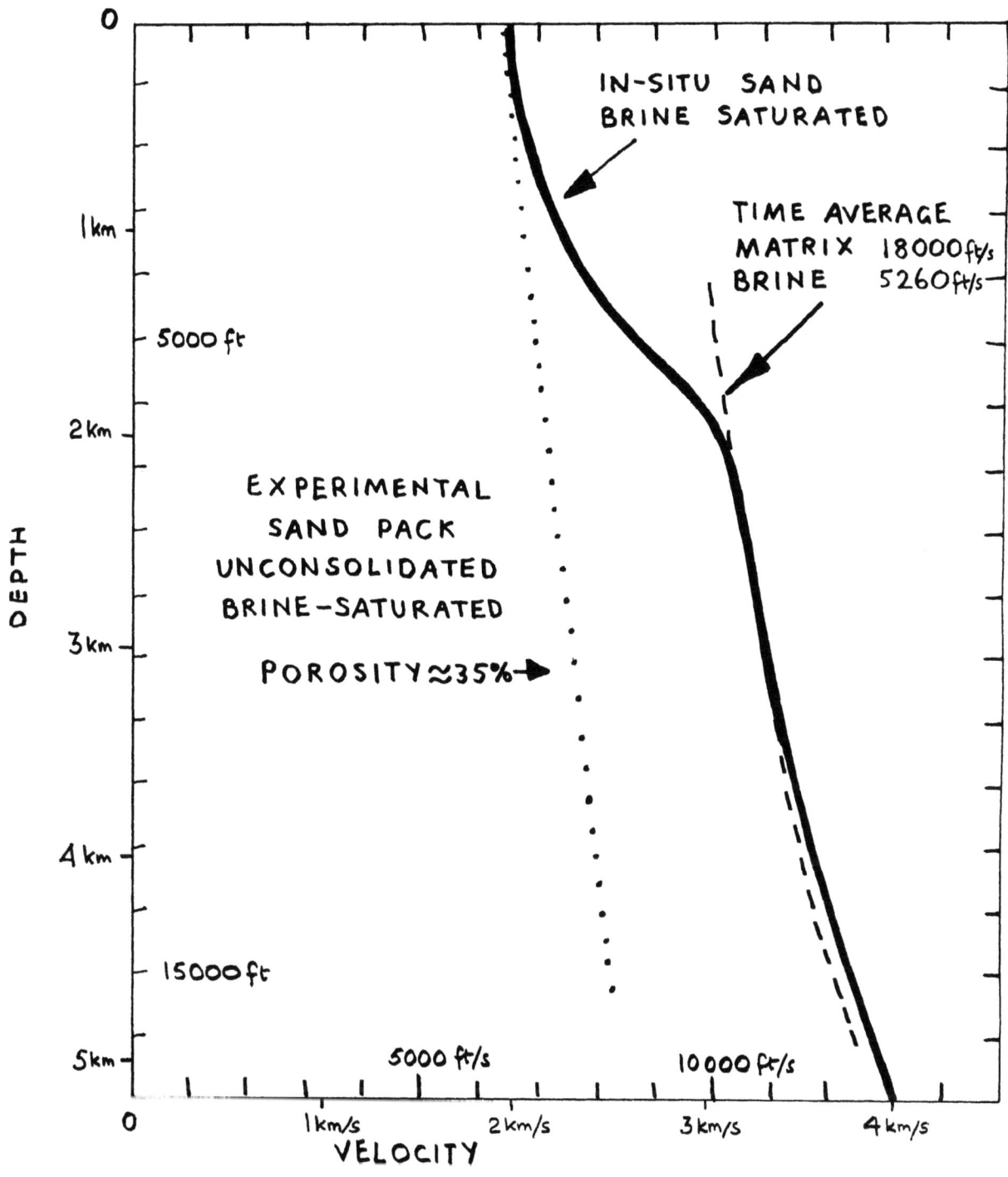

After Gardner et al., 1968

25   This allows the construction of a general relation between velocity, depth and porosity in sands (p.2-42). We see an expectation that, at any particular depth, there must still be some physically-reasonable relationship between velocity and porosity, though it is not the time-average equation; the latter relationship applies (in the case illustrated) only to the cemented region beyond about 1800 m of depth.

We should stress the distinctions which are being made here. Specifically:

- For deep-sea shales, the dominant influence on velocity is that of depth, to which velocity can be related by a compaction law. Porosity is also related to depth (p.2-16), but the velocity and porosity may not be interconnected using the time-average equation.

- For sandstones, the dominant influence on velocity is that of porosity. This need not have any connection with depth (being, as it is, related to the sediment and conditions of deposition in a very local manner); a coarse sand may overlie a coarser sand.

- For sandstones which are at least slightly cemented, the time-average equation may be used as the relation between velocity and porosity,

- These observations apply only to liquid saturation and normal pressures.

26(B)   The time-average equation is also generally held to be applicable to carbonates; typical values for V_m are:

$\quad$ 6400-7000 m/s $\quad$ (21-23,000 ft/s) for limestones and
$\quad\quad\quad$ 7900 m/s $\quad$ (26,000 ft/s)$\quad$ for dolomites.

Where limestones are massive and "well-behaved" (meaning that the porosity is of interparticle type -- not due to fractures or vugs) some sense can be made of their velocity-depth behaviour (Jankowsky, 1970). The Upper Cretaceous of N.W. Europe, for

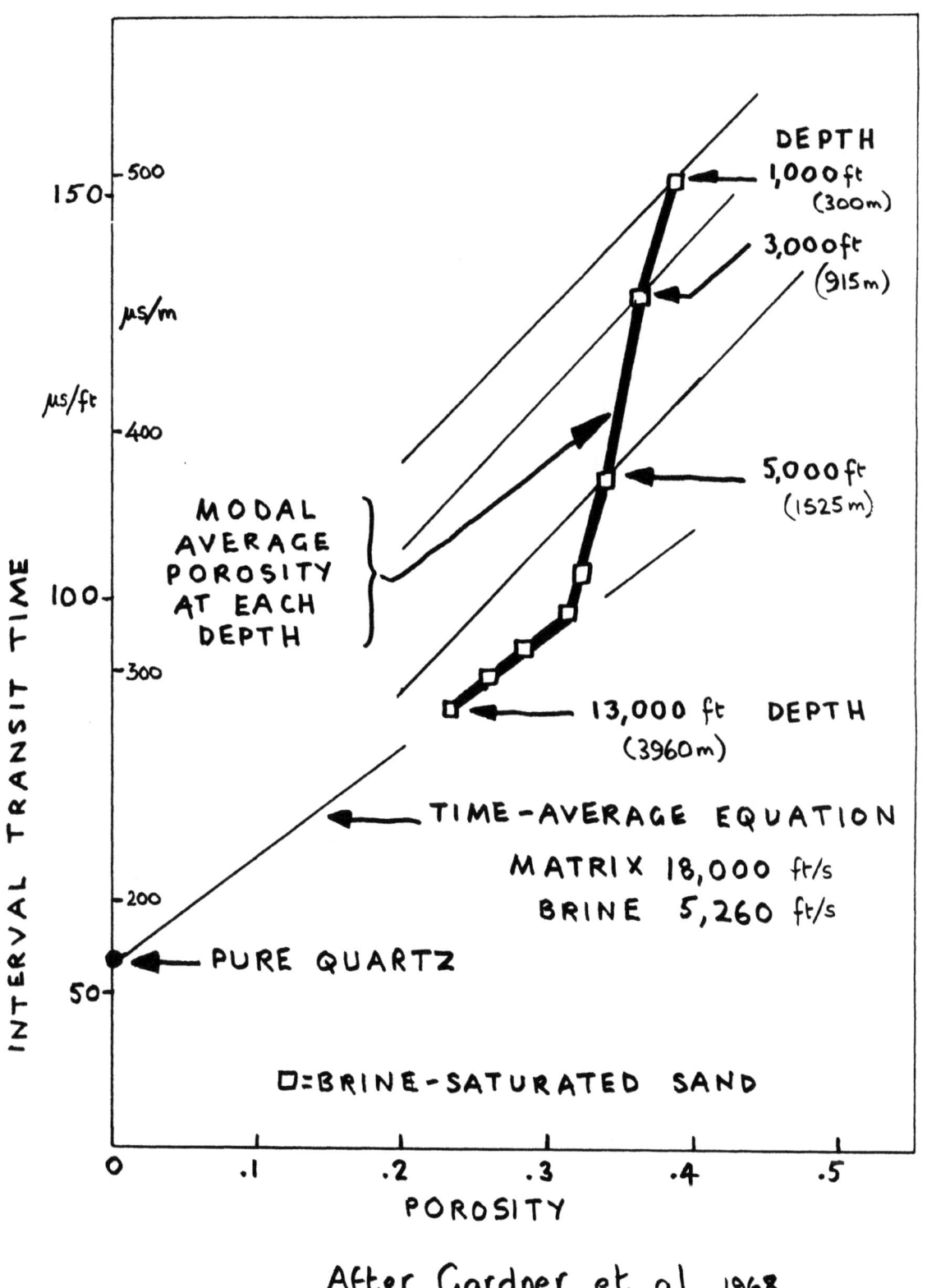

After Gardner et al., 1968

example, shows a semi-log gradient identical to that observed in shales; the figure on p.2-44 illustrates this. Although it would be attractive to infer a causal depositional connection, it appears that this identity is fortuitous -- the agency is cementation, not primarily compaction (Friedman, 1974). In carbonates deposited under less calm depositional conditions (or subject to subsequent tectonic or chemical modification), such ordered velocity-depth behaviour is unlikely.

This passing mention of carbonates leads us back to our discussion of fracturing.

The effect of fractures on seismic velocity is major. We remember our elementary physical model (p.2-30); as soon as there is a void, lateral deformation of the particles is possible, and the appropriate modulus of elasticity changes. We recall also how small is the displacement of a particle as a seismic wave passes across it -- typically at least two orders of magnitude smaller than the wavelength of light; consequently cracks far too thin to be seen under an optical microscope can produce major changes in the elastic properties. Indeed, our simple model would suggest that a certain fracture porosity realized by a great number of sub-microscopic microfractures might well produce a greater change of velocity than the same fracture porosity realized by a small number of large cracks. Thus there would be a difference of velocity in the two cases, but no difference of bulk density. This has been demonstrated convincingly by Kuster and Toksoz (1974) and by Toksoz, Cheng and Timur (1975). The practical warning to us is very clear: although we are delighted to be able to detect the presence of fracturing by a velocity measurement, we must appreciate that this significant velocity depression need not imply good porosity if the cracks are all very fine.

We should note that though these conclusions undoubtedly apply to the bulk velocity measurement made by the seismic method, sonic-log measurements may not agree. The sonic log selects the first arrival, whose velocity is depressed by general intergranular porosity or a system of fine fractures but not by local major fractures or vugs.

The figure on p.2-46 shows the observed change of velocity introduced by micro-fracturing a substantially non-porous rock; the change is very large (to about 60% of the original value).

Obviously the effect of applying increased overburden or lateral pressure to a fractured rock is to close the voids. However, we see from the figure that the velocity of the unfractured rock is never recovered entirely; this is because the fractured edges of the rock never fit together exactly when

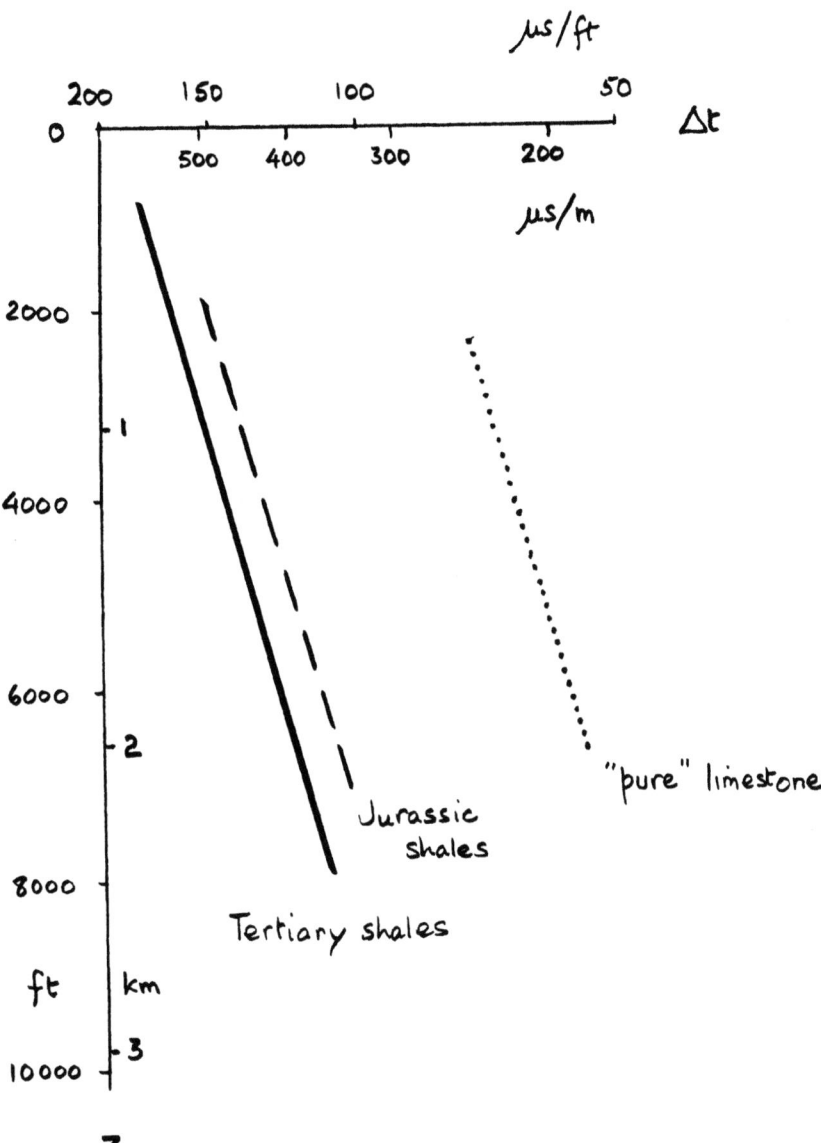

2-44

the pressure is applied, and some voids remain. This is relevant to the situation where a carbonate is fractured by uplift (for example, over a salt swell), and then sinks to a greater depth (for example, by withdrawal of salt elsewhere).

We remember that while the effect of the overburden pressure is to close the voids, the effect of the pore-fluid pressure is to hold them open. The effect of fracture porosity on velocity is therefore very dependent on the differential pressure.

Therefore, if we postulate a fractured carbonate reservoir which by reason of a massive overlying shale has become overpressured, we would hope to be able to detect the fractured zones seismically. As we shall see, this can be done. Its importance is major; several of the world's giant oil fields possess overpressured fractured reservoirs.

Returning to the normally-pressured situation, we stress again the lithologic distinctions. The "norm" of velocity in the case of a shale is defined by the shale compaction curve. In the case of sandstones (in the absence of fracturing) it is defined, to a first approximation, by the time-average equation. Now we include carbonates in this last category also, provided there is no fracturing and no chemical change. We emphasize that the time-average equation can be no more than an approximation, because it takes no account of the shape of the pores (which must be significant in terms of the elasticity).

The time-average equation for normally-pressured coarse-grained materials can be extended to include matrix, cement, water, oil, etc:

$$\frac{1}{V} = \frac{a}{V_a} + \frac{b}{V_b} + \frac{c}{V_c} + \frac{d}{V_d} \; ; \; a+b+c+d = 1.$$

As before, this implies that the bulk velocity is the same as that which would be obtained if the wave passed in sequence through the separated components. There is some suggestion that the time-average equation is less good for oil saturation than for water saturation; in any case, of course, the time-average equation for liquid saturation remains a matter of good fortune, a happy approximation.

So much for velocity-porosity-depth relations in general; as we did in the case of density, we must now see what the geophysicist can do to obtain better definition of the relationships if he is working in a mature area replete with well logs. The technique is set out by Sarmiento, 1961 (see also Pickett, 1960; Faust, 1951, 1953). Basically the logs are searched for segments which, from cores or cuttings or the comparison of the

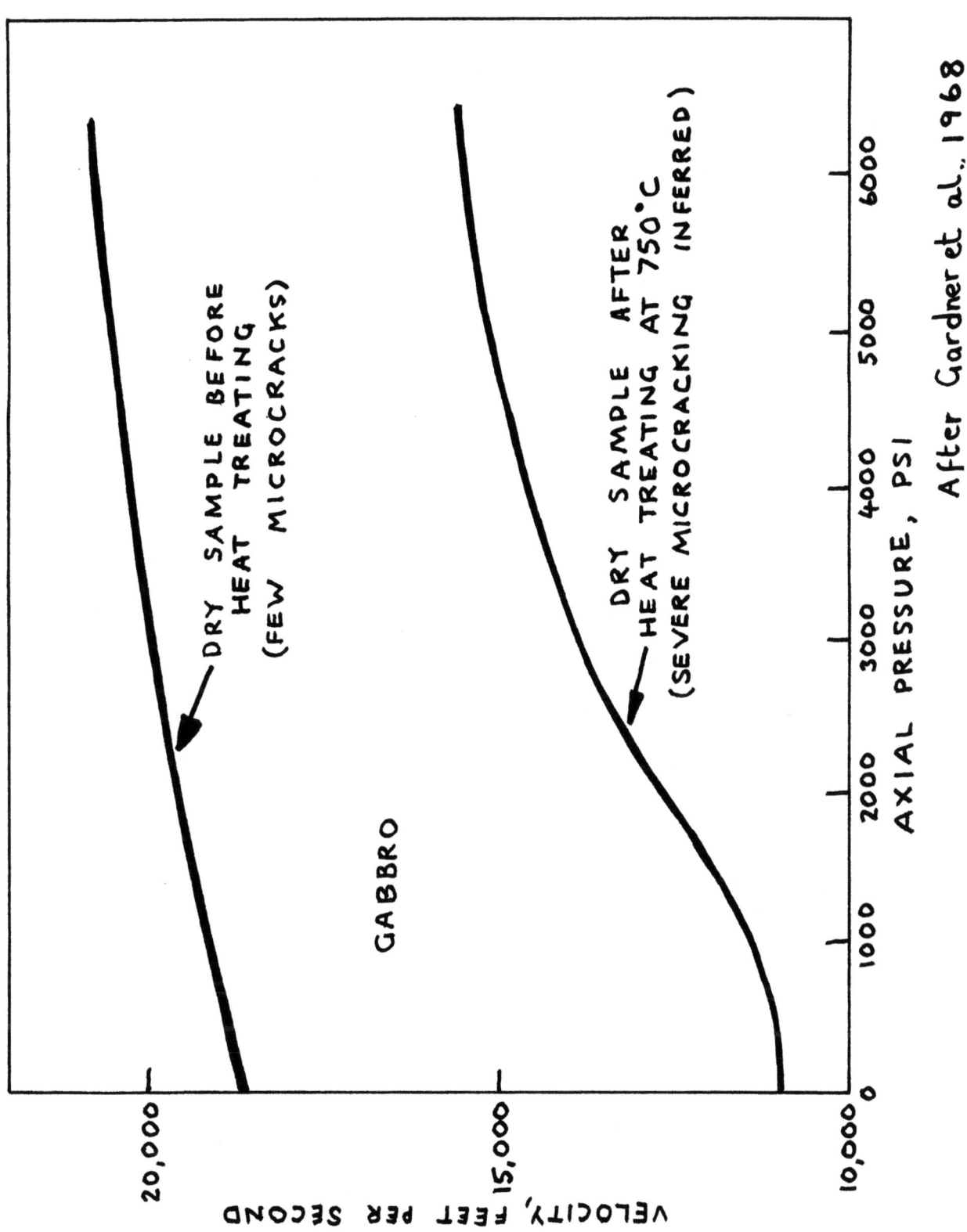

logs, can be uniquely identified lithologically; then the velocity (or more usually the sonic interval time) and the porosity and depth values are analysed for some sort of normal inter-relationship. Typical of the sort of relationship which may emerge as appropriate to a particular area are these from Sarmiento (who is, we should note, a log-logger):

For clean, well-sorted, cemented Paleozoic sandstones:
$$t = 50.6 + 5.525 z^{-0.16} + 5.249 \phi z^{-0.16} + 0.0359 z^{-0.16} \phi^2 - 0.0031 \phi^2$$

For dirty impure Paleozoic sandstones grading into graywackes:
$$t = 54 + 4.465 \phi z^{-0.16}$$

For Paleozoic limestones generally:
$$t = 50.9 + 1.23 \phi - 0.000296 z.$$

Of course there is no suggestion that these equations apply outside their area. However, they illustrate that the complete specification of the inter-relation between **velocity**, porosity and depth, for a particular rock type, may require terms in z, terms in ϕ, terms in ϕz^n, terms in $\phi^2 z^n$.

What is our reaction to equations like that — so physically unsatisfying? We are left feeling that, while such equations may yield a satisfactory empirical fit to the data, they do not in themselves represent the consequences of physical mechanisms; it is likely that the equations would be more meaningful, physically, if we could include a variable (such as the void aspect-ratio, or the shape of the pores) which would represent the type of porosity present. Our reaction is that we have fitted the observed data to equations with too few variables, and that one of the missing variables — probably the most important one — is the shape of the pores.

The equations (with the others given in the reference) also show the tendency for the effect of depth to be represented by the one-fourth power for Tertiary and Cretaceous, the one-sixth power for older rocks, and by virtual independence at great depth. This, of course, we are prepared to expect from the foregoing material. The one-sixth power is represented as well established for the medium range (Faust), though we must remember that the full significance of porosity, pore shape and pore pressure was not appreciated when many of the earlier analyses were made.

Before we proceed to gas-saturation, let us take a few moments to summarize the situation in the general case of liquid-saturation.

2.3.5 The overall picture of velocity and density variation

Let us summarize under three headings — density, velocity, and relationships between the two.

Density

As the depth of burial increases, the density increases (mainly by reduction of the porosity). The relation between density and porosity is well established and positive (p.2-17).

For shales, a composite experimental relation between porosity (or density) and depth (p.2-16) is widely accepted. Most of the compaction represented by this relation involves both the exclusion of water and the rearrangement of the clay particles, and is not recovered on relaxation of the pressure; only the minor quasi-elastic compaction is recovered. The total compaction effect with depth is major. The permeability of the shale decreases toward zero as the compaction becomes complete. In the zero-porosity state the shale density is typically 2.6 – 2.65.

For sands, the variability of sediment sizes and depositional mechanisms means that no generalizations can be made about the porosity or density relationships of different sands at different depths. However, for a hypothetically constant sand at different depths, we can say that the increase of density with depth is much less marked than for shales. Of that which does occur, part is due to rearrangement of the grains and to cementation; this is not recovered on relaxation. Again, part is simple elastic compression, which is recovered on relaxation.

For limestones, the increase of density with depth is generally very variable, although isolated well-behaved examples are found. There is a tendency for the compaction to be less marked if the concentration of calcium carbonate is high.

Velocity

Velocity is a function not only of rock type, porosity and depth, but also of rock condition and the shape of the pores. Attempts have been made to express it as an explicit function of geologic age; however, there is no reason to expect such a relationship to be strict, since the physical property of velocity can only be a function of the physical and chemical processes to which the rock has been subject, rather than of the geologic age as such.

For shales, the velocity increases smoothly with depth; physical arguments suggest that this increase must be rapid at first, and then progressively less rapid. In some particular areas, the curve can be established rather well (p.2-34).

For sands, the depositional variability is the main factor, as in the case of density. For a hypothetically constant sand at different depths, there is a shallow region of doubt until some cementation occurs; this is probably because the elasticity is very dependent on the number of contacts per grain, and so on the angularity of the grains. After cementation, the behaviour ceases to be critically dependent on the angular contacts. Beyond the shallow region of doubt there is a medium (and very important) range in which the time-average equation applies fairly well. At greater depths, or if a great wealth of log data is available, the use of a more sophisticated relation may be justified.

For limestones, smooth and well-behaved velocity-depth relationships (akin to those in shale) have been reported from thick and homogeneous limestones. This degree of homogeneity is

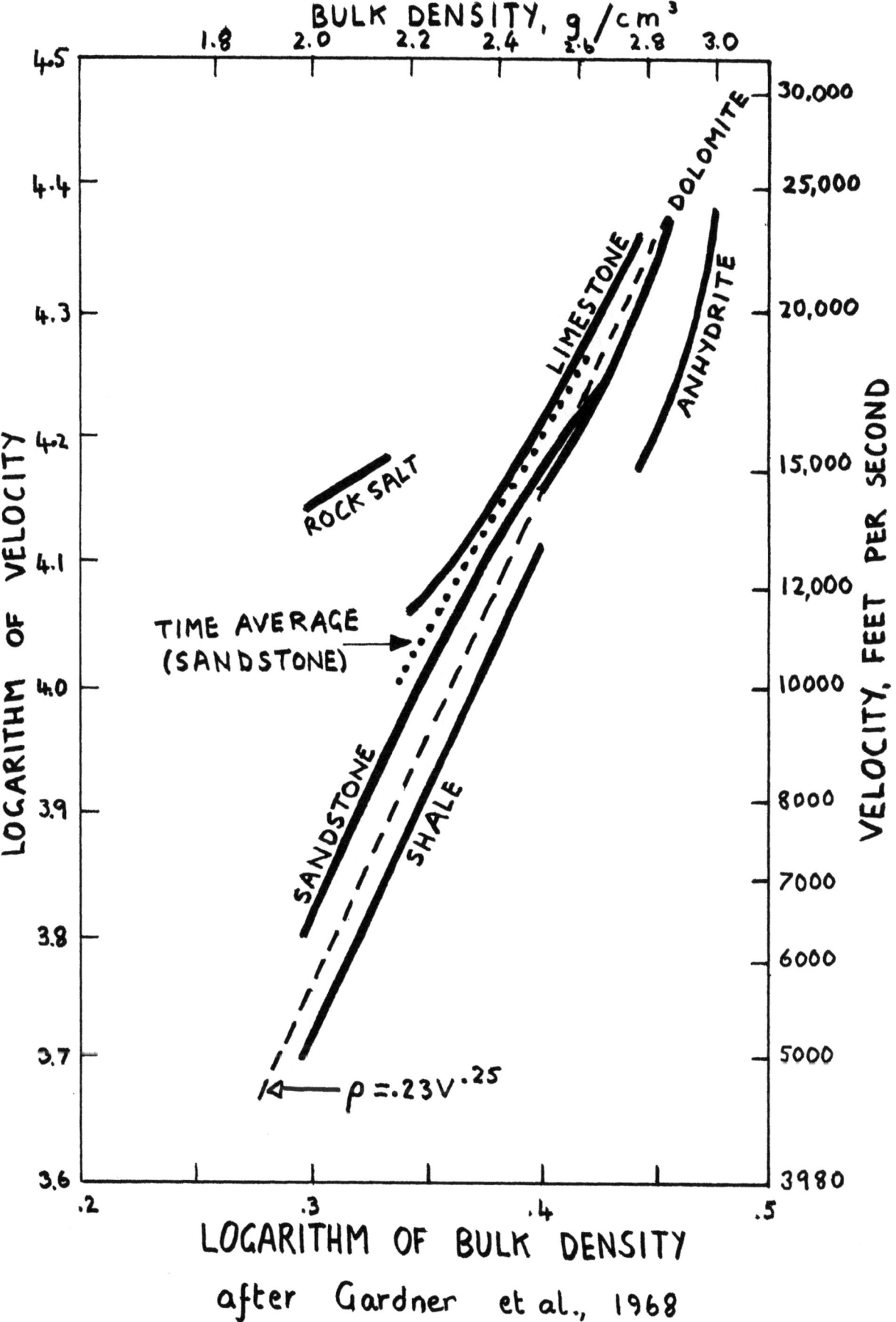
after Gardner et al., 1968

30 not universal, nor even general. Under homogeneous conditions,
31 the time-average equation applies reasonably well.

Velocity-density relations

32 Since both density and velocity are closely related to porosity and depth, it would be reasonable to hope for a simple velocity-density relation (at least for porosity of intergranular type). The figure on p.2-50 suggests a simple power relationship between velocity and specific density for many rocks, and shows that a fair approximation for almost all rocks except salt and anhydrite is given by:

$$\rho = 0.23\, V^{1/4} \quad (V \text{ in ft/s})$$
$$\text{or} \quad \rho = 0.31\, V^{1/4} \quad (V \text{ in m/s}).$$

33 The figure on p.2-52 shows another version of the velocity-density relations, this time with the porosity corresponding to the density noted along each curve. The two figures are reasonably harmonious, except at very high porosities. The curves therefore represent useful generalizations, of value in an area where we do not have sufficient well control to be more specific.

We must note very carefully that all of this summary section 2.3.5 has assumed water saturation and <u>normal fluid pressures</u>. If the pressures are highly abnormal, the velocity and density are amost independent of depth, and have very low values — typically 2150 m/s (7000 ft/s) and 1.8 g/cm^3 for shale.

Also we should stress and stress again that the density-velocity relations suggested above must become invalid in zones of extensive micro-fracturing, where major reductions of velocity occur <u>without</u> much reduction of density.

34 Finally, we note that these velocity-porosity-density-depth relations are concerned with the instantaneous (or local) velocity. Seismically, however, we can measure only the gross interval velocity across a major interval or layer. In deep thin layers the change of velocity and density across the layer

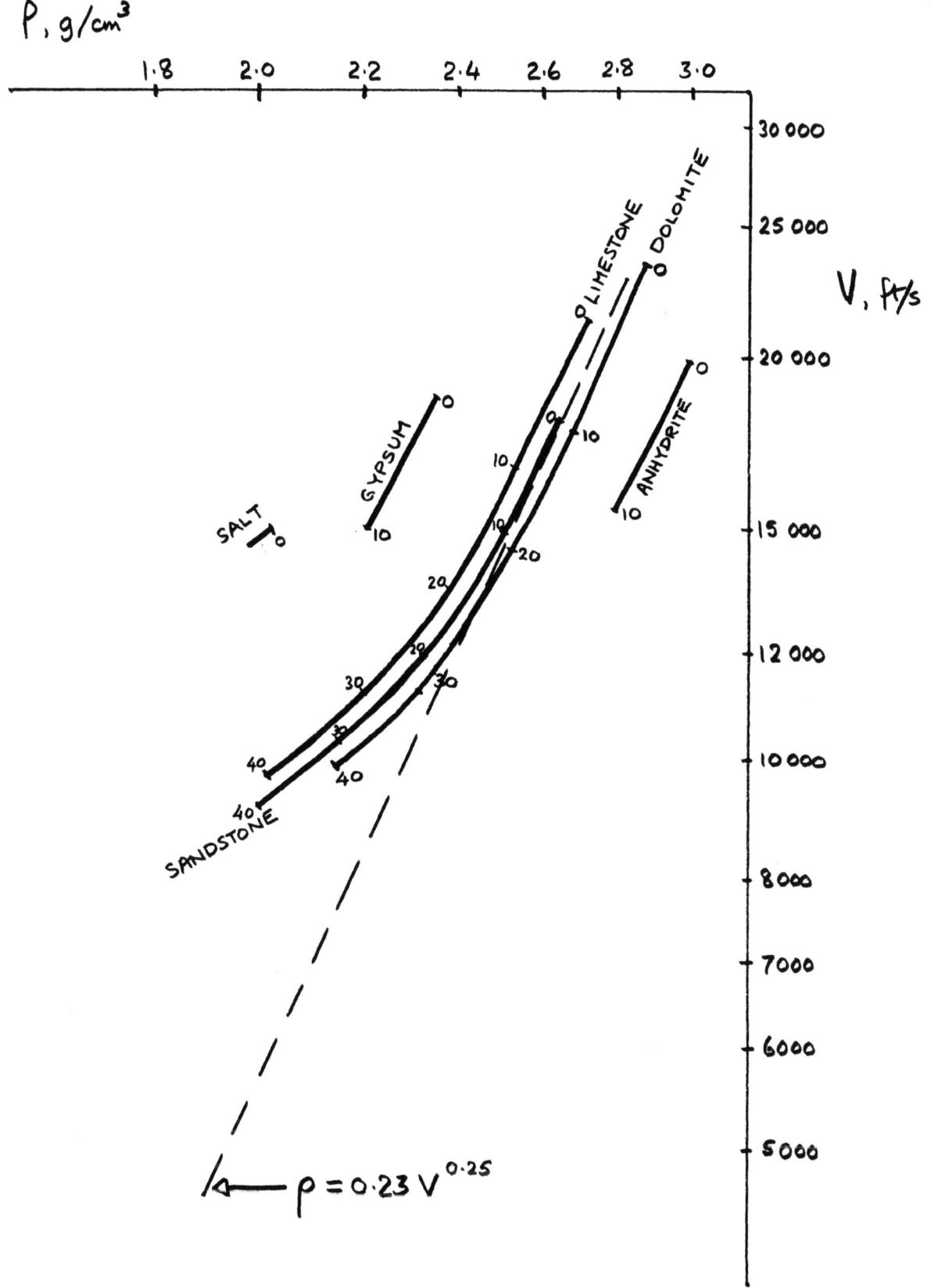

After Pirson, after Schlumberger

2-52

(by reason of additional compaction) is very small; therefore we can accept the seismically-measured interval velocity everywhere with the layer. The same applies to layers which have been deeply buried, but are now uplifted. However, generally, in thick or shallow layers this is not true. If we are to compute from the seismic interval velocity the instantaneous velocity at the top and bottom of the layer — and we shall see later this is critically important to porosity estimates in gas sands — then we must know the law of variation of velocity with depth. This is illustrated on p.2-54. In a case where we can accept a velocity-depth relation and a velocity-density relation, we can calculate from a seismically-measured interval velocity the local velocity (and hence density) just at the top and bottom of the layer. This is very important. The formulas for the calculation (for shale of the types discussed) are on pp.2-35 and 2-37.

Earlier, we raised hopes of narrowing the range of velocities to be ascribed to a particular lithology in a particular association, and so of using velocities for lithologic identification. What has happened to those hopes now? For the case of water saturation, we can say this:

- The shale compaction curve is essential to any attempt to narrow the range.

- Only two situations are really clear — the highly overpressured shale and the unfractured carbonate; the first has a velocity markedly less than that of the shale compaction curve, while the second is markedly greater.

- If a carbonate is clearly identified by its high velocity, and if that velocity is reduced significantly in an uplifted zone, we expect <u>fracturing</u>. This is often a useful conclusion, in a reservoir sense, except that even a major change of velocity need not imply a large porosity. Only a density measurement can confirm that the fracturing has yielded significant porosity.

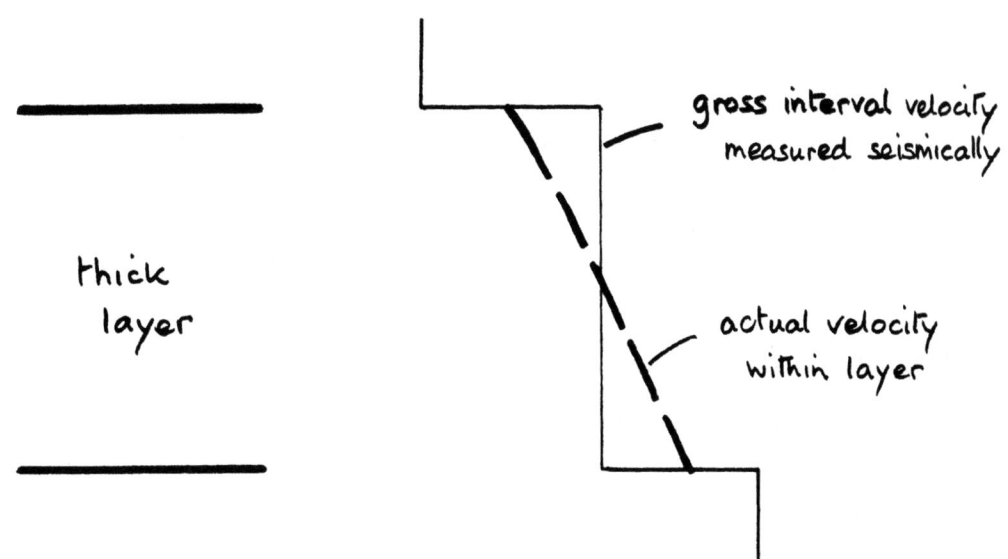

- For many clastic rocks, the condition of the rock is more material than the lithology, in determining the velocity. The porosity, the degree of cementation, the presence and type of fractures, the degree of recementation, and the resulting shape of the pores — all these may combine to make a sandstone have a higher or a lower velocity than a shale at the same depth.

- Obviously the continuum of grain sizes and the possibilities of contamination militate against clear-cut lithologic distinction from physical properties.

- The sandstone-shale abiguity is not resolved by linking a density measurement to the velocity measurement (since the zero-porosity densities are much the same). At depths below a couple of kilometres, however, a clastic density of 2.4 or less (indicating a porosity of more than 15-20%) suggests a coarse-grained material.

- The velocity and density of salt are not much changed by depth. In a young predominantly-clastic section, salt is often recognized positively by its velocity (4300-4600 m/s; 14-15000 ft/s); in an older section it is recognized by its low density (2.05 - 2.15 tonnes/m^3).

All of this means that we shall often be unable to resolve the lithology from our velocities, and sometimes not from velocity and density together. In many practical circumstances, the velocity of a clastic rock at a certain depth is a better indicator of porosity than it is of lithology; if the lithology can be established by considerations of depositional environment (as by the techniques of seismic stratigraphy discussed in section 5.2) then the velocity indications of porosity may be very valuable in a reservoir sense.

The above comments apply to the immature exploration province. In a mature province, of course, the knowledge from boreholes often allows us to be completely positive about the local association of velocity and lithology.

Our summary has been premised on water-saturation. Although the velocity and density of oil are 10-20% less than those of water, the effect of oil-saturation in a material of normal porosity becomes very small. We shall compute later that there is very little hope of detecting the difference between oil and water saturation by seismic measurements, except perhaps under very well-controlled conditions.

We turn now to the matter of gas-saturation.

TAPE 6 ### 2.3.6 The problem of velocity in porous gas-saturated materials

First we recall our simple physical model (p.2-30) and its conclusion that the velocity of a gas-saturated porous rock must be very much depressed at shallow depth but substantially unchanged at great depth (p.2-32). This conclusion follows from the ease with which the solid material can deform into the pore space, until that pore space becomes closed by external pressure. Further, in the discussion of our simple model we said that these would be a negligible difference between having vacuum in the pores and having a gas — the ability of the gas to resist the deformation of the solid material into the pore would be insignificant. Physically, this seems very plausible (unless perhaps the gas was at very high pressure). And if we do not expect much difference between vacuum and gas, we certainly do not expect much difference from one gas to another.

However, if we set up the time-average equation for a gas sand of 30% porosity, and start to insert some numbers, it is immediately obvious that the velocity of the porous sand depends critically on the velocity of the gas, and hence on the particular gas considered. This is unthinkable. It therefore braces us for the unwelcome but necessary conclusion that the time-average equation cannot apply to gas-saturated materials.

In the early days of bright-spot technology, much work was done, and many porosities were estimated, by applying the time-average equation to gas saturation. All this work was erroneous. Sometimes it produced plausible results, but that was just good chance — the time-average equation cannot apply to gas saturation.

Several noteworthy attempts have been made to arrive at an equation properly connecting velocity and gas porosity. The resulting equations are complicated (Gardner et al., 1974; Brandt; Biot; Geerstma). Basically, they do not give us what we want — a direct relation between velocity and porosity, as given by the time-average equation — because they require to know not only the elasticity of the rock grains but the effective elasticity of the matrix in its porous state, which we have no way of knowing. So we have to accept that the situation is not soluble unless we have some other input.

An approximate equation is given on p.2-58. This is useful not for computing gas-saturated porosity from velocity, but for calculating the effect on velocity if, in a water-saturated sand of constant porosity, the water is replaced by oil or gas. Thus we imagine a case where we can identify positively a water-saturated sand giving way to gas-saturation in a manner which makes it reasonable to assume that the porosity has not changed — and where we càn measure the velocities of the sand in the two conditions. Then we can set up two such equations, in which the only unknowns are the porosity $\emptyset$ and the porous matrix elasticity k_s; this allows us to solve for the porosity.

The figure on p.2-60 shows the result of this exercise on the water-saturated sand previously considered on p.2-40. This is a most important illustration. In particular, we note:

- The velocity distinction between oil and water saturation in a sand is very small except at very shallow depth.

- The velocity distinction between gas saturation and water or oil saturation is major down to about 1600m (5000 ft); at greater depths it still exists, but may be smaller than we can hope to measure.

$$V_r = \sqrt{\frac{1}{\rho_r}\left\{2k_s + \frac{(1 - k_s/k_m)^2}{\frac{\phi}{k_f} + \frac{1-\phi}{k_m} - \frac{k_s}{k_m^2}}\right\}}$$

where:

V_r = velocity in fluid-saturated sand

ρ_r = bulk density of fluid-saturated sand

   = $\rho_m(1-\phi) + \rho_f \phi$ (p. 2-17)

k_s = bulk modulus for the porous rock skeleton

k_m = bulk modulus for the rock grains

   = 38×10^9 N/m² for quartz sand

k_f = bulk modulus for the fluid

   = $2.4 - 2.75 \times 10^9$ N/m² for brine

   = 0.133×10^6 N/m² for methane at NTP

ϕ = porosity

ρ_m = 2.65×10^3 kg/m³ for quartz sand

ρ_f = $1.025 - 1.07 \times 10^3$ kg/m³ for brine

   = 0.0007×10^3 kg/m³ for methane at NTP

2-58

- The velocity distinction between the gas-saturated sand and the shale illustrated (approximately, $V_z = 1585z^{1/5}$) is very major at shallow depths, but non-existent deeper.

- The velocity distinction between shale and uncemented water-saturated sand is small at shallow depth.

These are <u>essential conclusions</u> to carry forward to future studies. The details will vary from area to area, but the basic tendencies are valid generally. In particular, the conclusions are not dependent on the odd shape of the velocity-depth curve for the sand; the same depressions of velocity would occur whatever the curve used.

36(B) All of the above applies to the ideal case where the pores are completely filled with gas. The usual practical case involves part water-filling and part gas.

Density has already been considered (p.2-18); there is a smooth increase of density with water-saturation.

Elasticity, however, is different. The ease with which the solid material can deform into the pore is scarcely affected by the presence of some water; all the deformation is readily absorbed by the gas. This remains true whether the proportion of water in the pore is 10%, or 40%, or 70%; the remaining gas absorbs the deformation. Over this range of water-saturation, therefore, the elasticity remains substantially constant, while the density increases; it follows that the velocity <u>decreases</u> with water-saturation. When the proportion of water rises to 100%, however, the velocity must rise considerably; there is no gas left to absorb the deformation, and the deformation is resisted appreciably by the water (p.2-31). All the change between gas-saturated velocities and water-saturated velocities therefore occurs with the <u>very first bubble of free gas within the pore</u>.

The computed effect of water-saturation on velocity is illustrated, for three depths, on p.2-62 (Domenico, 1974, 1976).

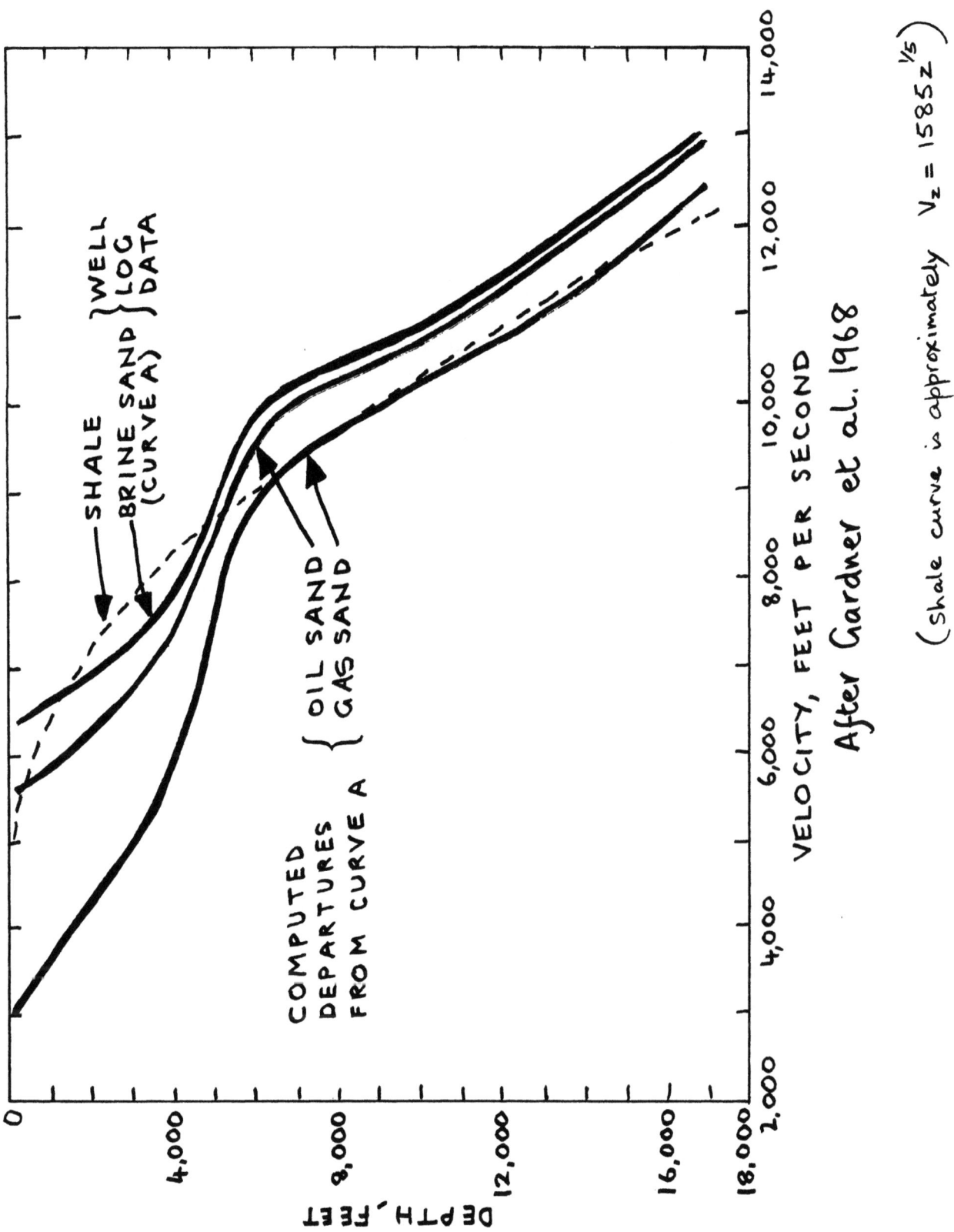

The effect as measured in the laboratory is not quite as dramatic, but all the change of velocity still occurs in the final 20% of water-saturation. (We note, incidentally, that no such effect occurs with water-saturation of an oil reservoir.)

There are two important conclusions from this.

- In any attempt to calculate the porosity of a gas sand from seismic measurements, the approach must be through density, not velocity.

- We are now much less disappointed by the news that we cannot use the time-average equation to calculate the volume of gas present, because the insensitivity of velocity to water-saturation means we could not have used the equation anyway.

In the real world, it is clear that very low velocities are indeed observed in very shallow gas-saturated sands. It may be true that these velocities are below water velocity, and as such undetectable by the sonic log. However, these velocities are not to be expected at commercial gas depths, where the sands are consolidated; in these circumstances a value of 2000 or 2500 m/s (say 7 or 8000 ft/s) may still represent a very attractive situation (or, of course, it may not — depending on the water saturation).

In the past, petroleum geologists might well have disputed whether it was possible to have just a few percent of free gas in the pores, on a widespread scale throughout a reservoir. The evidence of recent years, when many unquestioned bright spots have produced mostly water, establishes that it is indeed so.

Finally, let us acknowledge the incompleteness of our knowledge in many matters relating the seismically-measurable properties of a rock to its reservoir characteristics. A great deal of research is being done currently, and we must be ready to adapt to and utilize new insights as they are published.

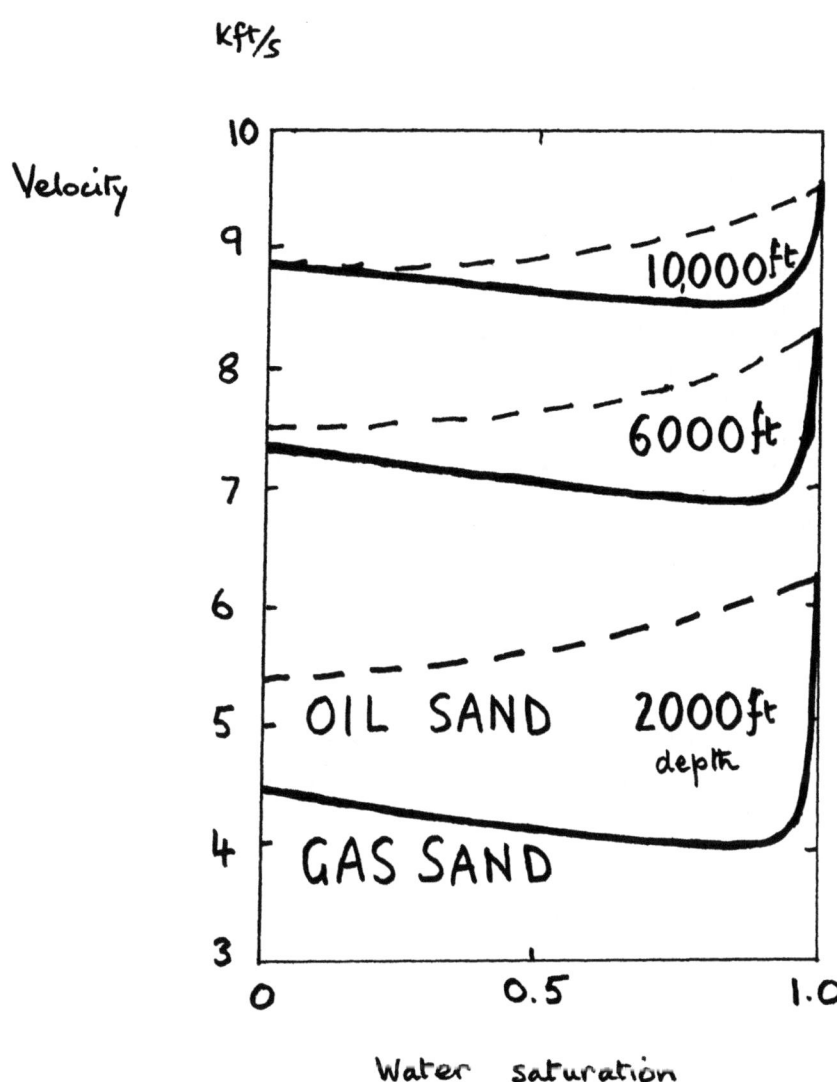

After Domenico, 1973

2.4 THE PARTITION OF PLANE SEISMIC WAVES AT PLANE INTERFACES

On p.2-9 we considered two acoustic wave packets of the same energy propagating in two materials of different acoustic impedance. We took the example of a clay and a granite, and concluded that the particle velocity would be large in the clay and small in the granite.

If a single wave packet propagating in the clay is incident on the interface between the clay and the granite, and if all the energy in that packet is transmitted across the interface, it follows that just on the clay side of the interface the particle velocity is large, while just on the granite side the particle velocity is small. This is impossible if we are to insist (p.2-2) that the materials should stay in contact across the interface, and no void is to appear between them. Consequently we are led to the necessity for reflection; not all the incident energy in the clay can be transmitted into the granite, and the amount which is reflected is just that which provides equality between the particle velocities on the two sides of the interface.

The process of reflection does not involve a loss of energy; the sum of the transmitted and reflected energy is equal to the incident energy.

2.4.1 Reflection of plane waves at normal incidence

As we established in the example of p.2-12, it is important to be specific about the type of measurement which we make. In seismic work we do not ordinarily measure energy; we measure amplitude — pressure amplitude if we are at sea, particle-velocity amplitude if we are on land. Let us start with a definition of the reflection coefficient for measurements of pressure amplitude.

If a plane wave of pressure amplitude P_i , propagating in a material of acoustic impedance r_1 , is incident on the interface into a material of acoustic impedance r_2, then a wave of pressure amplitude p_r is reflected from the interface, where

$$\frac{p_r}{p_i} = \text{reflection coefficient } R = \frac{r_2 - r_1}{r_2 + r_1}.$$

It is important to note that the measurements p_i and p_r are made in the <u>same</u> material — the r_1 material. It is also important to note that measurements of pressure amplitude are independent of the direction of travel of the wave (p.2-11); a squeeze is a squeeze, whichever way it is going.

Clearly, the reflection amplitude p_r is decided primarily by the <u>numerator</u> of the reflection-coefficient expression — the <u>contrast</u> of acoustic impedance across the interface. If r_2 is greater than r_1 (as with our example of clay followed by granite) the reflection coefficient is positive, and an incident compression is reflected as a compression. If r_2 is less than r_1 (which would be the case if the wave is incident from the granite side) the reflection coefficient is negative, and an incident compression is reflected as a rarefaction. If the reflection coefficient of an interface is positive viewed from one side, it is negative viewed from the other side.

What happens, physically, to cause a compression to be reflected as a rarefaction? The wave in the first (harder) material is maintaining a certain balance between the pressure acting and the motion produced — in harmony with the acoustic impedance of the first material — when suddenly it comes to the second material, in which motion is much more readily produced. The wave "falls flat on its face", and a rarefaction opens up behind it as it does so; the rarefaction then propagates in the backward direction.

2-64

Obviously, there are two limiting values implicit in the reflection-coefficient formula — when r_2 is zero (a "free" surface) and when r_2 is infinite (a "fixed" surface). The first is approximated by the interface between earth (or sea) and the air; its reflection coefficient is substantially -1, so that an upcoming seismic wave is sent back down into the earth by the free surface — substantially undiminished in amplitude but inverted in sign. As a corollary of this, to a wave in the air the earth appears as a fixed surface; it is fruitless, therefore, to attempt to generate a seismic signal in the earth by shooting in the air.

It may be helpful to draw the parallel between free and fixed seismic surfaces and the behaviour of a rope suspended from the hand.
If the bottom end is left free, slow vibration of the top end of the rope produces an equal and opposite motion at the bottom end. There are no forces exerted at the bottom end, but there is a maximum of motion. But if the bottom end is anchored, the inevitable zero of motion at the anchor point is accompanied by significant forces exerted by the rope on the anchoring mechanism.

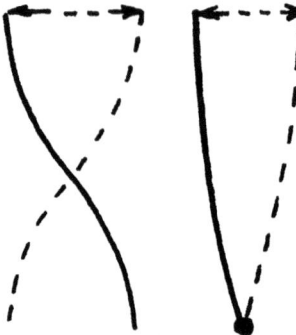

In the same way, when we work on land — with our geophones at the free surface — it is essential that we use geophones sensitive to <u>motion</u>, not to pressure. (As we noted earlier, the normal land geophone measures motion — in the form of particle velocity.) When we work at sea, the free surface is too noisy; to escape this noise we place our detectors deep in the water, and it is this fact that allows us the practical convenience of using geophones sensitive to pressure, not to motion.

TAPE 7 ### 2.4.2 Transmission of plane waves at normal incidence

The proportion of the signal reflected is defined by the reflection coefficient R; the proportion transmitted is defined by the transmission coefficient. If the pressure amplitude of the transmitted wave is p_t, and with the other symbols as before,

$$\frac{p_t}{p_i} = \text{transmission coefficient} = \frac{2r_2}{r_2 + r_1}.$$

By simple algebraic manipulation, this is equal to 1 + R. Our first reaction to the expression 1 + R is to check our algebra, because obviously the plus sign means that if the reflection coefficient is positive the transmitted signal is larger than the incident signal (which looks odd). However, our algebra is correct; on consideration, we see that the behaviour implied by it is also physically reasonable. If there was no reflection, and the whole of the incident energy was transmitted from the clay into the granite, then we know that the pressure amplitude in the granite would be increased to a large notional value (though it would be offset by a corresponding reduction of particle velocity). In the real world there is reflection, and not all the incident energy is transmitted; the consequence is that the pressure amplitude of the transmitted signal is less than the above notional value, but still greater than the incident amplitude. The key point about the transmission-coefficient formula is that the measurements are made in different materials.

We note also that the process of transmission cannot change the sign of the transmitted signal — cannot turn a compression into a rarefaction.

If instead of measuring pressure amplitudes (as we would with a marine hydrophone), we choose to measure particle-velocity amplitudes (as with a land geophone), then the reflection and transmission expressions change. The sketch at the left reproduces what we have already agreed for the pressure measurement; that at the right shows the corresponding situation for particle-velocity measurement. In both cases, the output from the appropriate type of geophone placed in the upper material is illustrated for a seismic pulse in the form of a simple spike, for the case where the reflection coefficient R is positive.

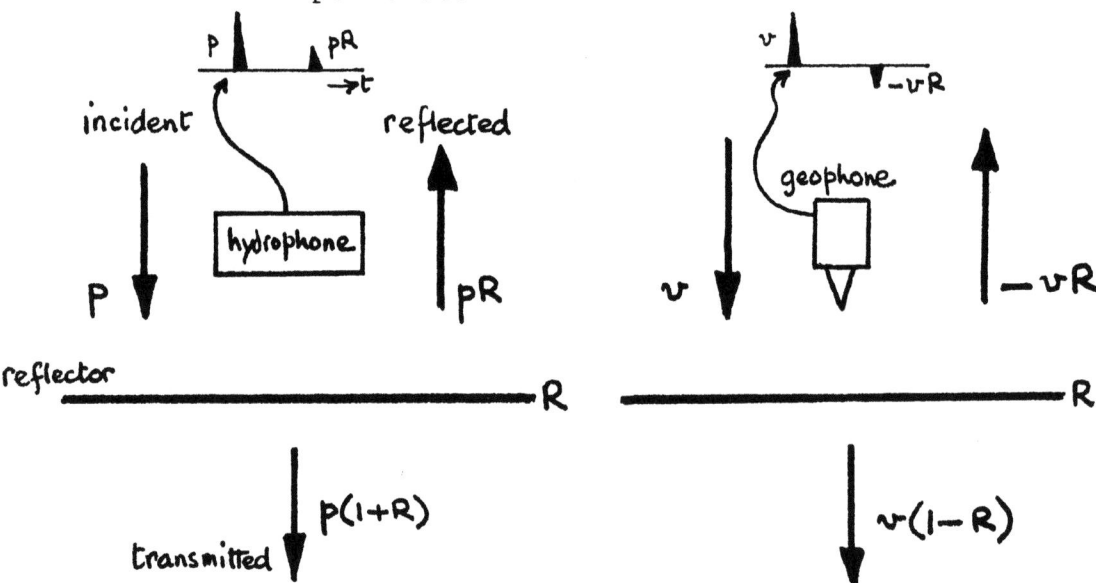

From the right-hand sketch we note that the reflection as detected by a velocity-sensitive geophone appears to be negative -- not because the reflection coefficient is negative (which it is not) but because the wave has changed direction (p.2-11). We note further that the signal transmitted into a harder material has a particle velocity smaller than that of the incident signal (which confirms that the reverse situation obtained with a pressure geophone does not constitute a problem).

Beware authors who do not say what they are measuring!

Finally, let us take full note of the fact that — although the sum of the reflected energy and the transmitted energy is equal to the incident energy — the sum of the reflected and transmitted <u>amplitudes</u> is <u>NOT</u> equal to the incident amplitude.

2.4.3 <u>Measurements with a borehole geophone</u>

If we assume a constant source, what happens to the amplitude of the check-shot signal as the borehole geophone passes from a formation of low acoustic impedance to one of higher acoustic impedance? The answer depends on the type of geophone. If it is a pressure-sensitive geophone (like the Gulf pressure 'phone), then the output <u>increases</u> as the geophone passes into the harder material. If it is a velocity-sensitive geophone (like the SSC or Geospace 'phones), then the output <u>decreases</u> in the harder material.

2.4.4 <u>Measurements with a surface geophone</u>

In the normal practice of reflection work, our reflection paths to any stated reflector always pass <u>twice</u> through every shallower interface. The transmission coefficient for the downward path is $2r_2/(r_2 + r_1)$, while that for the upward path is $2r_1/(r_2 + r_1)$. <u>The two-way transmission coefficient</u> is therefore $\dfrac{4r_1 r_2}{(r_2 + r_1)^2}$, which is equal to $1 - R^2$.

We note that the two-way transmission coefficient implies measurements made <u>in the same material</u>; it is always less than 1, and so (as we would expect) always represents a diminution of amplitude introduced by transmission. Further, the two-way transmission coefficient is <u>unaffected by the type of geophone used</u>.

2-68

In this context, the apparent reflection coefficient of a deep reflector as viewed from the surface is the product of the actual reflection coefficient and the two-way transmission coefficients of all the interfaces above. In the sketch illustrated, it is

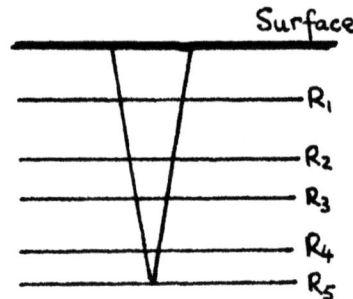

$$R_5 (1-R_1^2) (1-R_2^2) (1-R_3^2) (1-R_4^2).$$

Clearly, it is a general principle — consequent on the conservation of energy — that a geologic section which returns many strong reflections introduces a transmission-coefficient loss. We shall see that this principle (in four words: More up, less down) is useful in simplifying later concepts.

2.4.5 Some questions, to help it settle down comfortably in the mind

1. Verify that the equations for reflection and transmission are compatible with the principle of conservation of energy at an interface, using either a velocity-sensitive or a pressure-sensitive geophone as you wish (or both, if you are irrepressibly keen).

<u>Discussion</u>: Consider unit area of the interface, and use the intensity relation of p.2-9.

2. Compute the pressure-amplitude reflection coefficients and two-way transmission coefficients for the following important interfaces:

	ρ_1	V_1	ρ_2	V_2
Sea floor, for downgoing wave	1	1500	1.8	1667
Sea surface, for upcoming wave	1	1500	~0	330
Typical normal strong reflector	2.2	3000	2.475	4000
Top of gas-saturated sand	2.4	2667	1.8	2000

(ρ in tonnes/m^3; V in m/s)

<u>Discussion</u>: This is more than a trivial exercise in arithmetic, because it is important that we have a good feeling for the actual magnitudes of reflection and transmission coefficients

2-69

in the real earth. Thus a reflector of reflection coefficient 0·2 is a strong one, within the body of the earth; most geological interfaces are less or far less than this. Even for a strong reflector, however, the two-way transmission loss is only 4%. Among the few reflectors which can be expected to have larger reflection coefficients, the most important are the free surface, the sea floor, and the boundaries of a highly-porous gas-saturated sand shallow in the section. Note also that the sea floor (in the example given — recent sediments) is primarily a density reflector; so, for that matter, is the free surface. From p.2-16, the sea-floor reflection from recent sediments is likely to be a transitional reflector, occurring over several tens of metres; this is relevant to attempts to use the sea-floor reflection as representing the source pulse shape.

3. Assume a single gas-saturated sand between the surface and the "normal strong reflector" quoted above. The reflection coefficient of the top of the sand is -0.3, while that of the bottom is $+0.3$. Will the presence of the gas sand produce a visible weakness in the deep reflection? What if there were three such gas sands?

Discussion: The two-way transmission coefficient of the top and bottom of the gas sand, in cascade, is 0.91^2; this represents a "shadow" loss on the deeper reflections of 17%. It is doubtful whether this would be clearly identifiable by eye. Three gas sands, however, would yield a two-way transmission coefficient of 0.91^6, and so a shadow loss of 43%; that we should notice. The message is obvious, and important.

4. In one place successive layers have acoustic impedances of

2-70

3·5, 4·5, 5·5 and 6·5. In another place the first material lies directly on the last. Which condition transmits the seismic signal best?

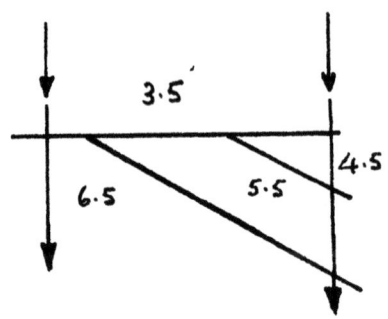

Discussion: Nature, like most of us, abhors the big jerk. It is a smooth transition between the limiting impedances which transmits the signal best; those of us with electrical training will recognize this as impedance matching, while musicians will see it as the reason why wind instruments have flared horns.

5. A slightly-cemented sand of 20% porosity is capped by a shale of density 2·4 tonnes/m^3 and velocity 10,000 ft/s. Part of the sand is oil-saturated and part is water-saturated. What is the ratio of the reflection coefficients at the top of the sand in the oil-saturated and water-saturated conditions? What would it be if the reservoir characteristics remained unchanged but the velocity of the overlying shale increased to 11 500 ft/s?

Discussion: The stated conditions allow the use of the time-average equation, with the values for matrix velocity and saturant velocity previously given in the text.

At first sight, a reflection amplitude ratio of nearly 3:1 would make us hopeful that we could recognize the change from water-saturation to oil-saturation. But the actual values of reflection coefficient in this condition are very small, and so prone to measurement error. Further, small changes in almost anything (including particularly the uninteresting overlying material) can change the ratio drastically, and even make it change sign. However, the concept of measuring a ratio of two reflection coefficients remains a useful one, as we shall see later.

6. A plane reflection changes its strength by 12dB (4:1). What is the most likely cause? Geologically, over what minimum horizontal distance could such a change occur?

Discussion: A question inserted to remind us that the choice of a smoothing operator (to smooth the measurement of some seismically-measurable property along a line) is a geological decision — not primarily a mathematical or statistical one.

TAPE 8   Before we pass to our next subject, let us note the consequences of combining our calculations of the reflection coefficient at the top of oil-saturated and gas-saturated reservoirs (questions 2 and 5 above) with our earlier discussion of the effect of water-saturation on velocity and density within such reservoirs (sections 2.3.4 and 2.3.6). The effect on velocity was given in the figure on p.2-62; the effect on reflection coefficient is shown on p.2-74 (Domenico, 1975).

37
38
39   The diagrams illustrate forcefully that the magnitude of the acoustic contrast at the top of the reservoir is sensitive to depth (which we knew already, from p.2-60) but remarkably insensitive to water saturation (up to 85-95%). As far as reflection coefficient is concerned, there is little difference between a highly commercial gas accumulation with 15% water saturation and a totally non-commercial gas accumulation with 80% water saturation. Needless to say, this fact explains many dry holes drilled in the early '70's; also needless to say, we shall return to the topic later.

2-72

2.4.6 Reflection and transmission of plane waves at inclined incidence

The first consequence of inclined incidence is, of course, the phenomenon of refraction, as quantified by Snell's law:

$$\frac{V_1}{V_2} = \frac{\sin \theta_1}{\sin \theta_2} = \sin \theta_{critical}.$$

Physically, refraction may be viewed as a necessary consequence of nature's preference for the quickest path between two arbitrary points; Snell's law is the solution of the everyday problem which asks, "How much is it worth going out of my way, if thereby I can do more of the journey by a faster mode of transport?"

We note that the process of refraction depends on the contrast of velocity across an interface; this is in contradistinction to the process of reflection, which depends on the contrast of acoustic impedance (product of velocity and density).

In the compressional waves we have considered so far, the motion of the particles occurs along the direction of the seismic "ray". If a plane wave of this type encounters a horizontal reflector at oblique incidence, the particle motion may have a significant component in the horizontal direction. This makes it reasonable to expect the generation at the interface of upcoming and downgoing waves in which the particles are shaken by the wave, rather than compressed; these reflected and refracted shear waves are generated in addition to our desired reflected and refracted compressional waves. The energy represented by these mode-converted waves is inevitably abstracted from the desired compressional waves; consequently we must expect the reflection and transmission coefficients at inclined incidence to be

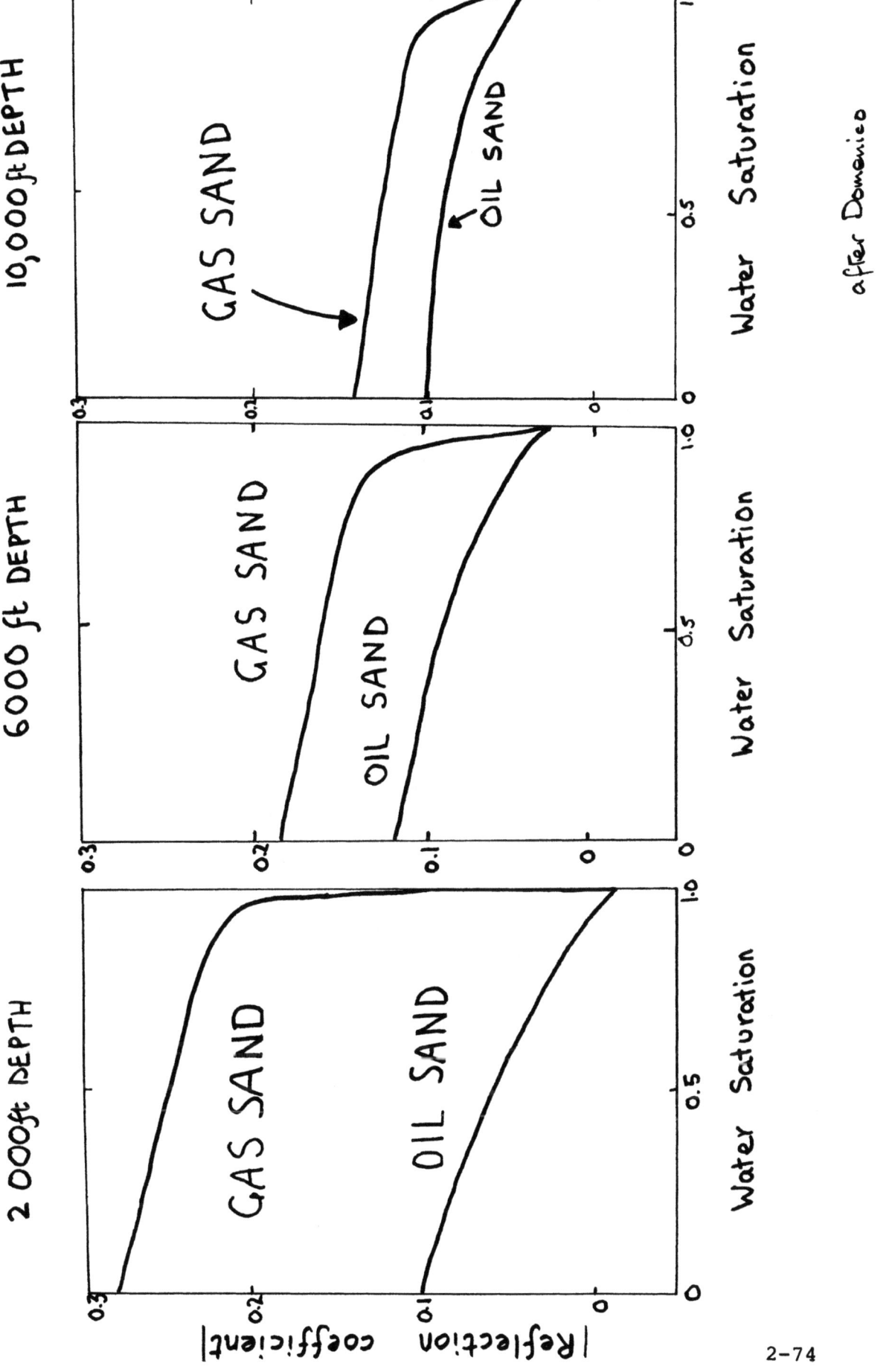

a function of the degree of mode conversion occurring, and so of the angle of incidence.

The actual expressions for the reflection and transmission coefficients become very complicated, and we shall not consider them here. All the useful conclusions, for present purposes, can be illustrated very neatly from Sheriff's diagram (1975) reproduced on p.2-76. As a function of angle of incidence, the diagram plots the proportion of the incident compressional energy reflected as compressional energy. Since the diagram comes from Bob (Webster) Sheriff, we can take it that the meaning really is energy; the proportion reflected must therefore be related to the square of our previously-defined reflection coefficient for amplitudes.

We consider first the full curve, representing an interface across which there is a velocity contrast of 3:1 but no density contrast. The amplitude reflection coefficient at normal incidence is therefore $(3-1)/(3+1)$, which is $+0\cdot 5$; consequently the percentage of the energy reflected at normal incidence is 25%. As the angle of incidence increases towards $20°$, the proportion reflected as compressional energy falls slightly (because a little of the energy is now being converted into shear waves). However, at about $20°$ there is a dramatic increase in the reflected signal; this angle is the critical angle for the velocity ratio of 3:1 — the angle beyond which (in a simple ray sense) there is no transmission into the lower material. And, of course, "less down" means "more up".

Beyond this critical angle the curve shows large swings (associated with favourable and unfavourable conditions for the generation of shear waves) before it finally rises to 100% at grazing incidence — when, again in a simple ray sense, no energy is transmitted into the lower material.

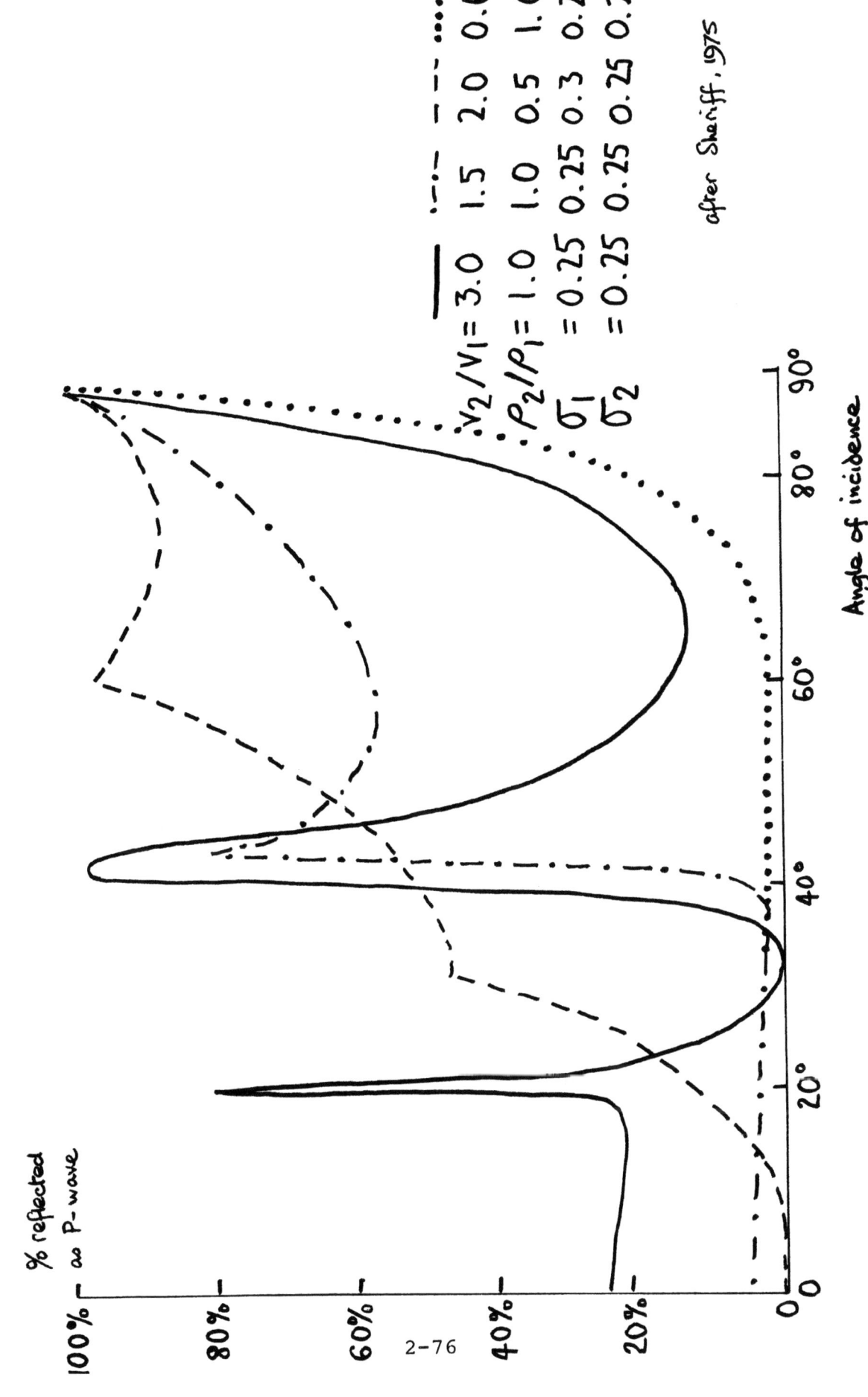

A velocity contrast of 3:1 would be rare in real geology. Much more frequent would be a contrast of 1·5:1, which takes us to the curve —·—·—. In this case the critical angle is about 40°, and the consequences of shear-wave generation are less complicated.

If the contrast has the same magnitude but opposite sign (curve........) there is no critical-angle effect; the two curves are coincident out to about 40°, but the curve associated with the negative reflection coefficient continues to decrease until eventually it rises (as it must) to unity at grazing incidence.

One other case is of interest. Curve — — — represents the situation where the velocity increases by 2 but the density decreases to ½. The reflection coefficient at normal incidence is therefore zero. The reflection coefficient can increase to significant values away from normal incidence.

What are the conclusions of all this for practical seismic work?

- Normally we avoid angles of incidence which, as defined by the velocity contrasts existing in the area, would bring us into the region of the critical angle. Doing this, we can regard the reflection coefficient as effectively constant, at the value given by the simple formula $(r_2 - r_1)/r_2 + r_1)$.

- We deliberately adopt a technique which give us critical-angle reflection if an important target reflector is very weak, but known to be positive; we then obtain a several-fold increase in the effective reflection coefficient. This technique is often valuable in obtaining reflections from the basement.

- We may also adopt such a technique as a means of identifying the polarity of a reflection; positive reflections increase in amplitude at the critical angle, while negative reflections do not.

- In practice, such techniques mean choosing the effective spread length appropriate to the target depth, to the velocity contrast at the target level, and to the variation of velocity with depth; a spread length equal to the target depth implies an angle of incidence of $26°$ if the velocity is constant with depth, and generally rather larger angles if it increases with depth.

- Interfaces across which the velocity increases but the density decreases (occasionally, shale to salt) may sometimes produce a reflection which is very weak on the inside traces of the spread but significantly stronger on the outside traces.

On land work a caution is necessary: the vertical sensitivity of the geophones means that the increase of reflection amplitude at the critical angle does not occur exactly where we would expect it (see Poley, 1964).

2.5 THE LOSS MECHANISMS

To what extent is it possible to apply the facts we have discussed so far, to give us a measure of the physical properties of a reservoir?

Let us consider the situation depicted in the sketch, in which we see on our seismic cross-section a reflection of amplitude 3mm from the top of the potential reservoir. Let us suppose that we also see a featureless interval directly above the reflection, and that over this interval we can measure a seismic interval velocity of V. What can we say about the reservoir material?

If it is geologically reasonable and prudent (we shall discuss this in Part 5) to identify the featureless interval as a marine shale, and if we have (or dare assume) a compaction law for that shale, then we can use the combination of the gross interval velocity and the compaction law to give us a velocity in the shale just above the reservoir. The same assumptions (or, less desirably, the $V - \rho$ relation of p.2-57) yield a value of density corresponding to that velocity. We therefore have the acoustic impedance above the reservoir.

To compute the acoustic impedance of the reservoir itself, we must convert the 3mm of reflection amplitude into a reflection coefficient. As soon as we have done this, we can compute the acoustic impedance of the reservoir material directly; if we are able to measure the interval velocity in the reservoir by seismic means, we can divide the one by the other to yield the reservoir density. This would be a useful prize, because an assumption of reservoir lithology and saturant would then define the porosity.

The next chapter of our course is concerned with the matter of turning a reflection amplitude into a reflection coefficient. However, we shall find that one of the possible methods (and one which we are sometimes forced to use) computes not the actual reflection coefficient but the apparent reflection coefficient as viewed from the surface. It follows that before using this to compute the properties of the reservoir, we must convert the apparent reflection coefficient into an actual reflection coefficient; this means that we must estimate all the losses which occur on the way down to the target reflector and back.

The most obvious of the loss mechanisms we have already discussed: that represented by the two-way transmission coefficients at all the shallower reflectors. This is not really a loss mechanism in the sense that energy is lost from the seismic wave — merely a consequence of the inevitable fact that if energy is reflected (as we hope it will be) then less is available to be transmitted to the deeper reflectors.

Before we look at this transmission loss in detail, let us turn to a mechanism which really does represent a loss of seismic energy — absorption.

2.5.1 Absorption

Absorption loses energy from an acoustic wave by irreversible conversion into heat. The mechanism is frequency-selective, introducing a greater loss at the high frequencies than at the low frequencies. The general effect is well known in everyday life — the sound of thunder or gunfire is a sharp crack if we are close, but a dull rumble if we are far away; a foghorn has to have a low-pitched note, because the higher frequencies would be absorbed in the fog. Similarly, with some types of rock we have an instinctive feeling that a sharp blow applied at one point will sound like a dull thud some distance away.

Absorption is, of course, regrettable. It means that, even without the effect of other agencies, the seismic pulse (from which alone we can make measurements) is doomed to occupy a rather low and narrow band of frequencies, doomed to have a long wavelength in the earth. The longer this pulse, the less our ability to see thin layers, or to detect small geological features, or indeed to make measurements of any rock property at the scale which the petroleum geologist would wish. To a geologist, a rock has meaning as a hand specimen, or even as a thin section; to a seismologist it has meaning only as a sample whose dimensions are measured in tens or hundred of metres. And one of the reasons for this sad fact is absorption.

However, in modern seismics we do not just bemoan the earth's bad manners; we explore the possibilities of turning them to good account. Granted we would have preferred no absorption — but given the absorption, can we measure it and regard it as a diagnostic rock property?

First we must ask what is the mechanism of absorption, so that we can hope to associate the effect with the nature of the rock.

How satisfying it would be if the mechanism were piezolectricity! For then we could say that electrical voltages are generated as the rock is compressed by the seismic wave; these in turn cause currents to flow through the rock and heat it, and so introduce a loss of energy to heat. The mechanism would be linear (a desirable property, obviously), and the absorption would be concentrated in the highly piezoelectric materials (for example, quartz). Unfortunately, however, it seems that piezoelectricity is not a major contributor to the absorption mechanism. It is now accepted that the dominant agency is friction.

TAPE 9   Friction is more complicated, but still not hopeless. If we accept that the loss to heat is caused by particles rubbing together as they participate in the propagation of the seismic wave, what relation would we expect between the degree of absorption and the condition of the rock? For there to be friction, there must be actual motion of one particle relative to another, at the point or points of contact. At the scale of rock grains, this is difficult to visualize in a rock which is unfractured and in which all the grain contacts are well cemented. However, it is easy to visualize in a rock where the grains are poorly sorted, or where not all the grain contacts are in pronounced compression, or where movement of the rock after fracture has created across the fracture a pattern of contacts of which some are in intense compression but others are not. And it is the poor sorting, particularly if existing between rock fragments of different velocity, which would provide the <u>cause</u> of relative movement between the grains.

45
46

This represents, of course, only a very crude and qualitative picture. But, to the extent that it is physically reasonable, it suggests that absorption is likely to be much more marked in a poorly-sorted, uncemented material (particularly one with angular contacts), or in a shallow fractured material, than in a uniform fine-grained or totally cemented material.

This, obviously, is not without interest; it means that many possible reservoir materials are likely to be frictional, and so seismically absorptive. But not all; a lightly-cemented porous sandstone without subsequent fractures is not likely to be highly frictional. So an absorption measurement would be a desirable corroboration of possible reservoir-type materials, but not a necessary one.

And it is quite clear that significant degrees of absorption are not to be expected at great depth, where the overburden pressure (coupled with the expansion due to the elevated temperature) has acted to close the voids and to ensure very intimate contact across all grain boundaries. The only exception to this would be the same as the one discussed in our consideration of velocity: that overpressure can act to hold the pores open and to maintain some of the grain contacts capable of sliding motion.

We shall therefore be interested in measuring anomalies of absorption, as possibly indicating potential reservoir materials, but we shall not be too disappointed if we find none (particularly at depth).

So we need to be able to measure absorption. As we shall see later, this has many practical problems. But let us just review first how we would set about making absorption indications quantitative.

47 (B) At the outset we must ask whether frictional absorption is linear. This is something to worry about, because non-linear stiction is obviously possible (or even likely) at individual grain contacts, and because non-linearity would weaken many of our standard seismic processes. However, it seems that frictional absorption can still be substantially linear as an effect observed in the mass, and so we adopt our usual attitude of declining to consider complications until someone proves that we must.

Then we must ask about the frequency-dependence of frictional absorption. The usual way to make us feel comfortable with the accepted view is to say: It is intuitively satisfactory that the loss should be <u>constant per cycle</u> of vibration.* Thus each successive cycle is attenuated to such-and-such a percentage of the last cycle. The wave therefore loses amplitude exponentially with distance, and the attenuation coefficient is proportional

---

*<u>Question</u> (for the physicists only): Are you comfortable with this, for a <u>frictional</u> loss mechanism? (The easy course is to say <u>yes</u>. If no, search the literature in the names of White, O'Brien, McDonal, Wuenschel, Walsh, Futterman, Strick, and their references — and then ask the question again.)

to the first power of frequency.

Let us go back to the big block of earth material which we considered in section 2.2.1. In that case we vibrated the left-hand edge to generate a plane seismic wave in the block, and concluded (at that stage, in the absence of absorption) that the wave would propagate without change of form or loss of amplitude. If we used a sine wave to shake the left-hand edge, and plotted the signal amplitude as a function of distance, it would appear as in the first sketch. In the presence of absorption, however, each peak is so-many percent less than the last (or, to make the calculations easier, so many dB less than the last); this is suggested in the second sketch. At a higher frequency (third sketch), we see the same dB loss from cycle to cycle, but (since there are more cycles in any given distance) we see a greater loss over any defined path length. The seismically-observable effect, therefore, is of a progressive loss of high frequencies (which implies a progressive broadening of a seismic pulse) over the seismic path.

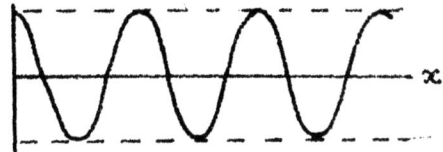

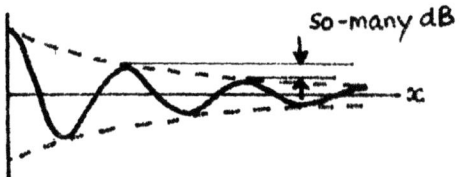

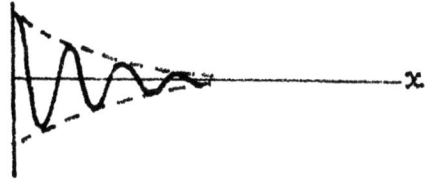

Let us quantify it. The natural measure of absorption, from our point of view, is already obvious: decibels per cycle (or per wavelength). The earthquake seismologists, however, use a different measure, called Q. This measure has an honourable tradition in vibration physics and electrical engineering, but is less immediately meaningful in our context than dB/cycle; the best course for our purposes is just to convert the one into the other, using the fact that the absorption in dB/cycle is about $27/Q$.*

---

*In fact, some immediate physical significance attaches to Q in that the amplitude of our sinusoidal wave falls to 4.3% of its original value in Q cycles.

Thus a typical rock at medium depth is usually ascribed a Q of say 135 (which would represent 0·2 db/cycle); deeper or less absorptive rocks are said to have Q's of several hundred, while weathered rocks at the surface are said to have Q's of 10 or 20. We shall see later that real absorption is probably less severe than suggested by these figures, but we will accept the figures for the moment. What is clear, physically, is that we have no difficulty with the observation that Q increases (absorption decreases) with depth. Friction can occur only where there is relative movement, and for many reasons (closure of the very thin voids under overburden compression and temperature expansion, cementing, the overburden pressure itself, reduction of particle motion with increasing acoustic impedance, and others) we would expect the frictional loss to decrease generally with depth.

The effect of an absorption of 0·2 dB/cycle, for different lengths of seismic path, is shown on p.2-80. Let us see how this is derived.

Question 1: The reflection path to the top of a gas sand is through material of Q = 135. The reflection time is 1·5s. Calculate the dB attenuation at 10, 20, 50 and 100 Hz.

Discussion: A Q of 135 is equivalent to 0·2 dB/cycle.
The number of cycles in 1·5s at 10Hz is 15. So the attenuation
is 3 dB
20   30   6
50   75   15
100   150   30

If now we plot attenuation against frequency, for this path length of 1·5s, we obtain an amplitude-frequency response for the path — similar to those of p.2-80.

2-79

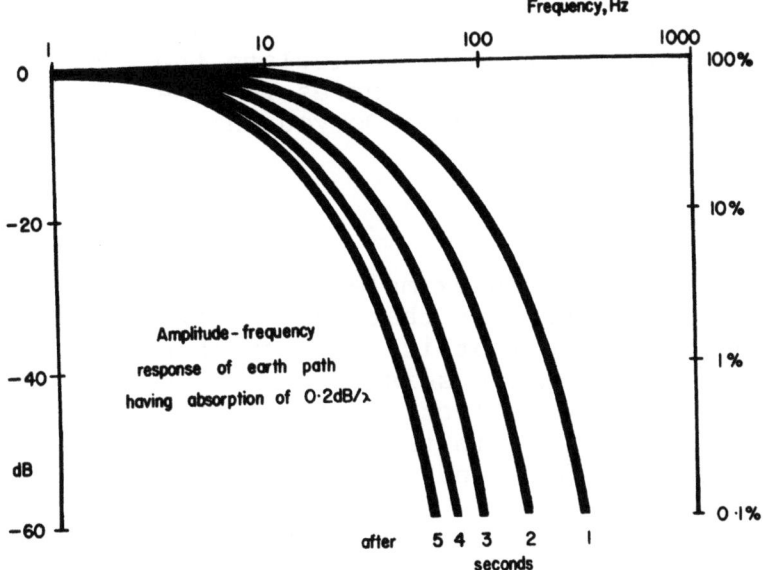

Question 2: The gas sand is 25m thick; it has a Q of 13·5 and a velocity of 2000 m/s. Calculate the additional dB attenuation at the above frequencies for the reflection from the bottom of the gas sand. Do you think this would be detectable in practice?

Discussion: Obviously, this example is an important one. We are asking whether the fact that the target material has a very low Q — and hence that it is a possible reservoir — could be established from the seismic data by comparing the frequency content of the reflections from its top and bottom. The solution is obtained as in the discussion of Question 1, for a two-way time of 25 ms and an absorption of 2 dB/cycle.

From the solution, we note that the differential attenuation over the "usual" seismic band of 10-50 Hz is just 2 dB. Would we be able to detect this, in the presence of the differential loss of 12 dB already present (Question 1) in the reflection from the top of the sand? Perhaps there is a possibility it could be done by machine — but it would be most unlikely by eye. And this is for a sand thickness with which most of us would be more than content, and a Q which is certainly at the bottom end of what is possible in the body of the real earth. We can say that the identification of reservoir-type material by studies of frequency content is likely to be at best difficult and at worst impossible.

As we remember from our pre-course revision (sections 1.1.5 and 6), the calculation of the amplitude-frequency response of a seismic path does not define the effect of that path completely — we need the phase-frequency response also. Here again the usual course is that of assuming simplicity until someone proves otherwise. The simplest plausible assumption is that of minimum-phase behaviour (section 1.1.11).

Although the concept of minimum phase is easy to define mathematically, it is less easily visualized physically. Perhaps it is helpful to say that it is the natural behaviour; most frequency-selective mechanisms in nature are minimum-phase. We may also note that, if a system is to make a distinction between low frequencies and high frequencies, it is reasonable to expect that the system will involve some delays in doing so — the discrimination cannot be made instantaneously. If the distinction is to be small, nature will need not less than a certain delay to effect the distinction; if it is to be large, nature will require a different minimum delay. It is possible to visualize an infinity of systems which will take longer, but none which can take less. For present purposes, we assume that the natural process of absorption in the earth involves this minimum value.

By combining the minimum-phase assumption with the amplitude-frequency responses of p.2-80, we can compute the progressive change of shape of the seismic pulse as it propagates. This is done on p.2-84, for a spike as input.

Since the process of absorption is a high-cut effect, the pulse remains one-sided; it would need a low-cut effect to add further half-cycles. Since longer travel involves a greater loss of the high frequencies, the pulse broadens as it goes. Since the behaviour is minimum-phase, the maximum of the pulse remains as far toward the front of the pulse as the amplitude spectrum allows (section 1.1.11).

Inspection of p.2-84 reveals a fact which has been cheerfully (but wrongly) ignored by a generation of seismic processors: although the decay of amplitude imposed by absorption on a single-frequency sine wave is exponential, the decay of amplitude of a multi-frequency pulse is not. (From the input pulse to the 1-second pulse we see an amplitude reduction to perhaps one-third; from the 1-second pulse to the 2-second pulse we see perhaps one-half; from the 4-second pulse to the 5-second pulse we see perhaps four-fifths.) The old practice by which processors claimed to compensate "inelastic attenuation" with an exponential expansion is quite wrong. The exponential expansion may be fine for other purposes (as discussed in Part 8), but we delude ourselves if we think it compensates absorption.

There is worse yet. Implicit in the minimum-phase concept is the necessity for seismic velocity to be a function of frequency. To the interpreter, this is unwelcome and irritating; he would wish a rock to have a velocity which could be ascribed uniquely to the rock, and which did not depend on the frequency content of the pulse used to measure it. However, that cannot be; it is incompatible with any plausible phase behaviour. The differential delays required for the rock to produce a specified frequency-selective action mean that the seismic signal takes longer to "get through" the system at some frequencies than at other frequencies; in other words, velocity is a function of frequency.

We should keep the effect in perspective; it is not very large, and seismic interpretation has come a long way while cheerfully ignoring it. Perhaps the only message we need to take forward with us is its answer to the question, "Whereabouts on the seismic pulse should the interpreter pick?" The answer is that the part of the pulse which travels through an absorptive medium with a velocity which is most nearly constant is the high-amplitude part; when an interpreter picks a peak or trough somewhere near the "middle" of his reflection, that is the best he can do. We shall see later that this has other merits also — in addition to being just easy to do.

The illustrations on pp.2-85 and 2-86 are an attempt to show how the progressive loss of the high frequencies gradually changes the velocity of propagation; a spike pulse starts with the nominal velocity appropriate to an

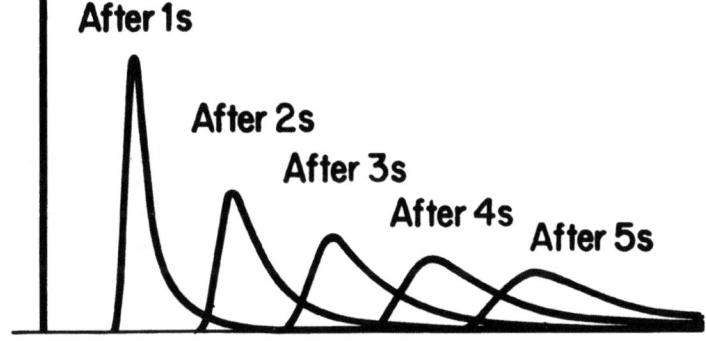

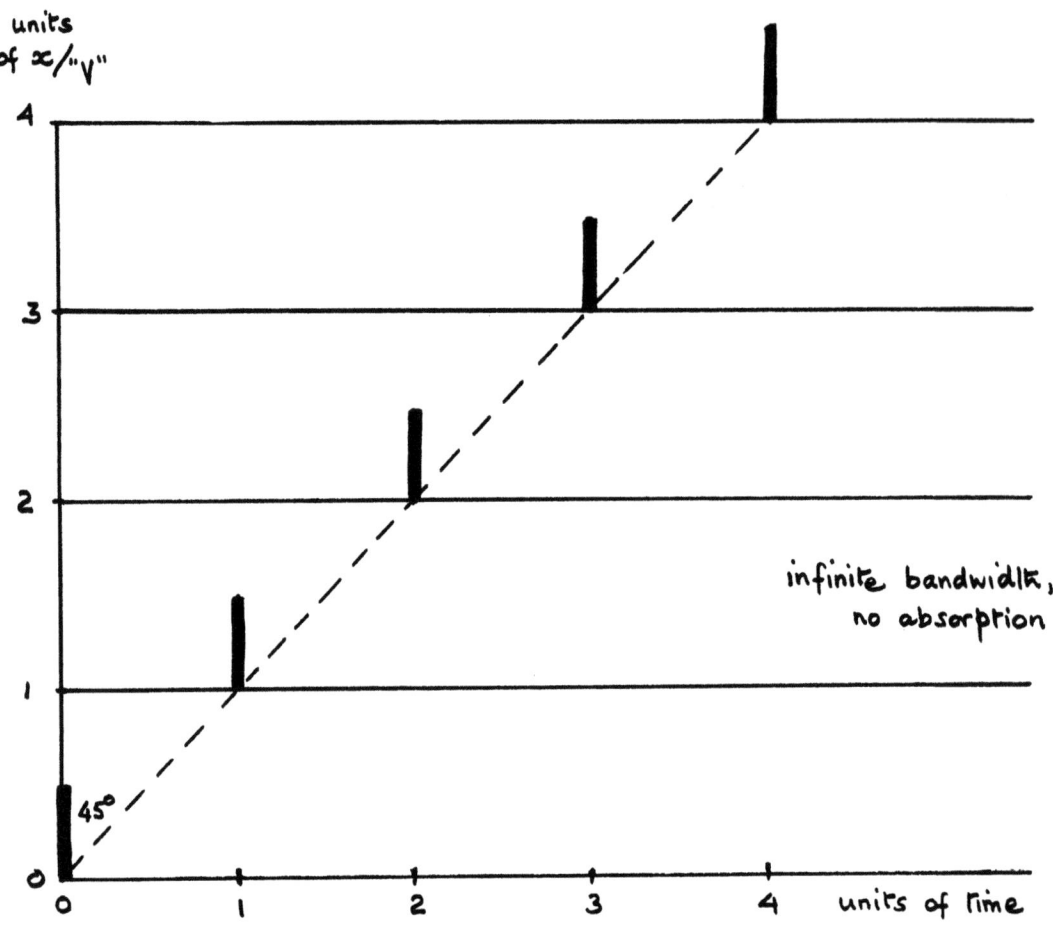

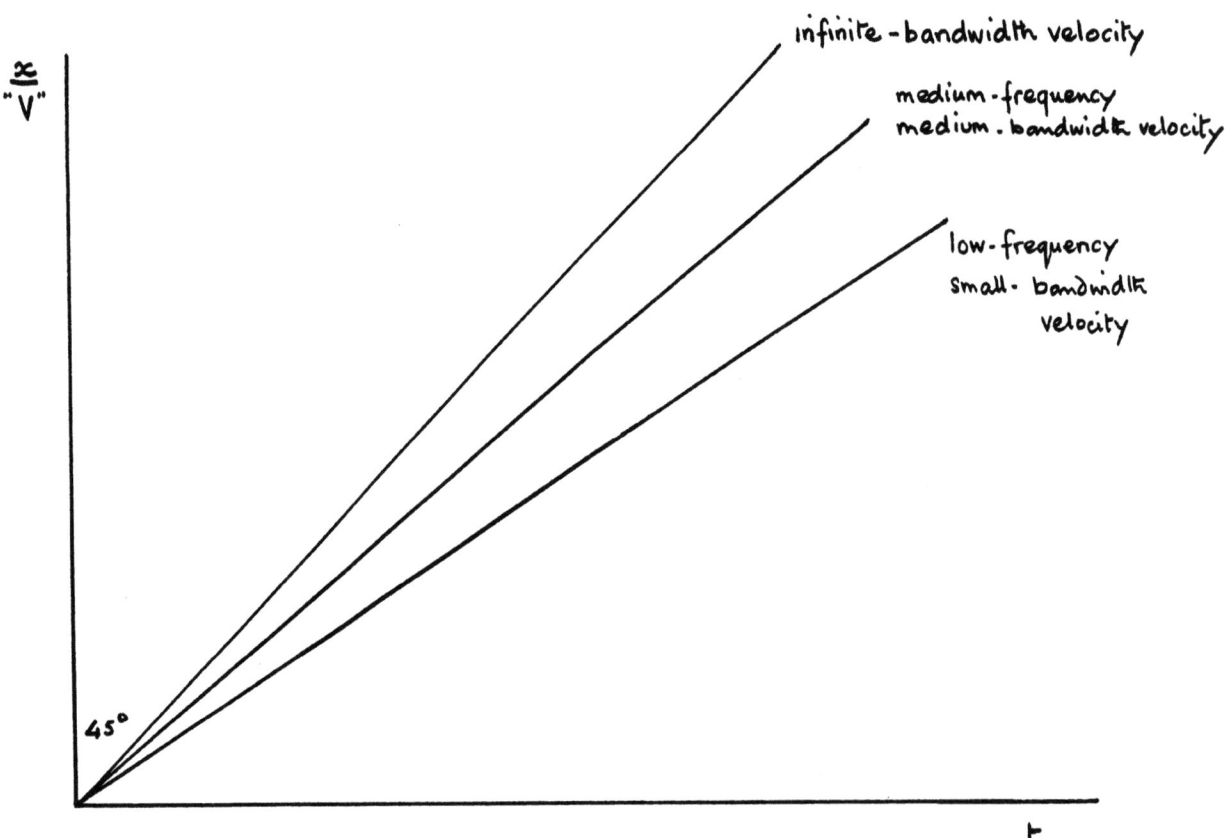

2-85

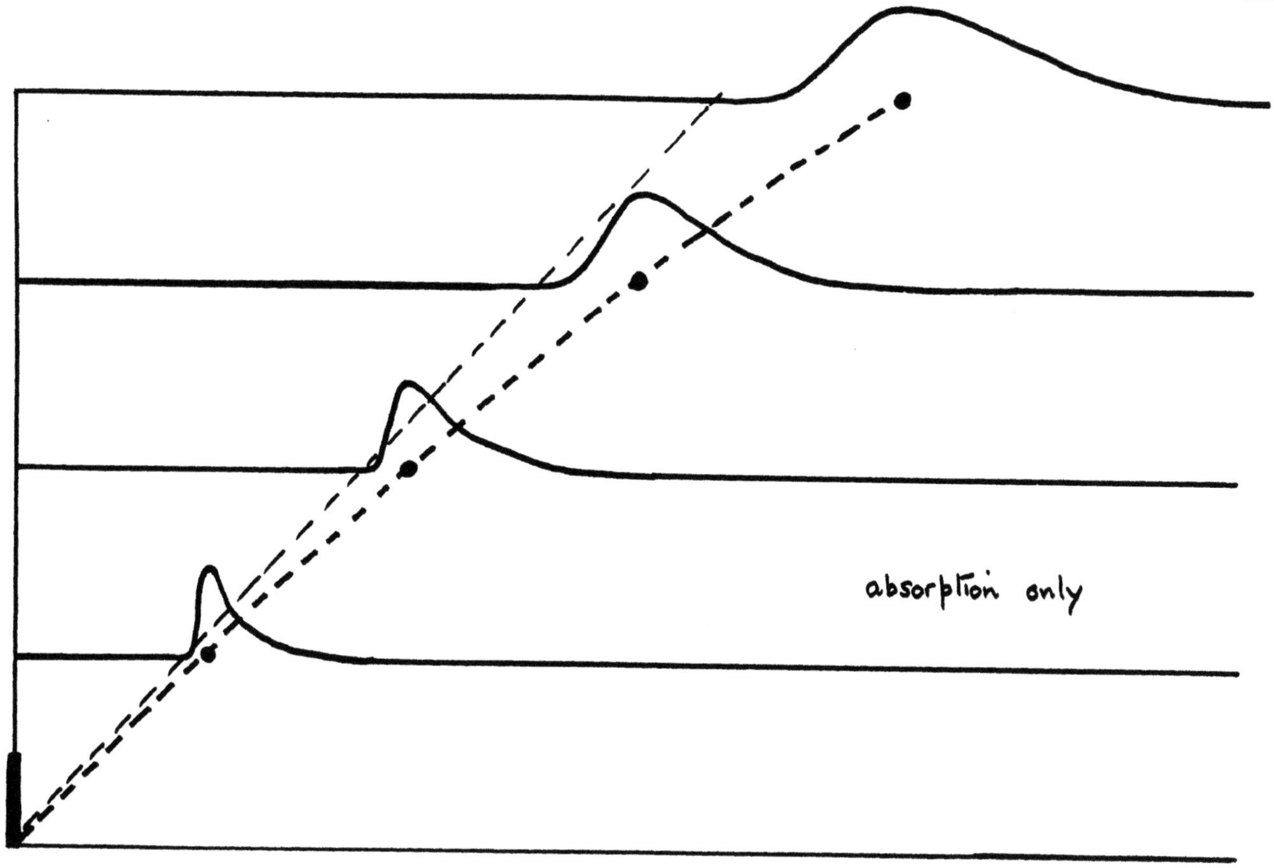

absorption only

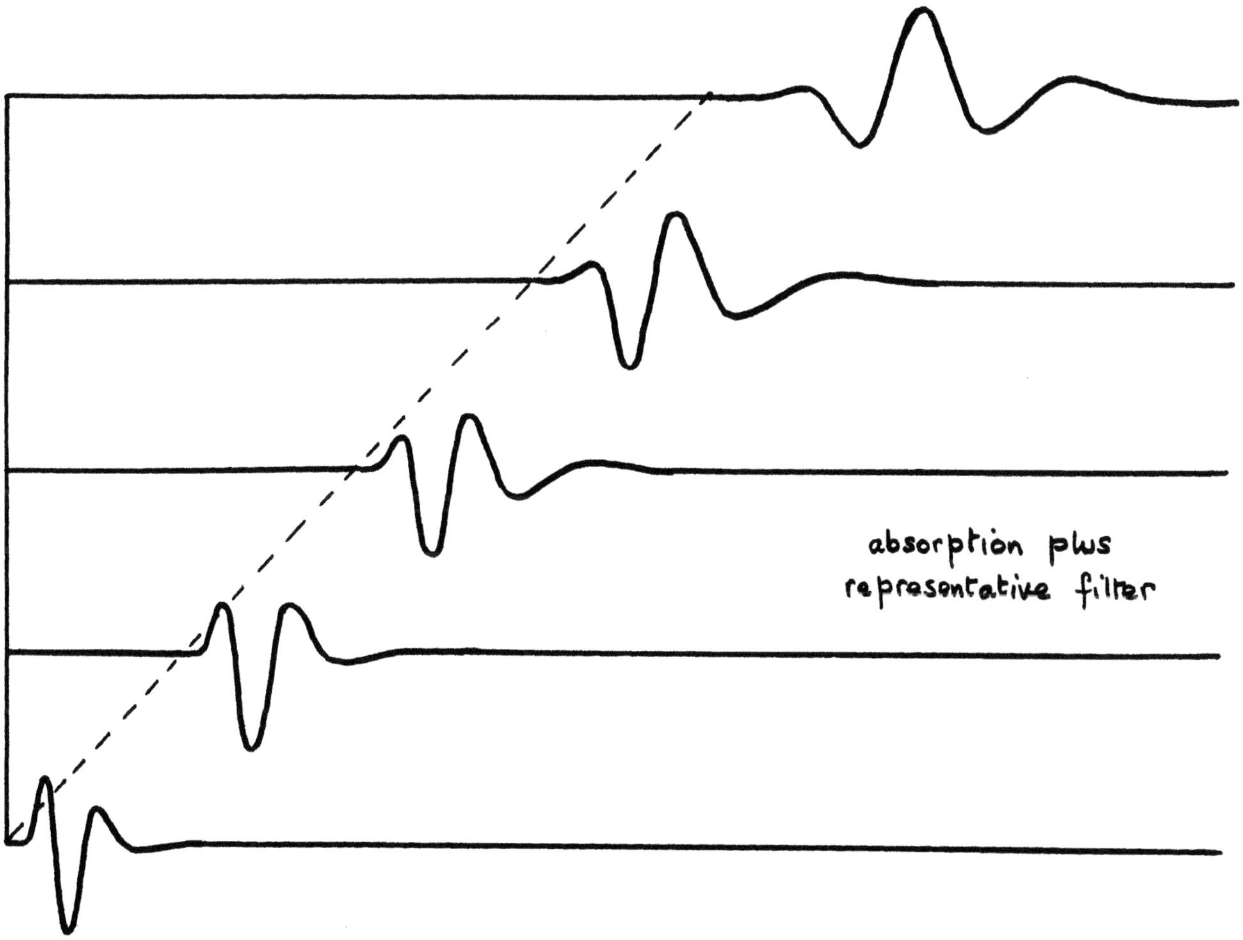

absorption plus representative filter

infinite bandwidth, but gradually slows to a low-frequency velocity as the high-frequency components are removed and the pulse broadens. No definable part of the pulse — onset, peak, trough or any other — travels with a constant velocity; the velocity of the high-amplitude part (black blobs at the top of p.2-86) is not stricly constant, but is better than any other.

48
49
50   Ever since Ricker, it has been well known that the so-called onset (or first break) of the pulse is not defined; it can be moved forward or backward just by changing the
51   display amplitude. The effect (which is very familiar to interpreters of check-shot surveys) is a related consequence of absorption-type effects; in practice it is not possible to increase the gain sufficiently to move the first break back to the position representing the infinite-bandwidth velocity, and so the velocity of the first breaks as picked is doubly unsatisfactory — because of the absorption, and because of the amplitude dependence (p.2-88).

In contradistinction, the time of an individual peak or trough, or of what we have called the high-amplitude part of the pulse, is not affected by gain.

52(B)   Now let us go back to our attempt to extract some <u>usefulness</u> from the regrettable fact of absorption. Let us ask, in particular, whether the degree of absorption in a porous rock is affected by the saturant — oil, gas or water.

First, we note with some interest that, whereas the absorption in rocks is believed to be related to the first power of frequency, that in fluids is quite definitely related to the square of frequency. However, our interest declines as we realize that the actual magnitude of the absorption in fluids is very small; no one ever saw a low-frequency record caused by operation in very deep water.

But how about the situation where the fluid is in the pores of the rock, and where it is possible that the passage of the seismic wave induces relative (viscous) motion between

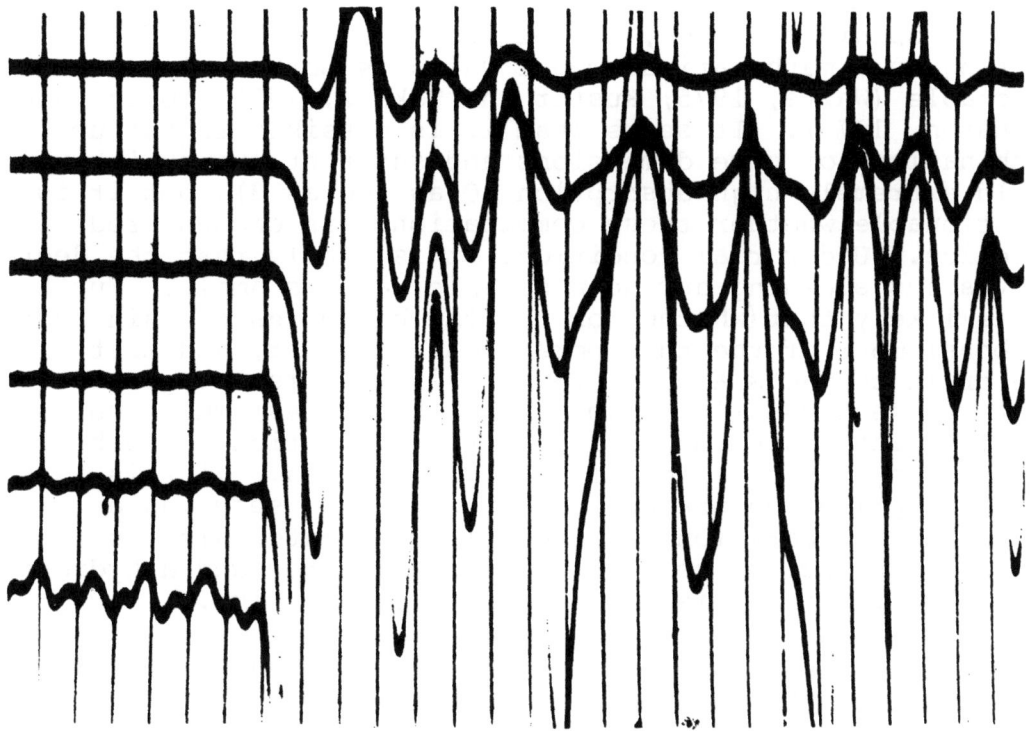

2-88

the rock and the fluid?

Needless to say, these "sloshing losses" are receiving a great deal of research attention; any way to distinguish the saturant using the seismic data would be a prize indeed. However, present opinions differ, largely because of the difficulty of setting up a mathematical model which is both realistic and tractable (White, 1975; Kuster and Toksoz, 1975; Toksoz and Johnston, 1975). It seems that it is possible to set up combinations of pore dimensions and saturant properties which yield extremely high absorption (Q as low as 3), but it is questionable whether these combinations are encountered in practice. Under real conditions it is likely that the degree of viscous sloshing during a seismic wave is small. Further, it is likely that (whether or not the big pores contain hydrocarbons) most of the thin pores contain water, and most of the grain contacts are water-wet; since it is the grain contacts (and usually those in the thin pores) which define the friction, this suggests that the absorption would be little affected by the presence of the hydrocarbons. If this is so, the absorption measurement continues to be useful as a possible corroboration of reservoir-type materials, but the seismic distinction between a water-saturated and a hydrocarbon-saturated reservoir is not possible at this time.

The practical geophysicist might ask, "Why mess with all that mathematics? Why not just take a lump of rock, and measure the absorption for various saturants?" Alas, it is not that easy. The first problem is to get the lump of rock from reservoir depth into the lab, without damaging the fabric, or introducing new fractures, or changing the differential pressure. The second is that we would prefer to make measurements not at the usual ultrasonic frequencies but at real seismic frequencies, using real seismic waves; for this it is desirable to have a cube of rock perhaps 100m on the side. Needless to say, no one is proposing to do a test on such a lump of rock in a lab. Perhaps, if we cannot crack the problem positively with mathematics, we ought to be using mineshafts and boreholes.

But enough of that; let us accept that absorption exists, that it is a significant limitation on our vertical resolution (and as such a confounded nuisance), that there is some small hope of using anomalous absorption as an indicator of reservoir-type materials, and that at this time we dare not say much more.

Earlier in the discussion we left open the possibility that values for absorption are ordinarily quoted too high. In this we were not disagreeing with measured losses of the high frequencies, but acknowledging that there might be other agencies at work besides absorption. It is feasible that the experimenters have observed a loss of 0·2dB/wavelength, for example, and wrongly assumed that all this loss was contributed by absorption. After all, the only observations we can use are:

- a loss of high frequencies, and a consequent broadening of the pulse,

- a loss of amplitude because of the loss of the high-frequency components, and

- a slightly reduced velocity on any particular part of the waveform.

Let us explore other loss mechanisms which might produce very similar effects.

TAPE 10   2.5.2   Short-path multiples

In section 2.4.4. we concluded that the two-way transmission coefficient through an interface of reflection coefficient R is $1-R^2$. In section 2.4.5 we computed that two-way passage through a strong reflector introduces only a few percent diminution in the signal reflected from below, and thus removed from our minds the surprisingly popular error that strong reflectors cast a marked "shadow" of weakness below them. The truth is that single reflectors in the body of the earth, even if they are strong, do not cast a significant amplitude shadow.

However, we also observed that the effect of passing through <u>many</u> reflectors involves multiplying together all the two-way transmission coefficients. On p.2-90 we plot the two-way transmission loss as a function of the number of interfaces, for reflection coefficients of 0·2, 0·1 and 0.05. Now we see that while single interfaces produce virtually no loss, large numbers of interfaces — even of small reflection coefficient — can produce a major loss. (The effect is a commonplace in the laboratory: although a single microscope slide appears perfectly clear to the eye, a stack of spaced-apart microscope slides quickly becomes opaque to the eye.)

From p.2-90, for example, we see that 1000 interfaces of rather weak reflection coefficient (0·075) involve a two-way loss of 40 dB, or a diminution of transmitted amplitude to 1%. That, of course, would be very serious indeed.

So we scurry off to look at outcrops again. Asking, this time, "What is the balance, in real geology, between the thickness of a layer and the reflection coefficients at its upper and lower boundaries? Does it happen that reflection coefficients of 0·075 are centimetres apart, or metres apart, or hundreds of metres apart?"

53 Quickly we see (whether we look at exposures or at well logs) that we must make an important distinction between types of sedimentation. At one extreme we have the classical low-energy deep-sea deposition of shale or carbonate, in which the unit is of substantially uniform composition, and where those interfaces which do exist represent very small reflection coefficients. At the other extreme we have highly stratified deposition, in which short cycles of sedimentation provide thin layers of markedly different acoustic hardness. In the massive material, transmission losses at interfaces may be very small. In the material characterized by highly contrasty stratification the losses may be enormous; one famous example of a thousand metres of interbedded shale

2-89

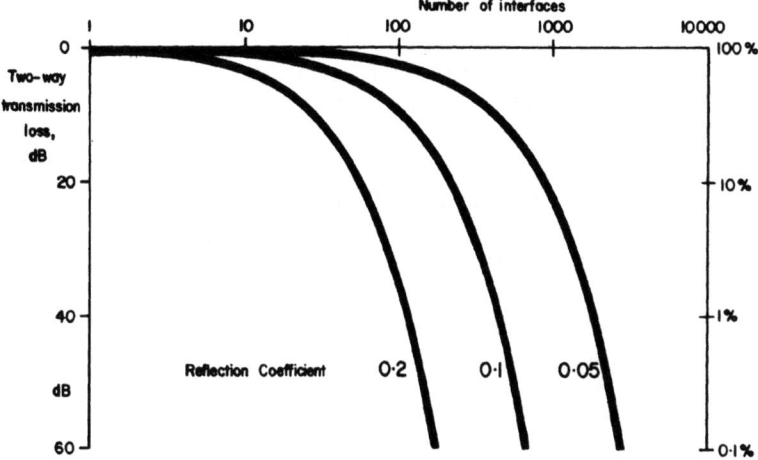

and anhydrite produced a computed two-way transmission loss of several hundred decibels.

This is ridiculous, of course; the seismic method could not possibly work in the presence of transmission losses of several hundred decibels, and such decays are never observed in practice. So we have to find an explanation. The explanation proves to be concerned with multiple reflections in thin layers (O'Doherty et al., 1971).

The essential feature of a geologic column which produces a large interface-transmission loss is that it must involve a tendency for successive reflectors to alternate in sign. A column which goes smoothly from a shallow velocity of 2000 m/s to a deep velocity of 4000 m/s produces negligible interface-transmission loss. A column which goes between the same shallow and deep velocities, but which introduces a large loss, can only do so by including a large number of layers which are alternately hard and soft — and so a tendency for successive reflection coefficients to alternate in sign.

Let us consider just two of such interfaces, alternating in sign (top left, p.2-92). The direct transmitted signal is followed after a short time by a two-bounce multiple always of the same polarity as the direct signal; whatever the sign of the lower reflector, the sign of the upper reflector viewed from underneath is necessarily the same, and the product of the two signs is positive. Of course, the multiple reflection is very small in amplitude. However, the effect is entirely systematic, occurring at all pairs of alternate interfaces. The sum of these two-bounce multiples in a cyclic sequence (top right, p.2-92) soon becomes larger than the direct transmitted signal.

The lower figure on p.2-92 illustrates the computed form of the signal transmitted through a real-life geologic column, after two-way times of 0·7s, 1·4s, 2·8s and 5·6s. The direct transmitted signal is the value appearing at 0 ms; clearly this makes only a negligible contribution after a second or so.

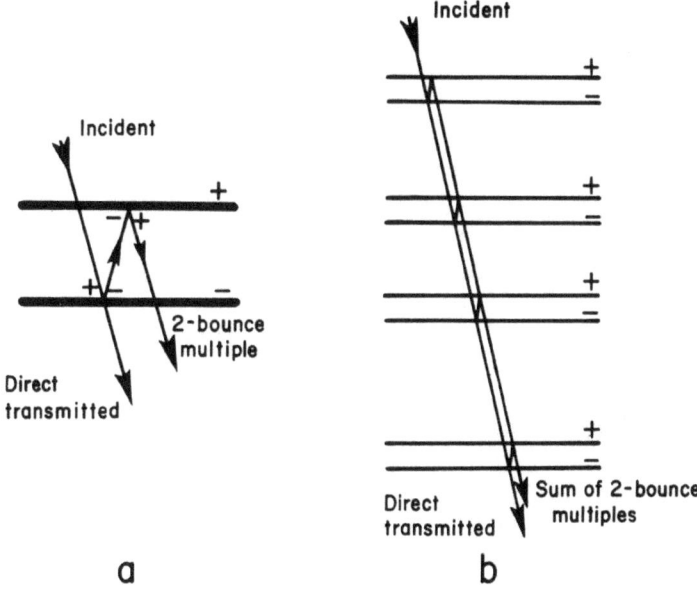

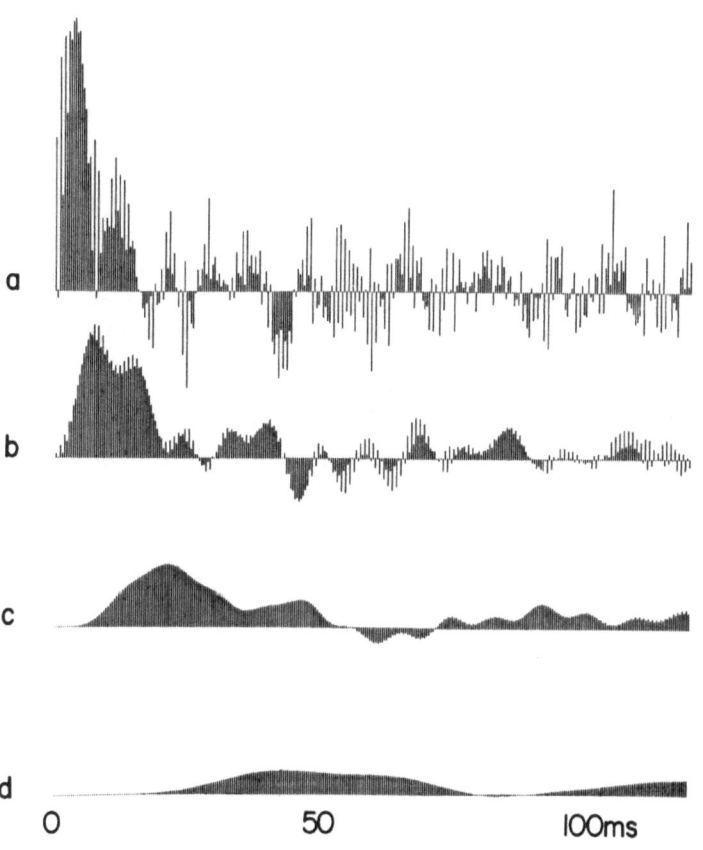

The bulk of the transmitted energy is carried <u>not by the direct arrival, but by the multiples</u>. Accordingly, the arrival time of the bulk of the transmitted energy is delayed; it is as though energy is taken from the direct signal and added back a little later, and this has the effect of <u>decreasing the amplitude loss expected from cascading the transmission coefficients</u>.

Originally this family of multiple reflections was termed "peg-leg" multiples. Since then, however, the term has been used to embrace all unsymmetrical multiples, and so we need another name. Some authors use "very-short-period multiples", others use "very-short-delay multiples", others use "short-path multiples" (which we shall adopt in this course) and yet others use the delightful and expressive "friendly multiples".

A general similarity is obvious between the pulse-shape changes of p.2-92 and those produced by absorption on p.2-84; we see immediately how the short-path multiple effect can be mistaken for absorption. The pulse becomes broader, losing its high-frequency components; the first break (although it exists in principle) cannot be detected, and the delaying of the energetic part of the arrival appears as a lowering of velocity. What is more, the effect is known to be minimum phase.

However, although the effect <u>looks</u> like the high-cut effect of absorption, we should have it <u>clear</u> in our minds that it is not really so. The effect of cascaded interface-transmission losses is not <u>really</u> frequency-sensitive; in contrasty stratification it appears to be so because the major frequency-insensitive transmission loss is compensated to a large extent by the short-path multiples — but only at the low frequencies. The frequency response of the short-path-multiple system is a function of the statistical relationship between the thicknesses of the layers and the products of the reflection coefficients bounding them.

We note a clear distinction from the absorption mechanism in this respect. The high-cut filter associated with absorption is a function of the rock itself, between interfaces. The apparent high-cut filter associated with short-path multiples

2-93

is a function not of a particular rock but only of the
statistical distribution of the interfaces — the degree to
which non-cyclic conditions dominate over non-cyclic conditions,
and the above-mentioned relationship between the thickness of
a layer and the product of the reflection coefficients at its
top and base.

(In passing, let us note that one of the basic premises
of classical deconvolution — that the effective sequence of
reflection coefficients is random, white and stationary — is
negated by the short-path multiple effect in real geology.
Sometimes we are luckier than we deserve.)

We should note also that there is a complementary effect
in thick gradational layers. While the pulse transmitted
through contrasty stratification appears to lose its high
frequencies, the pulse transmitted through a thick layer of
slowly increasing acoustic impedance appears to lose some of
its low frequencies. However, this latter effect is a very
small one; we shall neither see it nor be able to use it.

2.5.3 Scattering from inhomogeneities   (Miles, 1960)

Many common geological features are too small to be
detected as coherent reflectors, but yet produce local
distortions of the seismic wavefront and scattered seismic
arrivals; to the seismic method they constitute geological
"noise". Obvious examples are sediment slumps, cross-bedding,
turbidity effects, ancient erosional channels, small fault
blocks, small reefs, dykes and sills. Within otherwise
homogeneous layers, there can be zones whose acoustical properties
are changed by local tectonic pressures or chemical replacement
or crystallization. And continental deposits normally contain
erosional fragments having a large spread of sizes.

Much of this geological noise can produce an effect which has some similarity to the short-path multiple effect. In the latter, we remember, there was a systematic phenomenon by which every direct arrival had added to it a tail of multiply-reflected arrivals; these broadened the transmitted pulse and gave an appearance of high-frequency loss. The sketch shows how scattering inhomogeneities can have a rather similar effect — adding a tail of scattered arrivals to the direct path, and eventually, after a succession of occurrences, transferring considerable energy from the direct arrival to later times. Thus it happens that scattering from inhomogeneities becomes our third agency capable of producing the appearance of a high-frequency cut.

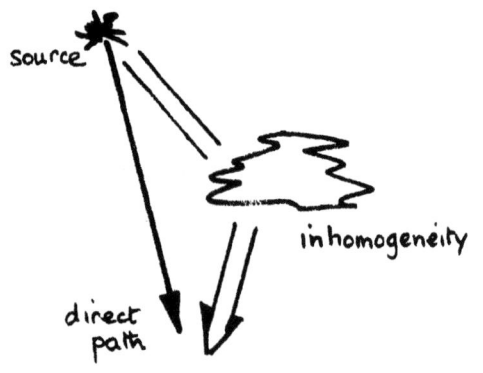

But how serious is the effect? From our sketch we would expect it to be variable with the distance from the inhomogeneity to the direct path, and with the dimensions of the scatterer, and with wavelength. Mathematical attempts to be more quantitative (Korvin, 1973; Nikolayev, 1973) show that for a random arrangement of scatterers, the scattering loss is generally proportional to the mean square of the scatterer dimensions, to the three-halves power of frequency, to the square root of the distance travelled through the scattering material, and to the inverse square of the average velocity through the scattering material. This is very scholarly, but leaves us wondering what to do with it. Perhaps the most interesting thing about the mathematical analysis is the suggestion that one day we may achieve a <u>unified</u> view of the loss mechanisms — a view in which the absorption associated with poor sorting melds with scattering, and in which short-path multiples become a special case of scattering between parallel or near-parallel bodies. This would better equip us to think about a wide variety of other circumstances which must occur in the earth — lateral velocity changes, reflectors of locally changing curvature, and a dozen other inevitable geologic features — which we feel intuitively must act as high-cut mechanisms.

Additional references: Spencer and Edwards, 1975,
                        Richards and Frasier, 1976.

2.5.4 The overall picture

72
We know that the shape of the reflection pulse which we pick on our sections is affected by many agencies. Some of these are concerned with the technology of the seismic method — the source pulse itself, the near-surface, the free-surface reflection above the source (and, at sea, above the hydrophones), the arrays, the recording equipment, the deconvolution process and the final time-variant filter. These are discussed and illustrated in some detail in section 8.2.4. In the present context, however, we are interested in those agencies which exist in the earth itself. Bringing together the conclusions of the last three sections, we see that these agencies are absorption, short-path multiples, and scattering from inhomogeneities. These three agencies affect the shape of the pulse <u>transmitted</u> down to depth and back again; superimposed on their pulse-shaping effect, of course, is the nature of the reflector itself — the details of the stratigraphy at the reflecting level.

73(B)
Apart from our concern with these agencies in determining the shape of the reflection which we pick, we have said that we are interested in assessing the geological message of changes in that pulse shape. We have said that:

- Absorption is a property of the rock itself. Several reservoir materials may be expected to show anomalously high absorption.

- Short-path multiples are related to the degree of contrast in the stratification (particularly within layers in the range 1–10 m in thickness).

- Scattering is related to the degree of inhomogeneity in the geologic section.

The geologic message is therefore quite different for the three agencies. Obviously, in the petroleum sense it is very important that we do not ascribe all observed high-frequency losses to

2-94B

absorption, becoming excited about reservoir possibilities which do not exist.

We shall return to this matter later. Let us say for the moment that we accept the caution, and that we acknowledge that the separation of the three effects on real data is not easy. What is clear is that sometimes one agency is dominant, and sometimes another — it all depends on the geology. Thus in one reported case (Schoenberger et al., 1974) the absorption and short-path-multiple agencies produced about equal attenuation of the high frequencies. In another (the interleaved shale-anhydrite sequence mentioned earlier) the multiple agency was clearly dominant. In a thick interval which shows little evidence of reflection "grain", the dominant agency is likely to be absorption; it is unlikely that highly contrasty stratification (even if fine) will not return at least a grain of reflections. However, a local loss of high frequencies on reflections from below obviously disturbed zones is most likely to be caused by scattering. And, as we computed in our question on p.2-81, a true absorption effect must persist over quite a thick interval if it is to yield a high-frequency loss which we can hope to measure.

2.6 THE DETERMINATION OF REFLECTION COEFFICIENTS

Having discussed the loss mechanisms, we revert now to the question we posed at the begining of section 2·5: How do we turn a reflection <u>amplitude</u> into a reflection <u>coefficient</u>?

Reading the processing advertisements, we really might believe that what we have on a modern seismic section is actually reflection coefficients. This is nonsense, of course; reflections are in microvolts, or in numbers of convenience on tape, or in millimetres of wiggle on a section, and an additional step is necessary to turn them into absolute reflection coefficients.

First we have to ask what we are going to <u>measure</u>, when we talk about reflection amplitude. And what happens to the definition of reflection coefficient, when we have real-world pulses instead of the spikes we illustrated earlier?

In section 2.4.2 we implied that a negative reflection coefficient would involve spike pulses as shown here — an incident erect pulse of amplitude A, and an inverted reflection pulse of amplitude -B. Then the reflection coefficient is obviously - B/A.

If now the incident spike is replaced by a real-world pulse, the reflected signal is unchanged in shape, but inverted in polarity. Further, its amplitude is decreased, in accordance with the same reflection coefficient. It remains true that the reflection coefficient is given by the ratio of the amplitudes of the reflected and incident pulses.
Any convenient measure of amplitude may be used — for example, a particular peak, or a particular trough. However, it is essential that proper account of the polarity inversion be taken;

thus we first determine the sign of the reflection coefficient by noticing whether the reflected pulse is erect or inverted, relative to the incident pulse, and then measure the amplitudes of the corresponding extrema (in the case illustrated, a peak on one pulse and a trough on the other).

If we have a leggy pulse, and a noise background, we may not be able to decide by eye whether the reflected pulse is erect or inverted. Any attempt to measure the magnitude of the reflection coefficient is then at risk. For such cases there is merit in making an amplitude measurement which is insensitive to polarity. Such measurements are computationally very easy; let us think of them as yielding the envelope of the pulse — the smooth shape which is obtained by connecting all the peaks and all the troughs. Since this shape is symmetrical about the axis, let us talk in terms of the one-sided envelope, and let us use the simple term strength to indicate the peak magnitude of the one-sided envelope.

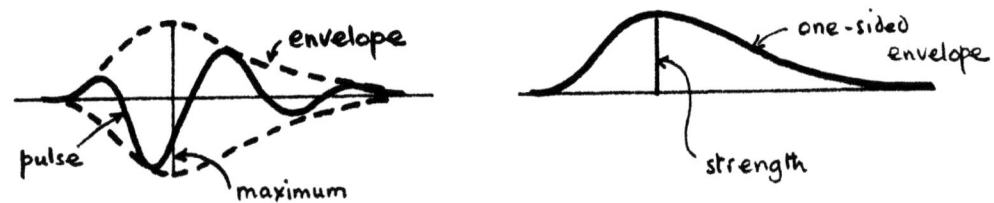

In the remainder of the course, the term "reflection strength" indicates this peak magnitude of the reflection envelope.

As we remember from our discussion of the meaning of velocity in the presence of absorption (section 2.5.1), the peak of the envelope is the point best suited for timing purposes. Now we are going to be consistent and to use it for reflection amplitude measurement also, noting as we do so that it necessarily has the highest signal-to-noise ratio.

TAPE 11   2.6.1 <u>The top-and-bottom method</u>

The first of our methods of turning reflection amplitudes into reflection coefficients is the top-and-bottom method. The method is particularly useful for determining the acoustic impedance of a layer of interest (for example, a reservoir) when the acoustic impedances of the layers above and below are known or can be estimated.

The principle is to measure the ratio of the reflection amplitudes from the top and bottom interfaces. As sketched on p.2-98, the acoustic impedance of the reservoir is r_2, and the reflection amplitudes are S_1 and S_2; setting up the equation for the ratio of S_2 to S_1 leads to a cubic equation in r_2, which can be rearranged as given.

More obviously meaningful than the equations is the graphical representation on p.2-98A. The horizontal variable — the ratio of the acoustic impedances above and below the reservoir — is assumed known; we shall work an example later showing how these impedances are derived. The parameter — the ratio of reflection amplitudes — is measured from the section or from a print-out. Thus we arrive at the vertical variable r_2/r_1, and so deduce the impedance r_2 of the reservoir.

The great merit of the top-and-bottom method is that the only amplitude measurements are made as a ratio, on quite closely-spaced pulses. Therefore the method is insensitive to most of the problems associated with the field work (particularly, source variations) and with the agencies operating on the pulse on its way down and back. It is also easy to ensure that the processing does not distort the ratio of these two closely-spaced amplitudes.

The great weakness of the top-and-bottom method is that it cannot be safely used under circumstances where the acoustic impedances of the overlying and underlying materials are about the same. Thus it is not suitable, in general, for the case of a sand reservoir encased in shale.

The reason for this is easy to see from p.2-98A. Wherever the ratio of r_3 to r_1 is near to unity, a considerable range in values for r_2 is possible without any significant change in the ratio of the reflection amplitudes. Thus if a shallow reservoir sand is saturated with water and encased in shale, the variation of acoustic impedance with depth might appear as in the left sketch; the ratio of reflection amplitudes is about 1. Although the depression of acoustic impedance produced by replacement of the water by gas is major (second sketch), the <u>ratio</u> of the reflection amplitudes remains about 1; the ratio is incapable of distinguishing the amount of the depression except by use of the <u>transmission</u> coefficient of the upper boundary, and we have agreed that in practice transmission coefficients show only weak dependence on acoustic contrast.

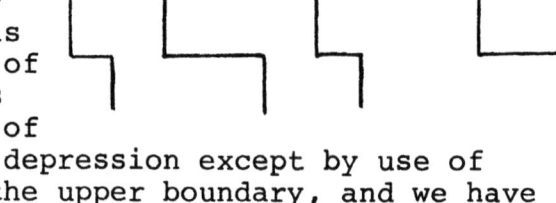

For the top-and-bottom method to be applicable, we must have impedance variations of the type shown in the third and fourth sketches, with the overlying and underlying impedances being strongly differentiated. The situation of the third sketch can occur when the top material is a shale, the second material is a gas-saturated sand and the third material is a water-saturated sand. The situation of the fourth sketch can occur in some reefs, or when a gas-full reservoir has formed on an ancient hard surface. We shall illustrate these situations later.

Because the top-and-bottom method is insensitive to error when overlying and underlying materials are similar, every opportunity should be taken to check its conclusions. Most reservoir configurations include the possibility of measuring amplitude ratios additional to those at the top and bottom of the hydrocarbon-bearing zone; some of these possibilities are sketched on p.2-98B. One extra caution is necessary, however; if the ratio is taken between reflection amplitudes on <u>different traces</u>, we must face the problems of ensuring that the <u>ratio of amplitudes</u> is valid. This matter is discussed in section 3.2.

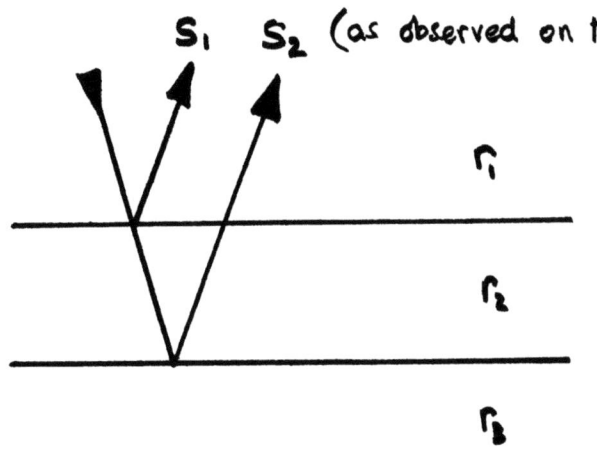
S_1, S_2 (as observed on the section)

$$S_1 = k \frac{r_2 - r_1}{r_2 + r_1} \qquad S_2 = k \frac{4 r_1 r_2}{(r_2 + r_1)^2} \cdot \frac{r_3 - r_2}{r_3 + r_2}$$

Or, rearranged as a cubic in the unknown r_2,

$$S_2 r_2^3 + (S_2 r_3 + 4 S_1 r_1) r_2^2 - (S_2 r_1^2 + 4 S_1 r_1 r_3) r_2 - S_2 r_1^2 r_3 = 0.$$

Or, rearranged in terms of the ratios,

$$\frac{S_2}{S_1} = \frac{4\, r_2/r_1}{(r_2/r_1)^2 - 1} \cdot \frac{r_3/r_1 - r_2/r_1}{r_3/r_1 + r_2/r_1}.$$

Or,

$$\frac{r_3}{r_1} = \frac{r_2}{r_1} \left[\frac{\dfrac{4\, r_2/r_1}{(r_2/r_1)^2 - 1} + \dfrac{S_2}{S_1}}{\dfrac{4\, r_2/r_1}{(r_2/r_1)^2 - 1} - \dfrac{S_2}{S_1}} \right].$$

N.B. Take care with the signs of S_2/S_1!

2-98

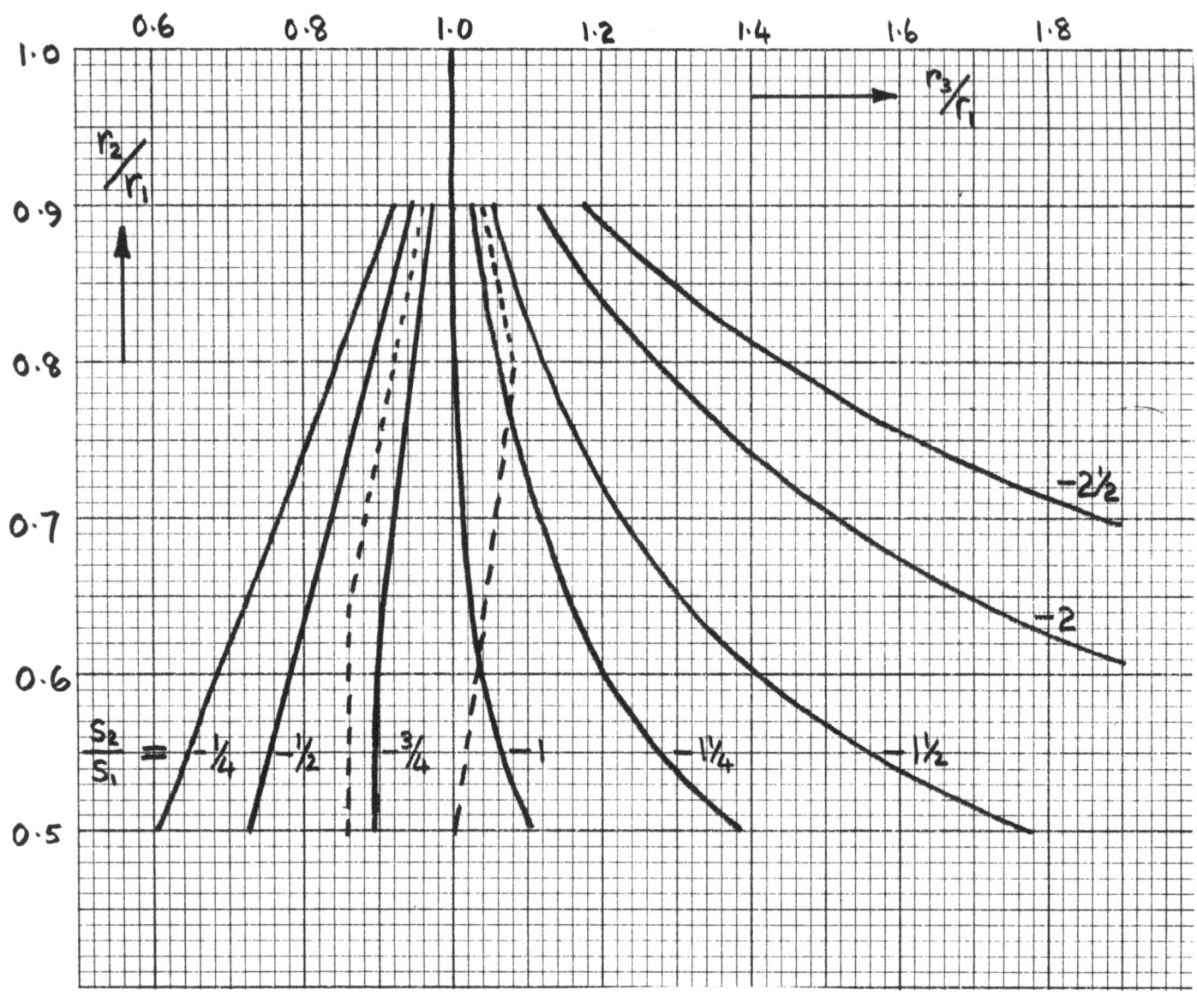

Note the insensitivity to error of the top-and-bottom method in the dashed region, where the acoustic impedances above and below the gas sand are approximately equal ($r_3/r_1 \approx 1$); a large change in the properties of the gas sand produces only a small change in the ratio of amplitudes (S_2/S_1).

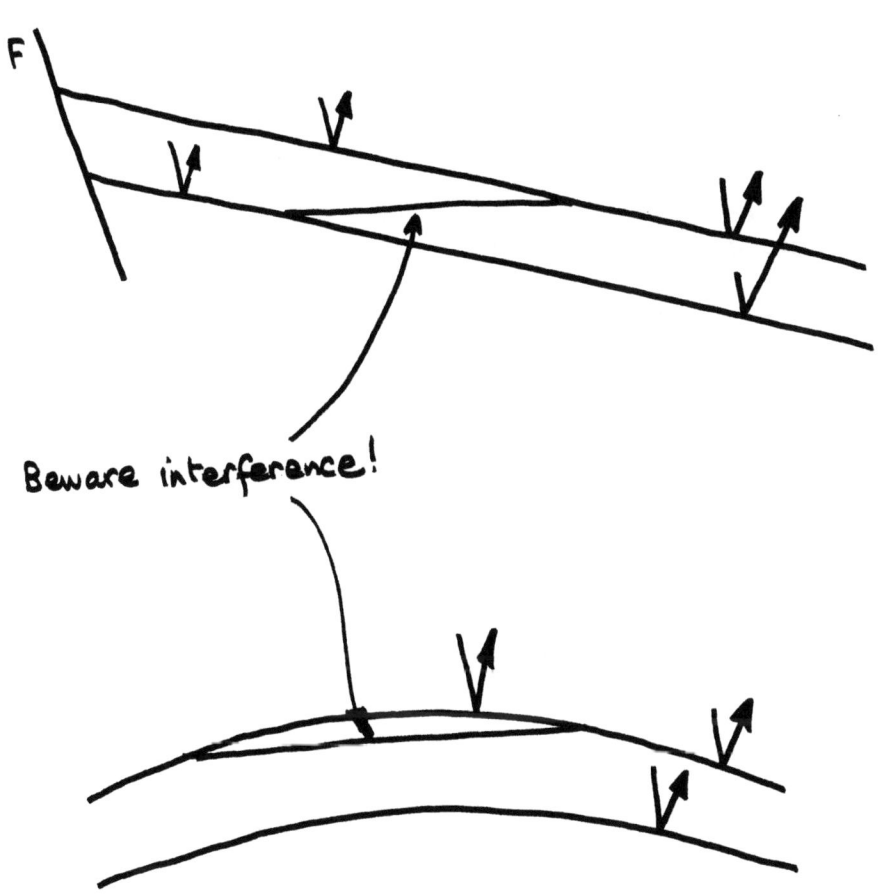

Beware interferance!

2-98B

2.6.2 Methods employing multiple reflections

Typical of these methods is the most basic version, in which we measure the ratio between a primary reflection and its surface multiple. Provided that we select measurements representing the correct geometry, and provided that the surface reflection coefficient can fairly be taken as -1, the ratio of multiple to primary is equal to minus the reflection coefficient of the primary reflector, as viewed from the surface; that is, we obtain the apparent reflection coefficient, which is the product of the real reflection coefficient and all the losses on the way down and back.

As we have agreed before, if we are able to estimate these losses, and to estimate the acoustic impedance of the material above our reservoir target, then this measurement allows us to infer the acoustic impedance of the reservoir using only reflections from its top interface; the multiple-reflection method may therefore be termed the top-only method, in distinction from the top-and-bottom method.

The merits of the top-only method are that we do not need to use the reflection from the bottom of the reservoir (which is often not an abrupt discontinuity, in practice) and that we do not need to know the acoustic impedance of the material below the reservoir. It is also true, as before, that the ratio feature protects us from most of the problems associated with field work.

In the case of the simple symmetrical multiple sketched, the processing risks are also minimized; as we shall see later, this is a consequence of the fact that the distance travelled by the multiple is known to be just twice that travelled by the primary.

The weakness of the top-only method, of course, is that it determines only the apparent reflection coefficient as viewed from the surface; we need to estimate the losses on the way down and back, and we shall find that this is seldom easy. For land work involving rugged topography and/or a thick weathering, there is an additional doubt about the effective reflection coefficient to be ascribed to the surface. However, the top-only method is forced on us when the top-and-bottom method is inapplicable. Further, the margin of error in both methods is such that whenever possible we compute both solutions, as a check.

So let us formalize the top-only method in a little more detail, and explore its variants. The discussion expands somewhat section 1.1.13 of the pre-course notes.

In any utilization of multiple reflections, we must remember the distinction between multiples involving only two reflectors (trapped reverberations) and those involving three or more (reflected reverberations). Let us look at these two classes in turn, identifying the physical system, the resultant pulse train (initially assuming spike pulses) and the autocorrelation function of that pulse train.

Trapped reverberations: These are exemplified by the simple reverberations between the surface (of reflection coefficient -1) and a single reflector (of reflection coefficient R).

The system —

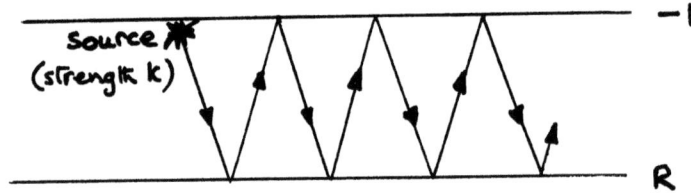

The pulse train —

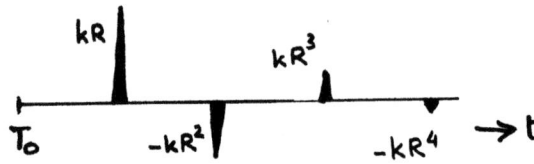

The autocorrelation function —

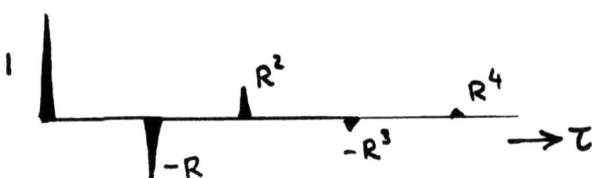

Clearly, we may derive the product of the reflection coefficient R with the surface coefficient -1 from the ratio of the first two arrivals on the pulse train (that is, the primary reflection kR and its first multiple $-kR^2$), or from the ratio of any other pair of adjacent arrivals, or directly as the value of the first negative value -R on the autocorrelation function.

The only value of the autocorrelation function in this context is that it averages all the ratios between pairs of adjacent arrivals. The use of the autocorrelation function is dangerous if dip or offset considerations mean that the reverberation time is not constant.

Reflected reverberations: These are exemplified by the reverberations between the surface (reflection coefficient -1) and a shallow reflector (reflection coefficient R), observed following the primary reflection from a deep reflector (reflection coefficient P).

The system —

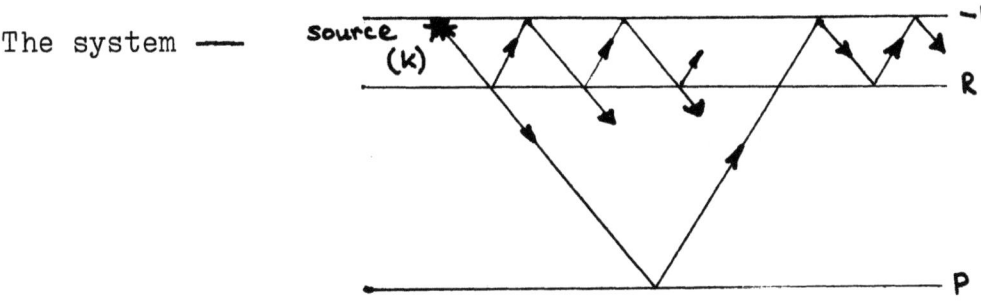

The pulse train —

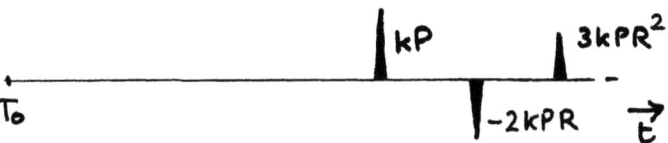

The autocorrelation function —

We may derive the desired product -R from half the ratio of the first two arrivals on the pulse train (that is, the deep primary reflection kP and its first shallow reverberation -2kPR), or from the first negative value $-2R/(1+R^2)$ on the autocorrelation function. This last expression is the one plotted as the upper curve on page 1-22 of the pre-course notes.

The use of the autocorrelation function to achieve averaging is safer in this case. Both methods of deriving the desired product -R assume that the reverberation time in the shallow layer is the same at the source and geophone ends of the path.

In practice, of course, we do not have spikes. It is no problem to work with real-world pulses if the reverberation time is longer than the pulse shape.

The pulse train is typically thus:

and the autocorrelation function is thus:

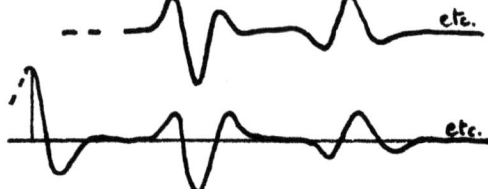

We can still make our measurements on these waveforms, provided we take due account of the inversions of polarity.

If the reverberation time is shorter than the pulse shape we have ringing.

The pulse train is typically thus:

and the autocorrelation function is thus:

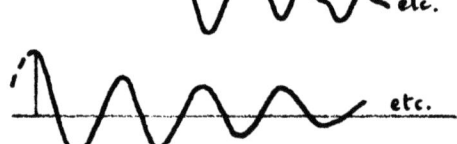

Measurements are now dangerous, because of interference. In the marine case, this danger begins to be significant in water depths less than about 60m (200ft).

It has been implicit in most of our discussion of reflection coefficients that our concern has been with single, abrupt and discrete reflectors. Of course this is often not true in the earth, and later we shall need to explore the complications in some detail. For the moment, let us just note that the demonstration as to whether the reflector is abrupt and discrete lies in a comparison of the pulse shapes of primary and multiple. If the multiple is clearly a simple inverted version of the primary, then the reflector is abrupt and discrete (in our bandwidth), and we are safe to compute the reflection coefficient. If the multiple is grossly changed in a manner whose meaning is unclear, then of course we would not delude ourselves that its amplitude ratio to the primary actually represents a discrete reflection coefficient.

Now let us apply the foregoing material to a simple case — that of finding the reflection coefficient of the sea floor in marine work.

The first method we might use is the study of trapped reverberations on unstacked records. The sketch suggests such a record after necessary corrections for gain and divergence (to be discussed fully later). Then if the near array is not too long, and not too far from the source, we can first make our check that the pulse shape of the first multiple is a simple inverted version of the sea-floor primary, and then compute the reflection coefficient as minus the ratio of multiple to primary on the near trace.

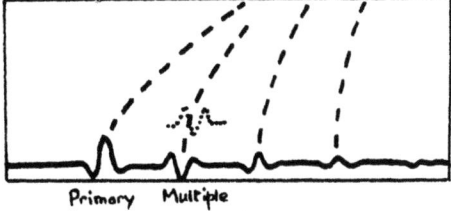

If the distance from source to near array exceeds the water depth, the primary (at wide angle) sees a greater reflection coefficient than the multiple recorded on the same trace (section 2.4.6). We can still compute this elevated reflection coefficient, by measuring the multiple on another trace selected to maintain the same geometry (dotted). But if we need the reflection coefficient at or near normal incidence, then we must either use the ratio between higher-order multiples, or go to reflected reverberations (as exemplified later).

If the near-trace offset is very small relative to the water depth, so that the reverberation time on that trace is constant, we can use the autocorrelation approach. The window might be as shown shaded; any primary or multiple reflections from deeper systems must be excluded from the window, as must be any noise.

In practice, clearly, we prefer not to use the auto-correlation approach in this instance.

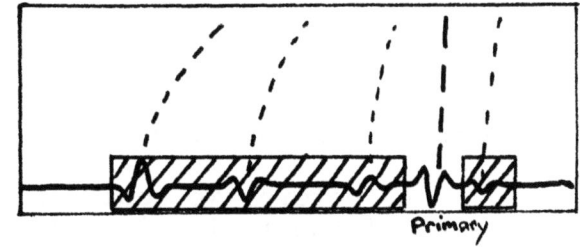

The second method we might use is that of reflected reverberations, again on unstacked records. The near trace, for example, might appear as sketched, where our interest focuses on the deep primary reflection and its first reverberation. Then the desired sea-floor reflection coefficient is equal to minus

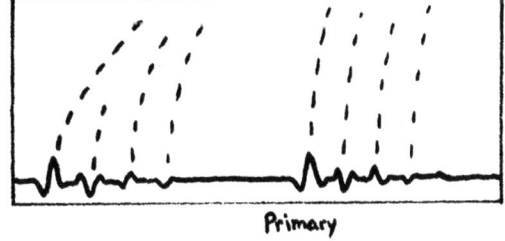

one-half of the ratio of reverberation to primary.

This method is necessary when the field technique employed did not provide for a near array at short offset from the source. Of course, the signal-to-noise available with this method is not as good as for trapped reverberations, but it is often possible to find a primary which is strong enough, even on unstacked data.

Where the signal-to-noise conditions preclude the use of unstacked data, we perform a special stack. In this context it is very important to realize that all the signal-to-noise benefits of stacking are available to help us study multiples, if we choose to do so.

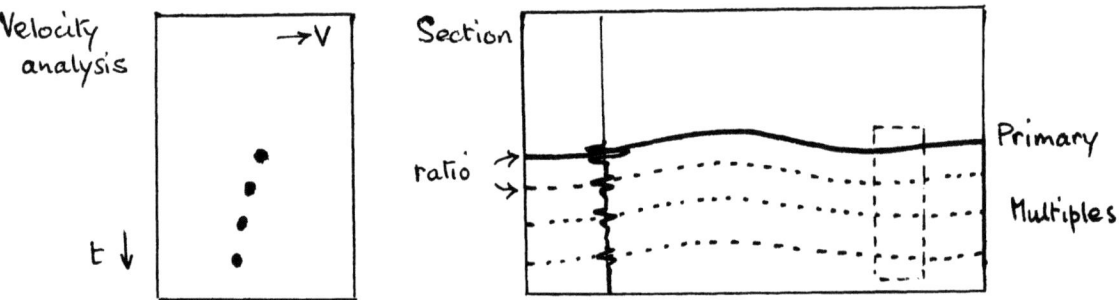

For this purpose we select a velocity distribution which follows the normal curve down to the primary we propose to use, and then branches back to follow the water reverberations of that primary. When we stack, therefore, the primary is enhanced, and so is the reverberation train. The ratio between first reverberation and primary can now be established with greater certainty than on the unstacked data (on the assumption, of course, that all the remainder of the processing has been done properly; we shall discuss this later).

As before, we could obtain also a conveniently averaged value of the reflection coefficient by computing the auto-correlation function over an aptly chosen window, as suggested by the dashed box in the sketch. Again, it is important to mute out any other primaries or noise entering the window.

When we use stacking, we are accepting some traces which may represent significant angles of incidence; judicious exclusion of the long traces may be desirable if the selected primary reflection is strong and positive (section 2.4.6).

For our present concern with the sea floor, we can accept the derived reflection coefficient as the actual reflection coefficient of the sea floor, because the losses on the way down and back through the water are negligible.

Now let us graduate to the case of a deep reflector in the body of the earth. In this case, we remember, what we shall obtain is the apparent reflection coefficient as viewed from the surface; later on, we shall need to estimate a correction for the losses.

The basic method is as for the sea floor. The only significant difference is that in the case of a deeper reflection we are unlikely to obtain more than one (or perhaps two) multiples; therefore we shall not be using autocorrelation techniques.

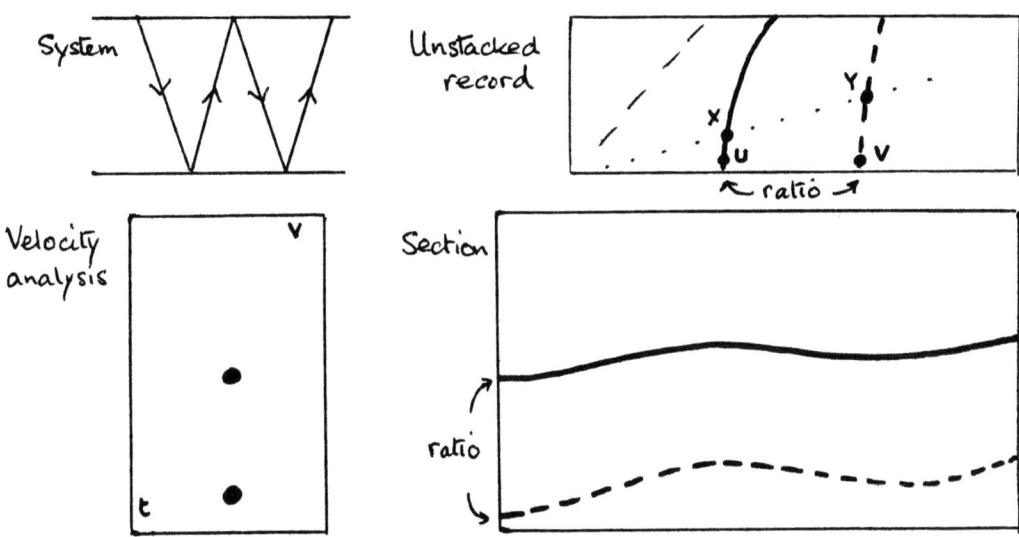

For the simple first-multiple case sketched at the top left above, we obtain the unstacked record at the top right. We can compute the reflection coefficient near normal incidence from the ratio V/U, and (assuming that our trace-to-trace validity has been maintained in the recording and processing) the reflection coefficient at inclined incidence from the ratio Y/X.

If we cannot find or measure the multiple with sufficient confidence on the unstacked data, we stack. We stack using the normal velocity distribution down to the primary reflection, and then that same constant velocity below. On the stacked section the desired ratio can be measured with greater assurance.

2-104

If the reflection whose coefficient we wish to measure is not sufficiently strong to generate a good simple multiple, we use a different reflector to provide the necessary second bounce. This second reflector can be any convenient strong reflector. Provided the dips are small, and provided we use appropriate muting of the long traces, the multiple is augmented by a factor of 2 because of the two reinforcing paths.

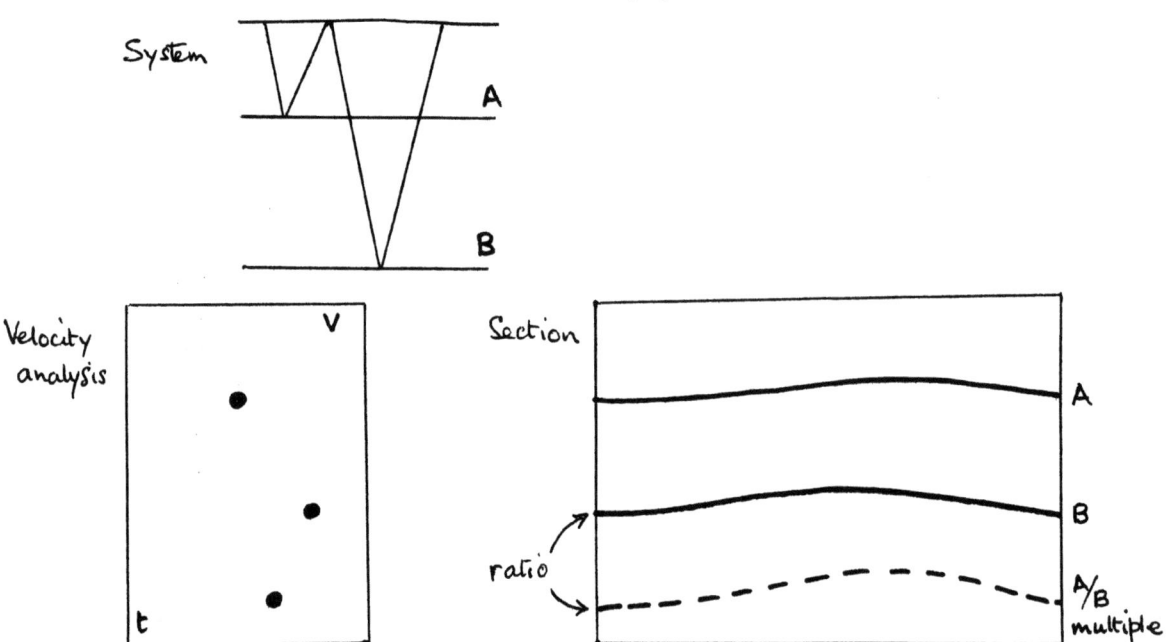

As before we may work on specially stacked traces if the desired multiple is not usable on unstacked data. The velocity distribution follows the primary function down to the second primary reflection, and then branches back to the multiple. The reflection coefficient of reflector A is minus one-half of the ratio of the multiple to the primary B, and the reflection coefficient of reflector B is minus one-half of the ratio of the multiple to the primary A.

Although we have talked in terms of special stacks, we should note that the necessary information is available at selected locations if constant-velocity stacks have been made as part of the regular processing sequence.

In anticipation of our future discussion of amplitude corrections for geometrical divergence, let us just note here that both stacked and unstacked amplitudes must be corrected, or a factor must be included in the arithmetic, to accommodate the decay arising as a function of the path length (which is known rather well in all these cases discussed). Obviously other processing stages involving amplitude scaling or adjustment must also be monitored carefully; we shall deal with all these points later.

TAPE 12   2.6.3 <u>The method of comparison with known values</u>

To the extent that the recording and processing can guarantee correct treatment of amplitudes, every apparent reflection coefficient follows, by direct comparison of amplitudes, as soon as one reflection coefficient is known.

For example, in fairly deep water the reflection coefficient of the sea floor can always be established (by the methods of the last section) with good accuracy. Then, subject to the above proviso, we can infer the apparent reflection coefficient of any other reflector by direct comparison of reflection strengths. In this sense the calculation and display of the sea-floor reflection coefficient provides a very valuable and authorative <u>calibration</u> of the amplitudes on the reflection section, and <u>this should</u> be standard practice on marine data.

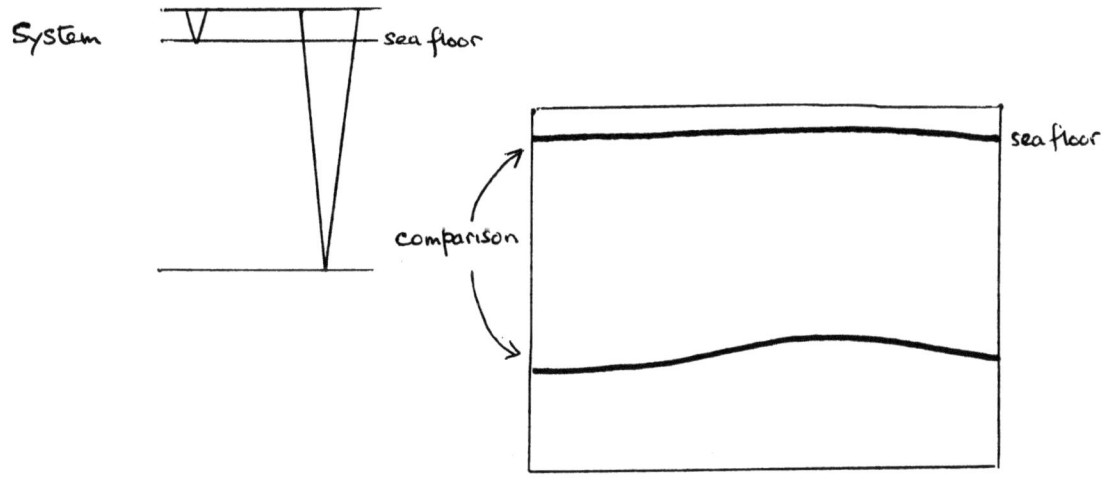

Obviously, all the methods we have considered for deriving quantitative reflection coefficients work best on good strong reflections. If our target is not a strong reflector, we compute the reflection coefficient of one or more strong reflectors reasonably close to it, and infer its own coefficient by simple comparison of amplitudes. Thus we should not think of the top-and-bottom method or
the top-only method solely in terms of reservoirs; it is often more convenient to use these methods on other reflectors bounding more-suitable geological units, and then to relate the reservoir reflection to the derived reflection coefficients of these other reflectors.

In passing, we might note that our technology has advanced to the stage where an absolute calibration of the entire seismic system (in millimetres of wiggle per unit of acoustic pressure or geophone velocity) is now easy. To the degree that we can also measure the outgoing source signal in absolute terms, and to the extent that we understand what happens near to the source, the apparent reflection coefficients in the earth follow from dividing the received signal by the source signal. We may be very close to this.

2.6.4   Removing the losses

We have said that if the geology allows us to use the top-and-bottom method to establish our reflection coefficients we would not normally need to worry about the losses on the way down and back. But if we are forced to use the top-only approach, using surface multiples, we determine only the <u>apparent</u> reflection coefficient; we need to remove the losses.

Only one part of this is easy. That is the compensation for the two-way transmission loss on passage through the sea-floor (or any other very strong single reflector). Since a two-way transmission coefficient is $1-R^2$, and since we are able to determine R by the methods of the last section, all we have to do is to divide the whole trace after the sea-floor reflection by $1-R^2$.

Because single two-way transmission coefficients in the deep earth differ so little from unity, this operation is worth doing only for the very strong reflectors. On the profusion of weaker reflectors, we remember, the very large transmission losses are largely compensated at the low frequencies (but not at the high frequencies) by short-path multiples.

So we have to find a way to assess the effect on our amplitudes of the three loss mechanisms of section 2.5.

First, it helps us to obtain some feeling for the general magnitude of the losses. The immediate check here is the comparison of the pulse shape of the primary and the multiple whose amplitudes we are measuring. If the pulse shapes are identical (only the amplitude and sign being different), then there are no frequency — dependent losses. This is the normal situation in the water, over a hard sea-floor. If the pulse shapes are very little changed, then these losses are small.

We may also take advantage of the fact that we know the losses are associated with a high-cut appearance. Thus we may compute the reflection coefficient of our target reflection on broad-band data, and obtain such-and-such a value; then we may make the same computation on filtered data (remembering, of course, that unless we have forced our data to be zero-phase, the reflection time changes somewhat with the filter). We might filter, for example, in bands 0-70Hz, 0-50Hz, 0-40Hz, 0-30Hz and 0-25Hz. The computed reflection coefficient diminishes, and then stabilizes close to its true value, as we reduce the upper frequency. Where it stabilizes, of course, the filtered primary and multiple have the same pulse shape.

At this stage of the discussion, we are assuming that the reflector whose reflection coefficient we seek is abrupt and discrete; in the next sections we shall see what must be done if this is not so. We must immediately note
the weakness of the above filtering method in this regard: as we lower the pass-band our filter operators become longer and longer, and even a reflection well separated from those above and below must eventually be corrupted by them. However, the method remains valuable where the target reflection is well spaced from others, and much stronger than they are. Further it has the merit that its message is self-evident, without any frequency-domain mathematics.

A method which provides compensation for all losses, including the transmission losses at interfaces, is based on a consideration of the multiple reflections from the topside and underside of the target reflection.

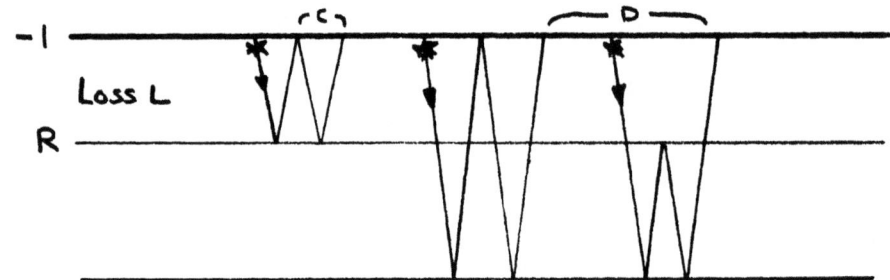

Here we identify separately the two-way loss L down to our target, and its true reflection coefficient R. Then in the simple system on the left, we know that the amplitude ratio C of multiple to primary is -LR.

Now comparing the two multiple paths at the right, we see that the only differences between them are the elimination of the loss L and the replacement of the surface reflection coefficient -1 by the underside reflection coefficient -R. The amplitude ratio D of these two paths is easily computed as $L(1-R^2)/R$.

Whence $\quad R = 1/\sqrt{1 - CD}$

and $\quad L = -\dfrac{\sqrt{1 - CD}}{C}$.

The method ordinarily requires a special stack, as discussed in the section 2.6.2, for each of the three multiple systems. The deep reflection, obviously, is chosen for its strength; it is not essential that it be discrete. However, we maintain the requirement that the target (upper) reflector should be discrete.

The method cannot estimate losses right down to the deepest reflection. However, we remember that the absorption loss is normally concentrated in the shallow section, and if we can do no more than estimate the losses down to 1.5 second (for example) this is still very worth while. If our target is at 2 seconds, and we know of no anomalous geological circumstances, then to a first approximation we can probably neglect the absorptive component of the loss in the intervening zone. In other words, if we have estimated the losses down to 1.5s, and so established the true reflection coefficient of a convenient reflector at that time, we obtain the reflection coefficient of our deeper target by simple comparison of its amplitude with that of the known reflector. Then we have only transmission losses and short-path multiples (and perhaps scattering) to worry about; if the intervening zone looks like a deep-sea low-energy shale (section 5.2) we ignore them, while even if it does not we have at least established a top figure (and perhaps a feasible range) for the reflection coefficient.

The method is strongest where the losses are indeed concentrated in the shallow section, where the reflection amplitudes are measured by an envelope technique (section 2.6, p.2-96A), and where the reflections are near zero-phase. Under these conditions the loss factor L represents the combined effects of both frequency-selective and non-frequency-selective agencies, summed over the relevant spectrum.

Without special attention, of course, the zero-phase item is not realistic; the method is likely to give losses a little too big and reflection coefficients a little too small if the reflection pulses are actually minimum-phase or mixed-phase.

Question: If we make the minimum-phase assumption for the source and for the earth processes, how would we attempt to correct the reflection-coefficient values derived from the zero-phase assumption?

The other method for separating apparent reflection coefficients into true reflection coefficients and losses is based on a spectral measurement of two or more of the reflections. Since we know that the loss mechanisms are high cut, having no loss at very low frequency, it follows that knowledge of the pulse spectrum before and after a particular loss allows us to compute the frequency response of the loss mechanism; further, we can then compute the effect of the loss mechanism on the observed amplitude of the pulse (for minimum-phase, or for our best alternative understanding of the phase behaviour).

At sea, for example, we may first compute the amplitude spectrum of the sea-floor reflection (having satisfied ourselves that it is discrete by comparing the form of primary and multiples) and accept this as the spectrum entering the earth. Then to the extent that we can establish the spectrum of a reflection zone deeper in the earth, we can divide the one spectrum by the other, frequency by frequency, to obtain the frequency response of the effective loss mechanisms along the deep path. Alternatively, we can divide the spectrum of a multiple by that of a primary to obtain the loss response of the multiple path.

In later sections (3.1 in particular) we shall consider the techniques, judgements and problems of such spectral measurements. Of course it would be foolish to pretend that there are no problems; in the past we would have considered such operations as hopeless. In recent years, however, we have become much more adept at frequency-domain transformations; further, we know how to recognize discrete reflections, and we are prepared to make a judicious selection of (possibly substantial) lateral averaging to take advantage of the slowly varying nature of the loss mechanisms.

Often, in this course, we shall find ourselves in danger of being unable to see the wood for the trees. The details of almost anything connected with the real earth are certain to be complicated; we cannot brush aside these details, because they are the difference between success and failure, but somehow we need to maintain a clear picture of the fundamental simplicity. So, in this cause, let us just restate our basic objectives in the context of this section:

- We seek to determine the quantitative reflection coefficient of our target.

- If our target satisfies the requirement of being a discrete and abrupt reflector, well and good. If not, we select a good discrete reflector just above or below the target, compute its reflection coefficient, and deduce the reflection coefficient of the target by simple comparison of amplitudes.

- We separate the true reflection coefficient from the losses either by selecting pairs of multiple paths which represent the same reflection coefficients but different losses, or by computing spectral ratios on pulses before and after travel to the target (or, preferably, by both methods).

- Phase assumptions affect the details. These manipulations may appear involved, but are readily handled by stock programs available everywhere.

- Whenever we can identify, at or near the target zone, a layer which is geologically likely to be of substantially constant composition and which has clear reflections of unequal amplitude from its upper and lower boundaries, we use the top-and-bottom method to check our conclusions.

The whole of this section has assumed a discrete reflector, and an abrupt surface reflector of reflection coefficient -1. Before we discuss in the next section what happens when reflectors are not abrupt and discrete, let us just note that abrupt and discrete reflectors do occur in the real earth. At one time, when we were flushed with the new insight of the synthetic seismogram, it was widely believed that all strong reflections were complexes; many are, of course, but many are not.

The assumed value of -1 for the surface is generally fair, unless there is a very deep, very absorptive and complicated weathering, or a mountainous surface. The water table and the sea floor may represent a measure of complication, but they can at least be pondered intelligently on the basis of the data we have already discussed.

The criteria for recognizing a discrete reflector are:

- Geological likelihood, based on knowledge of the lithology, the depositional history, and the major unconformities.

- Constancy of reflection strength over an area.

- Constancy of reflection character over an area; this we shall expand later in terms of measurements of frequency content and apparent polarity (sections 2.7.2, 3.4.5, 3.4.6 and 3.4.7).

- Similarity of form between the reflection and its surface multiple, any deviations being explicable as the effect of frequency-selective losses or of a varying reflection coefficient of the surface.

- Identical effects for reflection from the top and underside of the reflector.

2.7 INTERFERENCE BETWEEN REFLECTIONS

2.7.1 Interference in the time domain

Many important situations — wedge-outs, thin sands, fluid contacts, unconformities in general — involve two or more reflectors at a time spacing less than the duration of a seismic pulse, so that their reflections interfere. It is very important that we recognize the symptoms of interference, so that we do not start computing the reflection coefficient or the absorption at a reflection which is actually a complex (or at least, not without knowing what we are doing).

First and foremost, the major indicator of interference is geological likelihood, as evident on the section as a whole — can we see spaced reflections coming together and merging? And, of course, we know well that the assessment of this should be made on a section having the best possible bandwidth; events which are separately resolved at good bandwidth must become elided at narrow bandwidth.

74 Other symptoms of interference are clear from the Widess diagram on p.2-110, which shows the reflection train created by the overlapping of one erect and one inverted pulse, for a range of spacings expressed as fractions of a pulse length.

- If we have an estimate of the seismic pulse length, any reflection we see with a greater length is an interference complex. We shall learn in section 3.1 how to make such an estimate of pulse duration.

- The complex from the top and bottom of a thinning interval shows large amplitudes at spacings which yield constructive interference, and small amplitudes for destructive interference; these bright-dim amplitude variations are therefore a function both of the thickness of the interval and of the "period" of the pulse.

- The larger spacings (greater than one-half of a pulse length in the illustration) produce unexpected changes of slope ("hickies") in the waveform, and saddles in the envelope of the waveform.

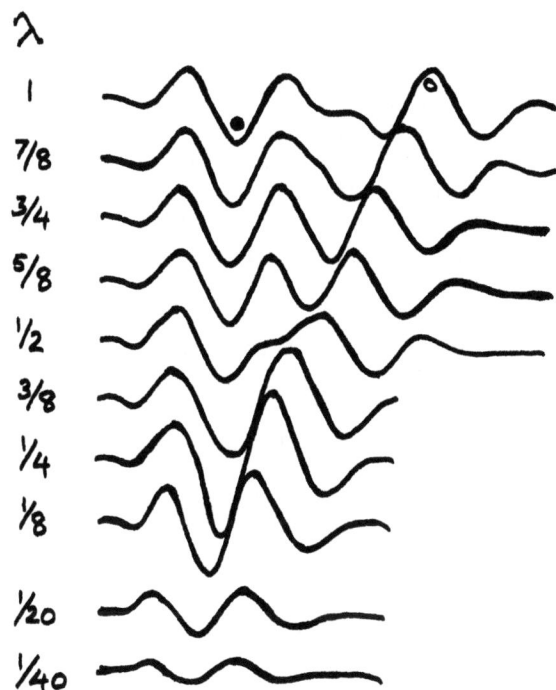

After Widess 1957 (1973)

- At smaller spacings (less than one-half of a pulse length in the illustration) a spurious dip becomes evident, intermediate between the two true dips. This occurs at large amplitude; in an inclined trapping situation, there is a risk of erroneous interpretation as a fluid contact.

75
76
- If the interfering reflections are of equal magnitude and opposite polarity (as on p.2-110, and as often occurs with shallow gas sands), a reflector spacing of one-eighth of a pulse length produces an approximate differentiation; the complex has <u>one more half-cycle</u> than the basic pulse, and a <u>higher-frequency</u> <u>appearance</u> (section 1.1.10). At spacings smaller than this the shape of the complex shows virtually no further change, but the amplitude decreases smoothly to zero; at such spacings our only hope for estimating the bed thickness lies in studies of amplitude, not in studies of reflection shape.

74
(repeat)
- If the basic pulse is or can be made to be zero-phase (so that it is symmetrical about its peak amplitude, as in the illustration) then the differentiation feature means that the signature of a shallow thin gas sand is the property of skew-symmetry about the central zero-crossing. Minor deviations from this are introduced if the top and bottom reflections do not have quite equal magnitudes.

- At thin-bed spacings the amplitude of the complex remains surprisingly large. We are accustomed to shrugging our shoulders at talk of a 3m sand (10ft); this would place us in the vicinity of the 1/20 trace, and we see that the amplitude of the complex reflection can still approach that of the reflection from top or bottom alone.

77(B)

If the interference is between two reflections of equal magnitude but the <u>same</u> polarity, the effect is just what we would expect: a tendency toward a <u>lower-frequency</u> appearance at small spacings, and double the <u>amplitude</u> at very small spacings.

If we have a <u>succession</u> of reflections of the same polarity, this must represent a staircase of acoustic impedance. In the limit this staircase may become a smooth gradation or transition between two limiting impedances. We shall call such a gradation a <u>transitional reflector</u>; obviously there are many geological processes which result in these gradations, and it is important that we are able to recognize and handle transitional reflections.

The seismic signature of a transitional reflector is an approximate integrated form — lower frequencies, loss of amplitude by loss of the higher frequencies (if the transition occurs over a distance longer than the wavelength of the higher frequencies), an appearance of one fewer half-cycles, and skew-symmetry if the basic pulse is symmetrical.

The higher-frequency appearance of a reflection sequence of alternating polarity and the lower-frequency appearance of a transitional reflection are, of course, strict corollaries of the principle "More up, less down", and our discussion of section 2.5.2.

2.7.2 <u>Interference in the frequency domain</u>

In the top diagrams on p.2-112 we see on the right the frequency-domain equivalents of the time-domain reflector spacings illustrated on the left. Whether the two reflectors are of the same polarity (a and b) or of opposite polarity (c and d), the effect in the frequency domain is a system of peaks and notches in the frequency response; if they are of the same polarity there must be a peak at zero frequency, and

2-111A

if they are of opposite polarity there must be a notch. The closer the two reflectors, the more widely spaced are the peaks and notches. The notches do not go to zero if the two reflectors are not equal in magnitude.

If then a seismic pulse having a certain spectrum is incident on a pair of closely-spaced reflectors, the spectrum of the resultant interference complex is obtained by multiplying the pulse spectrum by the appropriate peak-and-notch response (section 1.1.6). If sketch (a) in the lower illustration on p.2-112 represents the spectrum of the pulse, the spectrum of the complex typically takes one of the forms shown in sketches (b) - (e), depending on the reflector spacing. Sketch (b), for example, represents very close spacing of reflectors of opposite polarity; we see the high-frequency appearance associated with differentiation. The comparison of sketch (c) with sketch (a) reminds us that in such a case a narrow filter peaking at 50Hz would show no spectral difference between the single reflector and the complex. The same filter on the spectrum of sketch (d) would knock the hell out of the event.

At first sight, we might think that a suite of narrow-band filters could be used as a tool for unravelling interference complexes. Although it is true that an event classifies itself as a complex if it is present in one narrow band and weak or absent in another narrow band (both within the overall record bandwidth, of course), we shall come to the conclusion in later sections that the frequency domain represents a rather clumsy tool for these purposes. From the interpreter's point of view, it is certainly important that any such judgement should be made on a filter-panel type of display; this gives a type of frequency-domain insight while preserving the geological monitoring allowed by a normal time-domain section. We discuss this point further in section 3.4.7.

So, it may just be worth making some studies in the frequency domain, when the problem is interference. Another minor benefit is obtained by operating on our final time-domain reflection pulses to turn them into the symmetrical zero-phase shape, which compresses the signal into the shortest time allowed by the bandwidth (Schoenberger, 1974; Dedman et al., 1975; Barry, 1975); this is the process discussed in the last section in a more powerful context — as a means of recognizing a thin bed by its skew-symmetric signature. But when all this is said and done, it still remains true that the important answer to the problem of resolving an interference complex is the classical one — improve the bandwidth.

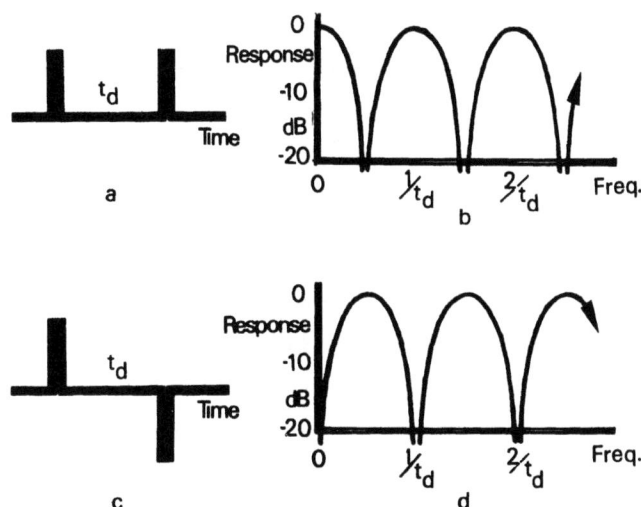

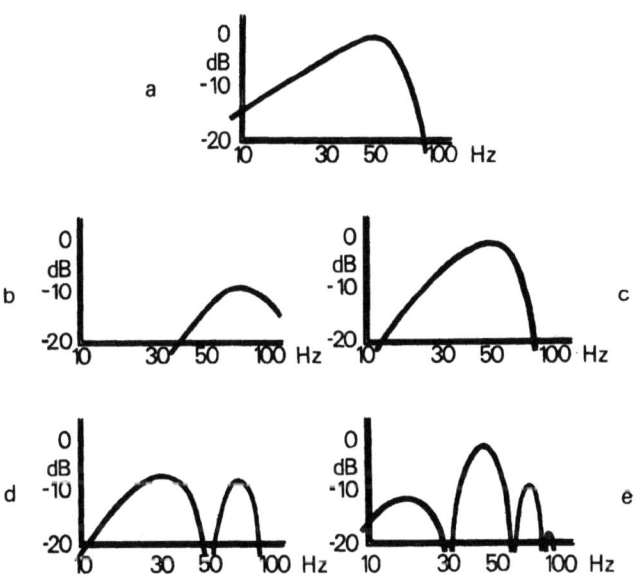

2.7.3 The consequences of interference for our measurements

In section 2.6, when we were discussing the problem of turning reflection amplitudes into reflection coefficients, we confined ourselves to abrupt and discrete reflectors. In the case of the sea-floor reflection, where we knew that there are no significant losses in the water, we said we would check the abrupt and discrete nature of the sea-floor reflector by comparing the form of the first multiple with that of the primary*.

Now we know that the sea floor is not always a simple abrupt reflector. In the case of recent sediments, for example, we know (p.2-16) that the zone immediately below the sea floor represents a gradation of density (and a weaker gradation of velocity) over several tens of metres. So although there is an abrupt reflector at the place deemed to be the sea floor by a sparker (or a fish), there is also a transitional reflector below it. The corresponding reflection must therefore be a complex, comprising a reflection of the basic pulse shape followed by another of integrated low-frequency form. Alternatively, the sea floor may be scoured and clean, but the materials below may be older consolidated stratified sediments; in this case the sea-floor reflection is a complex comprising one large and several small component reflections.

Only when the transition is rapid or the stratification is weak does the sea-floor multiple reproduce the shape of the primary. Otherwise, the primary reflection is the convolution of the source pulse shape with the transition or the complex of reflectors, and the multiple is the convolution of the source shape with the autoconvolution of the transition or complex. In such cases the sea-floor reflection coefficient computed from the strengths of primary and multiple ceases to be a

---

* Incidentally, this check is also necessary before we dare assume that the form of the sea-floor primary reflection properly represents the outgoing signal. Source signature deconvolution using the sea-floor reflection is conceptually dangerous unless this check is made.

strict reflection coefficient in the sense of previous
sections (being usually too small, numerically, if the
multiple is lower-frequency than the primary, and too
high if it is leggier); however, the value of the computation
for calibrating the section remains substantially unimpaired.

For deeper reflections we do not have the foreknowledge
that there are no losses above, as we have in the water; a
lower-frequency appearance on the multiple may be due to a
transitional reflector or to loss mechanisms along the path.
So if we are forced to use the top-only method, and our target
zone is transitional or a complex, we have no choice but to
do as we discussed just now: find a discrete reflector (or
preferably two), separate the losses from its true reflection
coefficient, and then scale the amplitudes at our target zone
using the known nearby reflection coefficient.

What can we do if there appear to be no discrete reflections
at all? In marine work, at least, we are not lost, though of
course our position is weakened by our inability to cross-check
between different approaches. We calibrate our section by
computing the nominal reflection coefficient of the sea floor
(even if it is not discrete), as above. Then we have an
apparent reflection coefficient at our target level, but no
way to estimate the losses by the study of amplitudes or
frequency content of discrete reflections at or near the target
level. So we set up a window straddling the target zone and
compute the average pulse spectrum within that window by normal
(lucky!) techniques used in deconvolution (sections 3.1.1.3,
3.2.8.3, 8.4); then to the extent that this is successful we
can compute the losses, on the assumption that they are all
high-cut, by referring this pulse spectrum to that of the
measured source pulse.

Question: Within a rather featureless shale section (for
example, the Tertiary of the North Sea) we see a reflection
develop locally in a manner and position which would be
compatible with interpretation as an oil-bearing lenticular
sand (for example, in the Eocene). Not far below is a strong
clear reflection known or easily shown to be discrete (for
example, the Tertiary-Chalk contact in parts of the North Sea).

Below that is another strong reflection. We wish to know what is the possible range of reserves in place, for the postulated oil sand.

The sequence goes like this:

- The reflections from the top and bottom of our postulated oil sand interfere to form a complex, which is fairly strong where the sand is thickest and which tails away to nothing at the edges of the sand. Therefore we cannot think of measuring the reflection coefficient of either the top or the bottom of the sand individually. So we seek to calibrate the sand amplitudes by reference to the reflection coefficient of the strong discrete reflector.

- We find the apparent reflection coefficient of the latter by comparing pulse shapes and then reflection strengths for the primary and its multiple on a constant-velocity stack (section 2.6.2). Preferably we subject the data to a program designed to bring the pulses to an approximate zero-phase condition.

- We separate the true reflection coefficient of the strong discrete reflector from the losses by one or other (or both) of the methods of section 2.6.4.

- We relate the reflection amplitudes at the level of the sand complex to the known true reflection coefficient of the strong reflector below it, on the assumption that the losses are the same to both levels.

- We compute the reflection coefficient of the top and bottom of the postulated sand for a range of possible values of porosity; for this we employ our best compaction curves for velocity and density of the surrounding shale, we assume the sand is clean and oil-saturated, and we use the time-average equation.

- Using the pulse shape associated with the discrete reflection, we construct an interference diagram (similar to p.2-110) using one negative and one positive reflector of the expected reflection coefficients, as a function of sand thickness.

- We find the range of combinations of porosity and thickness which provide a match with the observed sand complex, both in character and on an absolute amplitude scale of reflection coefficients. Since a poor porosity requires a thick sand to provide a match, the range of reserves per unit area is not large.

- We compute the range of total reserves in place by using the amplitude variations over the sand to map its relative thickness variations, and by assessing its areal extent.

The question, now, is a tough but important one: <u>identify the assumptions made along this chain of computation.</u>

Obviously this is a very important exercise. First, let us emphasize that numerical calculations of this degree of sophistication are now feasible under favourable (but realistic) conditions; indeed considerable work is already done at this level. Second, because of the positive exploration benefit obtainable, we would be derelict if we did not attempt such calculations. But third, we would be even more derelict if we attempted the calculations without establishing a check-list of the assumptions involved. We shall set up such check-lists on the examples we work in section 4.4.

2.8 SPHERICAL WAVES, CURVED REFLECTORS, AND DIFFRACTION

So far, almost all our attention has been to plane waves. We have said that a plane wave, in a non-absorptive material, propagates without loss of amplitude or change of form. We have defined the reflection coefficient for such waves, noting the distinctions between normal incidence and inclined incidence. We have discussed the phenomena of absorption and short-path multiples, again for plane waves, and we have noted the change of form and associated loss of amplitude which they entail.

But in practice we are not able to generate plane waves, and so we must ask what differences are introduced when we consider spherical waves.

2.8.1 Geometrical divergence   (O'Brien, 1971; O'Doherty et al., 1971; Newman, 1973; Shah, 1973)

The first consequence of the spreading nature of a spherical wave is that the energy transmitted across a certain area of the wavefront at a certain radius from the source must pass across a larger area at any larger radius. Thus if we visualize a pencil of rays radial from the source (top left diagram, p.2-114), we can see that the intensity (which, from p.2-9, is the flow of energy across unit area in unit time) decreases as the area increases — that is, as the inverse square of the distance from the source. This is the well-known inverse-square law that we remember from optics.

However, a seismic geophone does not measure intensity — it measures amplitude (p.2-10), which is related to the square root of the intensity. The pressure amplitude of a spherical seismic wave decays as the inverse $\underline{first}$ power of distance (that is, as $\frac{1}{tV}$).

This is the relation applicable to a constant-velocity medium (for a near-enough example, the sea) in which the rays remain radial and the wavefront remains spherical. In the earth, however, we know that real velocity-depth relations are such that the rays

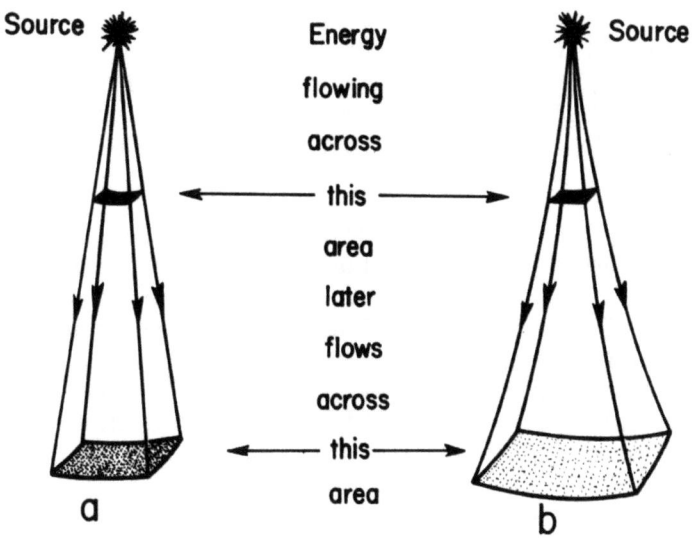

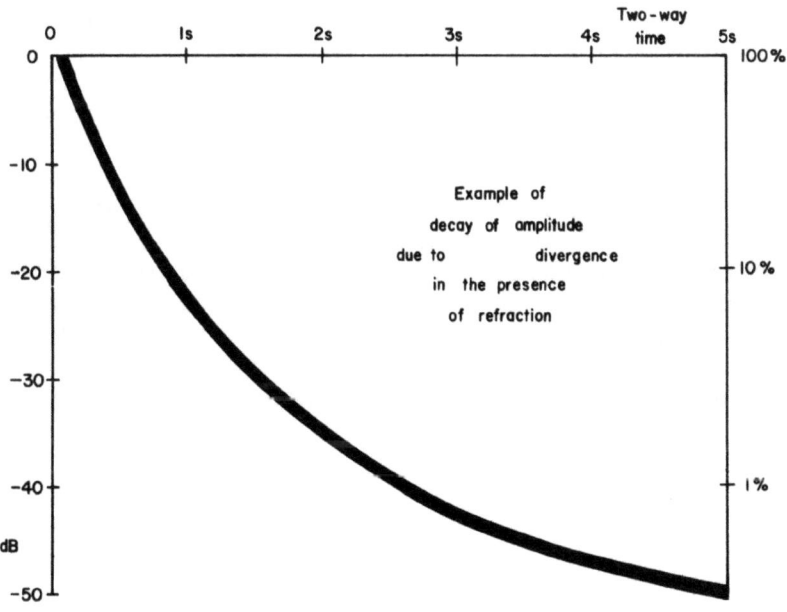

become curved by refraction and the wavefront is no longer spherical. As shown in the top right diagram, p.2-114, this implies a greater decay. For a near-normal-incidence path in a horizontally-stratified earth, the pressure amplitude decays as $\frac{1}{tV^2}$, where V is now the rms velocity.

How large a decay is this, in practice? The lower diagram on p.2-114 plots the decay of amplitude as a function of time, for a realistic velocity distribution. Relative to an event at 0.1s, an event at 5s is reduced by 50dB; 99.7% of the amplitude is lost. Clearly, this decay is very large; its cause, which we shall here term geometrical divergence, is (in normal bandwidths) the dominant reason for the observed decay of raw seismic traces as a function of time.

After the complexities of absorption, geometrical divergence is sheer delight. It is easy to understand, it is a definite effect which must be present, it is demonstrably a major effect, and — best of all—it is easy to compensate. For, clearly, all we need to do is to multiply each sample of the trace by a factor proportional to tV^2. Would that all earth effects were as simply understood, and as simply compensated!

There are a few side-comments which should be made:

- We remarked earlier (p.2-82 to 84) that an exponential expansion (as so-many dB/s) is not a proper compensation for absorption on a seismic trace. Now we see (from the lower diagram on p.2-114) that it is not a proper compensation for divergence either.

- Strictly, the simple expression tV^2 applies to the pressure amplitude, as measured in marine work. For land work, where we measure particle velocity, the theoretical expression is more complicated; near the source a spherical wave has a 90% phase shift between the particle velocity and the acoustic pressure, as a means of escaping the infinite amplitudes otherwise implied at zero distance. However, by the time of our earliest reflections of interest the particle velocity has come substantially into phase with the acoustic pressure, and it is standard practice to apply the simple tV^2 expansion to land data as well as marine data.

- Normally, the velocity distribution used for the term V in the expansion tV^2 is obviously that corresponding to primary reflections. This means that multiples, having smaller rms velocities, are over-compensated — they emerge larger than they should be. There is no escape from this.

- As stated earlier, the simple expression tV^2 applies strictly to near-normal incidence and to the Dix conditions of a horizontally-stratified earth; often we violate these conditions. However, the basic physical reality is an increasing area of wavefront enclosed within an expanding pencil of refracted rays; if the situation allows the tracing of this pencil of rays, the true correction for geometrical divergence follows immediately from the area within the rays. We shall see later that modelling techniques allow the computation of divergence corrections appropriate to wide-angle incidence and complex geology. However, we always start with the simple tV^2 expansion, and consider the complications only when we must.

- Those of us who have to write the specifications for marine surveys will recognize geometrical divergence (in this case, spherical divergence) as the basis for the usual method of specifying the output of a marine source. Thus we may state that, within a given bandwidth, the output must be at least so-many megapascal-metres (or tens of bar-metres in old-fashioned measure). If we specify 5 MPm, for example, we know that this would be demonstrated by measuring 5 MP at 1 m from the source, or 0.5 MP at 10 m, or 50 kPm at 100 m;

this is simple spherical divergence.

TAPE 13 ### 2.8.2 Absorption and scattering, for a spherical wave

In section 2.5.1 we agreed that we would assume absorption to be linear until someone proved it not so. If absorption is indeed linear, then there is no reason to expect any difference when we go from plane waves to spherical waves. We shall proceed on this basis. We might just note, however, that if absorption were non-linear, the effect would be that the absorption would be concentrated on the first part of the downward path.

Similarly, we do not expect any difference in the basic nature of scattering as we go from plane to spherical waves. However, when we are using point sources and individual geophone arrays, it is obvious that particular records and particular traces must become corrupted by scattering as the spread rolls across a shallow inhomogeneity. We shall see many examples of such localized scattering later in the course.

Sometimes a shallow scatterer can defeat reciprocity and even linearity (Balachandran, 1975). We shall assume such effects are not general, or are made insignificant by our field techniques.

2.8.3 Reflection and transmission at interfaces, for a spherical wave

The mathematical treatment of this real-life situation is complex (Dix, 1955; Bortfelt 1960; Spencer, 1961). We shall do no more than nibble at the physical aspects.

When a wave which is spherical or near-spherical is incident on a plane reflector, there is first a reflection at normal incidence, and then reflections at progressively increasing angles of incidence. We cannot limit the discussion, as we did in section 2.4, to normal incidence, or to a particular angle; incidence occurs at all angles. Consequently shear waves, boundary waves and refracted waves are inevitable. Further, each reflected and refracted shear wave generates both shear and compressional waves at every reflector subsequently encountered at non-normal incidence. This appears a formidably complicated picture; we would be dismayed if we did not know for a fact that the seismic method works. It works because, in general, we can select a spread geometry which imposes small angles of incidence for the signals we use, and because we can select for stacking only those signals which have compressional velocities. However, we are warned that there must be a level of geological complexity (for example, in overthrusts) where these tools are inadequate and interpretation becomes impossible.

We must now ask about the simple formula for reflection coefficient $(r_2 - r_1)/(r_2 + r_1)$. We introduced the formula when we were discussing plane waves and plane reflectors — does it apply to spherical waves and plane reflectors?

When a plane wave is incident on a plane reflector, the area over which the reflection occurs is theoretically infinite. But when a spherical wave is incident on a plane reflector, the area is a system of circles and rings. At the time that the outside ring illustrated here is reflecting the first peak of an incident pulse, the inner ring is reflecting the first trough and the inside circle is reflecting the second peak. The dimensions of the circle and rings depend obviously, on the wavefront curvature and the pulse "period."

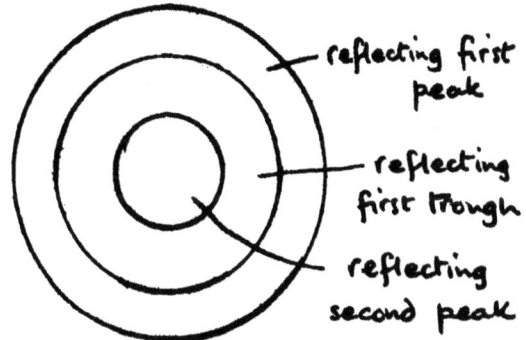

This prepares us for the observation that, if any simple formula is to define the reflection coefficient, there must be some proviso concerned with the area of the reflector in relation to the areas of these rings. Further, it is clear that the proviso is likely to be different for different distances up the spread.

We can visualize the proviso very easily in terms of an ordinary monitor record. On p.2-118 the alignment of a reflector on such a record is suggested by two hyperbolas representing adjacent zero-crossings of the pulse; the pulse itself is also shown on the near (X=0) and far traces. Then the vertical line as shown defines a distance up the spread which is equal to the diameter of the first zone of importance on the reflector. Most important is the inner (shaded) region of this zone, having say 60-70% of the nominal diameter. Then for the normal-incidence reflection (on the near trace), we ask that, <u>for the simple reflection-coefficient formula to hold, this shaded region of the reflector should be plane and of uniform acoustic contrast.</u> If the reflector is folded or faulted within this region, the reflection strength must be affected.

Since the dimension of the reflecting area (which the physicists among us will see to be related to the first Fresnel zone) is smaller for traces further up the spread, the area accepted for the near trace can be safely accepted for the cdp-stacked trace. The dimension — being related to the nmo and the pulse "period" — may be hundreds or even thousands of metres for low-frequency reflections deep in a high-velocity section, and smaller for shallow reflections.

2.8.4 Curved reflectors and velocity lenses

Now let us explore what happens if the reflector is not plane, over the relevant area.

If the reflector is concave-upward, it is clear that the system of rings is increased in size; the reflection amplitude is increased from the value given by the reflection coefficient, as by the focusing effect of a concave mirror. Conversely, a convex reflector decreases the size of the rings; the reflection amplitude is decreased, as by the defocusing effect of a convex mirror.

Since structural concavity and convexity are the stuff of exploration, we must ask the amount by which the reflection amplitude changes. The full solution is complicated (Shah, 1973), and so we must find some workable approximation. Such a "quick look" approach has been given by Hiltermann (1975). An even simpler approximation, for a stacked trace, is given by the graphs on p.2-118B, which are solutions of the following equations:

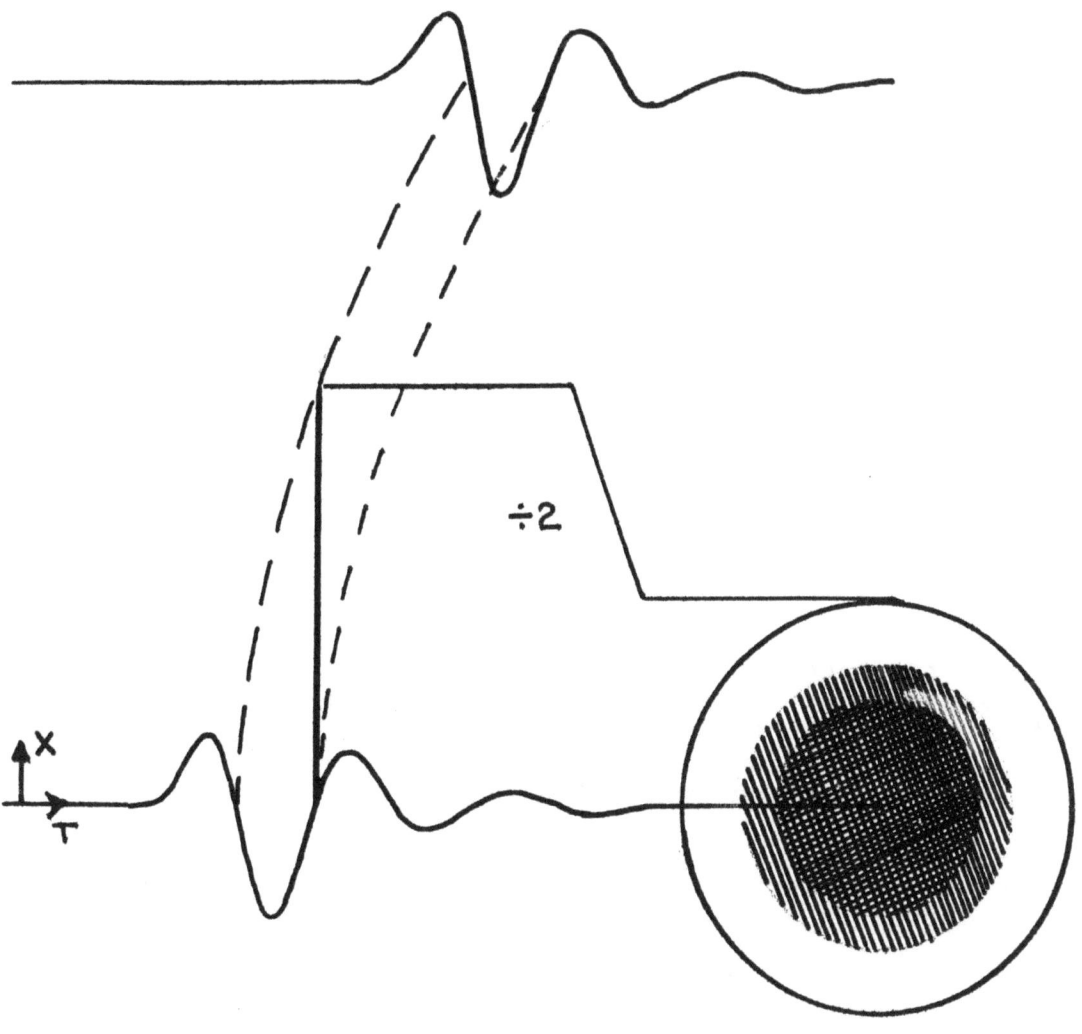

For a 3-dimensional structural feature (circular contours, dome or dish),

the bright-up ratio $\dfrac{S}{S_\infty} = \dfrac{1}{1 - R_W/R_S}$,

and for a 2-dimensional feature (parallel contours, ridge or valley)

the bright-up ratio $\dfrac{S}{S_\infty} = \dfrac{1}{\sqrt{1 - R_W/R_S}}$,

where S = amplitude in presence of focusing/defocusing,

S_∞ = amplitude as it would be in the absence of focusing/defocusing (if the reflector were plane),

R_W = radius of curvature of wavefront,

and R_S = radius of curvature of structure (both positive for curvature which is concave-upward).

The three-dimensional equation yields the solid line on page 2-118B and the two-dimensional equation the dashed line; structures with elliptical contours produce focusing/defocusing effects between the two lines. The 0 dB reference level represents the spherical wave incident on a plane reflector; above this we see the bright-up effect of focusing over a syncline, and below it the dim-down effect of defocusing over an anticline. The horizontal axis is the ratio of the two radii of curvature — that of the wavefront and that of the structure.

It is unfortunate that this defining ratio should be that of two <u>distances</u>; we would have preferred some solution which could have been implemented on a <u>time</u> section. The radius of curvature of the wavefront is no problem; this can be obtained by standard manipulation of the wavefront charts. The radius of curvature of the structure, however, must be expressed in metres or feet; for the level of operation at the interpreter's

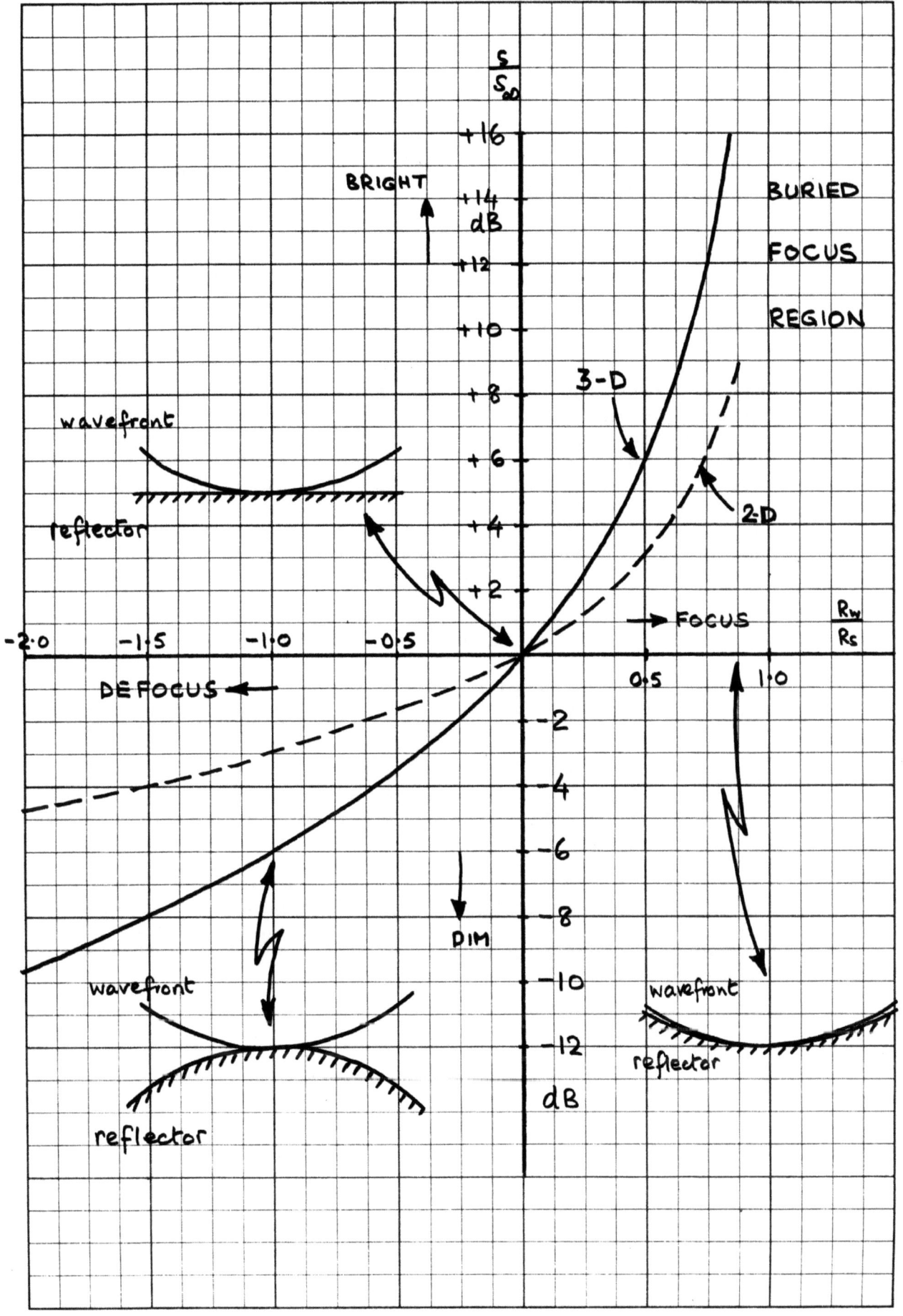

2-118B

desk, therefore, this requires an approximate hand migration and depth conversion.

Bounds can be placed on the focusing/defocusing effect without going through this migration exercise. On the defocusing side of the graphs, we see that when the radii of curvature are equal over an anticline, the dim-down ratio is -6 dB (down to 50%) for a dome or -3 dB (down to 71%) for a ridge. If the structure is tighter than this, as we shall illustrate later, it develops diffractions off its crest; only if these diffractions can be seen, therefore, can we be suffering more than -6 or -3 dB of defocusing. On the focusing side of the graphs, any syncline tight enough to produce a buried-focus bow-tie must take us beyond the point where the ratio of radii is 1; at this point the bright-up ratio (although infinite according to the approximate equations) is defined by the area of the reflector over which the radii are equal. In a conformable syncline the focusing effect can be seen increasing with depth until the buried-focus phenomenon appears; at this level we know the radius of curvature of the structure, and so we can calculate the focusing effect at all higher levels in the conformable sequence.

The reflection amplitude defocuses to half over a 3-D syncline when the ratio of radii is 1; it focuses to double over a 3-D anticline when the ratio is 0.5. It follows that focusing is a more obvious effect than defocusing, on real sections. Significant anticlinal defocusing is unlikely at shallow depth, and requires a rather tightly folded feature even at great depth. Significant focusing, however, is more frequent; it may occur both in synclines and in the concave curvature into piercement features and faults. Fortunately the synclinal focusing is seldom of quantitative interest, but curvature into the flank of a salt dome -- just where a genuine change of reflection coefficient may be anticipated by reason of a gas accumulation -- is an example of a situation requiring careful correction.

If we ever achieve perfect migration (which must be in three dimensions, of course) the amplitude effects of focusing and defocusing will be compensated in the program. Roll on the day.

In optics, focusing and defocusing are produced not only by specular reflection (from curved mirrors) but also by refraction (in convergent and divergent lenses). So also in seismics. Plano-convex low-velocity (convergent) lenses are exemplified by gas reservoirs in anticlinal structures and by overpressured shale swells; both produce local focusing effects. Evaporite lenses can produce defocusing effects.

More subtle, but still very significant, are the effects produced by refraction at an unconformity surface. We shall look at these in more detail in our later discussion of modelling (section 6.1, particularly p.6-6). For the moment, let us just note what happens when, in the sketch, the velocity V_1 of the upper material is intermediate between the velocities V_2 and V_3 below the unconformity. In the middle portion of the sketch, where the V_3 and V_1 materials are in contact, a normal-incidence-reflection ray from the deeper interface is refracted <u>anticlockwise</u> at the unconformity surface; to the right, a normal ray from the same interface passes through the upper dipping interface without refraction, and is then refracted <u>clockwise</u> at the unconformity surface. To the left, at the surface, there is a zone of "double vision" of the deep reflector; to the right there is a zone of blindness.

Of course this picture is oversimplified, but it does remain true that zones of local brightness and local dimness appear on reflections from below a truncation, and that the amplitude in these zones depends not only on the reflection coefficient of the truncated reflector but also on the <u>velocity</u> contrasts of the unconformity surface.

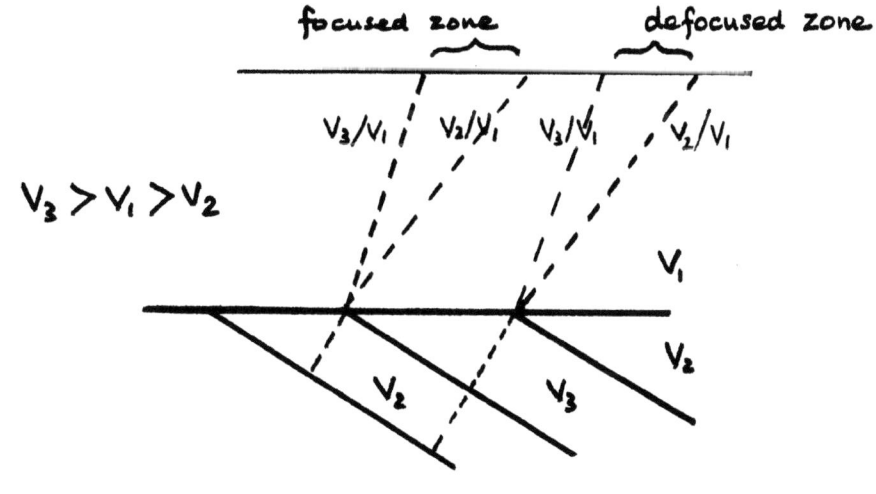

2-118D

2.8.5 Diffraction

Although not physically exact, Huygens' principle in its basic form gives a delightfully simple view of the diffraction process. Huygens' principle, we remember, says that each point on a wavefront may be viewed as the centre of a secondary wavelet, and that the position of the wavefront at a later time may be constructed as the envelope of such secondary wavelets.

One immediate use of this concept is to protect us from the fallacy of thinking that a source array, of normal dimensions, is more effective than a single source (of the same energy) because it generates a wave which is more nearly plane, and so less subject to geometrical divergence. We may regard the extended source — whether it is a Vibroseis baseplate, or an array of vibrators, or a pattern of shallow charges, or a length of detonating cord — as a number of Huygens secondary sources, each generating a secondary wavelet; it is true that after a few milliseconds the envelope of these wavelets has a plane front and curved edges (p.2-120, top left), but after a few hundreds of milliseconds the plane sector is inconsequential and the envelope is substantially spherical (p.2-120, top right). A source array of normal proportions does not generate a plane wave; it generates a substantially-spherical wave. However, the source array is directional in the sense that the amplitude of the wave in the downgoing direction is made greater than that in other directions by interference between the several wavelets. Thus a 100 m length of detonating cord generates a near-spherical wave whether the cap is at one end or the other; the only effect of the cap placement is a minor change of amplitude and high-frequency content in the "forward" and "backward" near-vertical directions.

Another, and very important, use of Huygens' principle follows from its message that the seismic wave returned from any geologic feature — no matter how complex — may be constructed as the envelope of secondary wavelets. Let us explore this, for a succession of situations of increasing complexity.

The first situation is that of a single point returning energy to the surface — one Huygens' secondary source in

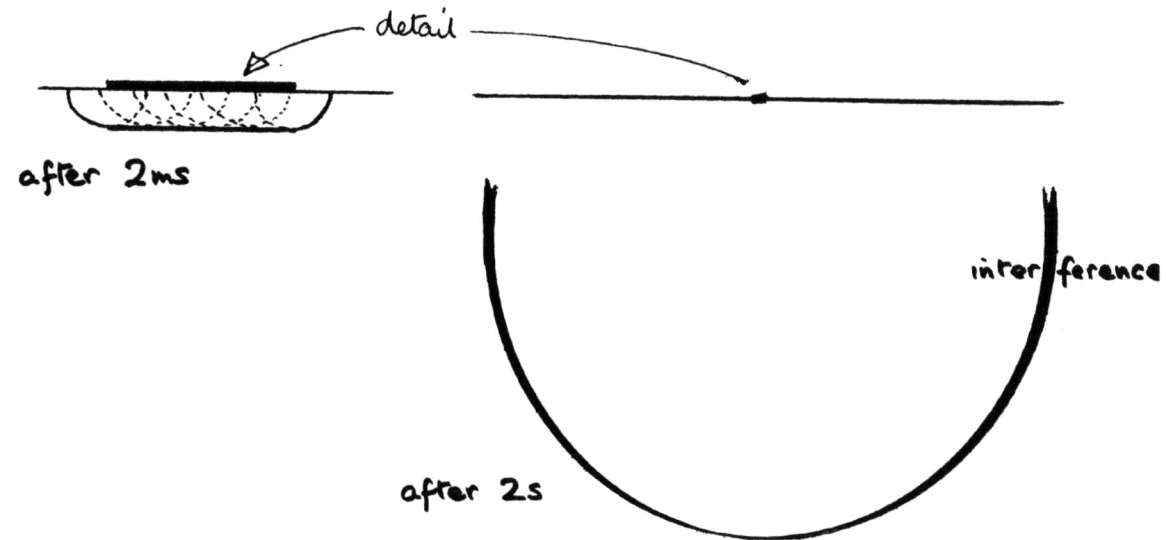

after 2ms — detail

after 2s — interference

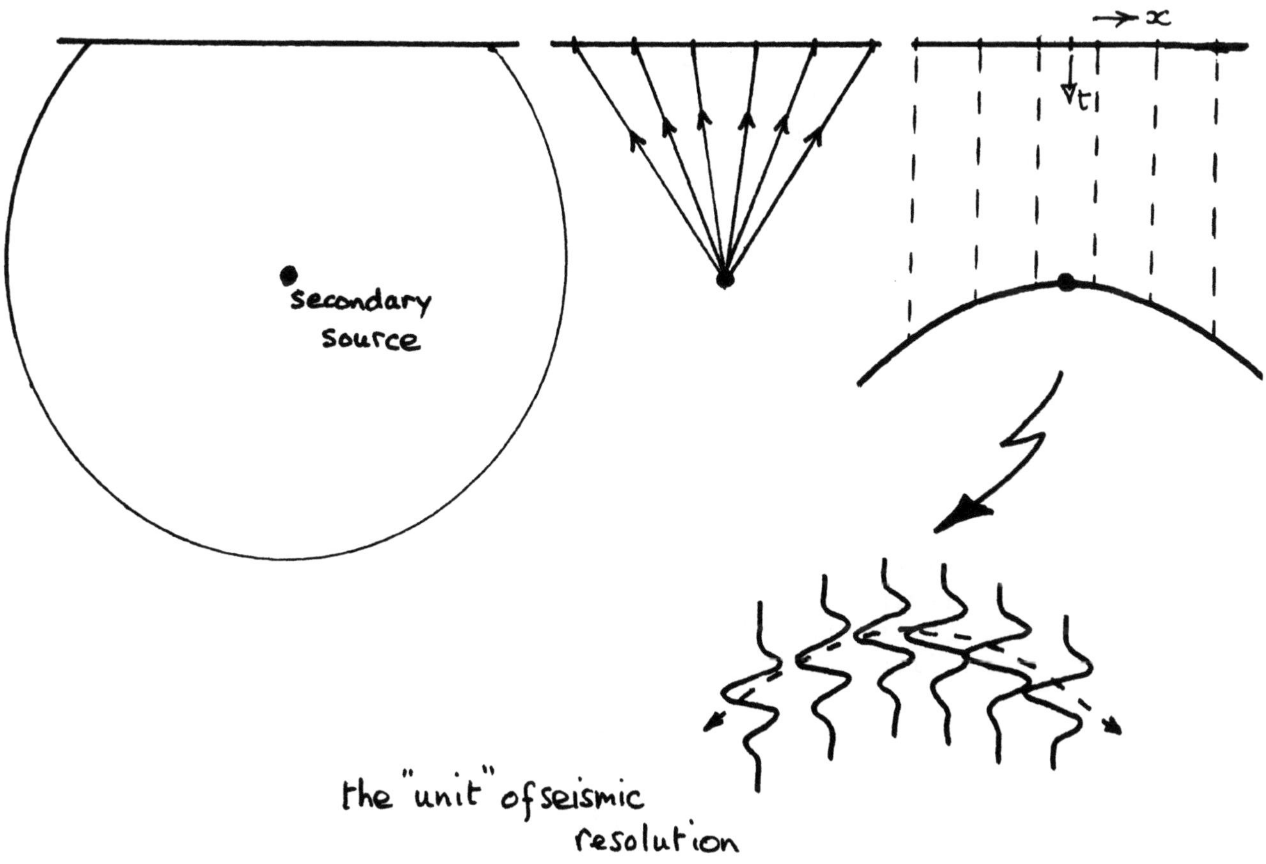

secondary source

→ x
↓ t₁

the "unit" of seismic resolution

isolation. The corresponding secondary wavelet is spherical (again we ignore the niceties of real velocity distributions, for the moment) and appears as at the left of the lower diagram on p.2-120. If we have a spread of geophones on the surface we detect arrivals, from this single source to each geophone position, along the ray-paths suggested in the middle sketch. These ray-paths represent cdp-type geometry, except that only one-way times are involved. If, as usual, we plot the arrivals on traces vertically below the surface positions of the geophones, we obtain a hyperbolic alignment of these arrivals (right-hand sketch). This hyperbolic alignment (which we shall call a Huygens hyperbola) therefore constitutes the trace-to-trace seismic signature of one individual Huygens secondary source in the subsurface.

At this stage, it is worth digressing for a moment to answer the question: What defines seismic resolution?

From Ricker we learn what defines seismic resolution in a vertical sense; our ability to separate two closely-spaced parallel reflectors is defined by the bandwidth of the reflected seismic pulse (preferably after phase-zeroization). From Huygens (through Hagedoorn, 1954) we learn what defines seismic resolution in a lateral sense; our ability to separate two geologic features spaced apart horizontally is defined by the curvature of the Huygens hyperbola (which is itself defined by the travel time and by a quantity which we shall recognize later as the migration velocity). Bringing together these two precepts, we can see that the "unit" from which the seismic record is composed is a hyperbolic alignment of pulses, of the form suggested at the bottom of p.2-120. Thus our ability to separate the top and bottom of a reservoir sand of constant thickness is a matter of reflection pulse bandwidth and phase; our ability to see a small reef (on an unmigrated section) is again a matter of pulse bandwidth and phase, but also of reflection time and migration velocity.

It is important that we have this straight. Reading the advertisements, we might be led to think that doubling our sample rate or halving our group interval will improve our resolution; although it is true that these variables must be properly chosen (we shall discuss this in sections 7.2 and 9.2), there is a point beyond which further change will increase the cost while doing absolutely nothing to improve the resolution. The resolution of an unmigrated section is <u>rigorously limited</u> by the reflection pulse, the reflection time and the migration velocity.

Now let us return to the concept of synthesizing the seismic response to a prescribed geology, using Huygens hyperbolas. In this we regard any geological interface which returns energy to the surface (note we do not insist on specular reflection here) as generating a number of Huygens secondary sources, each of which yields a Huygens hyperbola at the surface geophones. For illustrative purposes, we can regard the Huygens secondary sources as being of uniform strength and uniformly spaced along the interface; we shall consider deviations from this in Parts 6 and 7.

<u>TAPE 14</u>
85
86
87
88

Our second situation, then, is the <u>continuous reflector</u>. We may start with our secondary sources <u>widely spaced</u>, so that we can see the hyperbolas separately (p.6-13, upper). Then we bring the sources close together, and add all the hyperbolas at each sample position on each trace. The effect is that all the tails of the hyperbolas <u>cancel</u>, leaving only the continuous reflection that we would expect (p.6-13, lower).

By Huygens synthesis, then, as by any other means, a continuous reflector yields a continuous reflection. It is important to realize that the tails of the hyperbola generated by one secondary source are cancelled by the tails generated by many other secondary sources, and that if these latter sources are not present (or are not of equal strength) the cancellation is not complete.

This leads us directly to our third situation, which is the <u>terminating reflector</u> (Trorey, 1960). If a reflector of constant reflection coefficient terminates, the tails of the hyperbolas generated by secondary sources near the termination cannot be cancelled by tails generated beyond the termination, and so there must be a residual representing incomplete cancellation — an interference pattern of hyperbolas branching away from the termination point. As suggested in the upper diagram of p.2-122, the resultant diffraction pattern has a forward branch in which the reflection pulse smoothly assumes a lower-frequency integrated form (section 2.7.1), and a backward branch which is symmetrical except that the polarity is reversed.

The terminating reflector gives us an excellent example of the danger of good physics applied to an unreal geological model. There is no doubt about the physics — everything stated above is correct — but there is the gravest doubt whether a terminating reflector (of constant reflection coefficient up to the termination point) can exist in the real earth. We recall Question 6 of p.2-72: How quickly can a reflection coefficient change along a real geological interface? If it cannot change to zero abruptly, but requires a transition zone or some local curvature, then the interference pattern of the tails of the hyperbolas is changed. In particular, if the reflection coefficient changes slowly — say from 0.2 to zero over several kilometres — the cancellation of the hyperbolic tails remains substantial, and the residual interference pattern is too weak to be seen under real conditions.

But there is worse yet: in the literature, several authors have implied that what is a terminating reflector in physics is a fault in geology. This is emphatically not true. A reflector is a contrast between an upper acoustic impedance r_1 and a lower acoustic impedance r_2; it is not possible for this situation to terminate.

The simplest possible fault corresponds to a right-angled block of the material of acoustic impedance r_2, surrounded by that of r_1. This is quite a different situation, requiring the calculation of the strength of Huyghens sources down the vertical interface, and the inclusion of the corresponding Huygens

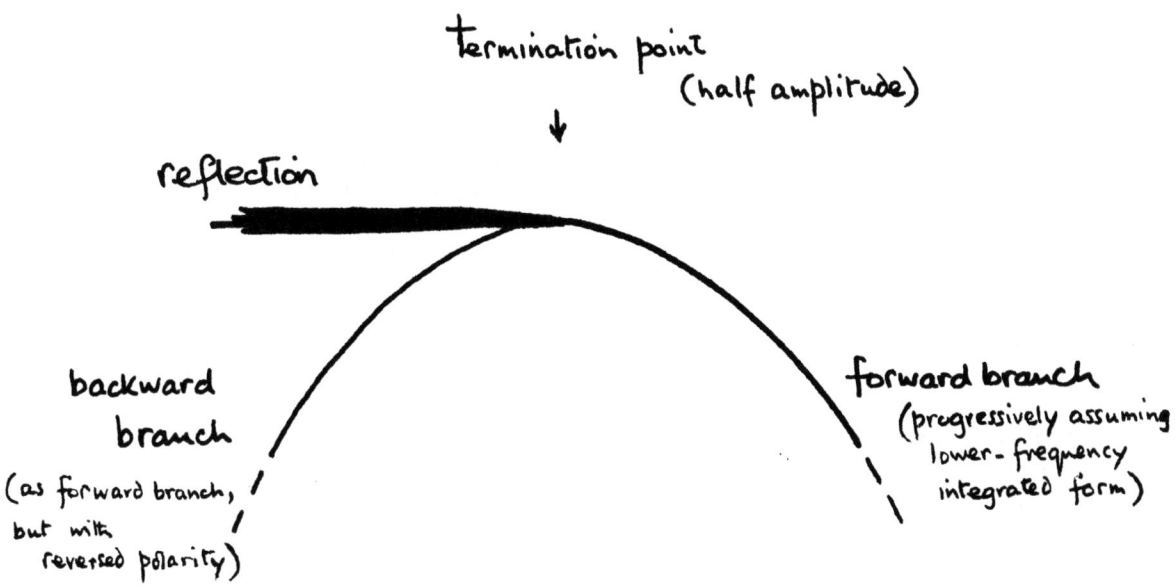

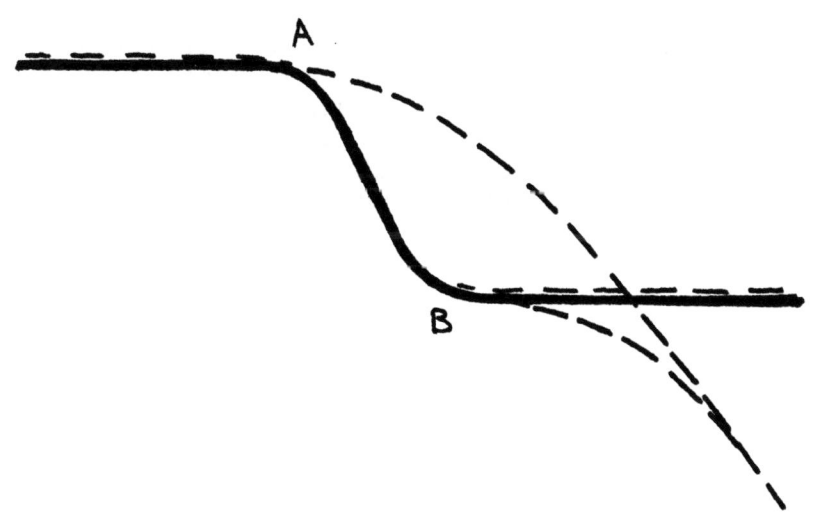

2-122

hyperbolas. The effect of adding these additional hyperbolas
<u>changes</u> the seismic signature; the signature of a reflecting
<u>corner</u> is not the same as that of a terminating reflector
(Kelly et al., 1976). The signature of the corner does not
show symmetry between the forward and backward branches of
the diffraction pattern, and indeed the backward branch
may not be visible at all.

So, when we are reading the literature, we must be on
our guard against equating a terminating reflector to a fault.
It is here suggested that a simple abruptly-terminating
reflector is impossible in real geology.

Our fourth situation is the flexure, illustrated at the
bottom of p.2-122. If we establish Huygens secondary sources
all along the geological interface represented by the heavy
line, and add together all the corresponding Huygens hyperbolas,
we find that the tails of the hyperbolas typically show
substantial cancellation everwhere except along the dashed
line. This dashed line — a forward-branch diffraction from
the top of the flex and a backward branch from the bottom —
therefore represents the seismic signature of this flexure,
at the appropriate travel time and migration velocity. At a
later time or at higher velocity the hyperbolas would be
flatter; at an earlier time or a lower velocity the hyperbolas
might be so sharply peaked as to yield a true reflection from
the flexed portion of the feature. In the case sketched there
is no specular reflection apparent from this steeply dipping
segment. The convex curvature over the upper bend of the
flexure yields defocused (weaker) amplitudes near A; the
concave curvature at the lower bend yields focused (stronger)
amplitudes near B.

Our fifth situation is the fault, in rocks sufficiently
plastic to show significant flexure before faulting. Then,
as we have just discussed, the upper bend of the flexure
generates a forward-branch diffraction (now ordinarily somewhat weaker than before) and the lower bend generates an
unchanged backward branch (upper diagram, p.2-124). In

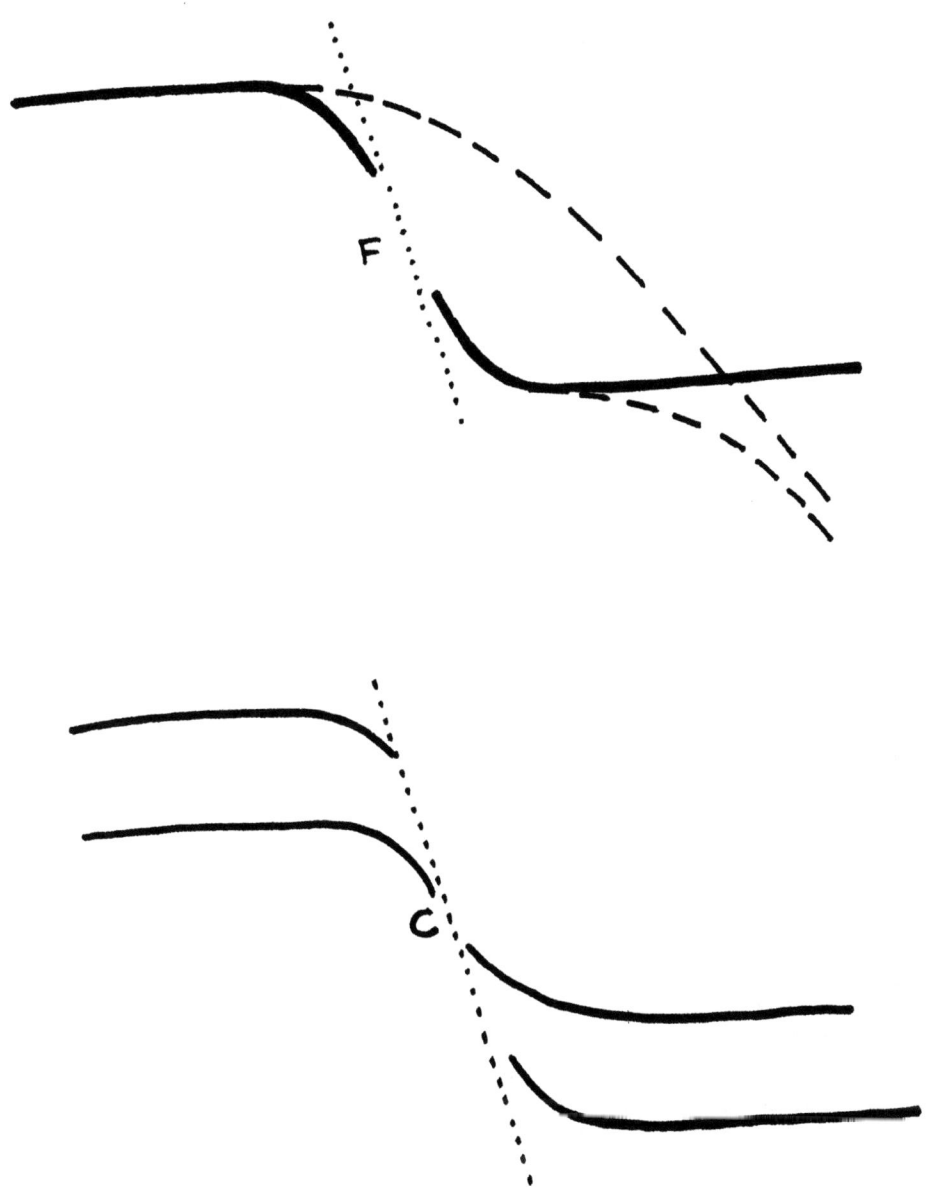

addition, there may be new diffractions introduced, because the faulting may introduce entirely new contrasts of acoustic impedance across the fault plane; such a new contrast is indicated at C on the lower diagram of p.2-124. As before, these new diffractions can be synthesized by adding together Huygens hyperbolas of appropriate strength. Obviously, there are copious opportunities for new interference complexes — even interference between the forward branches of one faulted impedance-contrast and the backward branches of another. It is suggested here that the published "examples" of mirror-image forward and backward branches (that is, the terminating-reflector signature) from the edge of a fault are in fact such multiple interference complexes, involving more than one contrast of acoustic impedance.

Clearly, real faults in multi-layered rocks become very complicated situations. Not only do we have the new contrasts of acoustic impedance across the fault plane, but in synthesizing the diffraction patterns by the addition of Huygens hyperbolas we have to ascribe strengths to the Huygens secondary sources along the fault plane. The latter problem depends on the hade, on refraction (and hence on velocity contrasts) above the secondary source, and on the position of the seismic source relative to the fault. The interrelation of these factors is not something which we can attempt at our desks, and so we have to accept that there is a limit to the usefulness of quick-and-dirty diffraction synthesis by the sketching of Huygens hyperbolas. However, for the less-complicated situations which we discussed previously, it is often true that we have only to sketch these hyperbolas roughly, with pencil and paper, to see the places where the tails must cancel and the places where they must reinforce into a strong diffraction.

Another rudimentary situation is worth specific mention — that of the corrugated tightly-folded reflector, or the reflector broken by extensive minor block-faulting. Clearly, we must add hyperbolas corresponding to secondary sources on the "up" parts of the reflector, and hyperbolas corresponding to secondary sources on the "down" parts. The result must depend on the spatial

92 wavelength of the corrugations, on the "throw" from top to bottom of the corrugations, relative to a **pulse length**, and on the curvature of the hyperbolas (that is, on the reflection time and the migration velocity). We are back to our definition of resolution. If the corrugations are of small amplitude and long spatial wavelength we do not see them; the reflector appears corrugated. If the corrugations are of small amplitude and short wavelength we do not see them; the reflector appears plane (but possibly of a changed character). If the corrugations are of larger amplitude and short spatial wavelength we do not see them; the reflector becomes an obvious complex, and some minor diffractions may be seen. If the corrugations are of large amplitude and longer spatial wavelength the diffractions may become very obvious.

It is important to remember the dependence of interference criteria on pulse bandwidth; we are working at a time when major efforts are being made to improve our reflected bandwidth, and to the degree that these are successful we must re-evaluate the changed consequences of interference. Thus we must re-evaluate our array lengths, since the response of an array represents the interference between the individual geophone outputs; we must also appreciate that diffraction patterns (which are the result of interference between Huygens hyperbolas) are similarly sensitive to bandwidth.

> Question: We have said that the sea surface has a reflection coefficient of substantially − 1. Do we need to make a correction for the waves?

Finally, we can graduate from the rudimentary situations discussed above to the case of arbitrary geological complexity. If we have a geological model defined, we can obtain a first approximation to the corresponding seismic section by constructing,

2-125A

93 for each of many closely-spaced points along each interface,
94 an appropriately-scaled version of the hyperbolic "unit" of
95 seismic resolution (p.2-120, bottom), and by adding together
96 all the values which fall in each cell of the section.

97(B)

As noted earlier, small corrugations may be invisible to the seismic method. It is as though the local deformations of the wavefront <u>heal</u> by reason of the diffraction phenomenon. This has led Sheriff (1975) to describe some manifestations of diffraction by the helpful term <u>wavefront healing</u>.

In particular, this view is <u>helpful</u> in guiding our expectations where small anomalies of transmission characteristics are concerned. The classic demonstration is to sit a few metres from a radio, and listen for the effect of interposing a transmission anomaly (say some heavy object half a metre across) between the radio and the ear. If the anomaly is close to the ear the sound can still be heard, by diffraction, but the insertion of the anomaly makes a significant difference to the volume and character of the sound. However, if the anomaly is in the middle region between radio and ear, its presence can scarcely be detected at all — the wavefront heals around it.

Translating this commonplace into seismic terms, we can see that a local transmission anomaly — whether it be an anomaly of absorption, or velocity, or transmission loss, or refraction, or scattering — may or not be seen on a seismic section, depending on its size and its position along the reflection path. An anomaly may be clearly detectable by a borehole geophone just below it, but quite invisible on the reflection section. If it is visible on the section, it is likely to be accompanied by two sets of diffractions from its ends: a broad flat set associated with the path from anomaly down to reflector and back to surface, and a tightly curved set associated with the shorter path from anomaly to surface. Often only the second set is visible, the first having become so flat that they have substantially healed the reflection. If the anomaly is small neither set may be visible.

And it is the <u>smallest</u> dimension of the anomaly which is the important one; a channel sand may be several kilometres in length but only a hundred metres wide — it is the hundred metres which largely determines the wavefront healing.

2.9 POST-DEPOSITIONAL GEOLOGICAL PROCESSES AS MODIFIERS OF SEISMIC PROPAGATION CONDITIONS

In section 2.3 we considered the relation between the geological circumstances of lithology and depositional environment and the physical properties relevant to seismic propagation. We also mentioned in passing that these physical properties (specifically, elasticity, density and possibly absorption) may be affected by geological processes occurring after deposition. We should explore this a little further.

We remember that the effect of compacting pressure is major in shales, and possibly significant in some limestones. The horizontal forces generated by tectonic activity can be just as important as the vertical forces due to gravity. In a tectonically-undisturbed zone (for example, the US Gulf Coast), the horizontal pressure is typically one-third of the vertical pressure, and the vertical pressure dominates. In a contorted zone (for example, the Rocky Mountain area), the horizontal pressure may be three times the vertical pressure, locally, and so be dominant. The interpreter therefore searches his sections for evidence of horizontal pressures which, by supplementing the vertical pressures, are likely to have increased the velocity and density of the shales beyond the normal compaction curves; in this context normal faulting suggests absence of such horizontal pressures, while folding and thrusting suggest their presence. Of course, both effects may be present on the same section, at different levels.

Dapples (1972) has suggested that the effect of tectonic stress can be greater than that consequent merely on compaction; the cementing material can actually be changed, as a chemical neoformation, by the stress.

Smirnova (1973) and Sakmuradova (1973) report measurements showing very significant changes of density (and presumably velocity also) in tectonically-disturbed zones. A material of specific density 2.0 - 2.2 on the flanks of a dome structure had a density of only 1.6 - 1.8 on the crest, but 2.3 - 2.4

in an overthrust.

We have previously stressed the importance of fracturing in depressing the velocity. We must also recognize that fracture systems can become cemented, so that the original velocity and density are restored or even exceeded. On the large scale, in the mélange zone of a fault, rocks of specific density 2.7 have been reduced to 2.45 and subsequently cemented to 3.5. On the small scale, in fine fracture systems, the volume proportion of cement can be minute (in a time-average sense), but yet sufficient to fill all the voids and so restore the original velocity and density.

We have agreed earlier that simple uplift yields a comparatively minor (elastic) decrease of velocity and density, whereas uplift producing fracture yields a major decrease of velocity and some decrease of density. Subsequent downward movement restores much of the density decrease, but less of the velocity decrease; this is because the increased pressure never fits the fractured rock together exactly. Cementation of a fracture system, consequently, can have an effect greater than that of deeper burial.

Anomalies of velocity and density are therefore to be expected where the possibilities for cementation are in some way changed. Such effects have been observed over reefs (Davis, 1972); the magnitudes are significant (upper diagram, p.2-128). Any occurrence which provides an upwards conduit for water excluded from deep sediments is likely to affect the mineralization of adjacent rocks; faults and piercements are obvious candidates. Conversely, any occurrence which provides a barrier to the upwards migration of excluded water is likely to provide an anomaly in the opposite sense. The most intriguing of these latter possibilities is the barrier presented by an accumulation of hydrocarbons; as suggested in the lower diagram of p.2-128 (after Pirson, 1970), it is quite likely that the presence of oil or gas can create an anomaly of cementation (and hence of velocity and density) in the sediments above.

Two other agencies affecting velocity and density may be worth mentioning. The first is that magmatic intrusions and contact metamorphosis can be expected to produce major changes. The second is that overpressured zones formed during the depositional processes may later be relieved; formations experiencing this change must certainly increase in velocity and density, but — since the packing is unlikely to be efficient, and since there has been no normal cementation — the resultant velocity and density are unlikely to match their normally-pressured counterparts.

As we approach the end of Part 2 — our discussion of the factors affecting seismic propagation — we should take a moment to appreciate the degree to which these considerations have led us beyond traditional seismic interpretation.

Let us think of it first as we stand looking at an outcrop, and then as we sit looking at a seismic section.

As we look at an outcrop, these might be the questions passing through our mind:

- What is the lithology?

- What was the depositional environment?

- What was the subsequent tectonic and burial history?

- How have these factors affected the lateral rate-of-change of physical properties?

- How have these factors determined the dominant compacting agency?

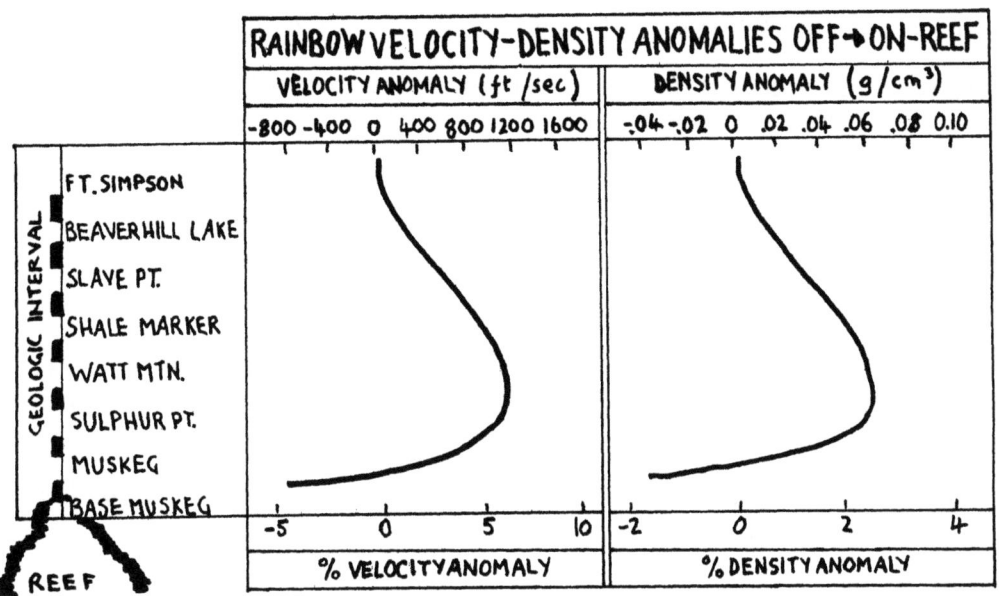

After Davis, 1972

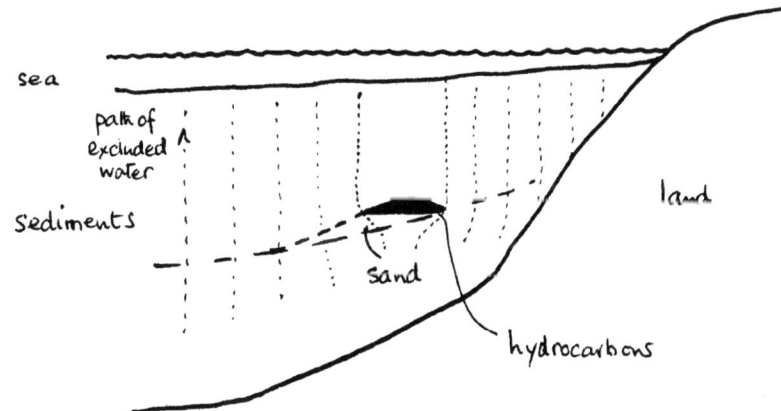

After Pirson, 1970

- What is the cement, and what is its proportion?

- What other solid contamination must be taken into account in determining the density and velocity?

- Has the material been fractured? Locally, or extensively?

- Has it been recemented after fracture?

- What is the present porosity?

- So what would be a likely value for velocity and density in the water-saturated state, taking into account the lithology, the supplementing of vertical compacting pressures by lateral pressures, the compacted porosity, the cementation and other solid contamination, the fracturing, the recementation, and the uplift-and-burial history?

- How much would the water-saturated velocity and density be changed by gas saturation?

- Are there any features which would cause a deviation from normally-expected relations between velocity and density?

- Having regard to the absence of any direct velocity-lithology relationship (by reason of the factors listed above), to what extent would a good seismic velocity measurement narrow the range of possible lithologies, in this case?

- Does the combination of grain sorting, angularity, fracturing and cementation allow the possibility of relative motion at grain contacts?

- Is there then a hope that the material might allow a further reduction in the range of possible lithologies, on account of anomalous absorption?

- Is the material thick enough to allow a measurement of anomalous absorption?

- Would the measurement be confounded by the short-path-multiple effect in contrasty stratification? Or by scattering?

- Are the bounding interfaces of the major units plane, over a likely range of Fresnel-zone dimensions?

- Shall we need to compensate any calculated reflection coefficient for reflector curvature?

- Is it safe to take the surface reflection coefficient as -1?

- Over what lateral extent would it be permissible to smooth measurements of interval velocity and/or reflection coefficient, for these formations?

Now let us exemplify the questions going through the mind of the modern interpreter as he looks at a seismic section. Of course he still thinks the traditional thoughts — marker reflections, and structural patterns, and faulting, and why-can't-those-idiot-processors-get-me-a-better-section-than-that — but now he is likely to be thinking other thoughts as well:

- Whereabouts on the section will the geology allow good interval-velocity measurements?

- What evidence is there of depositional environment?

- Can the shale compaction curve be established from nearby wells, or otherwise?

- Are we expecting zones of overpressure?

- Having regard to the above points, what are the chances of establishing the lithology from measurements of interval velocity?

- If the velocities suggest carbonates, can we detect local depressions of velocity (indicative of fracture porosity) over the uplifted zones? Are there reservoir implications?

2-130

- To what extent can we hope to supplement the lithologic indications derived from interval-velocity measurements by acoustic-impedance measurements derived from reflection strengths?

- How are we to turn reflection strengths into reflection coefficients, on this section?

- Which reflectors are discrete and abrupt, between fairly massive units?

- Are there one or more good pairs of discrete reflectors, for the top-and-bottom method?

- Is the reflection coefficient of the surface likely to deviate substantially from -1?

- Do the chosen reflectors have extents suitable for reflection-coefficient determination, having regard to the dimensions of the Fresnel zones, curvature, and velocity lensing above?

- Have the field work and the processing preserved valid reflection amplitudes?

- To what extent can we hope to supplement interval-velocity and acoustic-impedance measurements by absorption measurements?

- What evidence can we see defining the local agencies responible for changes of frequency content on the section?

- Do we see any possible gas-liquid contacts?

- Do we see any locally-anomalous strong amplitudes in a position which would be compatible with a hydrocarbon accumulation?

- How much of the apparent strength of this reflection is due to interference?

- Is there a risk the strength could be due to focusing, or velocity lensing or post-depositional tectonics?

In later parts of the course we shall explore the means for answering these questions in practice. In Part 3 we shall discuss the actual measurements and their validity; in Part 4 we shall enlarge on the direct recognition of hydrocarbons, and in Part 5 we shall be much concerned with establishing depositional environment. However, to conclude Part 2 let us just consider a light-hearted example to illustrate the extent of our release from exclusively <u>structural</u> thinking.

112
113

The sketch represents a classical river-delta system, complete with meanders, point bars, natural levées, flood plains, distributary channels, deltaic fingers, a beach, an offshore bar, and a prograding delta-front. Let us suppose that this system becomes covered by a large-scale transgression, that subsequent sedimentation and subsidence bury it to considerable depth, and that gas originating in the flood-plain deposits (or elsewhere) becomes trapped in the channel sands and the beach and bar sands. These accumulations require no structural component, so we shall assume no relief at all.

What would be the seismic manifestation of this system?

If we shoot an E-W line as suggested in the plan view, it shows no structure whatever (except at the delta front). The only anomalies are anomalies of amplitude, over the gas-saturated zones (lower sketch). These anomalies are uninterpretable on a single line. On a close grid of lines, however, they can be <u>mapped</u>, and their disposition on the map shows the outline of the river-delta system.

For such purposes, of course, the grid of lines needs to be very tight — an observation to which contractors have been observed to nod agreement.

114(B)

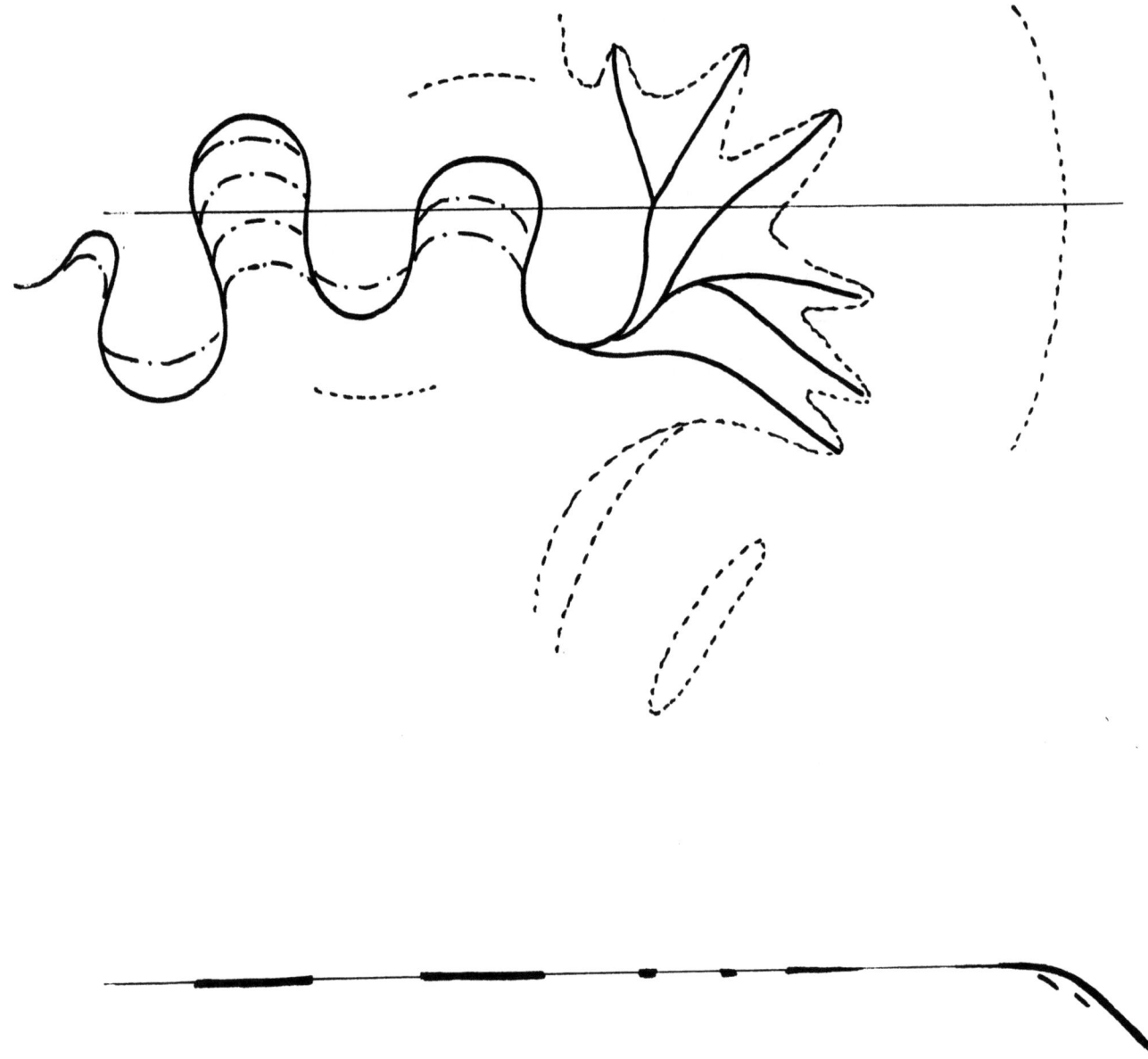

PART 3

SEISMIC SIGNAL MEASUREMENTS

IN DETAIL

TAPE 15   3.1   THE MEASUREMENTS WHICH MAY BE MADE

Let us look now at the measurements which may be made on seismic data — first on single traces, and then on groups of traces.

3.1.1   Measurements on a single trace

We may make the following measurements:

1. The trace <u>amplitude</u>; this means the sample values recorded on tape — values representing the acoustic pressure (at sea) or particle velocity (on land) as a function of time.

2. The trace <u>envelope</u> as described in p.2-96; this is a one-sided function which loses information on individual details of the wiggles, exchanging this phase-type information for a convenient measure of reflection strength. Without a measurement of this type, as we agreed earlier, we have some difficulty in defining what we mean by the amplitude and time of a reflection pulse.

3. The <u>frequency content</u>. To most of us, the display of frequency content which has the most immediate impact is the filter panel, made as part of a normal processing sequence. Where a visual appraisal of frequency content is sufficient, this remains a powerful — and safe — tool. For other needs, we may measure the frequency content in terms of a "window" (or "gate"), for which there are many choices including the following:

115

116(B)

- The whole trace, as suggested at (a) on p.3-2. This yields one spectrum per trace, the general shape of which represents the average form of the seismic pulse over the whole trace, and whose individual peaks and notches represent (in a complicated way) the thickness of the layers —

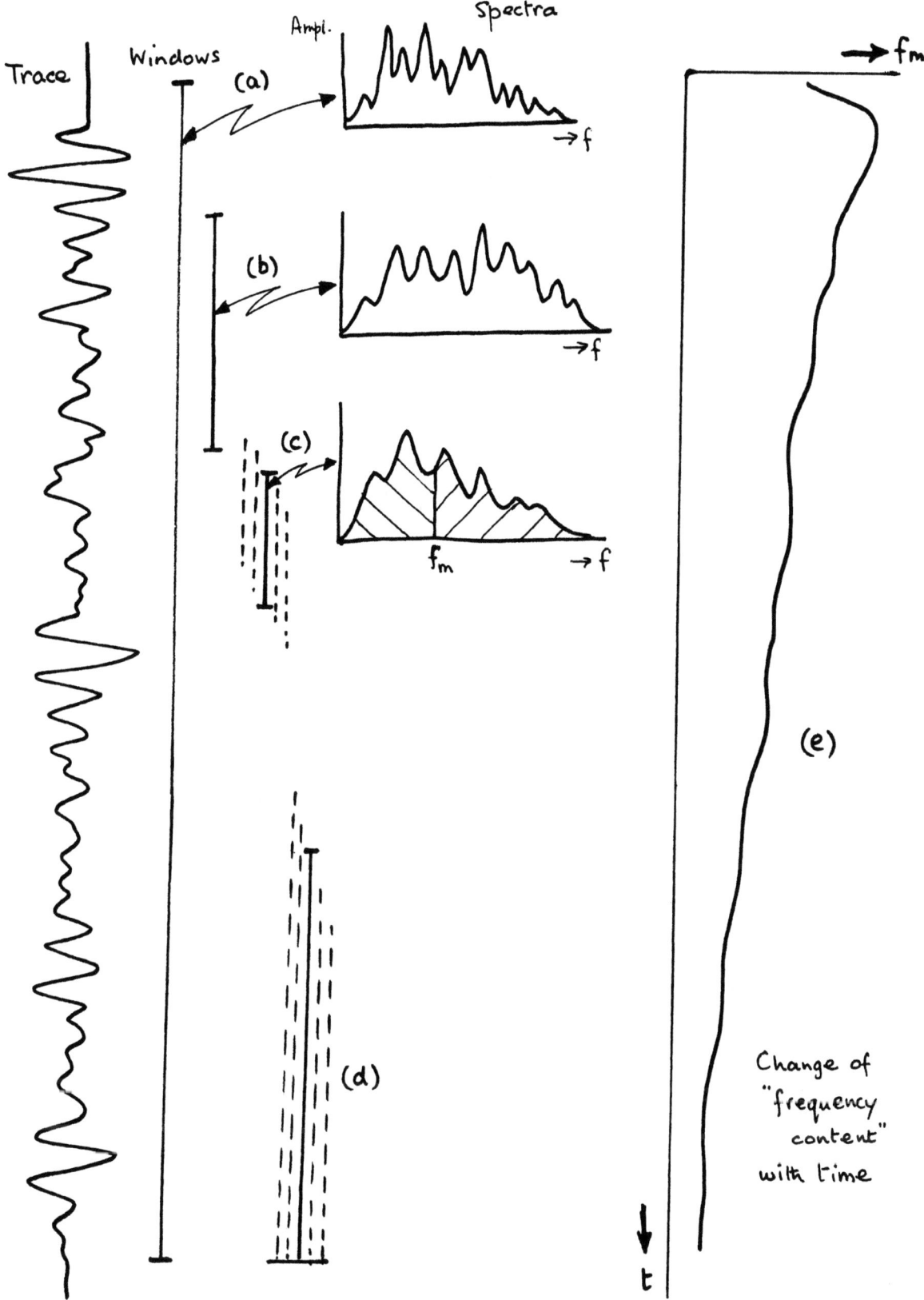

3-2

including multiple-generating layers — in the earth (section 2.7.2).

- A single shorter window, as suggested at (b). The general shape now represents the average form of the seismic pulse over this shorter time range; it includes more high frequencies than the long window if the short window is at early times, and fewer high frequencies if it is at late times. The individual peaks and notches of the spectrum now represent the layer thicknesses present within the shorter time range.

- A short _moving_ window. As suggested at (c), successive positions of the window may overlap, to any chosen degree. This gives us the means of studying the _change_ of frequency content with time. The change must have two components: a slowly-varying component representing the progressive effect of absorption and other high-cut effects, and a locally-varying component representing the consequences of successively dropping one layer thickness from the window while taking in another, as the window slides down the trace.

- A window of which only one end moves; this is illustrated at (d) for the case when the fixed end is late in the record.

Each of these choices of window is useful for particular purposes. However, those which use moving windows (and which are therefore most likely to give us information about changing geology) have the embarrassment of burying us in data; in the limit of case (c), we might obtain several hundred spectral values for each sample of the original data, on each trace. Therefore we are led to seek some measurement — preferably a single number for each calculated spectrum — which would contain the essence of the full spectral measurement in the context which interests us. If, for example, our concern is with the slow absorption-type effects, we are not interested in the individual spectral peaks and notches, and so might use a measurement such as that labelled f_m in sketch (c) — the frequency in hertz which bisects the area under the significant part of the spectrum. Or we might use the ratio between the area under the spectrum and its value at a particular frequency expected to have a stable content (Quarles, 1973). Or we might make a measurement of the instantaneous phase of the seismic trace, and compute its rate-of-change as a measure of the "instantantaneous frequency"; this would give us a new sample-for-sample frequency measurement which is conceptually rather different from Fourier-type analysis (Sheriff et al., 1977). Many other choices are possible; the important feature is the reduction of the volume of spectral data required for a defined purpose.

To the extent that this can be done, we can construct the graph of sketch (e), showing the variation of frequency content with time. For a frequency measurement over a moving window, this graph must contain rapid variations consequent on changes of layer thicknesses within the moving window, and slower variations consequent on absorption and the other high-cut agencies. Except for a spurious effect at early times (which we shall consider later) the slower variations must always represent a loss of high frequencies, as shown in the sketch.

Clearly this graph can itself be manipulated for particular purposes. We may smooth it, to enhance the slow variations, or differentiate it to enhance the layer-thickness content.

The problem of studying absorption in a particular formation (such as a reservoir) clearly requires that we make a careful choice of window length, window overlap and smoothing operator; in its nature as a local effect, it must be difficult to separate the effect of absorption (which we know to be small, from Question 2 of p.2-81) from the effect of changing layer thicknesses within the window. We shall return to this later.

Other manipulations of the frequency information are possible, as suggested (following White and O'Brien, 1973) on p.3-6. The case used as an illustration is that of two windows centred on times T_1 and T_2 (for example, two one-second windows centred at 1.5 and 3 seconds).

First, we may compute the two corresponding amplitude spectra A_1 and A_2 as shown. Then we may compute their ratio, at every frequency, and plot the logarithm of the ratio as sketched at the bottom. The individual points must have a considerable scatter, because of the different layer thicknesses (and hence spectral peaks and notches) in the two windows; however, to the extent that it is safe to draw a straight line through them, this line has a negative slope of $2\pi(T_2-T_1)/Q$, where the Q refers to the absorption occurring over the time range. (In using Q as a measure of the high-frequency loss, we are saying that either short-path multiple effects and scattering are negligible, or these other mechanisms are "lumped in" to a measurement purporting to be of absorption; the second would be the general case.) We note that this technique does not require that the A_1 spectrum be computed from a long window such as that suggested above; it could use the spectrum of the recorded outgoing source pulse, or that of the sea-floor reflection. We shall return to these possibilities later.

Second, we may make the assumption that all the processes contributing to the two calculated spectra are minimum-phase, and use this to construct an estimated pulse shape for the average wavelet in each of the two windows. This manipulation is sketched to the right of p.3-6. Later, we shall see several uses for these estimated pulse shapes; for the present, let us just note their value as one input to studies of interference; as soon as a pulse shape is available, different models for the interference mechanism can be readily tested (with nothing more than a desk calculator, if necessary).

Whenever our main concern is in identifying and measuring changes in shape or in spectrum, we are confronted with a

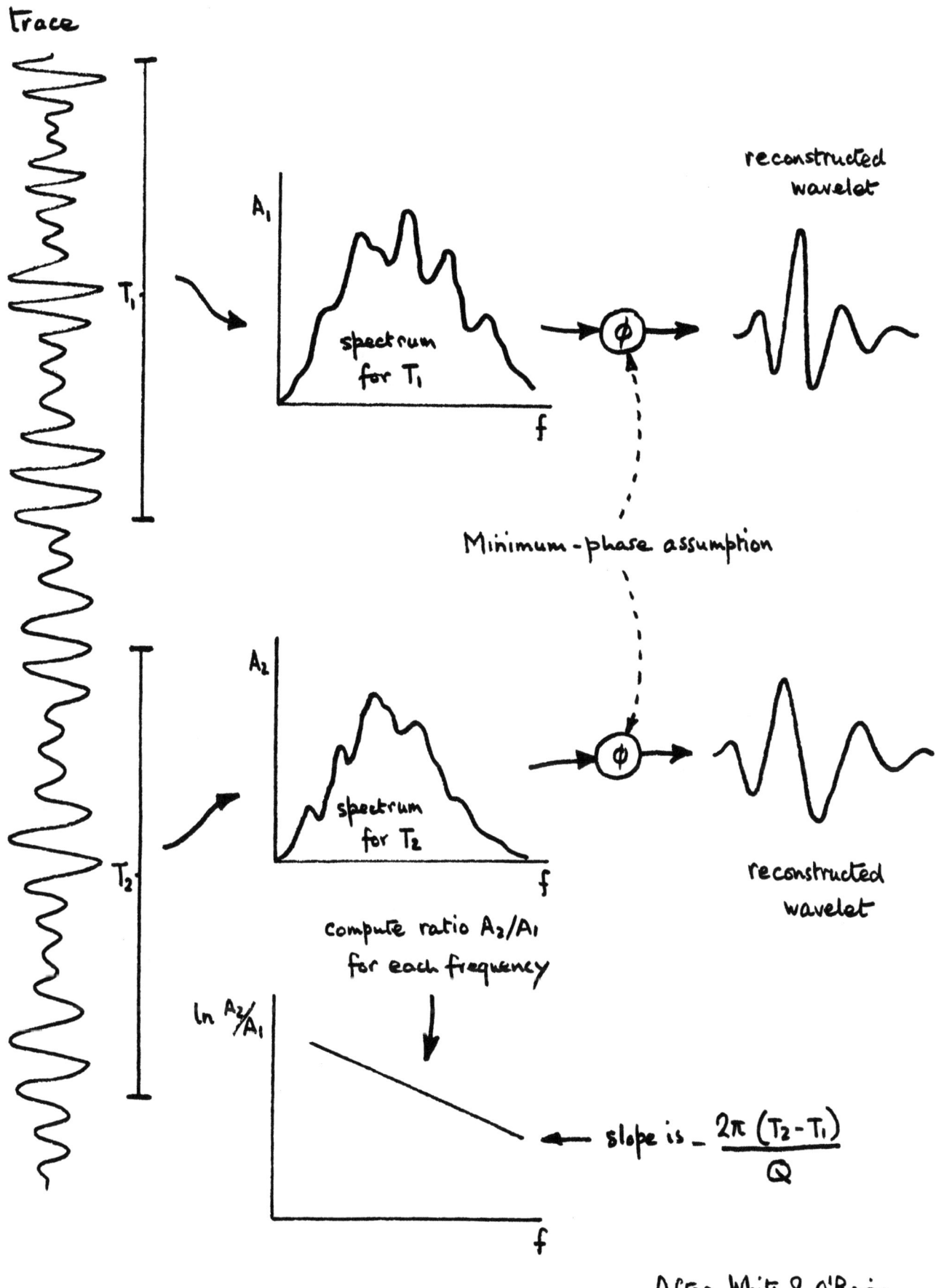

fundamental difficulty. This is that our desire to see the time-variation on a moment-by-moment basis necessitates short windows, but the shortness of the window decreases the certainty of the spectral estimates. There is no escape from this problem on a single trace, and the interpreter making studies of frequency content must be aware of it. Like the old problem of the conflict between signal-to-noise ratio and resolution, it requires a judicious compromise; the interpreter alone (after seeing the effect of various choices, and in the light of his defined problem) is qualified to make that compromise. (Incidentally, it is worth mentioning that this is much the same problem as that of choosing window lengths for a time-variant deconvolution; the degree of time-variation is too small if the windows are long, but the process removes primaries if a window is so short that it becomes dominated by a single reflection complex.)

That part of the pulse spectrum which does not change with time is the contribution of the source itself (and possibly of the free-surface reflections above source and/or geophones). Obviously it is of value to know this contribution deterministically whenever the source is such that its output can be measured; this point will recur many times in following discussions. Also obviously, if the source itself is narrow-band there is little chance of detecting changes of spectrum down the record; we shall never detect absorption if we are working Vibroseis with a sweep from 8 to 20 Hz.

4. The reflection _polarity_. First we note that this is not concerned with the convention for _display_ polarity; we are concerned to determine the _sign of the reflection coefficient_ (section 2.4.1 and examples 1 and 2 of section 2.5.4). We are asking: Which way up is the reflection pulse _as a whole_?"

If the outgoing signal could be a simple spike of compression and the reflected signal also a simple spike, this would be a trivial determination which could be done by eye (sketch (a) of p.3-10).

If the form of the outgoing signal is unknown, but we can see a certain pulse shape on a particular reflection and an obvious inversion of this shape on its surface multiple (sketch (b)), then we can be sure that the primary reflector is positive; if the surface multiple is obviously not inverted, then the primary reflector is negative. This is also feasible by eye.

The indications of sketch (b) would be clarified if the reflected pulses could be manipulated to be individually zero-phase, as in sketch (c). Such manipulations are highly desirable in all attempts at polarity estimation (Schoenberger, 1974). However (.... despite what we read in the advertisements!) such processes <u>do not allow the unique interpretation of a peak as a positive interface and a trough as a negative interface</u>; this would require that the whole seismic system have response to zero frequency, which is not possible with present equipment and techniques.

At the other extreme of the problem, we can imagine reflection pulses which (either by reason of a shot-cavity resonance or shallow-water reverberation) have the appearance of sketch (d). The "leggy" appearance of each reflection now defeats a polarity decision by eye.

Even in a case of medium complexity, such as that shown in sketch (e), it has been generally supposed that the visual identification of reflector polarity is not sufficiently definite to be useful. But nowadays we are not restricted to the power of the eye; if we can formalize the process by which the human identification of polarity is made, that process can be mechanized.

One of many methods by which this can be done involves a study of the position of the individual peaks and troughs of the reflection waveform relative to the position of the maximum of its envelope. Thus for each of the reflection pulses on p.3-10 there is a corresponding envelope, drawn at the right-hand side. The envelope is the same, of course, whether the reflection is positive or negative. It is a simple matter to pick the time of the maximum of this envelope, and then to ask whether the pulse extremum closest to this maximum (or just before it, or as we wish) is a peak or a trough. Clearly, we can devise a criterion of this type which will correctly identify the polarity of all the pulses shown. However, it remains true (as it always must, in the real world) that the chances of correct polarity determination are best on pulses having a broad amplitude spectrum and a zero phase spectrum — those closest to the spike of sketch (a).

If we do not have a recording of the outgoing signal, or if we do not know how to process it appropriately for these purposes (both of which aspects are discussed more fully later), our so-called polarity determination decides "which way up" the reflection pulse is, but does not tell us whether that means a positive or negative reflector. The absolute identification therefore requires a <u>reference reflection</u> whose polarity is known, and which can thus be used to calibrate the determination. This is conveniently the sea-floor reflection in marine work; otherwise it may be the top of a massive formation known to be of high acoustic impedance (for example, a limestone).

Since our reflection records are not (and never will be) composed of spikes, the determination of reflection polarity is always subject to error in the presence of interfering reflections. On a single trace, we can use some of our interference criteria from section 2.7.1 (particularly pulse duration, saddles in the envelope, and "hickies") to avoid making a judgement at all where interference means that the judgement must be in doubt. In following sections we shall see that the use of more traces adds to our powers of discrimination.

5. Other measurable variables are now feasible. In fact, the power of the computer means that we are now limited solely by our ability to devise and understand the seismic measurements, and not by our ability to perform them. Any anxiliary variable which has any hope of a geological meaning can and should be explored.

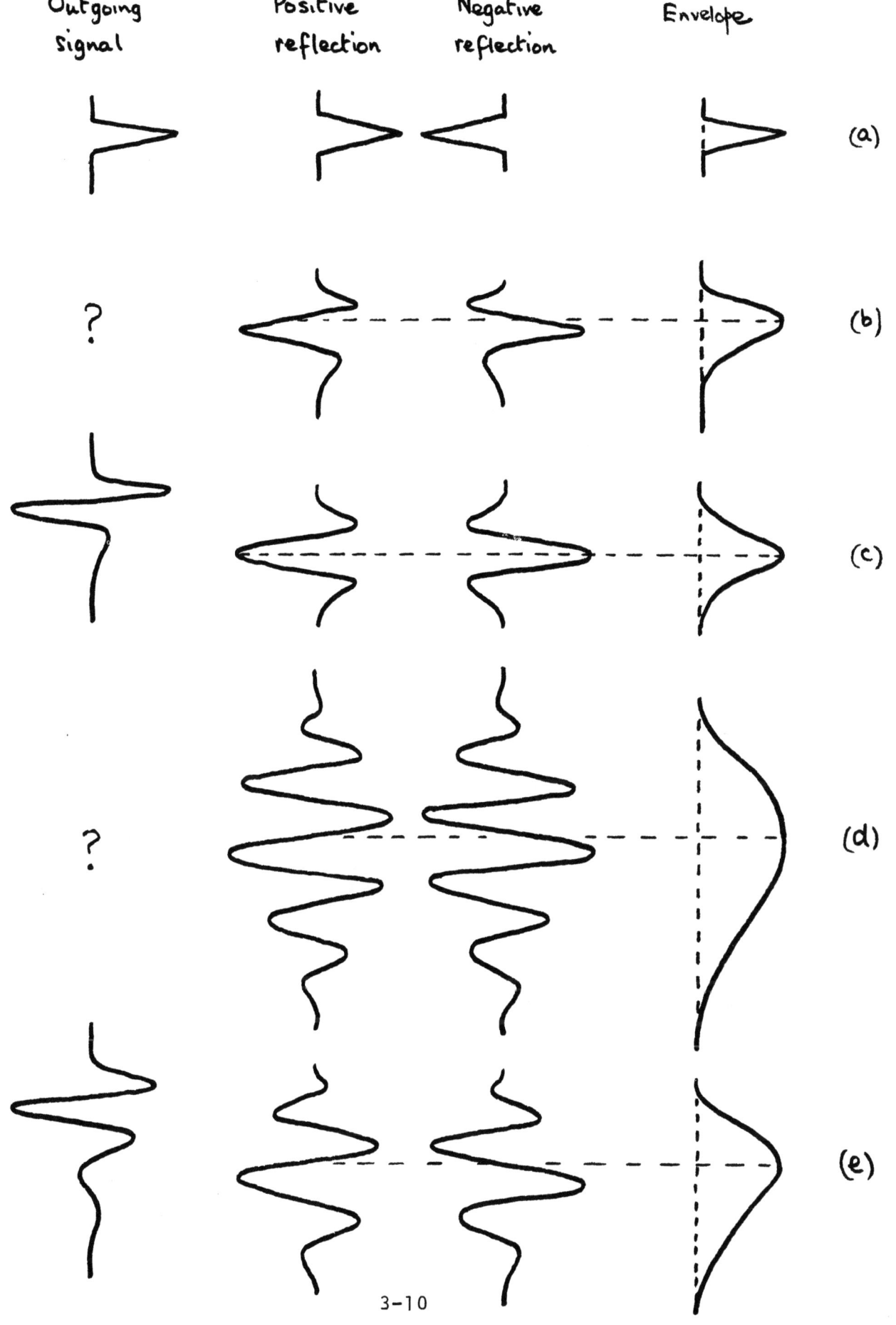

3-10

For example, we shall see in Part 5 that the amplitude distribution of a seismic trace has significance in terms of depositional environment — that a portion of trace which is characterized by continuing high-amplitude reflections suggests quite different depositional conditions from one characterized by a few individually strong reflections in a quiet background. A measurement based on the amplitude distribution can easily be mechanized to recognize this distinction.

Again, the mechanized recognition of "hickies" (as an unexpected increase or decrease in the slope of the reflection waveform) is easily implemented.

Whether such auxiliary measurements are valuable depends on the ease or difficulty associated with making the same assessments by eye and brain on normal sections. The important point is that the interpreter should be alert to see cases where the machine can help, and should not hesitate to ask that it be done. If we see something potentially useful, let us try it.

All the preceding material was concerned with measurements on single traces. We turn now to the additional information available when groups of traces are combined in particular ways. Important references in this context are Morgan (1970) and Taner (1974).

TAPE 16 ### 3.1.2 Additional measurements on groups of traces

117 No geophysicist requires to be told that pulse estimates are likely to be better if we use more than a single trace. The fact that so much processing of a statistical nature (for example, deconvolution) is done on a single-trace basis is therefore a great testimony to the marketing abilities of the contractors, who need to fill their machines.

The way in which traces should be grouped for better estimates of the auxilary variables depends on the objective. All the traces from one shot, for example, have the source signature in common, and so all these traces are available for selection if the objective is concerned with estimating that source pulse. However, we would not use all of them if either the deep geology or the near-surface zone suggests that some ray-paths are distorted by anomalous conditions (faults, scatterers, extremes of weathering or variations of water depth). The horizontal extent of the earth which may be taken into a single estimate is thus a matter involving geological judgement, and the interpreter should participate in it. Again there is a conflict requiring a judicious compromise: a large degree of horizontal averaging is available under conditions of undisturbed geology (where we need it least), and a very small degree near faults 118(B) and local anomalies (which may be our targets).

For a discrete reflector, the variation of reflection coefficient with angle of incidence is implicit in the variation of reflection strength from trace to trace of a single record (sketch (a) of p.3-16). Except as the angle approaches the critical angle, however, this effect is likely to be smaller than the observational error involved in having different reflecting "points" and different transmitting material along the ray paths (section 2.4.6). Near the critical angle, of course, the major increase of reflection strength associated with a positive reflector represents an additional and useful tool for ascertaining reflection polarity.

For a reflector which is not discrete, but is a complex involving several closely-spaced interfaces, the effect of changing angle of incidence is complicated. Perhaps there is some hope that this could be worthy of analysis (in both time and frequency domains) in cases of particular importance.

In principle, a discrete reflector allows us to separate the high-cut effects of absorption from the high-cut effects of short-path multiples. For the effect of absorption as a function of the source-geophone distance appears only as an increase of high-frequency loss consequent on the longer path in the earth; however, at inclined incidence the statistics of the short-path multiple-generating system move bodily to longer delays (apparently-thicker beds). Therefore both agencies are associated with a greater high-frequency loss at long offsets, but there is a chance that different degrees of loss would allow their separation. The measurement, of course, would be of frequency content in a short window straddling each discrete reflection, as a function of path length in the earth.

If instead of considering the traces from one shot we consider a cdp gather, we remove (or much reduce) the problem of different reflection "points", but introduce the problem that many different shots are involved (sketch (b), p.3-16). This is therefore the best course where the source is highly reproducible (either in fact or after appropriate processing, discussed later).

The most important measurement we derive from the cdp gather is, of course, that of stacking velocity; our discussion of this occupies the whole of section 3.5. While on the matter of general seismic measurements, however, let us observe here that velocity analysis can be effected in different frequency bands. In principle, therefore, it is possible to make direct studies of dispersion, or the variation of seismic velocity with frequency. The problem is then that the method requires discrete and well-spaced reflectors whose reflections are not so lengthened by the necessary filter operators that they begin to interfere significantly. However, encouraging results have been reported (Crowe and Alhilali, 1975).

We may also study the grouping of traces of a common-offset set (that is, all the trace 1's, all the 2's ..... along a line). The obvious measurement which comes from this is dip. When the dip measurement is made, the traces may be summed along the direction of dip to yield a single 2-D fan-filtered trace; measurements discussed in the last section are thereby improved, provided that the reflection alignments are plane over the aperture used.

One of the most powerful enemies of the interpreter (and the processor) is cross-dip. Three-dimensional field techniques now allow the grouping of traces having depth-points spaced laterally from the line of profile, and thereby allow estimation of the cross-dip. This in turn allows calculation of the true dip (Walton, 1971; Michon and Tariel, 1972).

Because of the possibility of variation of apparent reflection strength and frequency content with angle of incidence, the common-offset set is also the correct grouping to improve some of the frequency measurements discussed in the last section. Thus a Q measurement on a single trace is weak, for the reasons discussed in that section; a Q measurement averaged over many common-offset sets is quite strong. Indeed, we can visualize statistical measurements of frequency content for which the windows become, in effect, short and wide rather than long and thin (sketch (c), p.3-16). The requirement for random spacing of reflections within the window (previously discussed in terms of layer thicknesses) is now met by the changing geological patterns passing across the wide window.

The variation of frequency content with source-receiver distance can be studied by repeating the above exercise for all of the common-offset sets obtained in normal shooting. It becomes tedious to discuss all the possibilities and their objectives, but let us carry forward in our minds the general principle: that multiple-coverage (or redundant) recording gives us an altogether new power for making auxiliary measurements on the seismic signal. For any defined objective we can find a grouping of traces which will optimize the measurement, and good consistent results can now be obtained under conditions which we would formerly have dismissed as impossible. In particular, we can now afford to discard individual traces highly different from the norm; we shall conclude many times in this course that it is better to accept even only a small number of good traces, and to discard the bad ones, than to average them all.

3.1.3 Measurements on stacked traces

Since the cdp stacking process yields major improvements in the ratio of signal to noise and the ratio of primary reflections to multiples, we are concerned to know to what extent the measurements of section 3.1.1 (in particular, reflection strength, frequency content, and polarity) may be made on single stacked traces.

The first observation, obviously, concerns the quality of the stack. If residual static errors are present, or the geology causes deviation of the reflection alignment from a hyperbola, then the stacked reflection must have too small an amplitude and too small a content of high frequencies.

Further, the degrading effect of noise, though reduced by the stacking, still corrupts measurements of deep reflections.

The usual effect of these agencies is that the similarity of the reflection shapes across the stacking hyperbola is good over a middle range of the record, rather less good at early times (where deviations from the hyperbolic assumption occur) and definitely poorer at late times. All processing houses have some way of measuring this similarity, and so it is reasonable to ask whether that measure can be used to correct at least the reflection amplitudes for imperfections in the stack. This can be attempted (under the control of the interpreter) in particularly important cases, but it is difficult in the general case because this measure of similarity is also prejudiced when a multiple cuts across a perfectly coherent primary reflection.

The conclusion, therefore, is that the interpreter must first satisfy himself as to the quality of the stack before he commissions auxiliary measurements on stacked traces. And, when an effect is found and measured on a stacked trace, there is wisdom in going back to the unstacked traces in search of a confirmation.

A second problem arises when the interpreter wishes to measure the amplitude ratio between two reflections (for example,

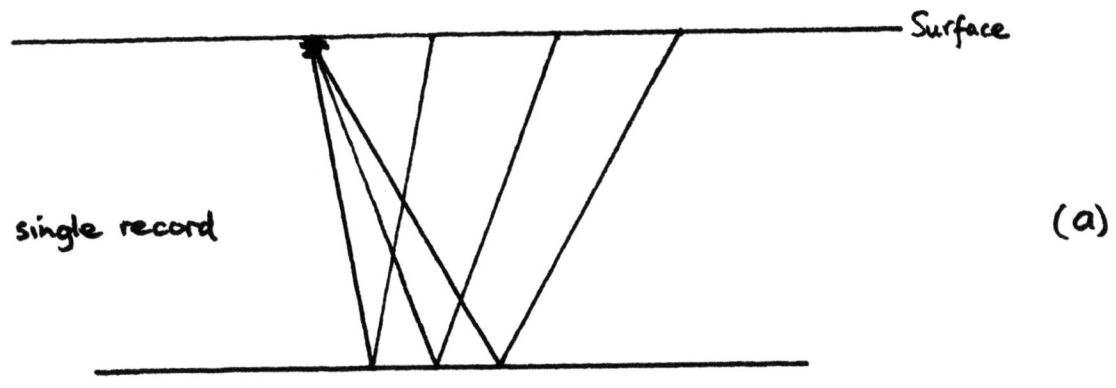

single record (a)

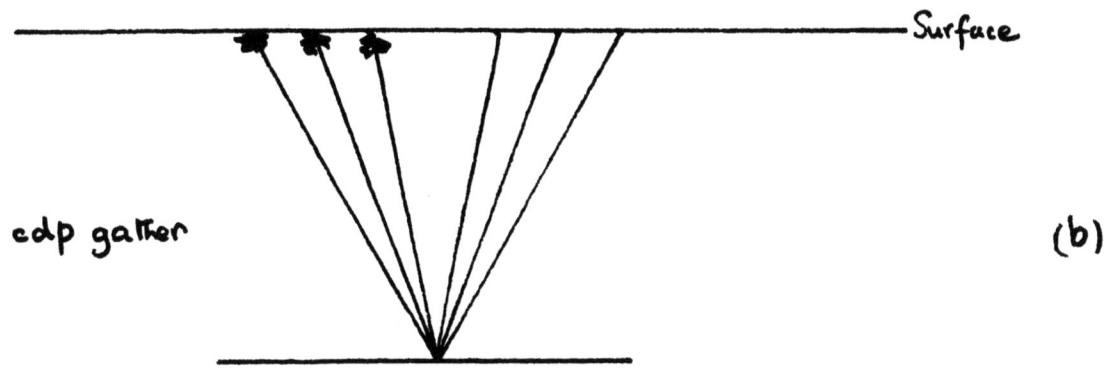

cdp gather (b)

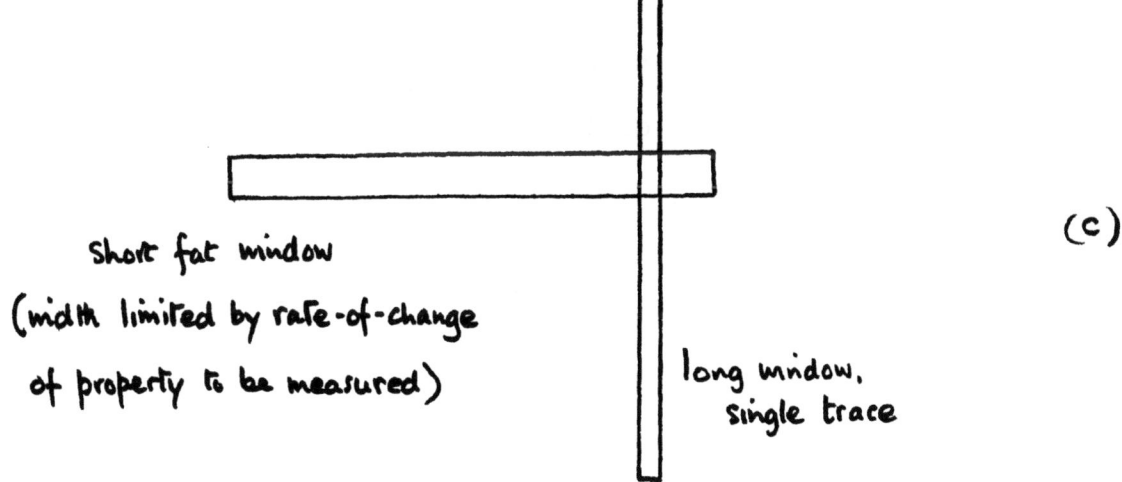

short fat window
(width limited by rate-of-change
of property to be measured)

long window,
single trace

(c)

3-16

in converting reflection amplitudes to reflection coefficients by the techniques of section 2.6), and when the stacked trace at these two levels is composed of different numbers of original traces. The problem is apparent, in extreme form, in the sketch.

In the last stages of investigation of a particular target, when the need is to eliminate as many sources of error as possible, the interpreter may be forced to confirm his ratio measurement on a special stack using no more traces on the deep reflection than on the shallow reflection. 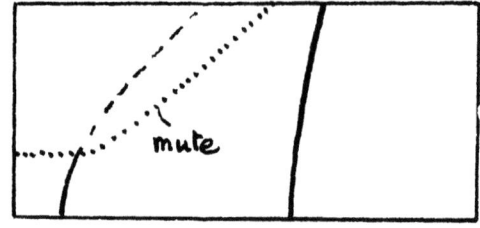 And, as before, he checks the evidence of the unstacked gathers.

Polarity estimation by visual inspection of the forms of the primary and its surface multiple (section 3.1.1.4) may be advantageously done on stacked traces, provided that a special stack is made (after the fashion of section 2.6.2) to optimize both the primary and its multiple.

Polarity estimation by the other methods of section 3.1.1.4 is almost always improved by normal stacking (provided, obviously, that the stacking is reasonably good). Reflections which maintain consistent polarity indications over many stacked traces (or, indeed, over many kilometres) identify themselves as discrete reflections associated with a stable pulse form. If the source reproduces well, those which oscillate in polarity indications (or for which no polarity judgement can be made) identify themselves as complexes whose local form varies with the local thickness of one or more thin layers at the reflecting level. These are useful distinctions in themselves; we have already seen, and will see again, that it is important to be able to recognize and distinguish discrete and complex reflections.

A problem in the measurement of frequency content on stacked traces is the distortion of frequency content by nmo stretch. The process of interpolation of the waveform after nmo correction adds low frequencies to the outside traces of the gather (Dunkin et al., 1973), and so lowers the general frequency content of the stacked trace at early times. This is the reason for the spurious <u>rise</u> in the median frequency f_m which we noted in the right-hand sketch of p.3-2. Again, the most sophisticated stage of analysis may require a special stack in which the mute pattern is selected on a criterion of only a few percent nmo stretch.

There is also a risk that the frequency content of the stack is distorted by the effect of reflection complexes observed at wide angle on the far traces of the gather. This can seriously prejudice the resolution of a thin reservoir (Larner, Mateker and Wu, 1974). Again a specially-muted stack may be necessary for this special purpose.

The above examples are just a few illustrating our changing approach to the relationship between interpretation and processing, and they give us a good lead into the material of the next section.

Basically we have three types of seismic survey:

- the reconnaissance survey, in which we try to assess the general prospects of a new area,

- the semi-detail survey (formerly called the detail survey), in which we delineate structure over features of specific interest, and make general studies of lithology and depositional history in the area,

- the real detail survey, in which we are concerned with hydrocarbon accumulations as such, with reservoir characteristics and porosity estimation, with the finest geological details visible to the seismic method — but with only a few dozen kilometres of line.

One of the present problems in the industry is that all processing houses are structured to handle primarily the reconnaissance survey. For this the machines are well adapted, the programs are optimized, and the staff are well trained.

indeed, the user gets excellent value. As we go to greater and greater degrees of detail, however, we find that the machines are poorly utilized, the programs are too inflexible, and the degree of understanding of the interpreter's problem is inadequate. Further, the work is very man-intensive, and unrewarding financially. Nobody wants to know.

The pressures to improve this situation will have to come from the interpreter.

3.2 CORRECTIONS REQUIRED BEFORE THESE MEASUREMENTS HAVE GEOLOGICAL SIGNIFICANCE

The interpreter is concerned solely with seismic measurements which relate uniquely to the deep geology. He wishes the complete removal of the effects of the near-surface, and the complete removal of all artifacts of the seismic method. This means that corrections are desired for all effects associated with:

- the source, the source coupling, the source environment, the source depth, and the free-surface (ghost) reflection above the source,

- the geophones, the geophone coupling, the geophone environment, the geophone elevation, and (in marine work) the free-surface reflection above the geophone,

- the near-surface and its variations,

- the offset distance, the spread geometry, and the array geometry,

- the recording instruments.

This is a formidable list. But if the interpreter is to use the seismic measurements as diagnostic of the real earth, <u>he must satisfy himself that these correction have been made.</u>

It is not sufficient for the interpreter just to ask the processors, "Say, you guys, did you do all those corrections right?" Of course the answer will be, "Sure!" — because the processors will be thinking in terms of the normal reconnaissance or semi-detail survey, and that answer would be acceptable as correct in such a context. But we have agreed that an altogether new standard — of precision in the field work and aptness in the processing — must be applied to the data when the modern seismic interpreter tackles a really detailed survey; only the <u>interpreter</u>, who knows the nature of the hydrocarbon problem, can define those standards and check that they are applied. The interpreter now finds himself asking questions about minute details of the manipulations applied to his data, and he often discovers that no one knows the answers except the man who wrote the program.

Let us look at some of the corrections which must be applied.

3.2.1 Geometrical divergence

The general nature of this correction is set out in section 2.8.1. There we decided that, for horizontal layering and near-normal incidence, the correction consists in multiplying each trace by a factor proportional to tV^2, where V is the rms velocity. This is straightforward, but there are a few cautions with which the interpreter should be familiar.

1. This very simple form of the correction is restricted to the conditions stated — horizontal layering and near-normal incidence. The correction can be calculated for any other defined condition, but this is normally done only for extremely detailed work shallow in the section (where the correction has its greatest effect).

2. Ideally, we would wish to apply the divergence correction at an early stage of processing, and to make the correction specific to the local value of V. The first point is vulnerable to the objection that the velocity is not as well known at that stage as it is likely to be later (though it is true that by the stage of the detailed survey the velocity should be already rather well known from the previous surveys). The second point — specific local variation of the velocity — may be inconvenient in practice; it is usually better to keep a constant average value for the velocity over the prospect at the early stage of processing, and to introduce final corrections at the time of stacking. If these are actually applied after stack, the interpreter must be concious of the fact that the time values

have been changed by the nmo process. There is nothing in these
considerations which is particularly serious (the divergence
correction is comparatively minor at medium and late times in
the record), and there is nothing for which the interpreter
cannot work out the safest course in any particular situation.
But he should be prepared to do so in those cases where it
might be significant — notably where amplitude ratios are being
measured, good deverberation is critically important, or the
study is of multiples.

3. We agreed in section 2.8.1 that it is not possible to
compensate both primaries and multiples, simultaneously, for
geometrical divergence. Obviously in the normal case we apply
the correction based on the primary velocity distribution, and
take our knocks on the multiples. It is only for the deliberate
study of multiples (along the lines set out in section 2.6) that
we consciously adopt any other course. Then we apply the
divergence exactly appropriate to the multiple train we are
studying — according to whether it is a trapped multiple or
a reflected multiple. All the references to the deliberate
stacking of multiples in section 2.6 require that <u>the velocity
distribution used for the unusual stacking should be used for
the divergence correction; this is important</u>.

4. If the velocity information which forms the basis for the
normal divergence correction is obtained on a continuous basis,
the interpreter must ask whether and to what extent it has been
smoothed (in the direction of the profile, or otherwise);
obviously he is concerned to know that the degree of smoothing
is compatible with the geological complexity. If the velocity
information is not continuous, the processors must apply some
form of interpolation between analyses; we will discuss this
topic more fully in section 3.5, and be content to note here
that the interpreter must be satisfied that the interpolation
algorithm is appropriate to the geology.

5. In practice, the divergence correction must be terminated, or it would produce an appearance of excessive noise at the end of the record. Ideally, as suggested in the sketch, this should be done at a time which represents the effective end of the record (unusable signal-to-noise ratio) on the <u>stacked</u> section — though the correction may actually be applied to the <u>unstacked</u> data. Subsequent choice of deconvolution windows must be made with due regard to the termination time, since it is important that the data entering the decon window should be entirely representative of the <u>reflection signal</u> — without any falsely exaggerated noise. Again these are judgements to be made by the processor, but of which the interpreter should be at least aware.

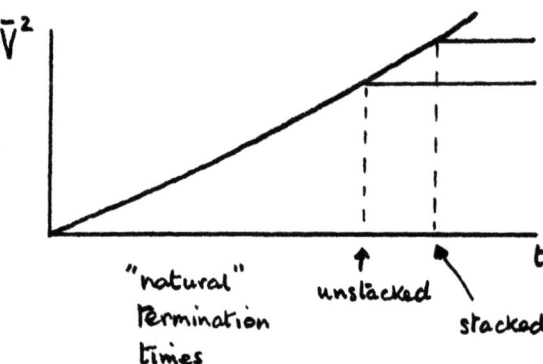

6. The above comments assume that the divergence correction is actually to be applied to the traces. Some processing houses, for practical reasons, prefer to use exponential expansions (which, as we have noted earlier, have no theoretical defence except in matters of dereverberation). True amplitude calculations can be done on traces treated with exponential expansions if we numerically change the amplitudes to what they would have been with a divergence correction — provided that the processors have preserved a record of every amplitude scaling (fixed and time-variant) applied during the processing.

3.2.2 Amplitude compensation for partial failure of an array of identical marine sources

This section considers the case exemplified by a ship fitted with a number of identical gas-guns — what do we do when one of the guns fails? (We leave to section 3.2.4 the problem when the sources are not identical — the case of the tuned air-gun array.)

The effect on the data is suggested by the sketches of p.3-26. Let us suppose that in the real earth we have two simple reflectors of constant reflection coefficient, as shown on the left. Then if the effective source output decreases for a period as we shoot the line, and if we make no correction for this, the appearance of the final section is as shown in the middle sketch. Obviously, then, we are vulnerable if we start to draw hydrocarbon conclusions from the observed amplitude variations; we must compensate the partial failure of the source.

One solution much used in the past was base-levelling, on a record basis. (In this course the term "base-levelling on a record basis" is used for the operation of multiplying each sample of each trace by a base-level factor such that the average amplitude of each record (within a single defined window) is constant from record to record. The term "base-levelling on a trace basis" is used for the operation of multiplying each sample of each trace by a base-level factor such that the average amplitude of each trace (within a single defined window) is constant from trace to trace and from record to record. Thus base-levelling does not change its effect with time; it compensates differences of average amplitude on the new data. The terms "agc" and "equalization" are used, interchangeably, in the classical sense of an automatic gain control — the operation of keeping constant the average amplitude within a moving window; this process obviously changes its effect with time.)

If we apply base-levelling on a record basis to the raw data of the middle sketch, each weak record is compensated, and we obtain the desired effect of the right-hand sketch. Of course the noise at the end of the section is made worse over the region where the gun failed — this is inescapable — but we are no longer at risk when we interpret the reflection amplitudes.

At this stage, then, we are well satisfied with the base-levelling process. The operation has detected a series of weak shots by noting a series of weak records, and has properly compensated them.

The problem with the base-levelling operation comes when a series of weak records (or strong records, for that matter) occurs for reasons not connected with the source.

For example, let us consider the two top illustrations of p.3-28. We have a salt dome, whose extent is large relative to a spread length. Let us say that the field work is accomplished without any variation in the output of the source; then the reality of the situation, and the corresponding raw data, are both represented by the left-hand sketch. When the raw data are base-levelled, however, a typical (and proper) window for the regions each side of the dome would find only one or two reflections within it over the dome; the base-levelling process would increase the amplitudes of the records over the dome (thinking that the paucity of signal there was the effect of poor shots), and so would cause the final stacked section to have spuriously high amplitudes at and and near the top of the dome. It would be entirely fallacious to interpret these high amplitudes in terms of acoustic contrasts or hydrocarbon accumulations.

Another deceptive situation is illustrated in the middle row of sketches on p.3-28. Here we assume that the first dipping layer is the sea floor, and that the depth at the left-hand end of the line is such as to yield well-separated reverberations; as the water becomes shallower, the reverberations turn into high-amplitude ringing and then eventually become mutually destructive in the very shallow water. Therefore, for

3-25

3-26

a reality represented by the left-hand sketch, the raw data
may have average amplitudes as suggested in the middle sketch —
with high amplitudes in the central portion and low amplitudes
to the right. The final consequences of this depend on the
order in which we apply base-levelling and deconvolution; if
the reverberations can be successfully removed first, and then
the data are base-levelled, the result is the best we can hope
for — as shown on the right. But ordinarily the base-levelling
has been applied fairly early in the processing; if the base-
levelling is done first and the deverberation second, the large
amplitudes in the central region are turned into small amplitudes.
We can see how important it is for the interpreter to
ask some pointed questions before he becomes excited about
amplitude variations near "up-dip pinchouts".

On the left of the bottom row of sketches we suggest a
real bright spot. As we learned in question 3 on p.2-70,
the very minor "shadow" caused by transmission coefficients at
the top and bottom of the bright zone is unlikely to be visible
by eye. Indeed the usual effect is of an increased general
amplitude level below the bright spot, because of the greater
strength of the multiples generated at the bright level (middle
sketch). When we base-level the raw data we decrease the
amplitudes of the records in the central zone, both because of
the bright spot itself and because of the locally stronger
multiples. We therefore obtain two undesirable effects — a
weakening at the bright-spot level itself (which may cause us
to lessen our interest in an important feature), and a weakening
of other reflections both below and above the bright spot.
The most likely explanation for the existence of so-called
"shadows" above bright spots is that the data have been base-
levelled. This matter is resumed, of course, in our full
discussion of direct-detection techniques (section 4.2.5).

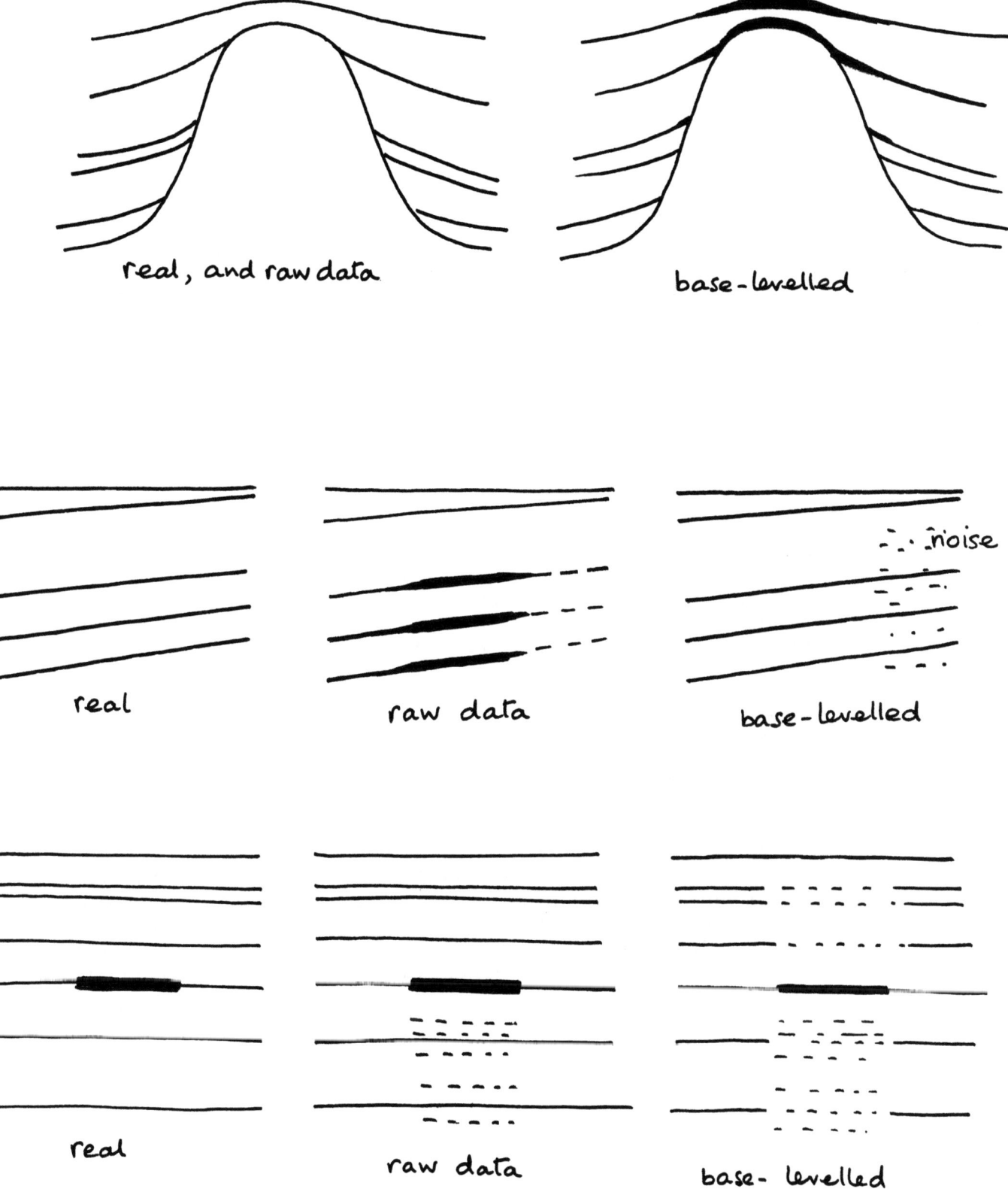
3-28

Question: What is the effect of base-levelling in a situation where the water is of constant depth but the sea-floor reflection coefficient increases locally?

TAPE 17

When base-levelling is applied, the processor must stipulate a window over which the average-amplitude measurement is made. In the case illustrated, it might be as suggested by the dotted line — the zone dominated by reflection signal. At first sight, it might seem attractive to vary the choice of window to "follow" just a few continuous reflections along the line — for example, just the first two reflections in our salt-dome example on p.28. But this has its own risks; in the limit, when we concentrate on just one dominant reflection, we force its amplitude to be constant, and so obscure possibly-interesting amplitude anomalies. In general, therefore, the window should include the whole of the reflection-dominated zone. The interpreter must be alert also to the distortion of true amplitudes caused by ground roll which varies from shot to shot, noise bursts (static, instrumental problems, etc.), and mud waves. The interpreter is also entitled to ask whether the "average" amplitude used in the base-levelling process is absolute or root-mean-square. The latter gives greater weight to the higher amplitudes; this is beneficial if (and only if) those higher amplitudes are reflections and there are plenty of them within the window.

It is obvious at this stage that base-levelling should be avoided if possible. Base-levelling has made us blind to highly significant changes of amplitude along reflections, and to highly significant areal variations in propagation properties; we are only just beginning to realize how great a disservice this has been. Floating-point machines (and simple programming tricks with fixed-point machines) have released us from any need to base-level for reasons concerned with numerical significance, and so we seek methods of compensating a failure or local weakness of the source while not using the base-level process.

The answer in the case under present discussion is to derive a measure of the outgoing source signal not from observations on the record (which also includes the effect of the geology) but from direct observations on the source. So if we are using a number of identical gas-guns as a marine source, we arrange a suitable calibrated hydrophone close to each, and record the outputs on tape, through normal recording channels with normal alias filters. ("Close" in this context means close enough to ensure that the free-surface and/or hull reflections, and the direct arrivals from the guns, are sufficiently attenuated by spherical divergence as to be insignificant.) Then we can sum the outputs and make some measurement (such as the envelope) of the resultant composite pulse. Finally we divide all samples of each record by a factor proportional to this source measurement, so that weak shots produce a small factor and so obtain a proper compensation.

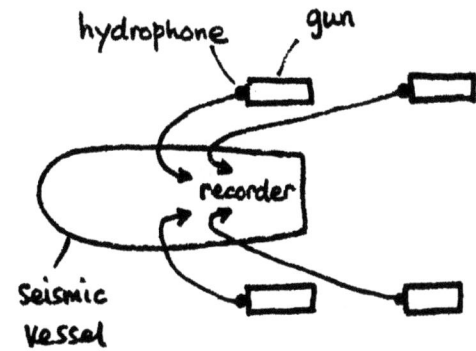

3.2.3 Amplitude compensation for weakness of a geophone array

Our first approach to the problem of individually weak geophone arrays might be to apply base-levelling on a trace (rather than record) basis. This accomplishes both functions — the compensation of individually weak shots (as in the last section) and the compensation of individually weak geophone arrays. Therefore it is open to the same objections as before, plus one or two new ones.

One new objection is that we should not expect the general amplitude level on the far traces (after appropriate divergence correction, of course, but before nmo correction) to be the same as on the near traces; it should be higher, because the same number of reflections is compressed into a shorter time range. Simple base-levelling therefore systematically depresses the outside traces (which is undesirable for purposes of multiple suppression in the stack). This objection can be overcome at the expense of more computation; however, the interpreter should be aware of how this has been done, because it is obviouse that local amplitude anomalies can be created, entirely spuriously, by an unfortunate combination of correction method, window selection, nmo and noise pattern.

Whenever there is some consistency to the usage of geophone strings on the spread, an additional tool is available which may remove any need for base-levelling on a trace basis. The prime example of this is in a marine survey, where a particular array is always recorded on the same channel. Thus a sum along the entire line of all the signals on channel 1 (after appropriate divergence correction and windowing) has a certain easily-established value, and the same is true for all other channels. Plotted as a function of channel number, we would expect a smooth curve. The actual form of the curve depends on many factors we have discussed (variation of reflection coefficient with angle of incidence, absorptive path length, short-path multiples, nature of reflection complexes, etc.), so we dare use no property of the curve except its smoothness. Individual poor arrays are recognized by their deviation from this smoothness, and so may be appropriately compensated.

Otherwise, the redundancy existing in multiple-coverage recording may be used to compensate the surface-consistent effect of local variations in the geophone coupling and the near-surface. A fuller discussion of this is postponed to sections 8.3.2 and 8.3.3. Whatever the system employed — whether base-levelling or some more sophisticated system — the interpreter must be aware of the vulnerability of the method to window selection and to the presence of ground-roll or other noise bursts within that window. Indeed the interpreter may be justified in saying, under some circumstances of weak geophone arrays, that he would prefer <u>inaction</u> to any attempt at compensation by automatic methods.

Nothing could be clearer, in this context, than the need to minimize the problem by good quality-control and good reporting in the field.

3.2.4. Compensation problems involving both amplitude and (stationary) frequency content

We now tackle the more difficult problems where variations introduced by the near-surface or the mechanics of the seismic method produce unwanted changes in <u>both the amplitudes and frequency content</u> of the records; initially we consider problems where the frequency content changes spuriously from record to record but does not change spuriously down the record.

Illustrative of this class of problems is the case of a tuned air-gun array when one or more guns fail, and the case where the source or streamer depth changes along a line.

Clearly, in these cases inaction is no solution, and base-levelling is no solution.

Indeed, the only clean solution is the deterministic one introduced in section 3.2.2 — to record the outgoing source pulse, to synthesize the effect of the free-surface reflections at the measured depths of source and streamer, and to use this determined signature for the compensation of both amplitude and frequency content to some chosen norm (Carroll, 1973; O'Doherty et al., 1975; Dedman et al., 1975; Michon et al., 1975). The technology now exists to record the signal from individual guns acceptably well, and the duplication between the "ghosted" source pulse and the signal observed with a deep hydrophone is surprisingly good. The message is therefore very clear: Record the actual source pulse, compute an inverse operator to normalize the record to what it would have been with some preferred source pulse and without any variations along the line, and avoid base-levelling.

3.2.5 Compensation problems involving time-variant amplitude and frequency effects

We have already noted (section 3.2.1.1) that in principle the divergence correction should be modified for non-normal incidence on the far traces. This modification, which we have said is reserved for the most critical problems, is obviously time-variant on any particular trace.

On land, the geophones are vertically sensitive. The sec θ compensation (which, again, would be considered only in exceptional cases) represents a time-variant amplitude correction. This is discussed also in section 8.3.1.1.

There are several agencies which produce time-variant effects on both amplitude and frequency content; in filter terms, they have an amplitude-frequency response which depends on the entry and emergence angles of the ray at the surface. These include the polar response of the geophone array, the polar response of the source array, and the delay time of the free-surface reflection system(s). All these problems are aggravated in the presence of dip; for the entry angle, the aggravation occurs on the updip side, and for the emergence angle it occurs on the downdip side. Thus differences in the amplitudes and frequency content may be observed on the updip and downdip sides if the source and geophone arrays are not the same (p.3-34).

The redeeming feature of these agencies is that they are all _calculable_. For the detail survey (and compensation at the present sophisticated level would be undertaken only for detail surveys), we know our dips and our velocity distributions rather well, and can consider computing the entry and emergence angles; then it is a straightforward matter to compute a time-variant amplitude and frequency compensation for the limited extent of

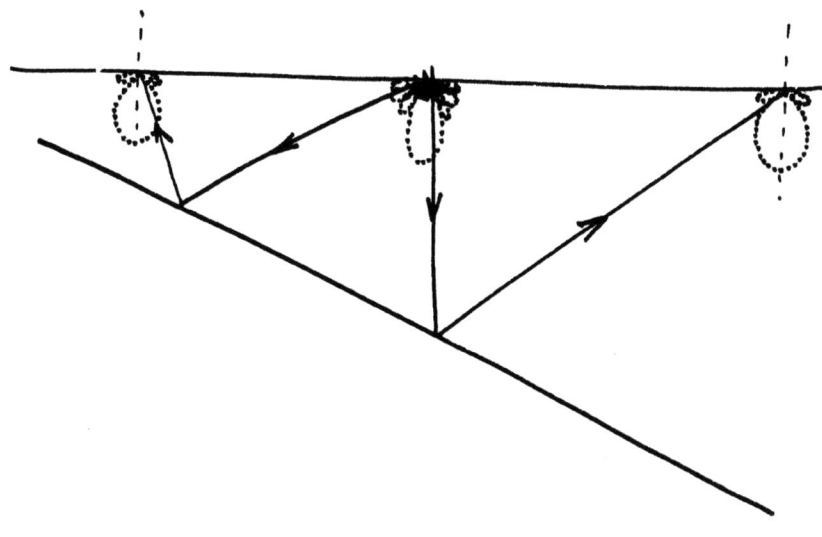

the data which interests us.

3.2.6 Amplitude compensation for transmission coefficients

In section 2.4.4 we computed the two-way transmission coefficient of a reflector of reflection coefficient R; it is $(1-R^2)$. Accordingly we can compensate any deep reflection for two-way passage through any such reflector above it, by dividing the deep reflection amplitude by $(1-R^2)$. For example, we can derive the reflection coefficient of the sea floor by the methods of section 2.6.2; then we may divide all the deeper reflections by $(1-R^2)$ to compensate the effect of two-way passage through the sea floor. As we calculated in response to question 2 and 3 on p.2-72B, this compensation is significant only for very strong reflectors such as the sea floor. The reduction of amplitude consequent on passage through more usual reflectors is significant only when there are many of them; this is the situation we discussed in section 2.5.2, and we remember that the loss is largely compensated at the low frequencies by the short-path multiples.

3.2.7 Corrections for time

The major problem of land statics is postponed to sections 8.3.2 and 8.3.3. Here we just note in passing that near-surface anomalies (including buried river-channels in marine work) can seriously affect our velocity determinations, and so impair the accuracy of our stacked amplitudes and our divergence corrections.

The other major correction for time is the correction of the origin, to accommodate the observation that the real zero time is not at the "timebreak" or the command-to-shoot, but some 20 or 30 or 40 or 50 or 60 ms later. This is the same as the oft-recognized fact that a simple multiple does not occur at double the travel time, but at some lesser time. The topic is discussed in detail in section 3.5.

3.2.8 General matters of concern

We have been discussing how we can correct the basic seismic measurements — reflection time, strength, polarity, frequency and, later, stacking velocity — for the mechanics of the seismic method and for the vagaries of the near-surface. We should also discuss a few matters of concern which arise in the earth and in processing.

1.   All the time we must worry about cross-dip. The major effects on the distortion of structure are well known, and are illustrated clearly in Tucker and Yorsten's monograph. But some of the consequences of cross dip can be quite subtle. For example, a reflection may dim suggestively over a small high — not because of a real change of reflection coefficient, but because an interpolation of stacking velocity has not taken into account the fact that one of the velocity analyses was affected by cross-dip. The important message is that the knowledge of true dip implicit in the interpreter's contour maps from the semi-detail survey should be injected into the processing of the detail survey.

2.   The proper muting of the raw records (on the basis of surface waves, refractions, noise, nmo stretch, wide-angle effects and thin-bed complexes) is a most important function; the interpreter should check it from time to time and in critical locations.

3.   Deconvolution is a problem. Let us just remind ourselves of the   processes in general use:

- Statistical spiking deconvolution, in which we try to sharpen each reflection pulse present on the record; the necessary information on the reflection pulses is obtained by statistical (auto-correlation) measurements on the record itself.

- Statistical dereverberation, in which we try to remove the effect of a shallow multiple-generating layer such as the sea; similarly the necessary information on the multiple system is obtained statistically from the record.

- Deterministic source deconvolution, in which we use actual knowledge of the outgoing source signal (from direct measurement on the source) to shape each reflection pulse present on the record.

We now have to ask what is the effect of these processes on our seismic measurements of reflection strength, frequency content, and even polarity.

First, we note that the deterministic process is safe. Not only does it correct for variations in the output of the source (sections 3.2.2 and 3.2.4), and so preserve true amplitude ratios along the line, but it also preserves true ratios between reflections vertically. Of course it may change the amplitude ratio between two reflections -- a 6 dB boost at 60 Hz may produce considerable benefit on a reflection at 0.5s but not on one at 5s -- but any such change is strictly according to its effect in the frequency domain. The final result is as it would have been if the source output had been constant and if the bandwidth had been that of the chosen norm.

Second, we note that no objection can be raised to any form of dereverberation which leaves the primary unchanged in form or amplitude. This is the case, for example, with predictive dereverberation operators having appropriately-chosen prediction lags (say 50 ms), for reverberations in water of depth greater than about 50m. It could not be true with any type of dereverberation operator which combined a pulse-shaping or spiking function with its dereverberation operator in shallow water.

Third, we have to admit disquiet about statistical spiking deconvolution, or indeed any process which changes the form of the primary reflection on the basis of a statistical measurement on the record itself. A statistical deconvolution does exactly what it is told — on average. The detailed validity of the process on specified reflections depends on whether the "average" (represented by the auto-correlation function) properly represents all the reflections present; this in turn depends on many practical factors (including the choice of window, the operator length, the aptness of the divergence and other corrections, and the degree of high-frequency loss present in the window). Therefore it is not possible to say with certitude that the ratio between two reflections has been properly preserved by a statistical shaping or spiking deconvolution. At this stage we begin to see more and more evidence for an argument which has been emerging steadily throughout Part 3: that different seismic measurements require different processing. Of course we must have statistical spiking deconvolution for the resolution of thin beds — this is beyond dispute. But obviously we cannot apply statistical spiking deconvolution before we study frequency content, and we shall be ill at ease if we do it before we study reflection amplitudes. In section 3.4 we shall discuss practical solutions to this dilemma.

4. We grow most concerned when it is proposed in the literature to measure the general decay of the signal after divergence correction (for example, as a best fit to an exponential), and then to read some simple geological significance into this decay as it varies over an area. Of course the measurement may have geological significance (and we shall consider a field example later), but the significance is certainly not simple. In the first place, the variation in decay may be caused entirely by a general weakening of the reflection coefficients with depth. Or it may be caused by a zone of anomalous high-frequency loss (a thick, shallow and highly absorptive unit, or a contrasty interbedded sequence). Or it may be caused by a change in the overall frequency response of the near-surface or the seismic tools (a change of water depth, or weathering thickness, or source bandwidth, or source depth, or streamer depth); anything which increases the high-frequency content (including spiking deconvolution) must increase the decay, while anything which decreases it (including normal filtering) must decrease the decay.

5. Finally let us stress the inevitable ambiguity of measurements of a change of frequency content, in the presence of both high-cut loss and bed-thickness effects. If two reflectors are discrete and well separated, we can attempt a measurement of the high-frequency loss occurring between them, even if their reflection coefficients are not the same; this is the situation at the top of p.3-40. But if one or both reflectors are doublets, the direct measurement becomes meaningless; two of the permutations include the cases where a low-frequency reflection is followed by a high-frequency reflection — a situation which (in the absence of artifacts like nmo stretch) can occur only when the upper reflector is a complex whose members interfere constructively and/or when the lower reflector is a complex whose members interfere destructively.

This ambiguity is, of course, a crying shame.

The <u>ambiguity</u> of a change in frequency content, in the presence of both high-frequency-cut effects (absorption, etc.) and bed-thickness effects:

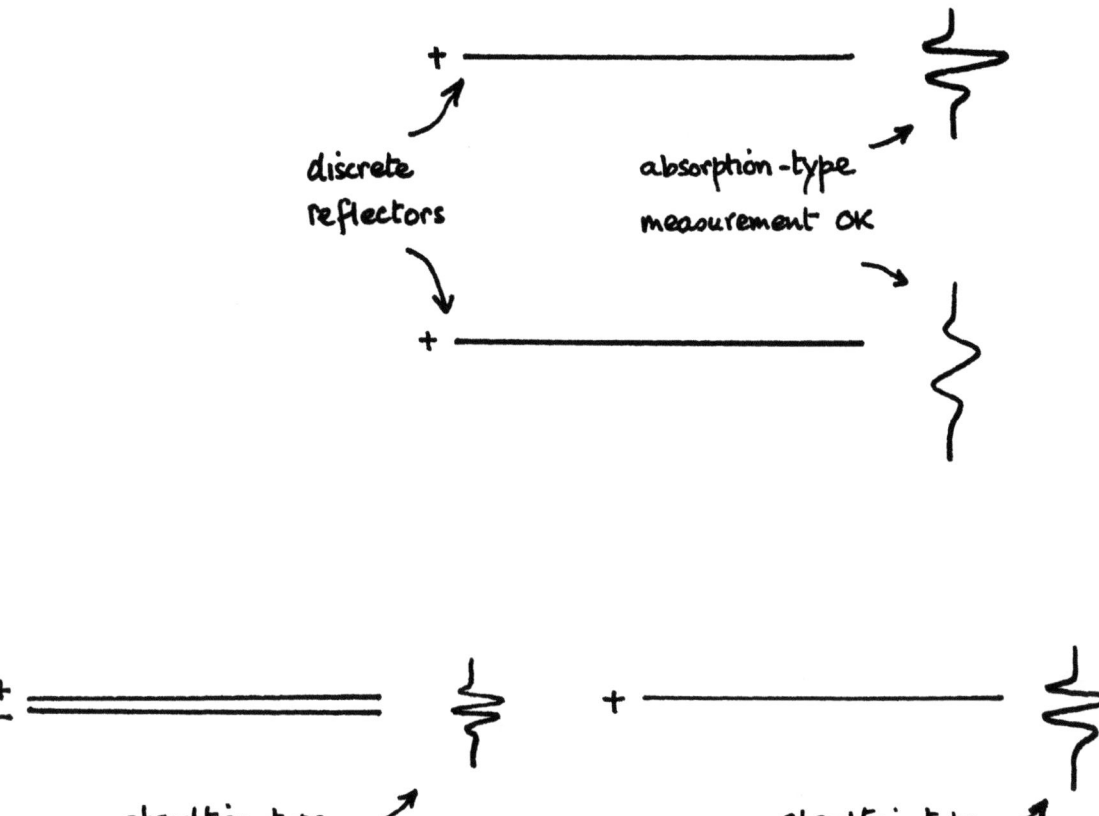

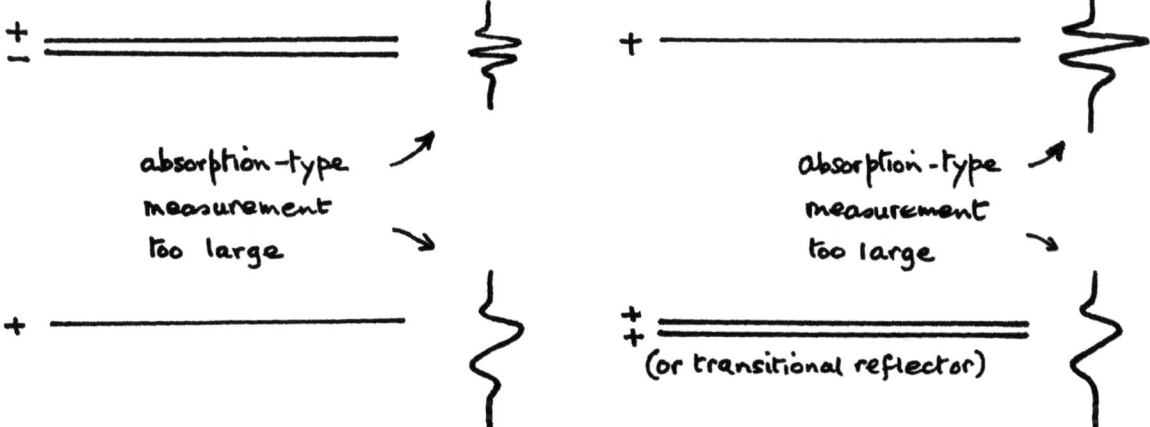

(N.B. Reversal of these situations can make absorption appear negative.)

TAPE 18 3.3 DETAILED STUDIES OF THE ATTRIBUTES OF A
              BOREHOLE-GEOPHONE SIGNAL

Now, having decided what measurements we can make, and having corrected the data for everything except the variables we wish to measure, let us consider some examples of real situations. We start with the situation of a geophone lowered down a borehole (that is, one-way travel), because of its contribution to understanding.

3.3.1 The amplitude of the direct arrival

The major factor affecting the amplitude of the down-going signal is geometrical divergence. For a downhole geophone we can correct this exactly, since we know the velocities and, usually, the geometry.

125   Having done that, we find that, here and there, the amplitude locally _increases_ with depth. The nature of such increases depends on the type of borehole geophone; if it is a pressure phone such an increase indicates a formation of _high_ acoustic impedance, and if it is a lock-in velocity-sensitive phone the increase indicates _low_ acoustic impedance (section 2.4.3). The illustration of p.3-42 was clearly made with a velocity-sensitive geophone.

In principle, since we have the sonic log, we can use knowledge of the density in any one formation (for example, in salt) to compute the density in any neighbouring formation, using the amplitudes of the direct arrival. The caution, of course, is interference between the direct arrival and reflections; this approach would not be used if obvious changes in the pulse shape are evident. The amplitude of the direct arrival is also subject to focusing and defocusing associated with velocity lensing; the likelihood of this could be assessed from the seismic section through the well.

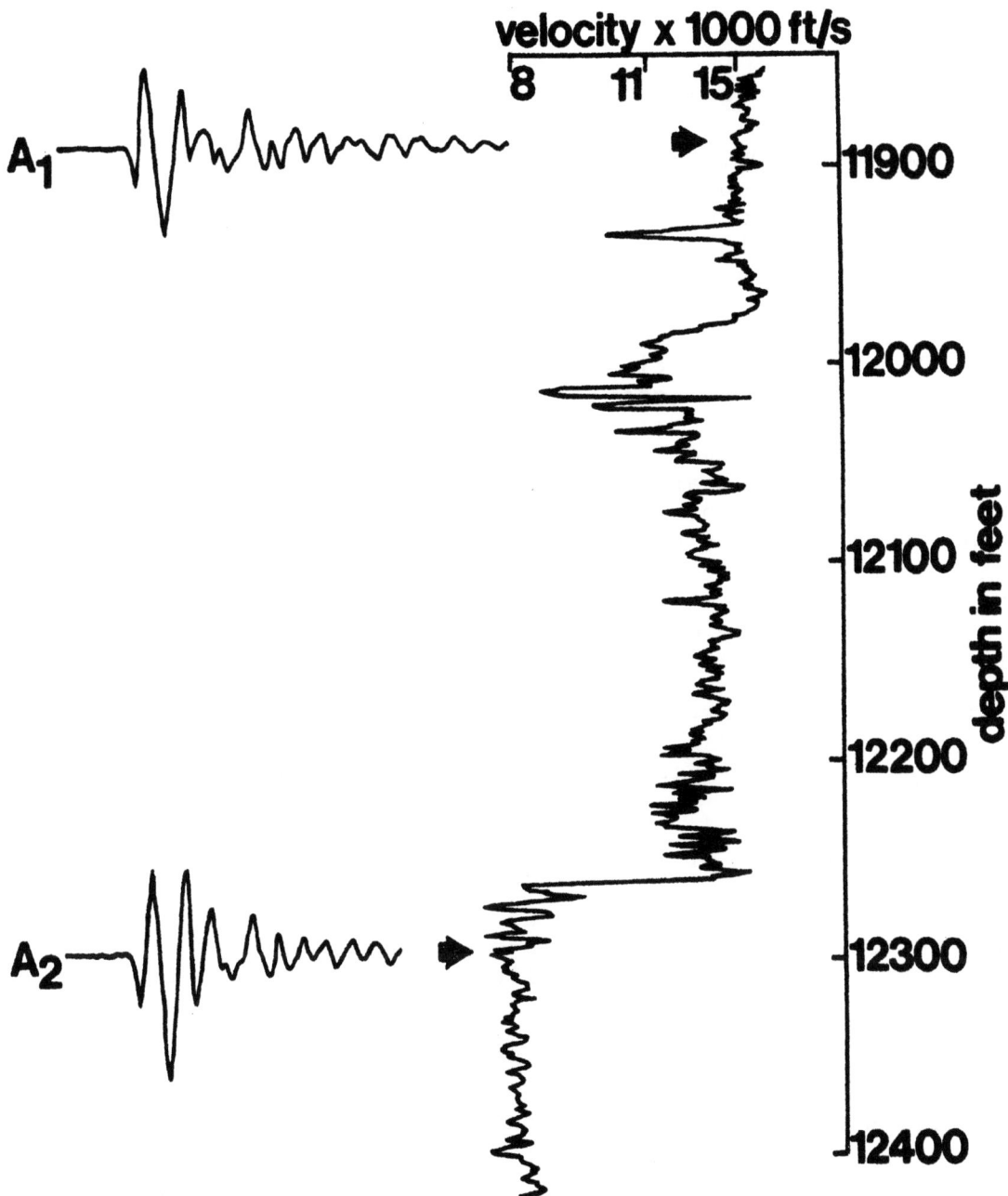

Courtesy of Kennett and Ireson, of SSL, 1973

3-42

3.3.2 The pulse shape of the direct arrival

Let us assume that the source is an air-gun in the sea or the mud-pit. Then the shape of the outgoing signal is monitored as a matter of normal practice, and we can refer the downhole shape to this outgoing shape (after synthesis of the free-surface reflection from above the source, if the conditions so require).

Our first observation is that the polarity of the downhole signal is always positive; the process of transmission cannot change the polarity (section 2.4.2).

Then we observe that if the source is rich in very high frequencies, these are quickly lost as the geophone goes through the shallow sediments. This extreme loss is not observed if the source does not emit the high frequencies, or if the recording system cuts off at say 60 Hz.

Thereafter we observe a generally small but systematic loss if the geophone passes through a major highly-absorptive unit or a very contrasty cyclic sequence. (In principle, we also observe a loss of the low frequencies as we pass through a significant slow transition of acoustic impedance. This effect is almost certainly undetectable (last paragraph of section 2.5.2)).

3.3.3 The effect of a gas-saturated highly-porous reservoir

As an illustration, let us consider the effects to be observed as the geophone approaches, enters and passes through a gas-saturated zone. We use the illustrations of Kennett and Ireson (1973).

On p.3-44 we see the borehole, the gas zone, the velocity log, and the relevant paths (direct and reflected) for two positions of the borehole geophone. On p.3-46 we see the signals observed from a velocity-sensitive geophone. First, considerably above the gas zone, we see the direct arrival (a) followed by the

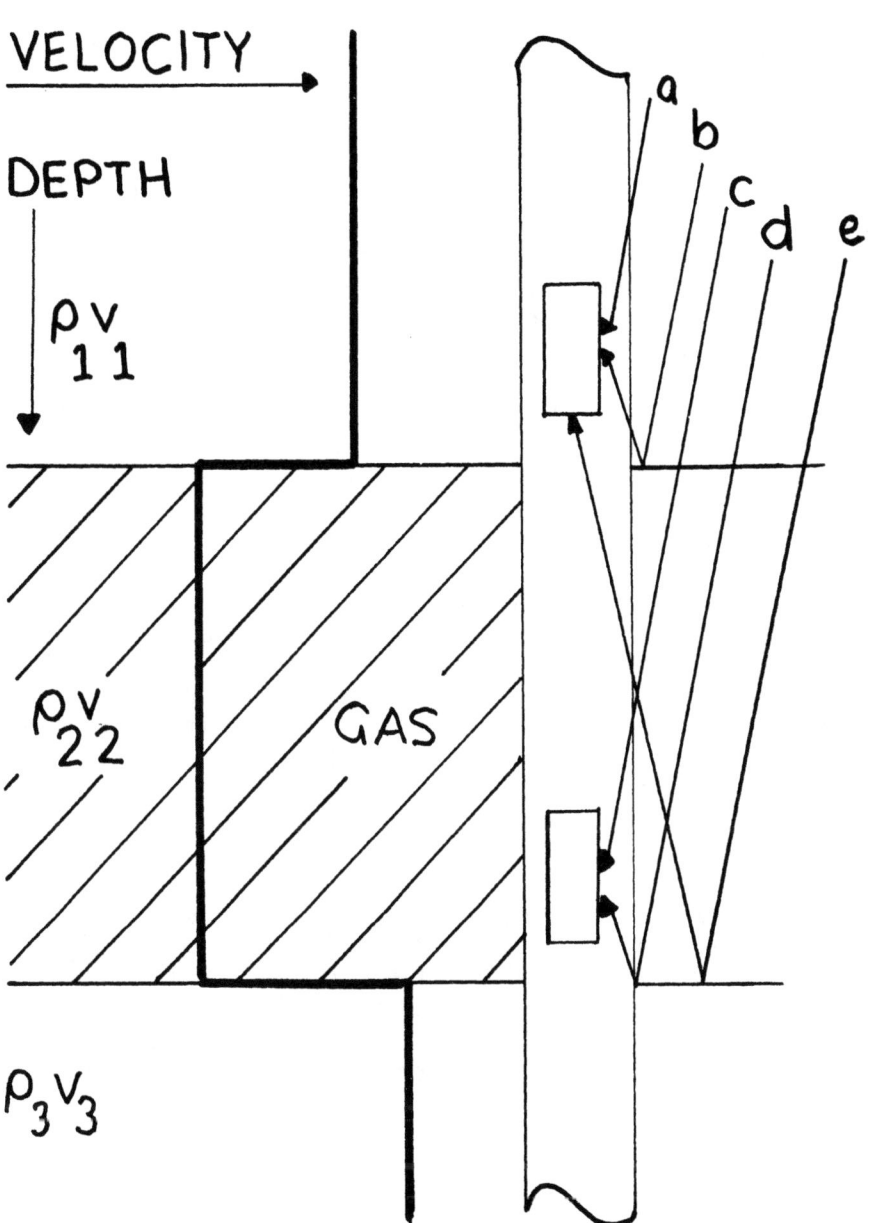

After Kennett & Ireson (SSL) 1973

top-gas reflection (b) and the bottom-gas reflection (e). The spacings are such that all three of these arrivals are well separated and individually measurable. The reflection coefficients of the top gas and bottom-gas reflectors are <u>directly measurable</u> as the ratios b/a and e/a (with minor adjustments for geometrical divergence and for the transmission coefficient of the top-gas reflector).

When the geophone is lowered to be just above the gas zone, but still in the overlying material, the amplitude of the direct arrival is modified only by slight additional divergence; however, the top-gas reflection now interferes with the direct arrival. When the geophone enters the low-impedance gas zone (and to the degree that the lock-in mechanism provides a good coupling to the porous and probably drill-damaged formation), the direct arrival increases in amplitude, and so does the bottom-sand reflection. Finally, when the geophone is well below the gas zone, the direct arrival is again seen clearly, at an amplitude appropriate to the impedance of the formation, the transmission coefficients of the top-gas and bottom-gas reflectors, and geometrical divergence.

If the absorption in the gas zone is very large and the layer very thick we might just see some high-frequency loss on the transmitted pulse. More likely, this would be detectable only by machine analysis. However, at a great distance below the gas zone, this minor effect would almost certainly be made quite undetectable by wavefront healing. This, of course, depends on the smallest horizontal dimension of the reservoir and on the depths of the gas zone and the geophone below it (section 2.8.5).

The great elegance of the borehole measurements lies in our ability to measure the reflection coefficients <u>directly</u>; this, obviously, is a very desirable check on any reflection coefficients which we are able to compute by the techniques of section 2.6. If many positions of the borehole are occupied. it becomes possible to stack the downgoing and upgoing signals

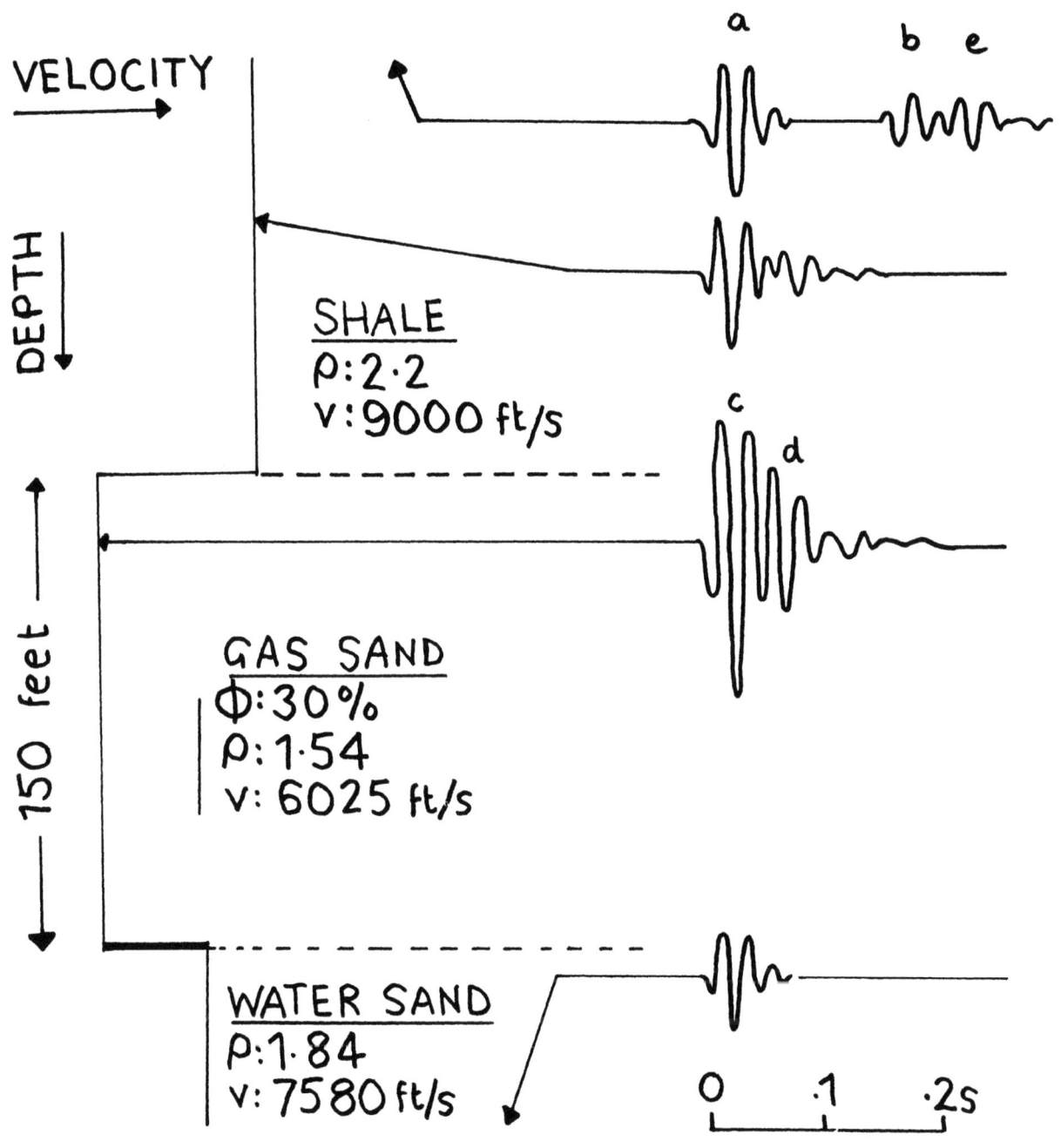

After Kennett & Ireson
(SSL) 1973

independently, and so to obtain the greatest freedom from doubts associated with interference. Where this is not possible, the known forms of the direct arrival and the sonic log allow us to attempt analysis of the observed waveform, by hand, in terms of its interfering components.

<u>Question:</u>   Would you modify the polarities of events (b), (d) and (e), having regard to the amplitude ratio of events (c) and (a)? (Remember p.2-67.)

3.4 DETAILED STUDIES OF THE ATTRIBUTES OF A TWO-WAY REFLECTED SIGNAL

Let us now add the return path to the borehole studies of the last section.

3.4.1 Consequences of the two-way nature of the path

The first point we note is that the return path always contributes an additional 6 dB of divergence correction; the signal at the reflector may have an amplitude only one-hundredth or one-thousandth of its amplitude at shallow depth, but the signal returned to the surface is always just half of its amplitude at the reflector.

Then we note that the transmission coefficients to be applied are now two-way coefficients (section 2.4.4). Consequently, the results are the same whether we use pressure-sensitive (marine) geophones or velocity-sensitive (land) geophones.

Then we observe that the high-frequency-loss effects of the absorption and short-path-multiple mechanisms are simply cascaded (auto-convolved) on the return path. The effect of scattering from inhomogeneities is probably not so simple; for a profusion of small scatterers the overall effect is simply cascaded, but for single scatterers of large dimensions the effect may depend on the size, depth and configuration of the scatterer.

3.4.2 The complete picture

On p.3-50 we reproduce the Sheriff summary diagram (Sheriff, 1975). This combines in one illustration all the effects we have discussed.

For convenience, some of the effects occurring between the surface and the reflector are shown on the return path only; in fact, of course, they occur on both downgoing and upcoming paths.

3.4.3 The processing problem

At this stage we wish to look at a large number of real-world sections, and see to what extent we can identify on them the numerous effects we have discussed. Before we do so, however, we must turn again to a topic we broached in section 3.2.8.3 — the observation that the authentic measurement of a particular signal attribute requires appropriate processing of the original data for that attribute, and that the processing appropriate to one measurement is not appropriate to another. In short, we must process by different means for different ends.

Every processing house has some solution to this problem. The solution proposed here is put forward as one of many possibilities, and not with any suggestion that things must be done this way.

In tackling the problem, we are not prepared to take any backward steps. Classical processing has developed many sophisticated techniques for the enhancement of the section in terms of its structural clarity; we are not prepared to throw away these advances. Thus, for purposes of structural interpretation, we wish to retain the benefits of time-variant spiking deconvolution, space-and-time-variant frequency filtering, and equalization (plus, if appropriate to the geology, space-and-time-variant fan-filtering, coherence modulation of the stack, diversity stack and a host of other options). We reject emphatically one view recently expressed in the literature, that "it is essential that we discontinue cosmetic processing". Cosmetic processing, in terms of the processes exemplified above, has served us well; for the purpose intended — clarification of the structure — we insist on retaining these benefits.

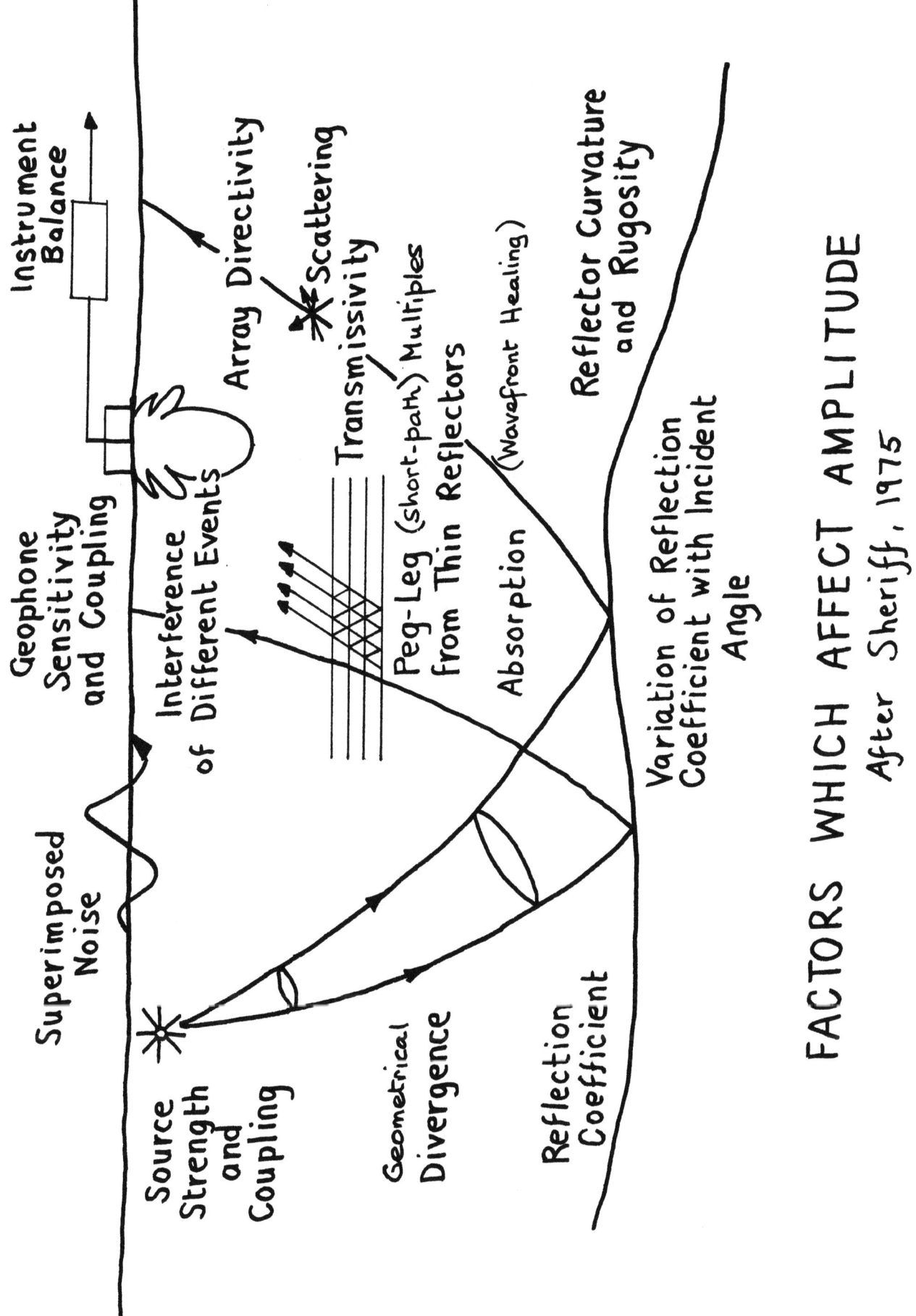

So we still plan to make a <u>structural section</u>, to which we apply every legitimate process <u>in our tool-box</u> for the visual enhancement of the reflection alignments. This gives us the best possible definition of the time, thickness and attitude of the layers in the earth.

But we also say that we now wish to study the <u>material</u> in those layers, and that to this end we must make <u>authoritative</u> measurements of the strength, frequency content and polarity of those reflections, together with the interval velocities between them. Some of these measurements are not compatible with data-dependent processes like equalization and statistical deconvolution, nor with filtering, and so we accept that some of the processing must be done two ways.

This leads us to forked or two-stream processing. For economic reasons, we wish to perform as much as possible of the processing <u>in common</u>, before we fork into two streams. The philosophy is sketched on p.3-52; a possible processing scheme derived from it is also illustrated on p.3-52, for a particular prospect. In the latter diagram, the order of individual processes (for example, deverberation and stack) depends on local conditions and data quality, and the order is therefore to be regarded solely as illustrative. The important feature is not the details of the scheme, which vary from prospect to prospect, but the broad concept of maintaining the integrity of the data (particularly as regards amplitudes and frequency content) for as long as possible, and then dividing into two streams — one a <u>cosmetic stream</u> for structural purposes only, and the other an attribute stream for the valid measurement of amplitudes, polarities, frequency content and any other of the auxiliary variables discussed in section 3.1.

3.4.4   <u>The matter of display</u>

Clearly, the output of the structural stream should be displayed in the classical manner — as a cross-section of traces presented in wiggle, variable-area, variable-density,

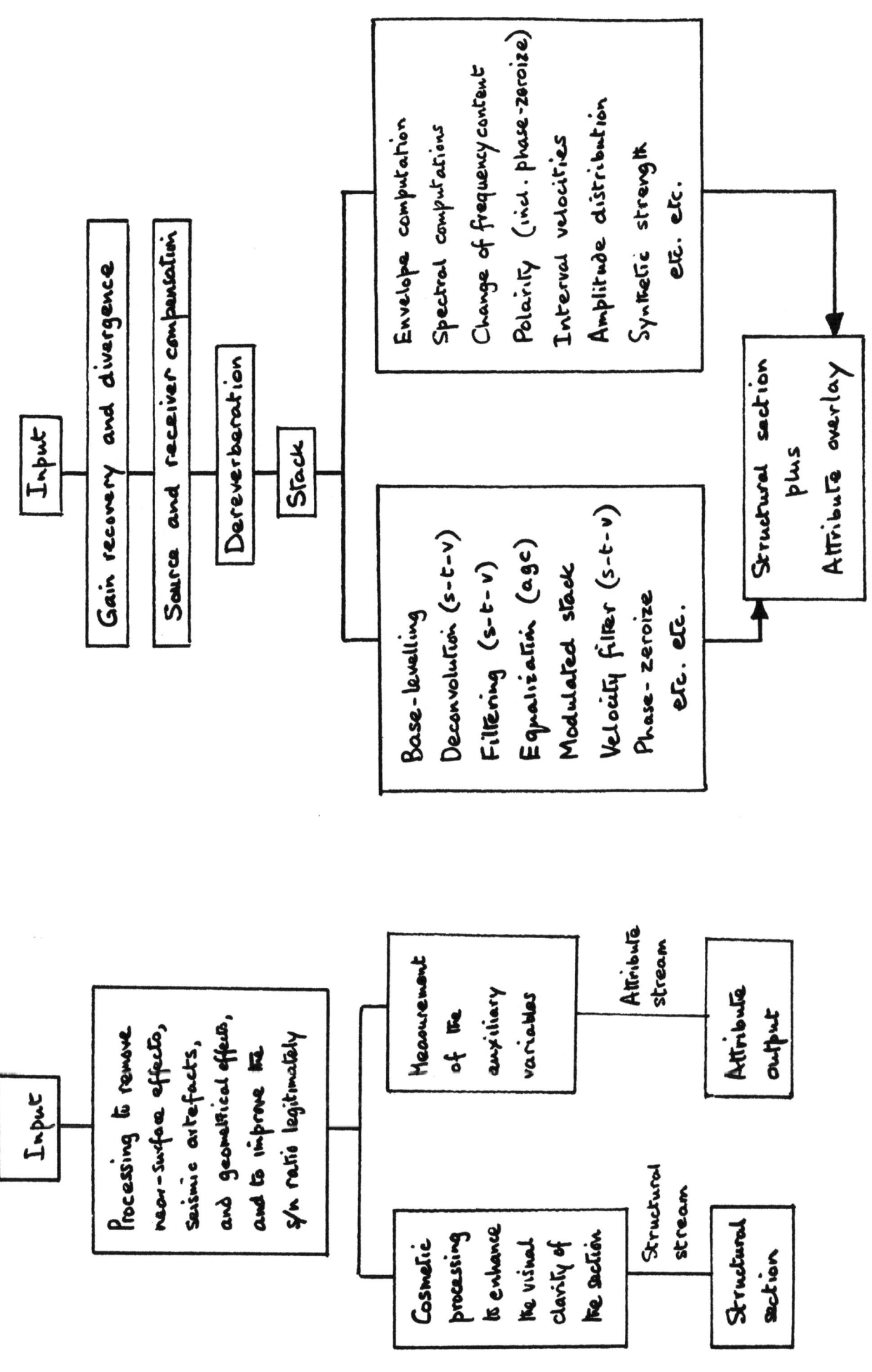

3-52

or some combination of these modes. At this stage of the discussion we can allow all possibilities; the objective is to convey the structural information through the eye of the interpreter to the mind of the interpreter, and the mode which does this best, for any individual, is the right one for him.

So the display of the structural picture is easy — but what should we do about the output of the attribute stream? Where the interpretation is tightly focused on a particular measurement and a particular feature, a simple print-out of the numbers may be all that is required. But before that, when we are taking a broader view, we seek a display which, like the structural section, yields the easiest transfer of the attribute information into the mind of the interpreter. Again this is a personal matter; the only clear limitation is that in general <u>it is not possible to interpret the significance of the auxiliary variables unless we know the structural situation also.</u> It is therefore folly to attempt interpretation of amplitude anomalies, or changes of frequency content or polarity, or velocity anomalies, without clear evidence of the structural environment which explains them.

We are driven, them, to two cross-sections — one optimized for structure, and one to present the selected attribute. Obviously they should be to the same scale, and preferably the attribute section should be some sort of overlay. The type of overlay is entirely a matter for personal preference and for personal eyeballs. For much of the remainder of this course we shall use overlays in which the variable is displayed in colour; this is convenient for teaching purposes because of the immediacy of the message conveyed, but it is emphatically not to be taken as limiting or in any way affecting the two-stream processing philosophy. Any method of display which conveys the message may be employed, according to taste; <u>the important feature is the two-stream processing, and not the display.</u>

3.4.5 Illustrations of factors affecting reflection strength (other than hydrocarbons)

Our object in this section is to look at a large number of slides which illustrate the degree to which variations of reflection strength are observed on real sections, for reasons <u>other than</u> the presence of hydrocarbons, and to attempt to recognize the reason for these variations. This is valuable because, when we come in Part 4 to study direct hydrocarbon detection, it is important that we should not misinterpret such variations as hydrocarbon indicators.

The displays we shall use are of the type shown at large scale on p.3-58. In this the black-and-white background (variable-area, in the example) is the output from the structural stream; to this may have been applied time-variant deconvolution and filtering, agc, and any other process which improves the clarity of the reflection alignments. The colour modulation, however, represents the truth of the matter — the reflection strength undistorted by any of these agencies. The eye readily find places where the variable-area amplitude is equal but the true strength obviously is not, and the colour key at the side allows
133   us to <u>quantify</u> the difference in strength immediately.

       On p.3-58 the variable displayed in colour is the amplitude of the reflection envelope — what we have
134   termed the strength (p.2-96A). It might equally well have
135   been the absolute amplitude after the fashion of the current Geospace literature; again we stress that the important feature is the <u>dual input</u> from the structural and attribute streams — however it may be displayed.

       With this as preparation, we now pass to our extensive suite of illustrations of the nature and cause of reflection-strength variations on real sections. Of these, we preserve
136L   in the manual, for record purposes, the section on pages
137   3-59 and 3-60; the former in its usual aspect and the latter
138L   with the true reflection strength superimposed in colour.
TAPE 19 This section is used extensively for later purposes also.
140
141
142L
143
144

3-54

From our suite of illustrations, we draw the following conclusions:

- Major changes of reflection strength (as much as 12 dB or more) occur along reflections. We did not fully appreciate the magnitude of these variations in the old days, because of the effect of agc and/or base-levelling. The variations are observed in situations where no hydrocarbons are involved. Some are due to major but entirely normal changes of acoustic contrast across a plane interface, because of facies changes or because an abrupt discontinuity becomes transitional. Some are due to the effects of compaction or relaxation after tectonic deformation. Some are due to the local introduction of a new material between the two formations across the interface. Some are due to velocity lensing. And many are due to interference.

- The signature of an angular unconformity, in terms of reflection strength, is a sequence bright-dim-bright-dim ..... This is caused in part by the fact that the unconformity brings different materials into contact across it, in part by the inevitable constructive and destructive interference associated with the wedge (section 2.7.1), and in part by velocity lensing (section 2.8.4 and 6.1.1).

- Velocity lensing is a significant effect recognizable by its position under a lens-type configuration; wavefront healing is often seen to modify the effect at considerable distances below the lens feature (depending in part on the return path also).

- A considerable degree of short-wavelength corrugation of a reflector can occur before the reflection strength is much affected; this is entirely in accord with the concept of the hyperbolic "unit" of seismic resolution (section 2.8.5).

- The maximum brightening when two equal and separated reflections come to interfere is obviously 6 dB. If the observed increase is more than this, an additional effect (perhaps associated with a change of reflection coefficient) is indicated.

- Focusing and defocusing are observed on real sections. However, major defocusing effects (greater than -6 dB) require very tight folding (section 2.8.4), and can be associated only with the condition that produces diffraction hyperbolas from the top of the fold. Significant focusing effects occur more often; most of the (syncline-type) situations which produce them are of little prospecting interest, but the problem of curvature into a salt dome flank is a very real one.

- In marine work, array lengths can be selected, having regard to the source bandwidth, to avoid any difference of reflection strength when shooting updip and downdip; on land, where surface waves may dictate longer arrays, and where practical considerations may dictate unequal source and geophone arrays, differences can occur. These effects are calculable (section 3.2.5)

- Often the cause of an amplitude anomaly can be seen by inspection; sometimes, however, several agencies are at work simultaneously, and the interpretation is difficult. We shall see in Part 5 that considerations of depositional environment are often helpful in untangling these problems.

TAPE 20 3.4.6 Illustration of polarity estimation

We recall that we have discussed three methods of determining the polarity of a reflection. Reduced to essentials, these methods take advantage of the following facts:

- The amplitude of a positive reflection increases at the critical angle; that of a negative reflection does not (section 2.4.6).

- The first surface multiple from a negative reflector has the same polarity as the primary; that from a positive reflector has inverted polarity.

- The distinction between an erect and an inverted reflection is present in the relation of individual peaks and troughs to the maximum of the envelope; this distinction can be transformed into a distinction of reflector polarity by observing the relation obtained on a reflection of known polarity (provided there is no major change of pulse shape between the known reflector and the others whose polarity is desired).

Here we illustrate the third method.

Two approaches are in current use. The first presents the peaks of the waveform in grey or in colour, and the troughs (rectified to appear as peaks) in black. The judgement as to polarity is then made by eye, on the basis of whether the reflection appears to be (for example) predominantly black-grey-black or grey-black-grey.

The second (section 3.1.1.4) asks where the individual peaks and troughs are, relative to the maximum of the reflection envelope. On p.3-60, for example, we can see that for the reflection which meets the left-hand edge at about 0.6s, the maximum of the envelope (the reddest colour) tends to be just after a strong trough of the reflection; the same is true for a reflection at about 1.4s. The unexpectedly bright zone at 1.1s, in the central part of the section, is different; the maximum of the envelope tends to be just after the strong peak.

When this judgement is mechanized, we obtain the display of p.3-63. In this, reflections judged to be of positive polarity are coloured red, and those of negative polarity blue (each to a saturation dependent on its strength). The unexpectedly bright zone clearly identifies itself as an inversion of acoustic impedance.

The performance of the method is surprisingly coherent on some reflections, but appears to be unrealistically erratic on others. Those reflections which maintain a constant indication are usually found to be discrete, and are often the geological marker horizons; those which are erratic classify themselves as complexes, in which minor

Color plates 3-58 through 3-64 follow page 349.

changes of bed thickness produce minor changes of shape in
the composite pulse, and for which these changes of shape
are falsely interpreted as changes of polarity. Bringing
together these illustrations with the previous suite showing
reflection strengths (together with the discussion of
principles in sections 2.7.2, 3.1.1.3, 3.1.1.4 and 3.1.3)
we can summarize the techniques available to us for distinguishing discrete reflections from complexes:

- Geological reasoning based on knowledge of the regional geological history and the depositional/erosional patterns present on the section.

- Borehole logs, if available.

- The fact that discrete reflections tend to maintain rather constant indications of strength, polarity and frequency content over many kilometres (provided, of course, that the source output is substantially uniform).

- If the frequency-selective losses are small, the fact that the surface multiple of a discrete reflector has a form much the same as the primary.

- If suitable multiples can be selected by stacking, the fact that a reflection from the top side of a discrete reflector has the same character as that from the under side.

As noted earlier, the use of knowledge of the out-going
source signal to force the final reflected pulse closer to the
zero-phase condition must be beneficial to all methods of
polarity estimation. Poor processing (particularly, bad stacking)
must be harmful. A highly variable source signal (which often
troubles us in land work) is also harmful.

It is always important to remember that the polarity of a
reflection is decided by the contrast of acoustic impedance
— not the contrast of velocity.

3.4.7 Illustrations of frequency content

Illustrations are given in the literature showing how the
details of the complete spectrum vary as a short window slides
down the trace (Mossman and Schoellhorn, 1973), and how the signal
may be passed through three fixed filters and displayed in colour
using the optical analogy (Balch, 1971). We can also display

the effect of sliding a short window by plotting only some derived frequency such as the value f_m discussed in section 3.1.1.

Consideration of all these displays drives home the point that we are now able to make any measurement we care to define, and to display it; the problem is understanding what the measurement means geologically. In the case of frequency, we have also to ask whether the quantification of spectral content really adds very much to the measurement of cycle breadth (which is done quite well — though not quantitatively — by the unaided eye).

We remember that three useful results have been suggested from frequency analysis: the improved resolution of thin beds, the detection of zones of anomalous high-frequency loss, and the identification of transitional reflectors.

Our discussion of interference in the frequency domain (section 2.7.2) prepared us for disappointment from the first of these — improved resolution of thin beds. The previous reference to Mossman and Schoellhorn (1973) confirms the pessimism, and an example given by Lindsey (1973) clinches the matter. For virtually all practical purposes, we may say that a detectable notch in the frequency spectrum appears at a reflector spacing identical to that at which a hickey in the waveform (or some other visual indication of interference) can be seen on the normal section, in the time domain. In the sense of giving us an additional tool for the study of a particular thin reservoir, therefore, the frequency domain is little or no help.

It must always remain true, however, that a thick zone of highly contrasty stratification yields a "high-frequency" reflection train. This follows inexorably from the "More up, less down" principle, and our observation that it is in such a section that the transmitted pulse loses its high frequencies. Below a "high-frequency" zone, therefore, we expect to see a loss of the high frequencies.

When we come to study this matter on real sections, and also when we attach our second and third objectives above (anomalous absorption, and transitional reflectors), we may use to advantage the display of the median frequency f_m, or the instantaneous frequency, or some other frequency measure. It is very important that we superimpose the frequency information on the normal section, so that it can be interpreted with regard to the background geology.

If we do not have the capability to do this, there is much to be said for the good old filter-panel, at least as a supplementary means of frequency analysis.

When we make a filter panel, we have two initial decisions to make, and these affect the type of conclusion which it is safe to draw. These decisions are: the type of bandwidth we shall use for the tightly-filtered outputs, and whether or not we shall normalize or base-level the outputs. We discuss these in turn.

The obvious polarization of choice of the narrow bandwidths is whether the bandwidth should be constant in hertz or constant in octaves. In the first case, for example, the narrow bandwidths we use might be the 10 Hz bands 0-10, 10-20, 30-40, 40-50, and 50-60 Hz; in the second they might be half-octave bands 7-10, 10-14, 14-20, 20-28, 28-40 and 40-56 Hz. The choice depends on the objectives; the important thing to remember is that in the case of constant bandwidth in hertz the filter operators all have the same envelope, whereas for constant bandwidth in octaves they all have the same shape (p.3-68). Thus for display in classical section form, the hertz-band filters yield very leggy reflections at the high frequencies; when it is the envelope which is displayed, this difference is not apparent. Further, the hertz-band filters do not present a longer and longer elided sample of the geology as the frequency is lowered, which the octave-band filters do.

The filter panel of pp.3-69 and 3-70 is included to allow study of the detailed behaviour of the interference complex at the unconformity level when filtered by half-octave filters; the bottom-left section on the first page is the unfiltered data, the top-right is 7-10 Hz (marred by a photograph fault), the bottom-right is 10-14 Hz, and then on the second page we see 14-20, 20-28, 28-40 and 40-56 Hz. The apparent position of the wedge moves considerably, of course, both horizontally and vertically.

The other decision we must make is whether to normalize the filters. If we do not, we shall see the real spectral content as a function of frequency; this is therefore the display which gives us the gross spectral shape of the seismic

274

Constant bandwidth in hertz (same envelope)

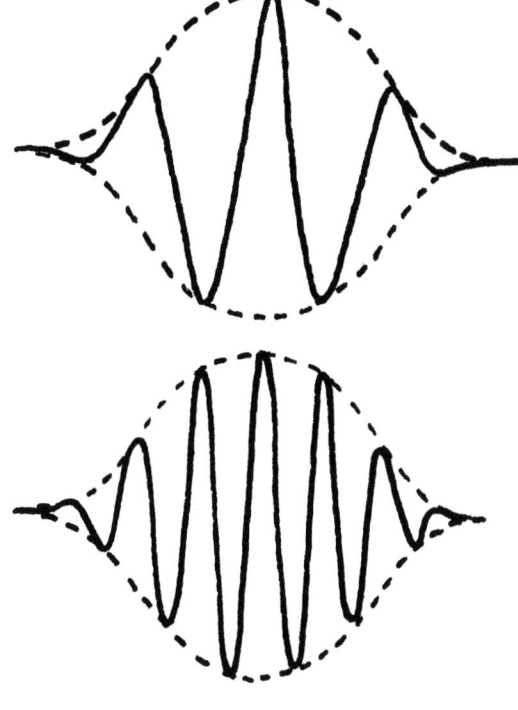

Constant bandwidth in octaves (same shape)

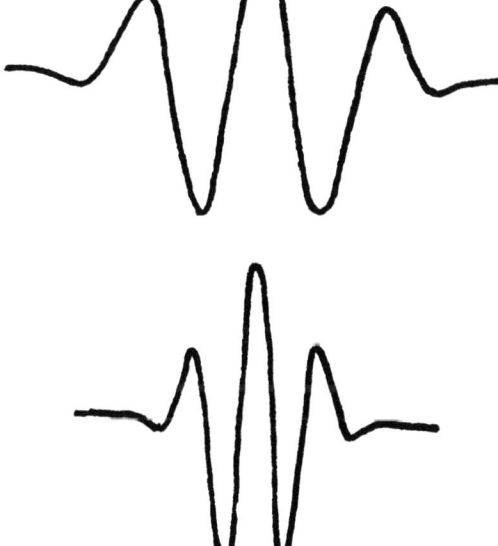

(zero-phase filter operators)

3-68

275

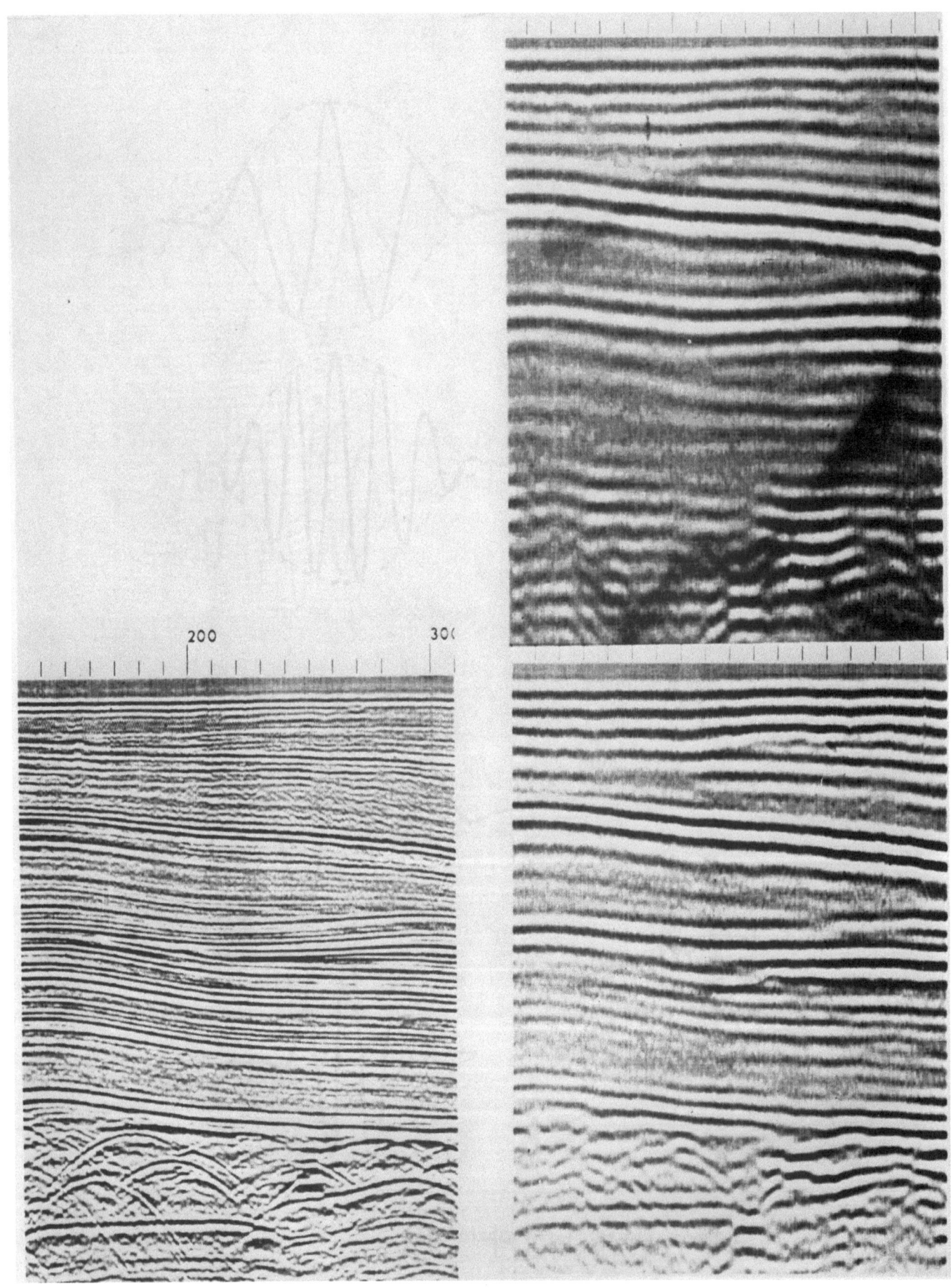

3-69

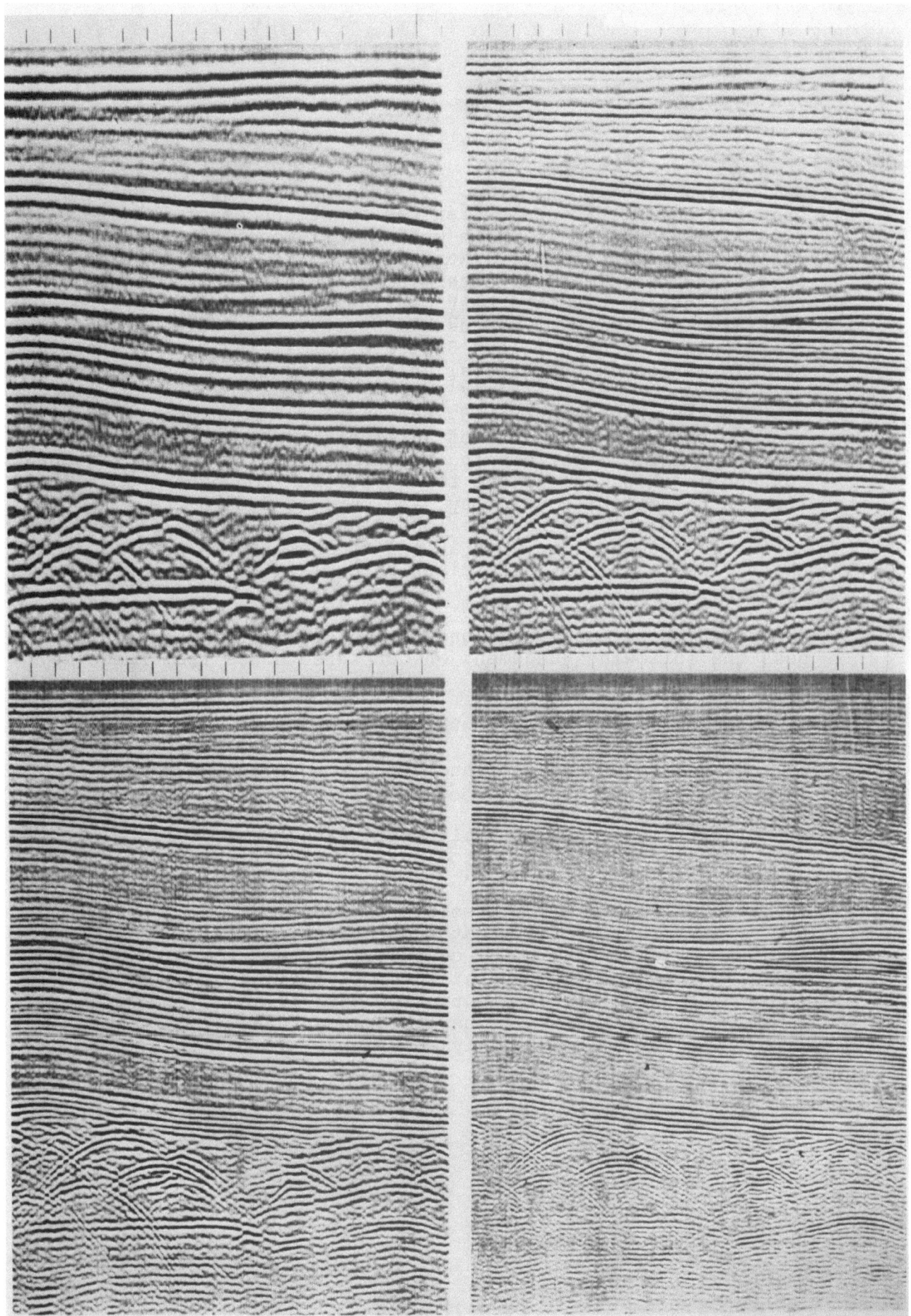

3-70

pulse as a function of time. If we do normalize, we shall probably be able to see better the consequences of interference, and the zones in which weak transitional (low-frequency reflections are present. (Normalization is also the standard processing technique for determining in which filter bands the reflections have the best continuity, and for observing what are the limits of the usable spectrum as defined by noise.) The displays of pp.3-69 and 3-70 are normalized, so they contain no information on the general pulse spectrum.

In the case illustrated, the envelope measurement can be used to show the existence of a transitional reflection just after the first major dipping horizon. Part of the low-frequency content in this zone is due to nmo stretch consequent on a rapid rise in the stacking velocity, and part is due to the minimum-phase nature which makes the low frequencies of the strong reflection appear late, but it can be shown by computation that these effects are not sufficient to explain the richness of low frequencies present.

With these considerations as background, it seems safe to pass the following judgement on our three hopes from frequency analysis:

- Improved resolution of thin beds — no.

- Detection of zones of anomalous absorption — maybe; we will postpone the decision.

- Identification of transitional reflectors — yes, provided they are not too near other reflectors.

Question: What are the possible explanations for the fact that the reflection at 2.45s in the centre of the section on p.3-60 (and at the bottom right of each filtered output on pp.3-69 and 3-70) has a higher frequency than those immediately above it?

3.4.8 Illustrations of change of frequency content

191
We have agreed several times that the measurement of the change of frequency content may be more useful than that of frequency content itself. Furthermore, it is the change of frequency content which is basic to the determination of Q or other measure of the high-frequency loss (section 3.1.1.3).

192L
193
On p.3-64 is the change-of-frequency-content display associated with the "standard" section of p.3-60. This represents the change in the value of the median frequency over a period of 240 ms. The measurement is much more sensitive than the eye; the highest values present represent a median-frequency shift of the order of 1-2 Hz. We notice that the change is unexpectedly negative in some places: the top of the section (where the explanation is nmo stretch), just below the first strong dipping reflection (nmo stretch plus a transitional zone), and just above the deep reflection specified in the above question. The explanation in the last case — which also answers the question — must be that the high-frequency reflection is a dipole doublet (section 2.7.1).

With such a display, we have the tool for detection of zones of anomalous absorption. The mechanized measurement of change of frequency content therefore raises our hopes a little, beyond our previous "Maybe". However, it still remains true that a local depression of frequency content may be brought about by complexes within the measurement interval (section 3.2.8.5) rather than by anomalous absorption; the new tool gives us no protection against this.

TAPE 21 3.5 <u>VELOCITY MEASUREMENT</u>

194   Here we assume that the general principle of velocity measurement (Taner et al., 1969) is well known.

196B For velocity measurement, as for seismic surveys generally, it is useful to recognize three levels of concern:

- velocities for stacking,

- velocities for lithologic purposes, migration and initial depth conversion, and

- velocities for porosity estimation and detailed depth conversion.

3.5.1 <u>The first level — velocities for stacking</u>

At the outset, we must remind ourselves that these are not <u>velocities</u>. The quantity we loosely call a velocity is just the variable defining a hyperbola which best fits the reflection alignment.

Perhaps we should take a moment to identify our different uses of the word "velocity". In sketch (a), p.3-76, we see the ray-paths comprising a cdp gather when the propagating material is homogeneous and of constant velocity; the corresponding reflection alignment is a hyperbola (sketch (b)) from which the constant velocity can be calculated. This situation, in which the variable defining the hyperbola has a true physical meaning as a <u>velocity</u>, does not occur in practice (except in the sea).

Sketch (c) illustrates the Dix conditions of horizontal uniform layering. The reflection alignment is still very close to a hyperbola, under most conditions, and we call the variable defining this hyperbola the rms velocity.

Sketch (d) illustrates the general case of a cdp gather, when the depth point is no longer common. Fortunately, the reflection alignment is still surprisingly close to a hyperbola; the variable defining this hyperbola is the stacking velocity. The stacking velocity has no immediate physical significance as a velocity.

Sketch (e) illustrates the ray-paths from a Huygens secondary source to a spread on the surface. In many practical cases this situation also yields a hyperbolic event; the variable defining the hyperbola is the migration velocity.

Finally, sketch (f) represents the average or depth-conversion velocity, which has obvious physical meaning as a velocity (even though the reflection path may not be physically realizable).

We shall discuss these situations in greater detail later. In the present context we need to appreciate very clearly that a stacking velocity need have no physical significance as a velocity. It is merely the variable defining the hyperbola which best fits the reflection alignment. We are free to choose it, therefore, to give a good stack — not, at this level, to be meaningful as a velocity.

It follows that our concern with stacking velocities may be of two quite distinct types. We may be conducting a reconnaissance survey, in which event our velocity analyses are likely to be spaced apart; in this case we wish to locate the analyses in such fashion that the stacking velocities do approximate to real velocities, in order that we can use geological rationales for the steps of picking and interpolation. Or we may be at the most detailed stage of investigating a prospective feature, where, over a short length of line, we require to place the greatest confidence in the amplitudes and waveform details on each stacked trace; in this case our velocity analyses are conducted at every depth point, and regarded solely as the means of producing a good stack (or of recognizing conditions where a good stack is not possible, and where the data should be discarded). The two objectives are quite different, and should be kept distinct in the mind.

Let us talk about the reconnaissance survey first. Then our initial problem is to locate the spaced-apart analyses intelligently. (We would not even consider <u>uniform</u> spacing, of course, unless the geology is also uniform.)

- In general, unless there is a previous survey in the area, the siting of analyses has to be done on some form of single-coverage section or brute stack. Our first criterion, then, is to avoid zones of excessively poor signal-to-noise ratio evident on this section; the reason for the poor signal-to-noise ratio, whatever it is, almost certainly militates against good velocities.

- We position an anlysis at the crests and troughs of folding (analyses 1, 4d and 5d on p.3-78), where the conditions approximate uniform horizontal layering and the stacking velocities come nearest to rms velocities and to physical significance.

- For reasons concerned with interpolation (discussed later), we position at least one analysis on the flank (analysis 2); we accept, of course, that this must yield "velocities" which are unrealistically high.

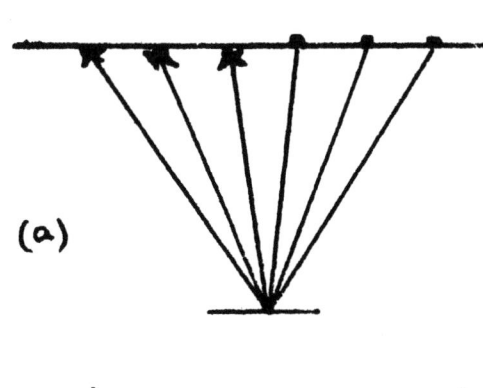

(a) true constant velocity

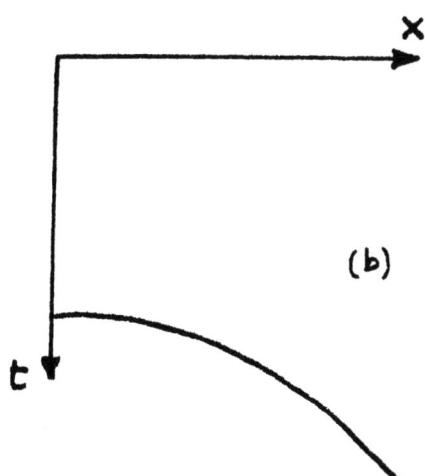

(b)

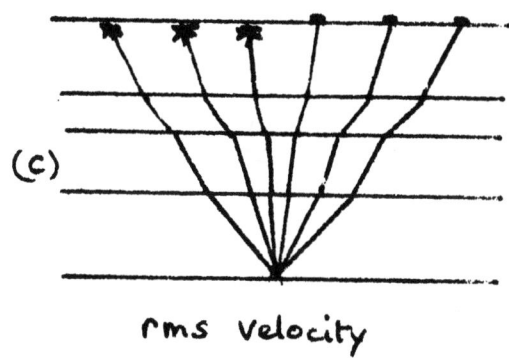

(c) rms velocity

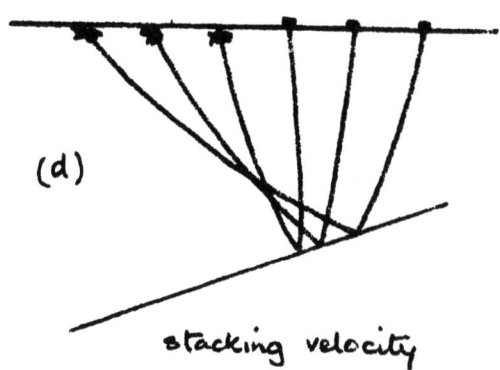

(d) stacking velocity

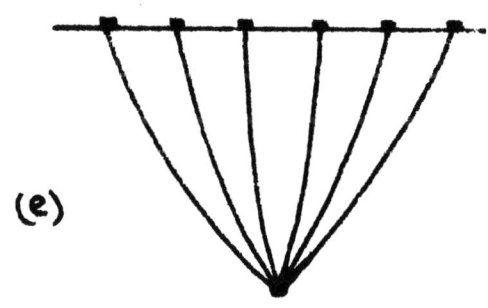

(e) migration velocity

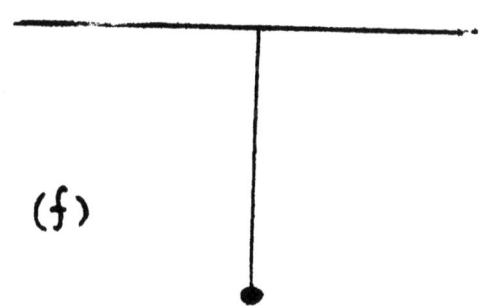

(f) average or depth-conversion velocity

- Where fragmentary reflections are visible at depth, we position an analysis on each of these (analyses 3 and 9e). It may happen that the analysis is unusable at all other levels, but the fragmentary reflections are the only evidence we have at depth, and we must use them.

- We avoid using any analysis at levels where there is a fault or where the ray-paths have passed through a faulted or otherwise disturbed zone; thus we use 5 c but not 5 d, and before using 6 d we check the likely position of the fault zone relative to the spread at the surface. Within these constraints, we try to position an analysis each side of a fault.

- We avoid using any analysis at levels where there is obvious interference. Thus analysis 9 is where it is because of the fragmentary horizon (e); in general, we would give ourselves only problems by forcing a pick in the zone of horizons (b) and (c). Analysis 10 is positioned to give this information as soon as the horizons become separated.

- We prefer to avoid siting an analysis where any ray-path passes through an obvious near-surface anomaly. This accounts for the gap between analyses 7 and 8. However, as with all other local conditions which preclude reliable analysis, analyses 7 and 8 are positioned as close to the anomaly as may be done without suffering its effect; we use no geophone positions located on the anomaly.

Let us develop this discussion of the near-surface. Of course the objection to siting an analysis on a near-surface anomaly is removed if we have a perfect means of removing the effect of the anomaly, and we may think that we have this in an autostatics program. But we must remember that autostatics programs are plagued by a vicious circle of inter-relation between the determination of the statics and the determination of the velocities (section 8.3.2), and that in any case the effect of a near-surface anomaly often extends beyond the mere introduction of simple _time_ shifts as calculated by an autostatics program.

284

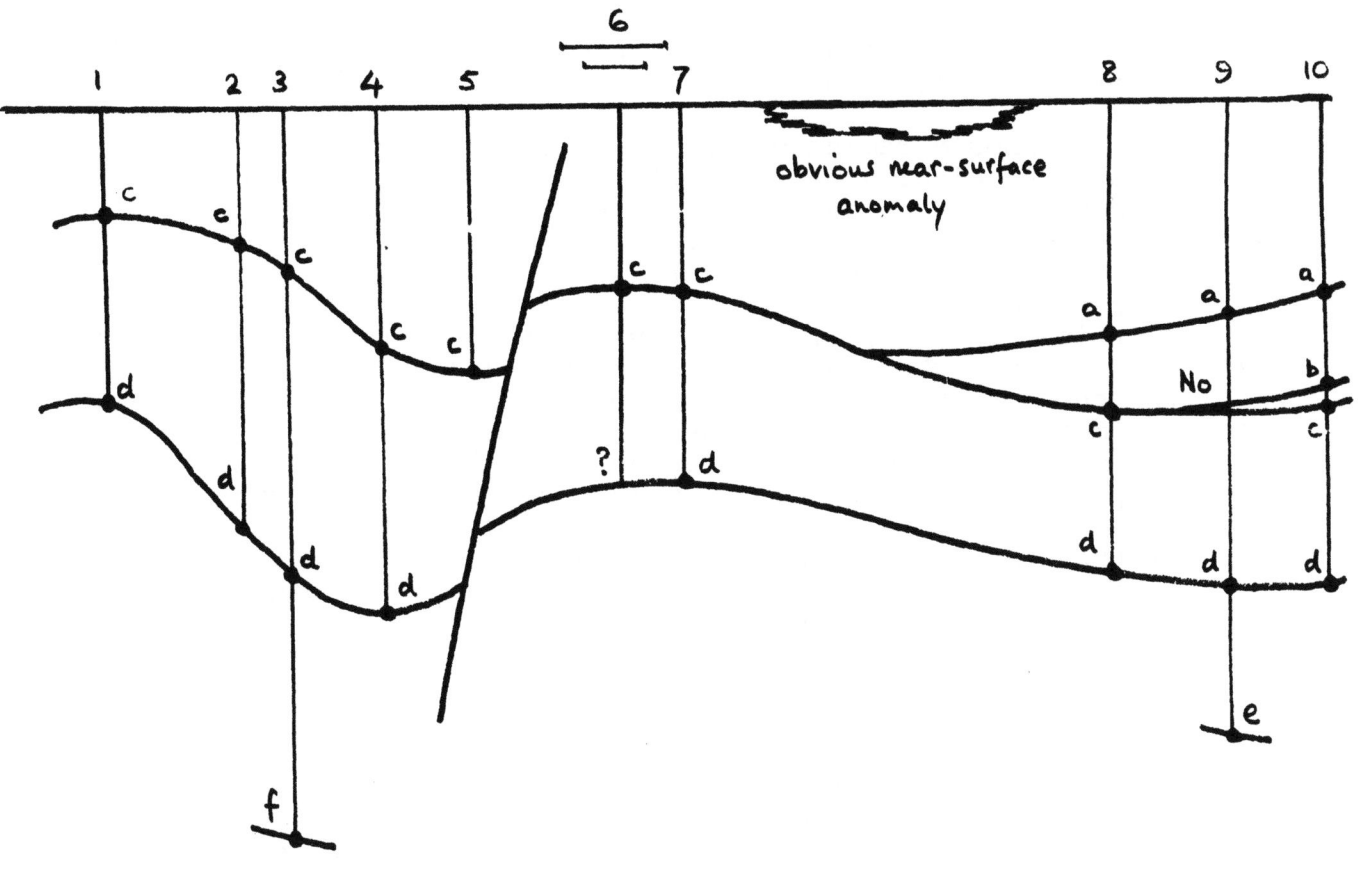

3-78

Therefore it is always best to avoid the near-surface anomaly if it is feasible to do so.

We should remind ourselves of the consequences of static errors on velocity measurement, using the illustrations of Prescott and Scanlan (1971). The diagram on p.3-80 shows the effect of a locally thickened weathering on the velocity measured for two deep horizons A and B. The true velocity is reduced by the thickened weathering (lines labelled "vertical velocity"), but the velocity measured by cdp analysis is locally <u>higher</u> under the anomaly and locally lower beyond the ends of the anomaly. As shown on p.3-81, the effect always becomes more serious with depth, because a simple time shift has more impact on a measurement at small Δt than at large Δt. On p.3-82 we see the way in which the velocity distortion is related to the ratio between the spread length and the extent of the near-surface anomaly; in terms of magnitude, the distortion is greatest when the extent of the anomaly is 0.6 — 0.7 of the spread length.

The effect of the near-surface is seen to be the greatest disruptive factor in velocity analysis.

The siting of velocity analyses is primarily the concern of the processor, rather than the interpreter. Nevertheless there is wisdom in having the interpreter review the siting on the stacked section, so that he becomes aware of zones where, for one of the above reasons, the stacking is not to be trusted. (When he finds an error of placement, he should remember that he has the advantage over the processor, who must make the siting judgement on the basis of less information than is present on the stacked section.)

There is also wisdom is being aware of other decisions which are made by the processor — such as the window length, the window increment, and the use of deconvolution, dereverberation and filtering. The window length is advantageously set to represent about 1½ cycles of the average reflection pulse; less than this is generally harmful to the signal-to-noise ratio, while more definitely increases the risk that a multiple will fall in the same window and so degrade the primary stack. The window increment is advantageously half the window length, for reconnaissance purposes. Dereverberation is obviously

DERIVED VELOCITIES

After Prescott & Scanlan, 1971

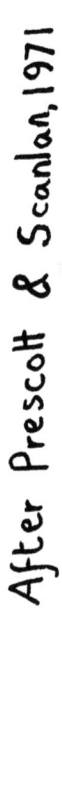

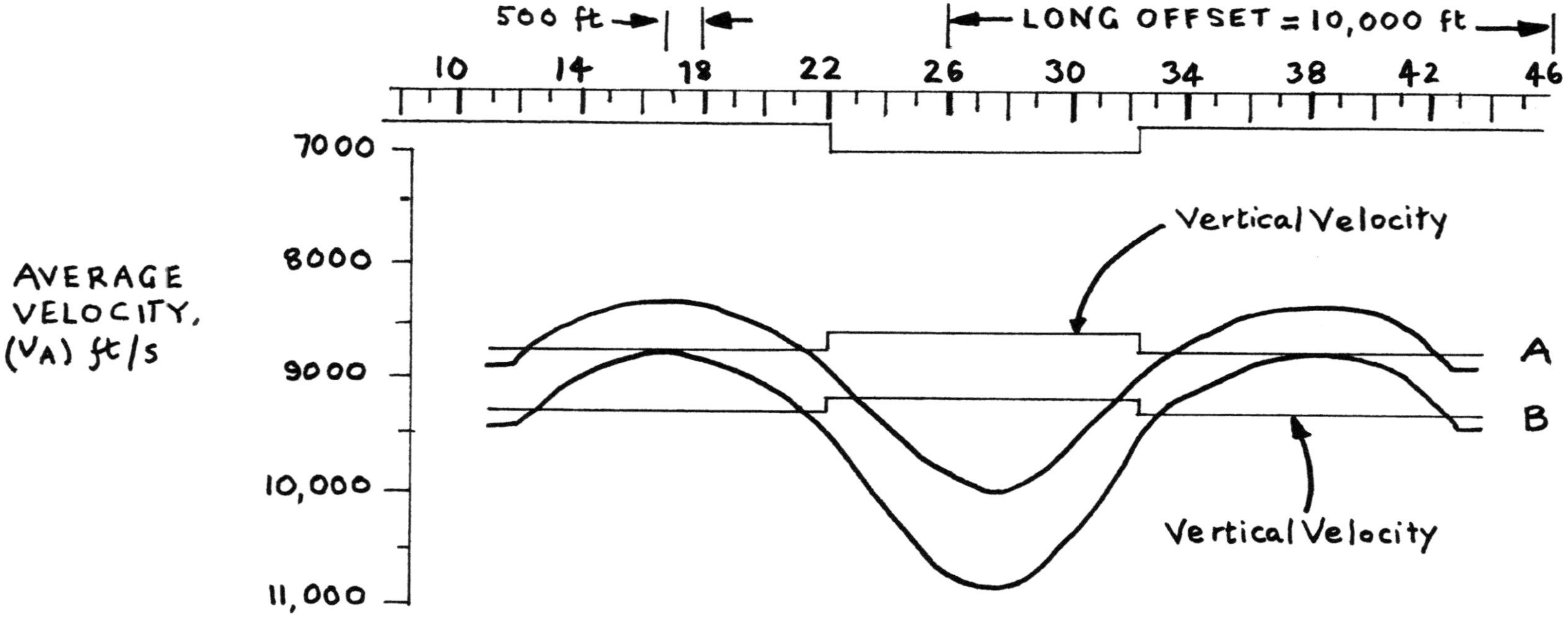

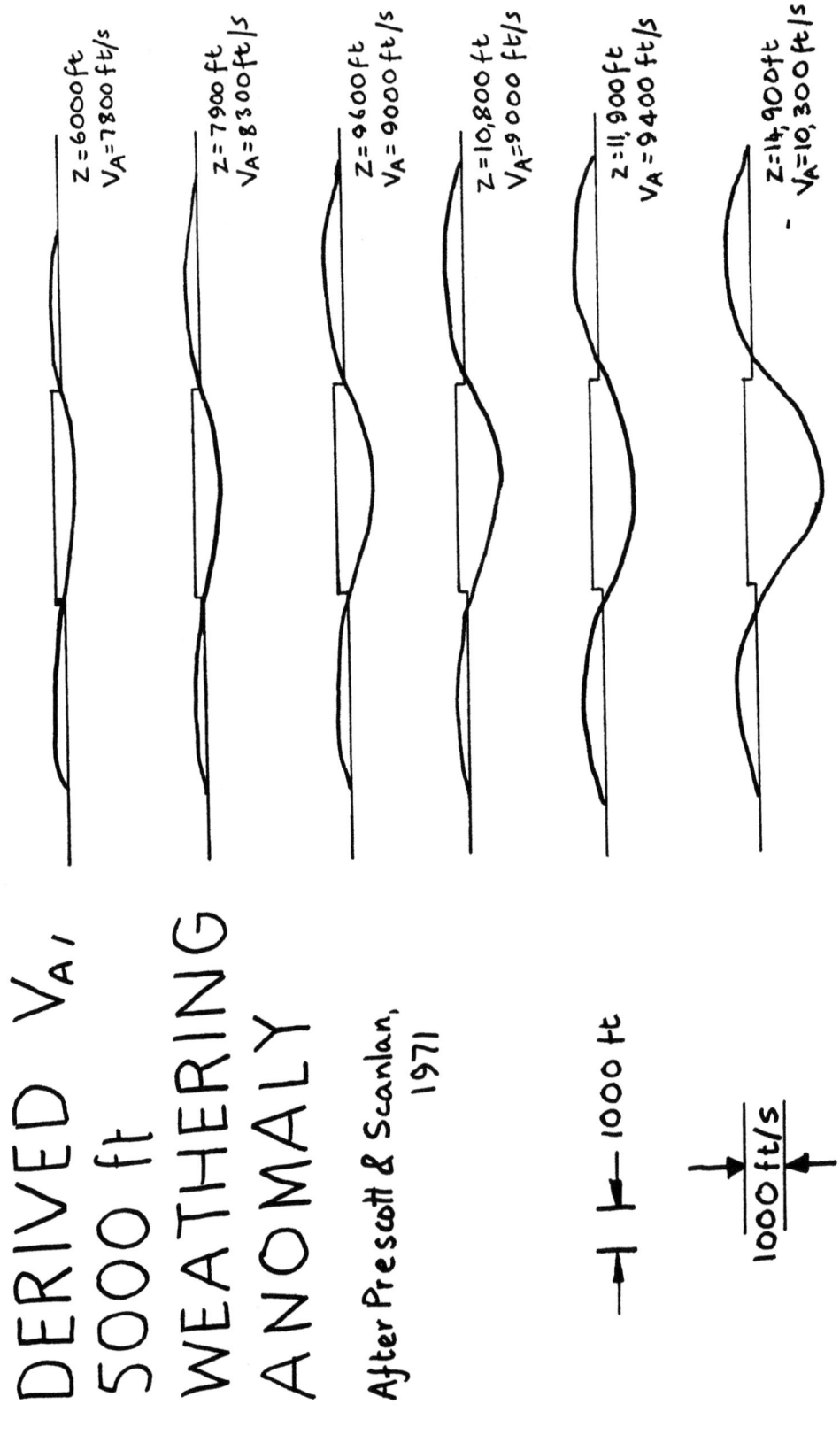

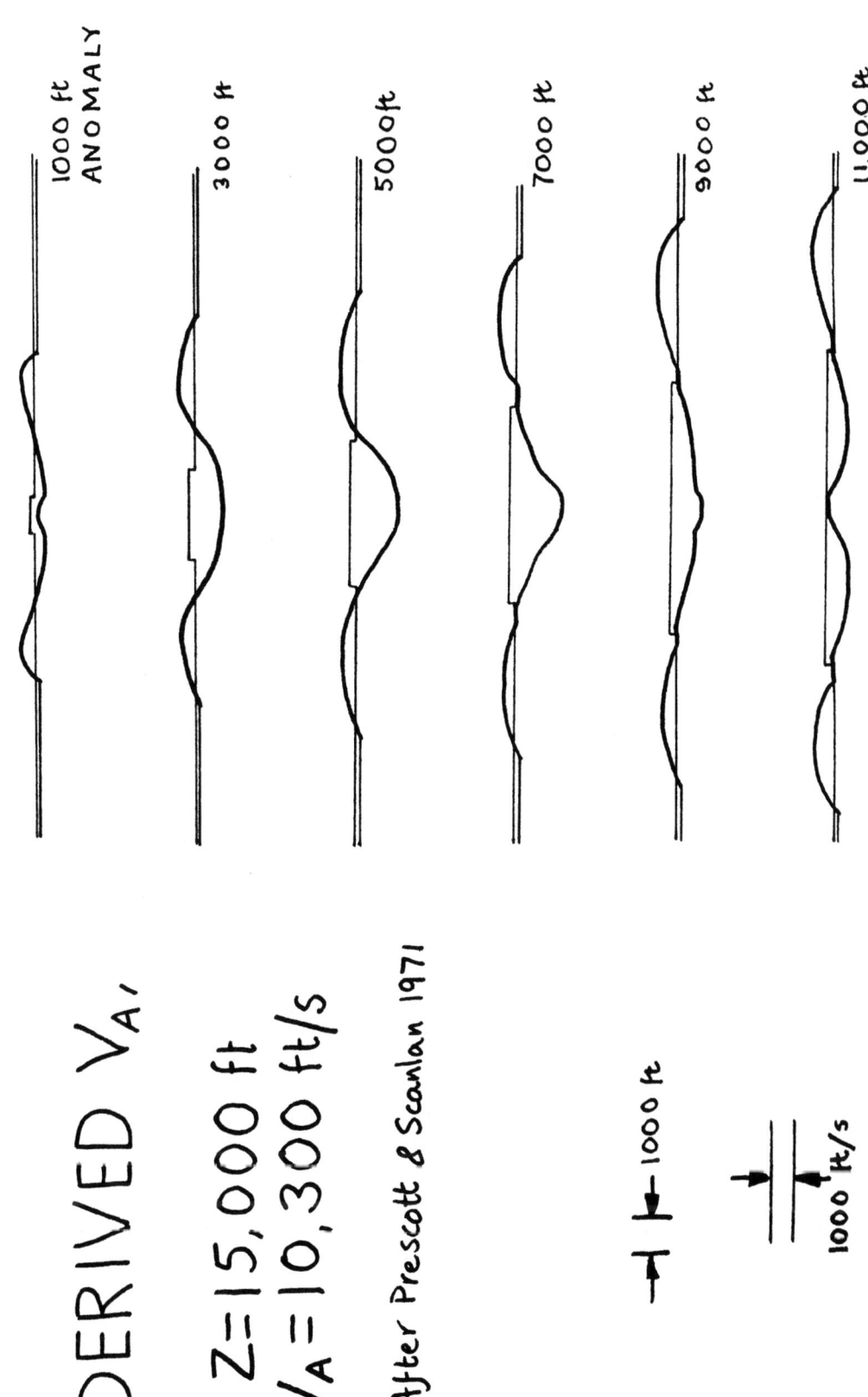

3-82

beneficial if effective — the fewer disturbing events on the analysis the better — and deconvolution and filtering are selected on the usual balance of judgement between resolution and signal-to-noise ratio (section 8.4.1).

The interpreter should also remember that the velocity analysis acquires a new importance (particularly at late times, where it was formerly of little significance) in connection with the geometrical divergence correction; we recall that the correction varies with the square of the velocity.

Our next concern is with another of those matters likely to be hidden deep in a program: what interpolation of stacking velocity is used vertically, and what horizontally?

Some of the choices available for the verticle interpolation are illustrated on p.3-86. The first is perhaps the obvious one — linear in stacking velocity (a). However, this is equivalent to a rather unrealistic variation of interval velocity within the layer bounded by the picks. If we assume constant interval velocity within this layer, we obtain the interpolation of sketch (b); if we assume some compaction law (for example, the $z\gamma$ of p.2-60) we obtain a corresponding interpolation exemplified by sketch (c).

(In passing, we note that for a specific problem in which we wish to minimize the problems of nmo stretch, we could interpolate as in sketch (d), where the interpolation function ensures constant ΔT throughout each reflection pulse, and concentrates the stretch between them. However, this would certainly not be adopted for general use.)

The matter of vertical interpolation has acquired a new importance because of the techniques of seismic stratigraphy to be discussed in section 5.2. Formerly it could have been argued that if there is no reflector present between the two picks the vertical interpolation is immaterial; now we see the wisdom of having the interpolation geologically reasonable if possible.

Having by interpolation arrived at a value of stacking velocity for every time sample, we must now ask how the interpolation is done horizontally, between analyses. It could, for example, be linear in stacking velocity, or linear in ΔT, or some more sophisticated interpolation. The situation is the one at the bottom of p.3-86, where analyses are shown at positions 1 and 2. Stacking velocities are deduced for times t_1 and t_2 on both analyses, and we now ask what velocity we ascribe to each of these times at position X, between the analyses. Clearly neither the linear horizontal interpolation nor any other mathematically-defined interpolation of <u>stacking</u> velocity can produce the right answer. And, of course, no interpolation at all can provide the <u>increase</u> of stacking velocity necessary to obtain a good stack on the dipping flank.

Every processing house has its own approach to these interpolation problems. Ideally, horizontal interpolation requires a full specification of the dips and interval velocities, and this is unlikely to be available, at least on a reconnaissance survey, at the time when the decision must be made. So the solutions are likely to be less than perfect; the important conclusion in this situation is that the velocity analyses should be sited — and there should be sufficient of them — to minimize the adverse consequences of inappropriate interpolation.

The interpreter is also concerned to ask about interpolation across a fault (should we interpolate, or should we maintain unchanged stacking velocities up to the fault plane?). And interpolation at line intersections — if the analyses on the crossing lines are at the locations shown, will the stacked sections tie at the junction, even if all the velocities are clear and correctly picked? They may not, and it may be that only 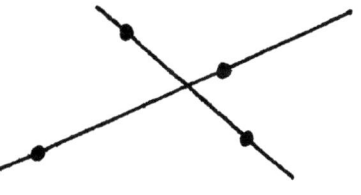 after the geological picture becomes evident on the stacked section can the answer be found.

Of course the interpreter cannot and should not be required to check every aspect of the processing of a reconnaissance or semi-detail survey; these comments are injected so that the interpreter should know the likely source of a particular

3-84

processing-related problem from its symtoms, and should also know where the results of such surveys are likely to be in doubt.

Perhaps we should insert a comment also on machine picking of conventional spaced velocity-analyses (Cochran, 1972; Beitzel and Davis, 1973). These are of value for on-line processing (particularly, for the preparation of a brute stack on board ship). However, we must not ask too much; we must remember that the picking of velocity analyses is at once a geological and a geophysical interpretation, and often a complex one. The geological component in the judgements is provided by the feasible lithology and tectonics; the geophysical component is provided by the variations of stacking velocity with dip and other complications, and by the fact that there should be at least a general compatibility between the breaks of interval velocity and the strength of the reflections. All these components are difficult to include in a mechanized judgement.

TAPE 22 3.5.2 The second level — continuous velocity measurements

The logical approach is to help the machine, by making as many as possible of the judgements in some other way which the machine can absorb easily. The answer is for the interpreter to pick the initial section (preferably an initial stacked section — though there is some risk in that), to give the machine these picks, and to ask it to find the velocities solely on these selected horizons. Since the volume of data considered is much reduced by the knowledge of approximate reflection time, it becomes economically feasible to compute the velocities on every depth-point; since the approximate dip and reflection curvature are known, it becomes ecomically feasible and conceptually sound to fan-filter the common-offset data (in a time-and-space-variant manner) before forming each velocity gather. The power of the velocity-analysis method is thus greatly increased; this is made possible by the entirely proper step of asking the interpreter to select the horizons to be considered.

The interpreter should terminate any identified reflection where it encounters any of the difficulties (at its own level,

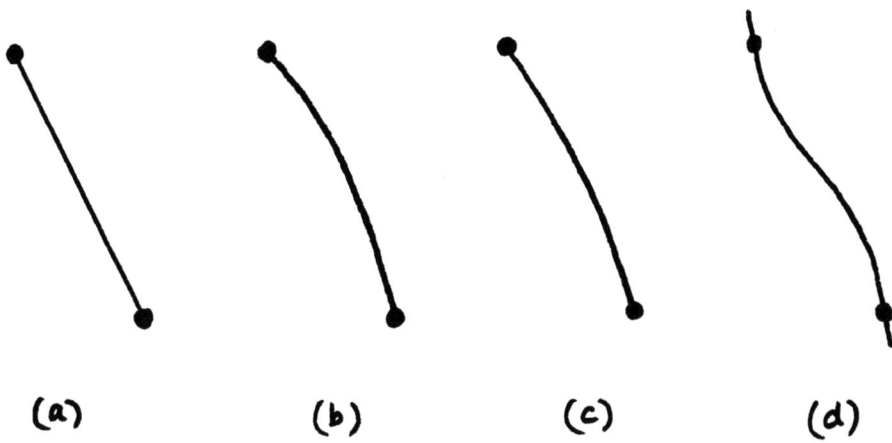

(a) (b) (c) (d)

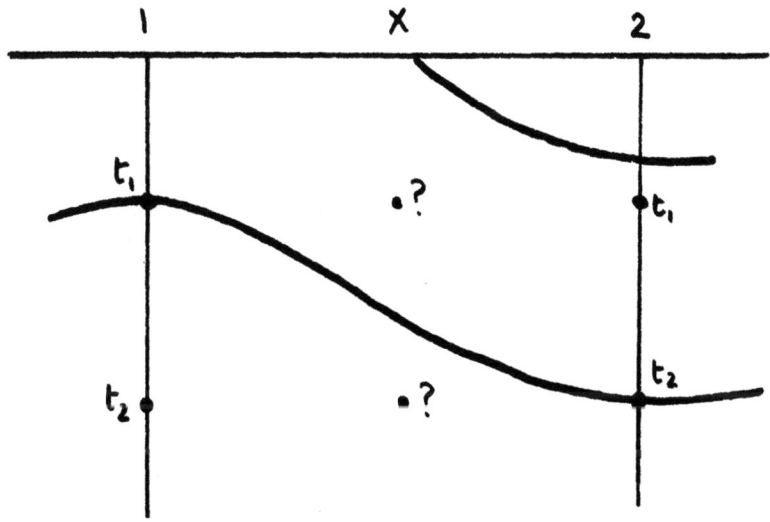

3-86

or above) which he knows must disturb or invalidate the computed results — interference, faulting, major changes of reflection strength, or obvious near-surface problems. It is far better to omit data which can be seen to be unsuitable, and thus to obtain consistent results with the minimum of smoothing, than to rely on a large measure of smoothing as normal.

For many purposes the smoothing is best done in V^2t (see, for example, Larner 1974, and his references). It is essential that the interpreter involve himself in choosing the extent of the smoothing operator; <u>it must not be chosen too long</u>, either for the faulting or for the expected lateral changes of interval velocity. (A processing house left without guidance will tend to use long smoothing operators, to make the results look "better" than someone else's, just as it will always strive for a generally "pretty" section rather than one which addresses the interpreter's particular problem.)

It is important to preserve, in some form, a measure of the scatter in the velocity values <u>before</u> smoothing; this is so that the interpreter can establish <u>his</u> level of confidence in the variations he sees — specifically, is this apparent lateral change in velocity greater or less than the observational error?

At the simplest level, the computed velocity values (which are still <u>stacking</u> velocities) can be used to calculate interval velocities and depth-conversion velocities according to the Dix equation, which assumes horizontal layering. At the next level of sophistication, we can calculate for the interval velocities and depth-conversion velocities a correction for dip and for refraction in the dipping layers above; the same process also yields the migration velocities. A fuller discussion of this is postponed to our discussion of migration in Part 7; let us just note here that the process requires the time-dip, travel-time and stacking velocity of all reflections, and hinges critically on the assumption that the reflections picked do represent in fact the real velocity breaks.

When assessing results expressed in interval-velocity terms, the interpreter is obviously alert to the problem which exists when he has terminated a reflection; left to itself, the machine will crank out an interval-velocity V_3, just to the right of the termination, which is obviously geologically inharmonious with the values V_1 and V_2; whatever the reason for the termination, it is unlikely that abrupt changes in the calculated values represent the real truth.

We mentioned above the importance of picking the right horizons (specifically, all the significant velocity breaks) if refraction in dipping layers is to be properly compensated. An error can affect all reflections below. It is also important to pick the velocity breaks if the interval velocities are to have any lithological meaning (though in this case an error has more local effects). For example, if the real velocity breaks are as shown in the left-hand sketch, but the interpreter picks those in the right-hand sketch, then clearly the final interval-velocity values are without significance lithologically. Therefore the picking exercise in the present context is not the same as the interpreter's normal concern. Usually he concentrates on the horizons which are useful as geological markers; in the present application he must consult the original velocity analyses first, and then pick those horizons which represent the major contrasts of interval velocity (not acoustic impedance).

3.5.3 The third level — detail velocity studies

In this section we are concerned with the final stage of seismic detail, in which the interval velocities of reservoir layers are required or in which we need the utmost accuracy of depth conversion. At this stage, it is wise for the interpreter to handle all the decisions himself, leaving the machine to do only the arithmetical tasks.

Because of the need to transfer the maximum information from the original data to the mind of the interpreter, the optimum display of the velocity analysis is important. Most interpreters seems to prefer a contoured display. The shape of the velocity-time contours contains within it the <u>character of the reflection pulses aligned along the hyperbola</u>, and is found to be a useful diagnostic tool.

For the preparation of the velocity analyses, the interpreter must make several decisions:

- He must decide how many analyses to run. If his concern is to measure the interval velocity in a thin target layer, he will need all the help he can get; he may even find himself calling for the shooting of more lines than he needs to delineate the feature structurally — just to obtain more ray-paths for velocity analysis.

- However, it would be quite wrong to think that mere numbers will guarantee success. The input traces must be carefully edited (having regard to everything that is known from the sections about the geology above the target) for near-surface problems, passage through disturbed zones, surface waves, poor shots, poor plants, anomalous frequency content, unexplained amplitude variations, and noise. The golden rule is to select the good traces — not to average all blindly.

- Then the interpreter must locate the analyses, and decide how many depth-points can be used in each (it is assumed here that some form of two-dimensional fan-filtering in the common-offset plane will be used before assembling each velocity gather). The number is a function of the bandwidth at the target level, of the reflection curvature, and of any local anomaly which it appears wise to avoid.

- If the problem is an interval velocity in a thin layer, the interpreter calls for the best bandwidth which the processing (including statistical spiking deconvolution) can provide. If the problem is depth conversion, he may make a more normal compromise between resolution and signal-to-noise ratio.

- The analysis window, and its increment, are then set to accord with the realized bandwidth. The window should be such that it does not include both the top and bottom of the thin layer whose interval velocity is desired; in cases of excellent signal-to-noise ratio (bright spots), it may be 20ms or even less. The increment must be such that, whatever the reflection times, a window position <u>squarely overlies</u> each important reflection.

- The velocity increment is then decided; for these detailed purposes it is likely to be 20 m/s or less.

If it is important to obtain numerically significant interval-velocity values in shallow layers, or to obtain correct interval-velocity variation in dipping layers, it becomes necessary to correct zero time. In fact, this can be defended in all applications — both on the section and for velocity analysis — just as a matter of good practice (Farrell, 1976).

The need to correct zero time is evident from p.3-92 and from our discussions and illustrations in section 2.5.1. In the velocity analysis at the top of p.3-92 we see three events as velocity contours; we also see them in the form of the trace at the right (which may, depending on the processing house, represent the stacked power, or the stacked envelope, or the semblance, or some other measure of the coherence across the gather). It is general practice to pick the velocities at the highest point of the contours, which is ordinarily also the maximum of the trace at the right. <u>This is entirely correct</u>; such a point represents the closest we can get to a part of the pulse which travels with the velocity characteristic of the propagating rock. It is unsound to try to correct this point to a so-called "onset" time; even if it were sound, the correction would be time-variant.

Picking of the maxima therefore yields time-velocity pairs which give the best estimates of interval velocities. However, when we come to consider the first layer, we see that past practice is inconsistent — it has accepted its first pick (the zero time) at the time-break, instead of at the maximum of the envelope of the pulse effectively generated by the source. If the source pulse — recorded and filtered like the reflection pulses — were displayed on the velocity analysis, it would appear as at the top left on p.3-92; by its side is shown its envelope. The time from the time-break or the command-to-shoot to the maximum of the envelope may be 20 — 60 ms, depending on the source and the filtering. <u>The real time zero is at this maximum.</u>

In cases where the source pulse is recorded, the correction may be calculated and applied automatically. This is certainly the best course. Where it is not recorded, the pulse may be estimated from the record (section 3.1.1); this usually leaves an uncertainty associated with the first half-cycle. Alternatively, the correction may be estimated as that necessary to make a simple multiple reflection, observed at normal incidence, arrive at double the travel time of the primary; in practice this is usually done by comparing the times of pulse maxima on traces and their autocorrelations (lower sketch, p.3-92). For land work using dynamite, the major part of the correction may be the response of the instruments and the processing, which can be established by recording and processing a spike. At sea, it follows from the argument of the last paragraph that the appropriate correction is that one which brings the interval velocity in the water to its proper value, and keeps it independent of depth.

So, with the zero-time correction applied (as so-many milliseconds) <u>before</u> the data enter the velocity analysis, we pick the peaks of the analysis, and all our interval velocities emerge correctly.

Thereafter, one of the first interpretation tasks is to revise the picking of the reflection times on the section. Ordinarily we shall have picked the times of the trough which

nominal T_o
V
source pulse
real T_o
t

nominal T_o
reflection trace
T_o correction
auto correlation
zero-lag

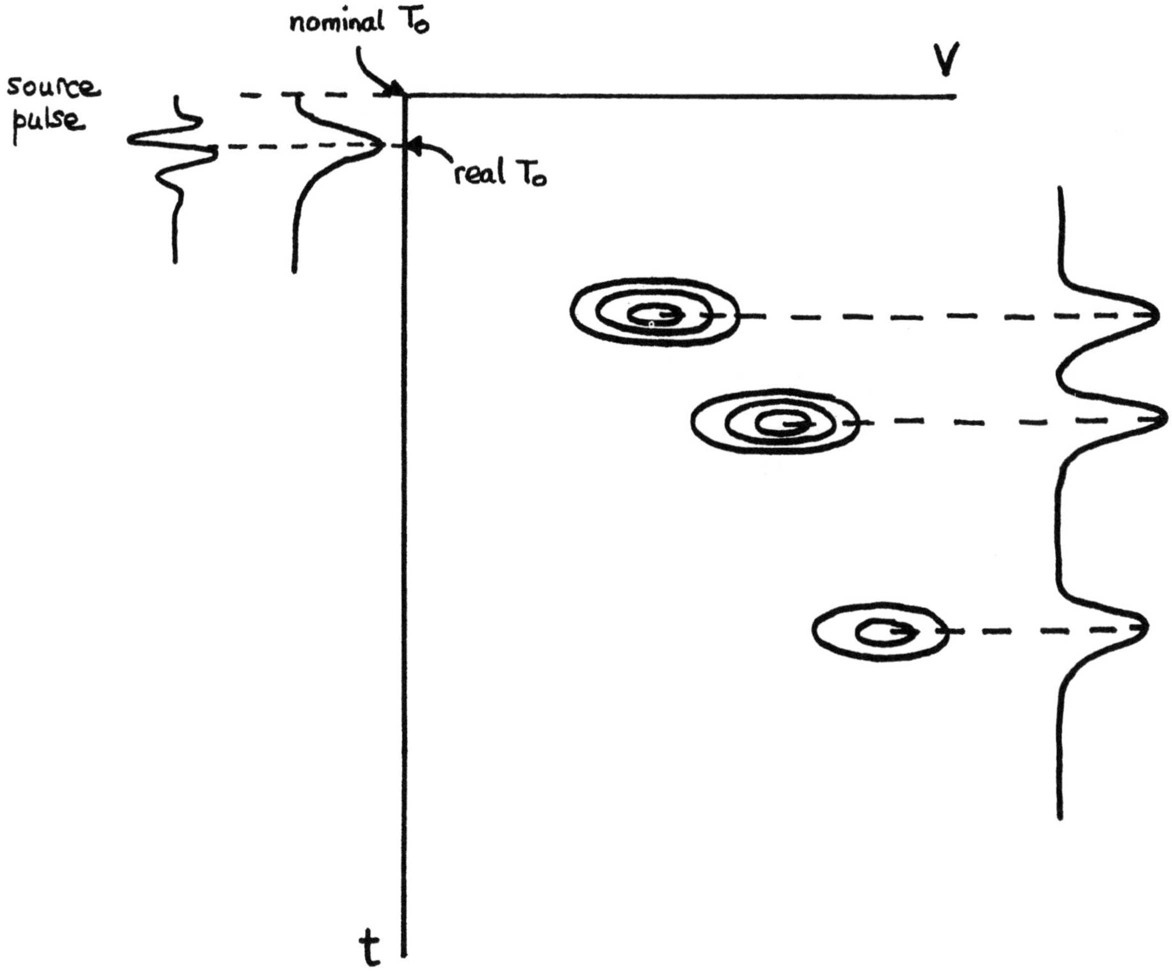

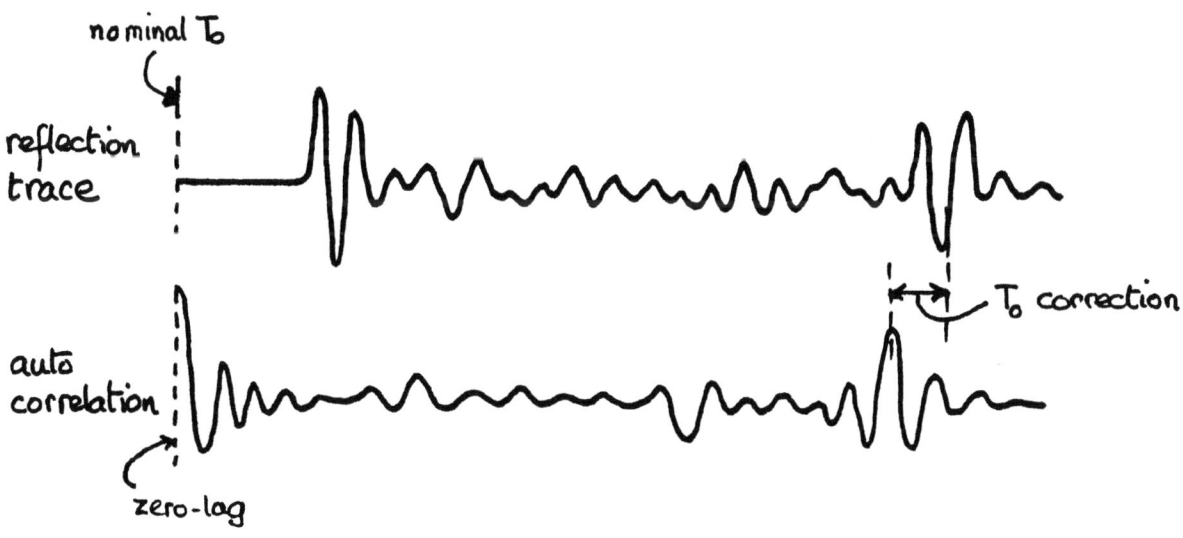

we coloured on the section; as suggested in the sketch at the top of p.3-94, this may or may not coincide with the reflection time observed and picked on the velocity analyses. For the present detailed purposes, it may be important that the times should be in general agreement, across the section and at all levels.

Then, having established a coherent scheme for picking velocities and reflection times, we start the detailed picking of the individual velocity analyses. We do this, like all other picking, by starting in the zones where the message is clear, and preferably where the section extends deep. This minimizes the interpretation problem when we come to onlap and offlap situations.

With the picture of velocity variation established in these zones, we start to work into the difficult areas. As we do so, we recognize the contour signature of interference (middle sketch of p.3-94); obviously it is important that we do not make a pick on this elided event. We may also encounter aliasing from a strong multiple (bottom sketch of p.3-94); the strong multiple event, recognizable by its low velocity, generates an aliased event which may be mistaken for a primary. At all times we keep track of the interval velocities consequent on our picks, ensuring that we keep within the constraints provided by geological plausibility; various proprietary displays and overlays are available to facilitate this interval-velocity monitoring.

Obviously the interpreter must define for himself a code of proper practice. One suggestion is that the picking should never be forced; a pick can be _discarded_ if some good reason for doing so is evident (interference, aliasing, residual near-surface anomaly), but a pick may never be _moved_. Neither may a pick be inserted where there is no closed contour.

Guidance in the picking is sometimes obtained by working in triplets. For an example we consider the top sketches of p.3- 96 where the left and right analyses are identical, but where the middle analysis suggests an anomalous high velocity on the second pick. First, of course, we would check whether this appears plausible on the sections. Then we would consider

3-93

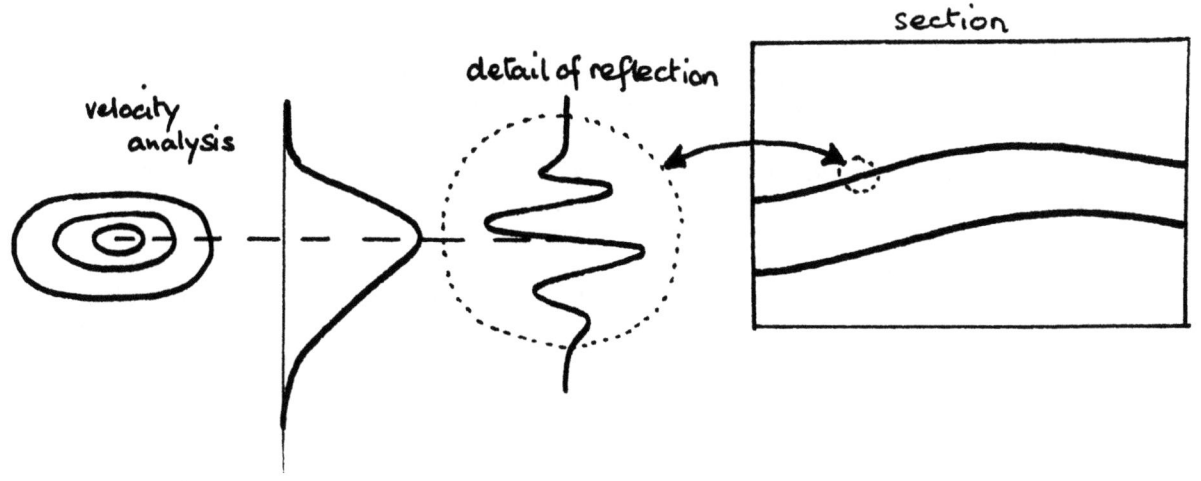

smear

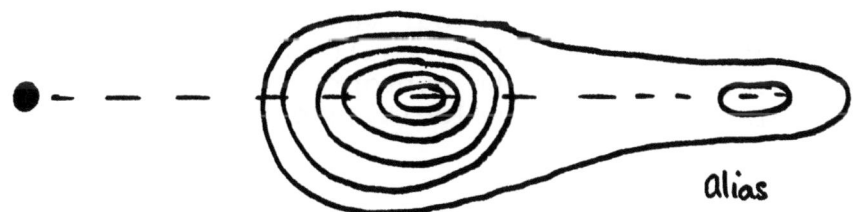

alias

3-94

the interval velocity between the second and third components of the triplet — does this velocity stay constant, and is that what we would expect from the sections? If so, then the likelihood of the high-velocity anomaly is enhanced.

This device becomes critically important when we are working with a low-velocity anomaly (for example, a gas-saturated porous sand) of such an extreme nature that it actually produces an inversion of stacking velocity; from the material of Part 2, this situation (which arises if the interval velocity in the reservoir is less than the stacking velocity to its upper boundary) is entirely feasible down to moderate depths. The problem, obviously, is that we must make a judgement whether the inversion is real or whether the pick we wish to make is a water reverberation or other multiple. To minimize the risk of improper interpretation of a multiple, we are careful to apply a closely-monitored dereverberation process before entering the velocity analysis. If the possible pick persists, we again check the reasonableness of the interval-velocity behaviour <u>below</u> our target (lower sketches, p.3-96).

The method of working in triplets may fail if the interval below the target is very thick. In this case, as suggested in the sketches, the stacking velocity of the deep event may remain almost unchanged, because of wavefront healing around the target anomaly (section 2.8.5).

O.PROJ.

212

213L

214

In part 4 we shall work an example of highly detailed velocity analysis. For the present, an illustration of the type of output obtainable is given on pp.3-61, 62, for our "standard" section. In this illustration, which is really more of a semi-detail example, an analysis using all the available ray-paths is conducted every 12 depth-points. The results are displayed as interval velocities, according to the colour key at the right of the section. On p.3-61 there is no smoothing; each analysis is represented by an independent vertical bar of colour 12 depth-points wide. A measure of the scatter in the results is given by the colour variations from one vertical bar to the next. On p.3-62 the results are smoothed with a space-variant operator of length appropriate to the geology at each level. Places where no velocity judgement could properly be

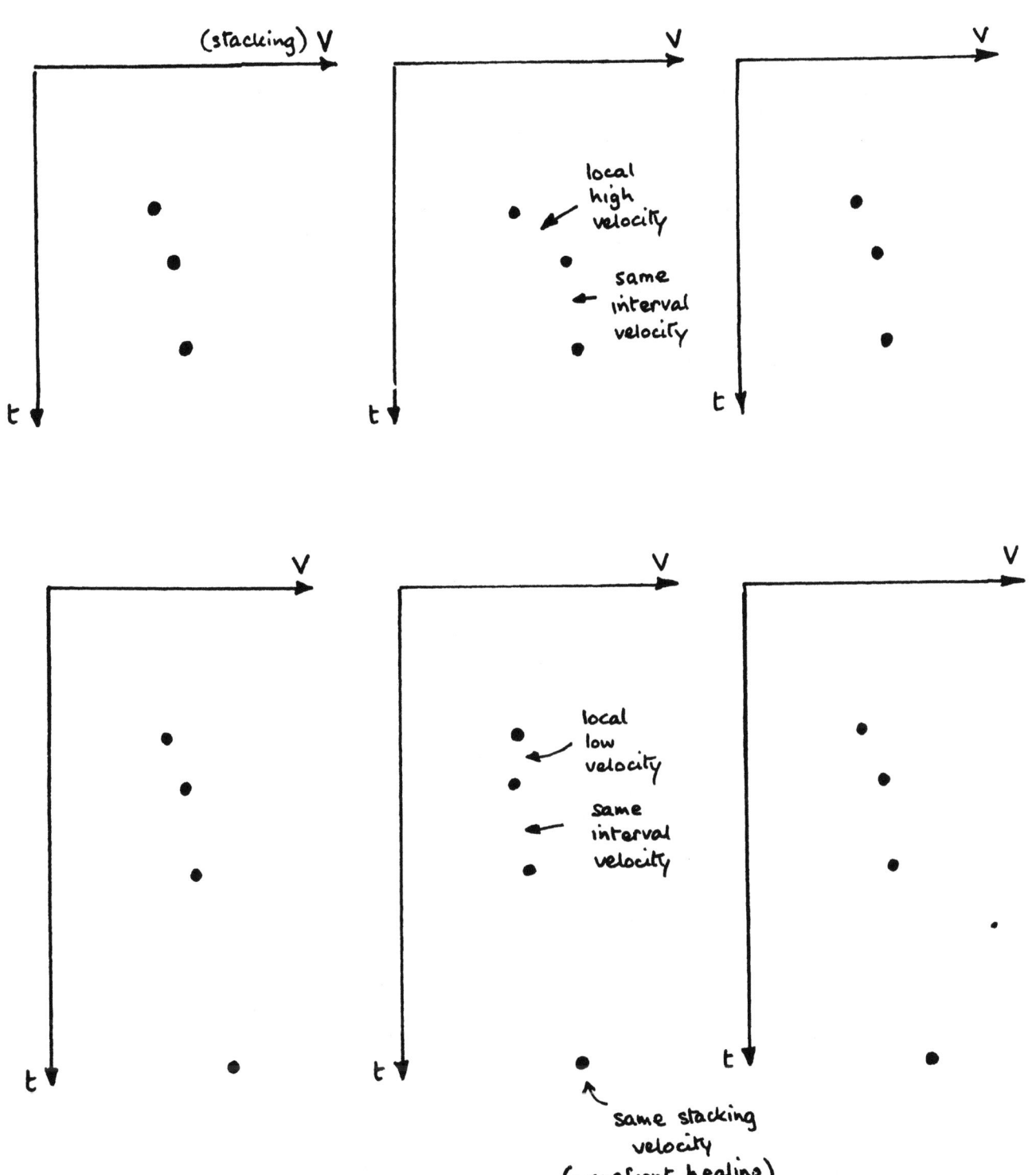

made (on account of interference, diffractions or noise) are distinguished by an absence of colour. We note that there is never any shame in saying "Here no valid measurement was possible."

215L We see clearly the reason for the negative reflection at about 1.1s in the centre of the section, which we previously discussed in connection with the polarity display on p.3-63. The reflection does indeed come from a major inversion of velocity (and probably of density).

On the matter of density, we note that no significant distinctions of interval velocity are detected to explain the strength of the reflection at about 1.4s (left edge). We are led to suspect a material of anomalous density in this region (probably salt).

If our concern is with interval velocities in thin layers, we may make the judgement that the results are not likely to be improved (and may be degraded) by correcting for refraction and dip in the layers above. If our concern is with depth conversion, and if dips exist above, then we probably would perform this correction; the matter is discussed more fully in Part 7.

If our concern is with interval velocities at shallow depth, we may need to apply a correction for deviation from a hyperbola (Al-Chalabi, 1973). The test is whether or not the stacking velocities obtained using only the outer traces of the gather are higher than those obtained using only the inner traces.

We recall that, in any case, the interval velocities we have computed are what we have called "gross" interval velocities. If the layers are thick or shallow, and if we are concerned to specify the velocity just above or just below an interface, then it is necessary to apply a compaction law (p.2-54). The sketch at the top of p.3-98 makes the point that, in an extreme case, this may change an apparent increase of interval velocity at an interface into a decrease.

TAPE 23 We have said earlier that, except for situations created by an anomalous density, there should be general _compatibility_ between the interval velocities derived from our _velocity picks_ and the reflection strengths observed on the true-

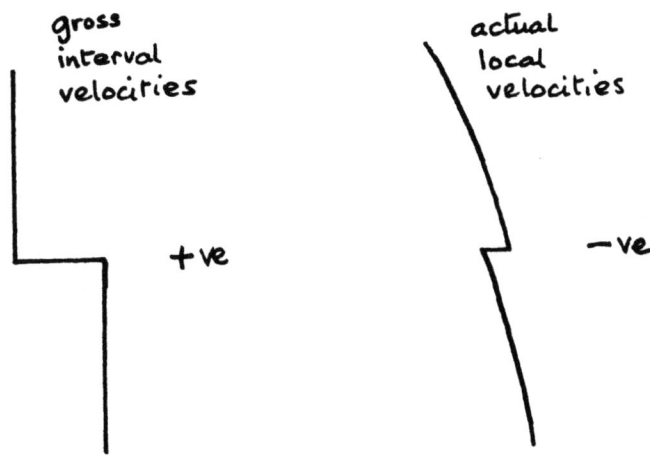

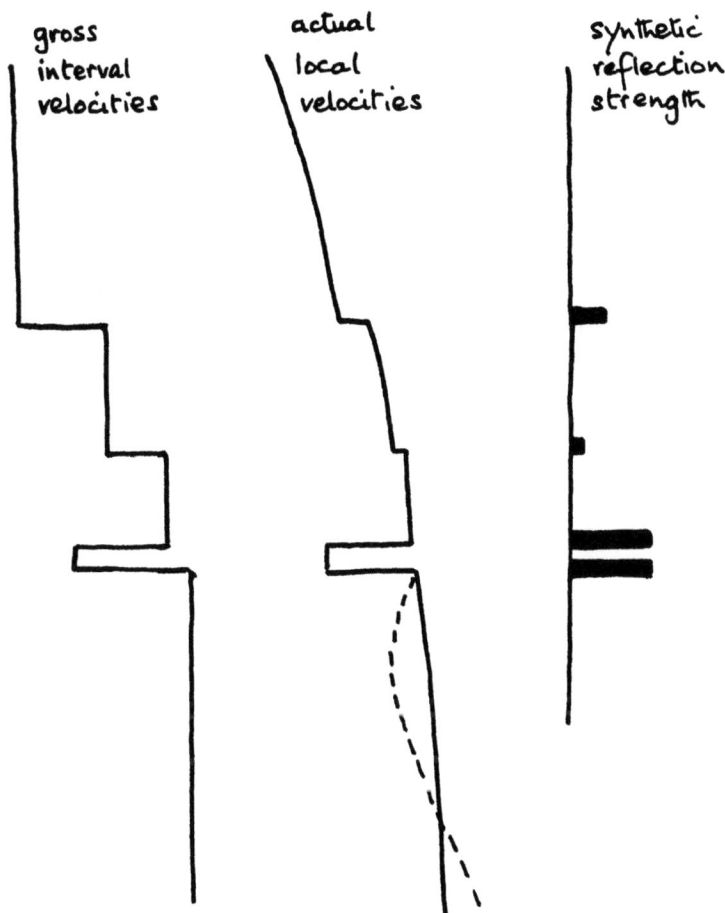

amplitude section. There are major limits on this, of course; thin beds may not allow an interval-velocity measurement, and may interfere to cause either strong or weak reflections. Yet where the layers are massive and the interfaces abrupt this general compatability should exist.

The presence or absence of such compatibility can be demonstrated by the construction of synthetic reflection-strength sections (Galloway, 1975). Following the course suggested in the lower sketch on p.3-98 we first construct the pattern of gross interval velocities at each analysis point; this might be the information displayed in the smoothed velocity section on p.3-62. Then we apply any knowledge we may have on local compaction laws (middle sketch). At this stage we may also introduce any density information we have; if we have nothing better than the $V^{1/4}$ relation this is not worth doing, since it just increases all reflection coefficients by a constant factor. Finally we compute the reflection coefficients at each analysis point, to derive the synthetic reflection-strength section (right sketch).

If the synthetic strength section does not match the observed strength section, we know that several possible explanations exist (including, for example, the transitional reflector shown dashed) and that these may be diagnostically useful; this topic is explored in section 4.5.2.3. If the synthetic strength section does match the observed strength section, we just smile.

PART 4

DIRECT

HYDROCARBON DETECTION

4.1 INTRODUCTION

Direct hydrocarbon detection is not really direct; it relies on the identification of an <u>acoustic contrast</u> associated with the presence of hydrocarbons. It follows from the material of Part 2 (and p.2-60 in particular) that identifiable acoustic contrasts are most likely if the hydrocarbon is <u>gas</u>. We remember (Question 5, p.2-73) that significant distinctions of reflection coefficient between oil-saturation and water-saturation can occur if the overlying material happens to have an acoustic impedance close to those of the reservoir rock in its water-saturated and oil-saturated states, but that in this case the actual values of the reflection coefficients are necessarily small, and measurement is therefore prone to error. Consequently the majority of our concern in this Part 4 will be with gas. That means, of course, <u>gas in the reservoir</u>; it says nothing (without additional information) about the results at the well head. It also means <u>free</u> gas — not gas in solution, and not gas liquefied by pressure. (In this context, we note that most of the useful constituents of natural gas cannot be liquefied by pressure at temperatures existing in the earth. In particular, methane and ethane cannot be so liquefied; propane, carbon dioxide and nitrogen may possibly be liquid in the earth, depending on the local temperature and pressure.)

4.2 THE CRITERIA FOR DIRECT RECOGNITION OF A GAS ACCUMULATION

In the literature (particularly in trade journals), there have been given several lists of criteria which are dangerously oversimplified. It is important to remember that the acoustic consequences of a gas accumulation depend on the overlying material, the porosity, the depth, the water-saturation and the reservoir configuration; further, these circumstances may be such that, in any given case, a particular criterion may or may not be satisfied. However, everything is calculable; black magic is neither appropriate nor necessary.

Let us discuss the seven criteria in turn.

4.2.1 The gas-liquid contact

This is certainly the indicator nearest to direct detection. The gas-liquid contact is always best seen at compressed scale; contacts which are quite clear at squashed scale are often not noticeable at natural scale, and this is one of the reasons why geophysicists were long sceptical about what is now observed as a fairly frequent phenomenon.

The gas-liquid contact must necessarily be a positive reflector.

Despite the popular tag "flat spot", it is obvious that a gas-liquid contact need not necessarily be either flat or horizontal in time. It may also not be quite horizontal in depth; when this happens the most likely explanation is probably minor faulting, but occasionally, in a simple trap, it may be related to changes in the nature of the permeability. In any case, obviously, our conclusions on this matter should be reported to the reservoir engineers.

Because of the importance of fluid contacts, any suggestion of an anomalous near-horizontal event in a possible trap location must be investigated very carefully. Techniques for enhancing such events are set out in section 8.5.6, and for being sure that they are real in section 4.5.

4.2.2 Anomalous reflection coefficients

Of course, at a superficial level our attention is attracted by anomalous reflection amplitudes. But we must always remember that the visual effect of amplitudes is subject to display gain and to frequency content (low-frequency events seem larger); further, anomalies can appear only by reference to the amplitudes of other events, which are at best irrelevant and at worst differently processed. It is only by the quantitative calculation of reflection coefficients that a proper measure of an amplitude anomaly can be obtained. Many dry holes exist as testimony to this fact (Post-mortem comment: "That's odd — it looked bright to me").

As we remember from sections 2.7.1 and 3.4.5, the most generally likely cause of large local anomalies of reflection amplitude is interference. The reflections from the top and bottom of a thinning unit of marked acoustic contrast can interfere, to produce at one separation an increase to approximately double the amplitude of top or bottom alone, and at another separation to produce vanishing amplitudes; certainly amplitude variations of 4:1 along such a unit are commonplace. Many dry holes exist as testimony to this fact also (Post-mortem comment: "That's odd — it was a nice little up-dip pinch-out, and bright as bright could be").

However, with that said, it is also true that major anomalies of reflection amplitude are often associated

with gas accumulations. In the case of a shallow gas sand
encased in shale, the anomaly takes the form of large
reflection coefficients (a "bright spot"); in the case of
a shallow carbonate reservoir encased in shale, the anomaly
may take the form of a small reflection coefficient (a "dim
spot"). As stressed above, it is very important not to
oversimplify this matter (to say, for example, that bright
equals sand, or dim equals carbonate), but to remember that
the nature and amount of the anomaly depends not only on the
reservoir material but also on the depth, the porosity, the
water saturation and the encasing material. A reflection
does not come from a rock — it comes from a contrast between
two rocks; it is distressing to read in the trade literature
that "a gas-filled sand is a better reflector than is a water-
filled sand". Such oversimplifications must surely lead to
dry holes (Post-mortem comment: "That's odd — we were
expecting to hit gas, but we hit the water").

4.2.3 Anomalous low velocities

It is always true that, other things being equal, the
presence of commercial gas must produce a significant depression of velocity in the reservoir rock, relative to the water-
saturated condition. The converse is not true; as we saw on
p.2-62, a significant depression of velocity in the reservoir
rock does not always indicate commercial gas (Post-mortem
comment: "That's odd — all we're getting is water, with a
little gas").

The depression of velocity may be measurable directly
as a low interval velocity if separated reflections are
obtained from the top and bottom of the gas-saturated zone.
Or a local depression of stacking velocity may be evident on
a reflection just underneath the zone; however, in section
6.4 we shall learn to be very cautious of end-effects in this
connection (Post-mortem comment: "It was bright enough, but
we missed drilling it because the velocity seemed to be high").
Or the consequences of the low velocity may be observed as a
pull-down of reflection times at the bottom of and just below

the zone; however, there is inevitably an ambiguity between velocity pull-down and genuine relief. Again it is important to stress that an identified effect must be <u>quantitatively</u> plausible; for example, a pull-down or a velocity pick is not defensible if it implies an interval velocity of 1000 m/s, even at shallow depth.

4.2.4 Inversions of polarity

If (and only if) the reservoir material in its water-saturated state has an acoustic impedance greater than that of the overlying material, and if (and only if) the local replacement of water by gas depresses the acoustic impedance of the gas-saturated zone to be <u>less</u> than that of the overlying material, then (and only then) does the reflection from the top of the reservoir show a polarity inversion over the gas (section 2.4.1). It is obvious that these conditions may or may not happen in typical gas accumulations; consequently the existence of a polarity inversion is to be taken as <u>corroborative</u> evidence for gas, but not as <u>necessary</u> evidence. Again it is important to remember the significance of the overlying material in contributing to the reflection response.

In practice, the identification of polarity inversions is often made difficult by interference; there is also a major problem of distinguishing such inversions from small faults (section 4.5.2).

4.2.5 "Shadows"

First we must remember that incorrect processing (specifically, base-levelling and similar techniques) can produce a general loss of amplitude below — and above — bright spots (pp.3-25 to 34). We remember also that bright spots, in their nature as strong reflectors, generate trains of multiples which increase the general signal level at later times. The processing weakness can be overcome, by avoiding data-dependent amplitude manipulations, but the multiples must inevitably tend to mask any amplitude shadow which may truly be present.

Then we must remember that (as we computed in Question 3, p.2-72B) the two-way transmission coefficient through the top and bottom of a gas-saturated reservoir is only about -1½dB, which is scarcely measurable on real data. From simple gas reservoirs, therefore, we do not expect an amplitude shadow to be obvious. If such a shadow is obvious, and if it is due to transmission coefficients, we conclude that multiple gas reservoirs are present.

However, there remains the possibility of shadows due to absorption in the frictional reservoir material; in contradistinction to transmission coefficients, this is frequency-selective, introducing a loss of amplitude strictly by loss of high frequencies. We computed (Question 2, p.2-81) that such absorption in a realistic sand 25m thick might well be undetectable unless the seismic spectrum could be extended above 50Hz. If an absorption-type shadow is visible on a normal section, therefore, we infer the presence of a considerable total thickness of sand. In this case, we note, we know nothing about the number of sands.

The two shadows — a simple amplitude shadow and a frequency-selective amplitude shadow — therefore have different (and useful) meanings. But the problem of distinguishing between them is not trivial (though, since their message must be good, we would wish we had the problem more often).

It is essential that we do not confuse a frequency-selective shadow (observable on all reflections below the gas zone until wavefront healing removes it) with a "low-frequency tail" on the bright-spot complex itself. The latter phenomenon is quite different in nature, and we return to it in section 4.3.

4.2.6 Diffractions

The generation of diffractions requires, first and foremost, significant contrasts of acoustic impedance;

these are provided in any situation which generates a bright spot. However, it also requires favourable spatial rates-of-change of acoustic impedance; the later illustrations of section 2.8.5 are examples of this, and we shall see others in Part 7. As a consequence, significant diffractions are to be expected in association with gas reservoirs where the reservoir is thick and the trapping mechanism is a fault, or in other situations where the lateral limits of the reservoir are abrupt. However, they are not to be expected in the case of a lenticular sand, whose thickness decreases smoothly to zero in all directions. Reefs, of course, may represent a middle situation (sometimes with diffraction-generating steepness on one side but not on the other).

4.2.7 The inter-relation of these criteria

The most useful of the above criteria are no more than the observable manifestations of one single property of a gas reservoir — a locally depressed acoustic impedance. As a consequence, these criteria must all be mutually compatible. For example, a large positive reflection from the top of a good gas reservoir is unthinkable; so is a clear polarity inversion without a major depression of interval velocity within the reservoir. The demonstration of compatability is a powerful means of validation, which we shall discuss in section 4.5.2; the vehicle for this demonstration is the model, which we shall discuss in section 6.2.

4.3 INFERENCES FROM PUBLISHED ILLUSTRATIONS

At this stage it is beneficial to review all the published bright-spot illustrations given in the references to Part 4, and thereby to become familiar with the general appearance to be expected of shallow gas accumulations in a Tertiary sand-shale sequence. A selection of these illustrations is reproduced (with acknowledgement and thanks to the authors and their companies) in the following pages.

From these illustrations we can exemplify several conclusions which, even if not general, are widely observed.

1. The clearest bright spots exist at times less than about 1¾ - 2 seconds, which at Tertiary velocities typically represents a depth of 2000m (say about 6500 ft). This is entirely in accord with the velocity-depth curves of p.2-60.
However, it must be remembered that the illustrations probably come from the same area as the measured velocity-depth curves, and that significant variations (either up or down) may be expected from area to area.
Indisputable Tertiary bright spots have been observed at a time of 2.7s, but in a definitely low-velocity section.

2. Fluid contacts are observed, though obviously only when the pay zone is thick. Fluid contacts have been interpreted on sections of pp.4-12, 13, 14, 15 and 19.

3. In the identification of fluid contacts (and reservoir boundaries generally), we are often concerned with fairly subtle indicators of interference. These indicators (section 2.7) are just as likely to appear in the negative half of the seismic waveform as in the positive half, and it follows that highly asymmetrical types of display which accentuate the peak relative to the trough (for example, variable-area-wiggle) are dangerous. If the interpreter is forced to work with such displays, he must call for sections with both display

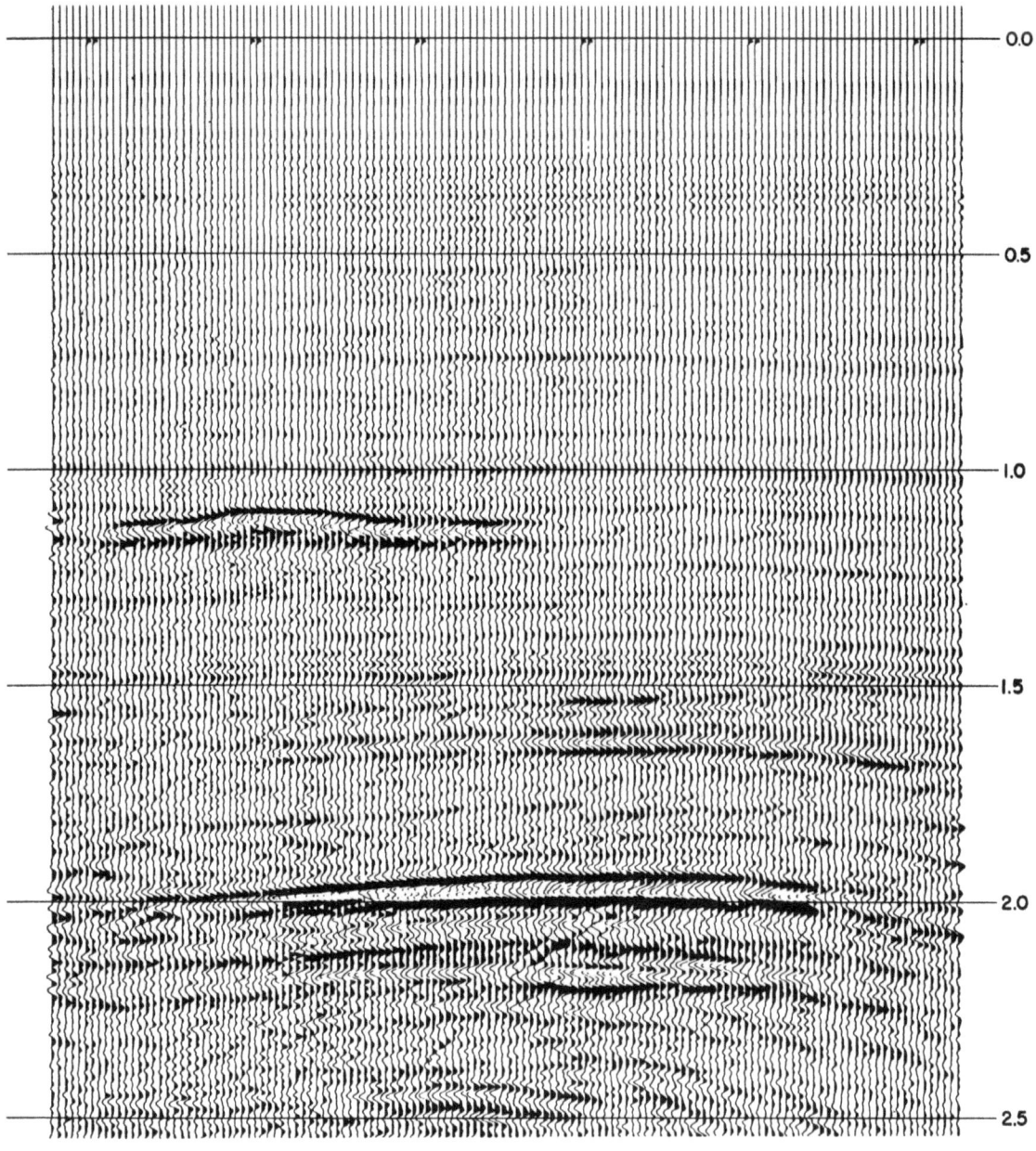

Courtesy of Barry and Shugart, Teledyne Exploration Company, 1973

4-9

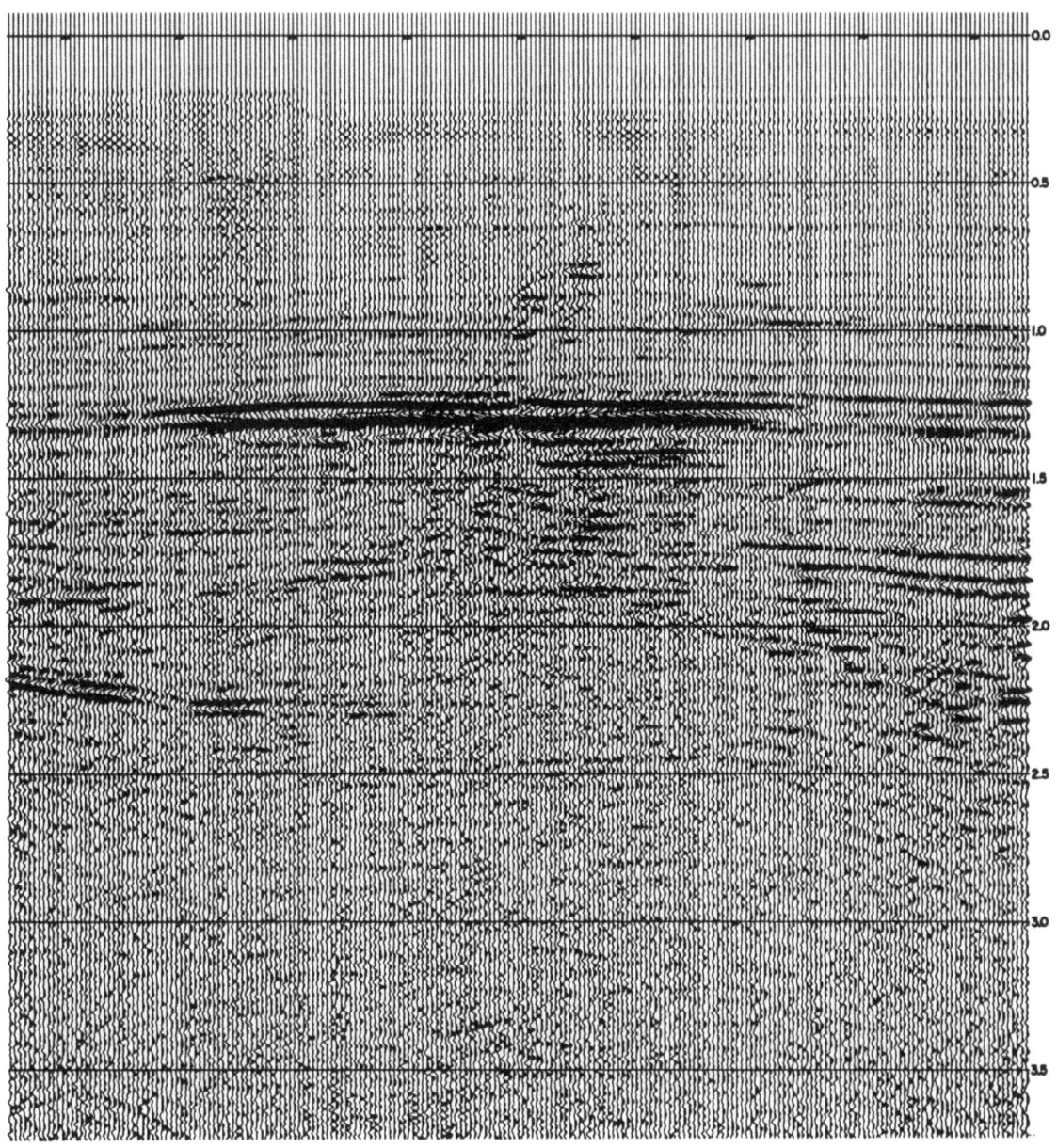

Courtesy of Barry and Shugart, Teledyne Exploration Company, 1973

4-10

317

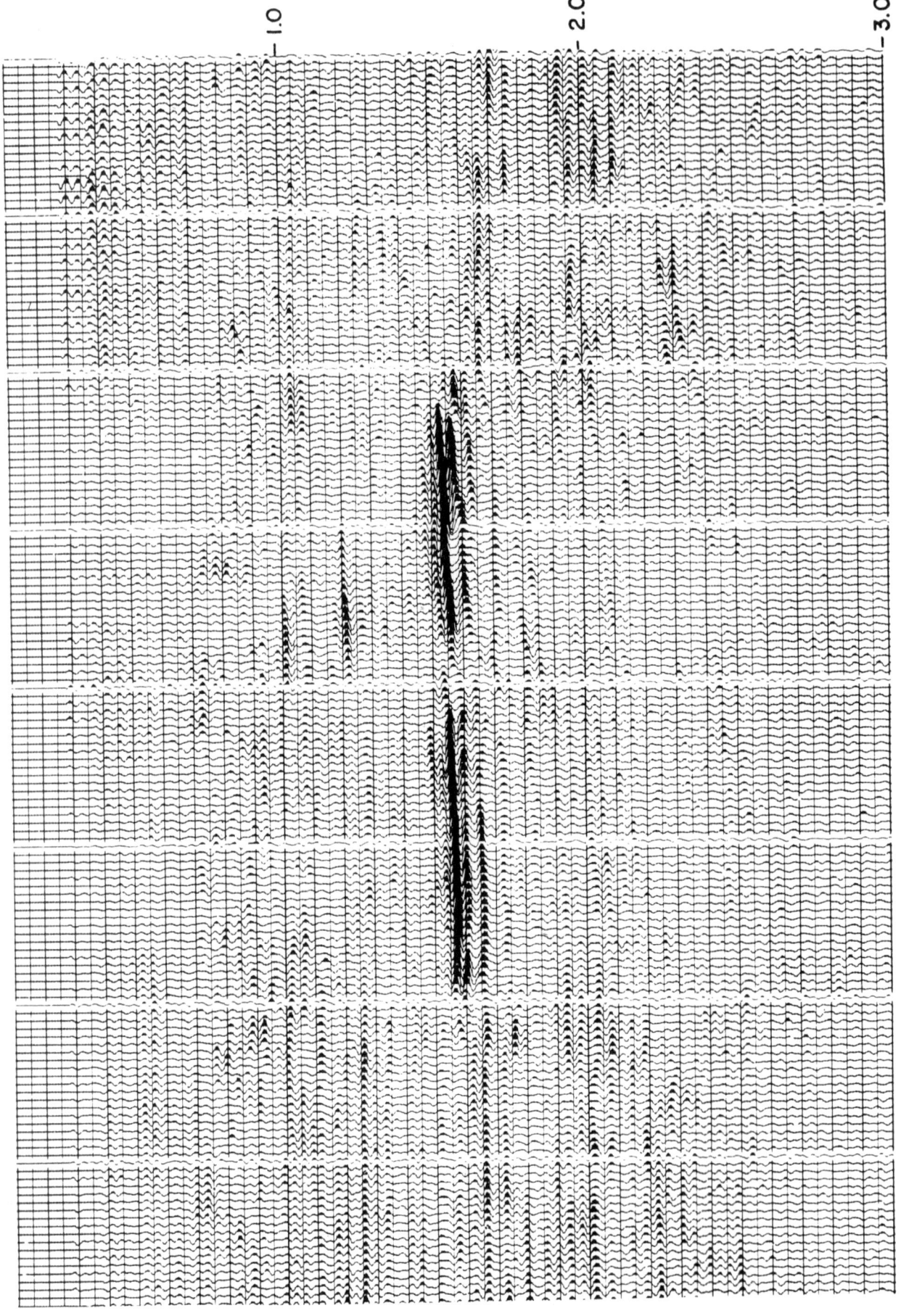

Courtesy of W.T. Prescott, Continental Oil Company, 1973

4-11

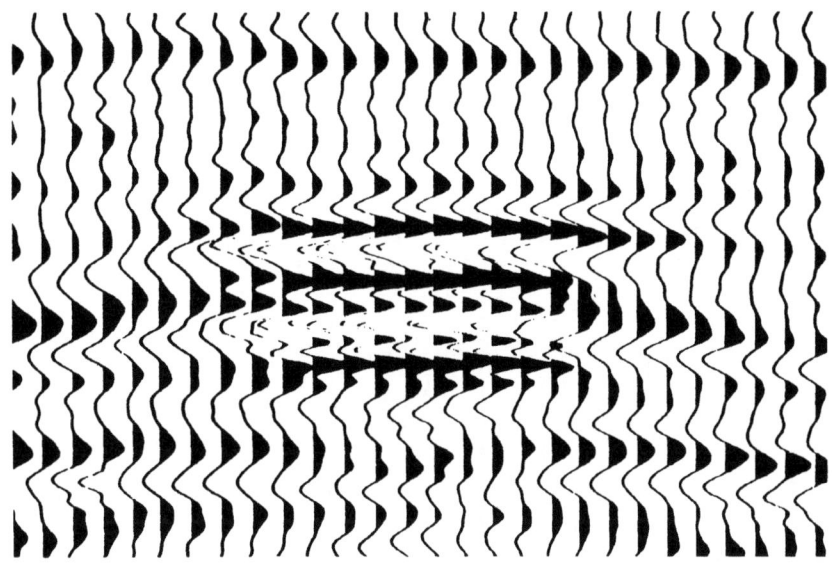

Courtesy of C.H. Savit, Western Geophysical Company, 1974

4-12

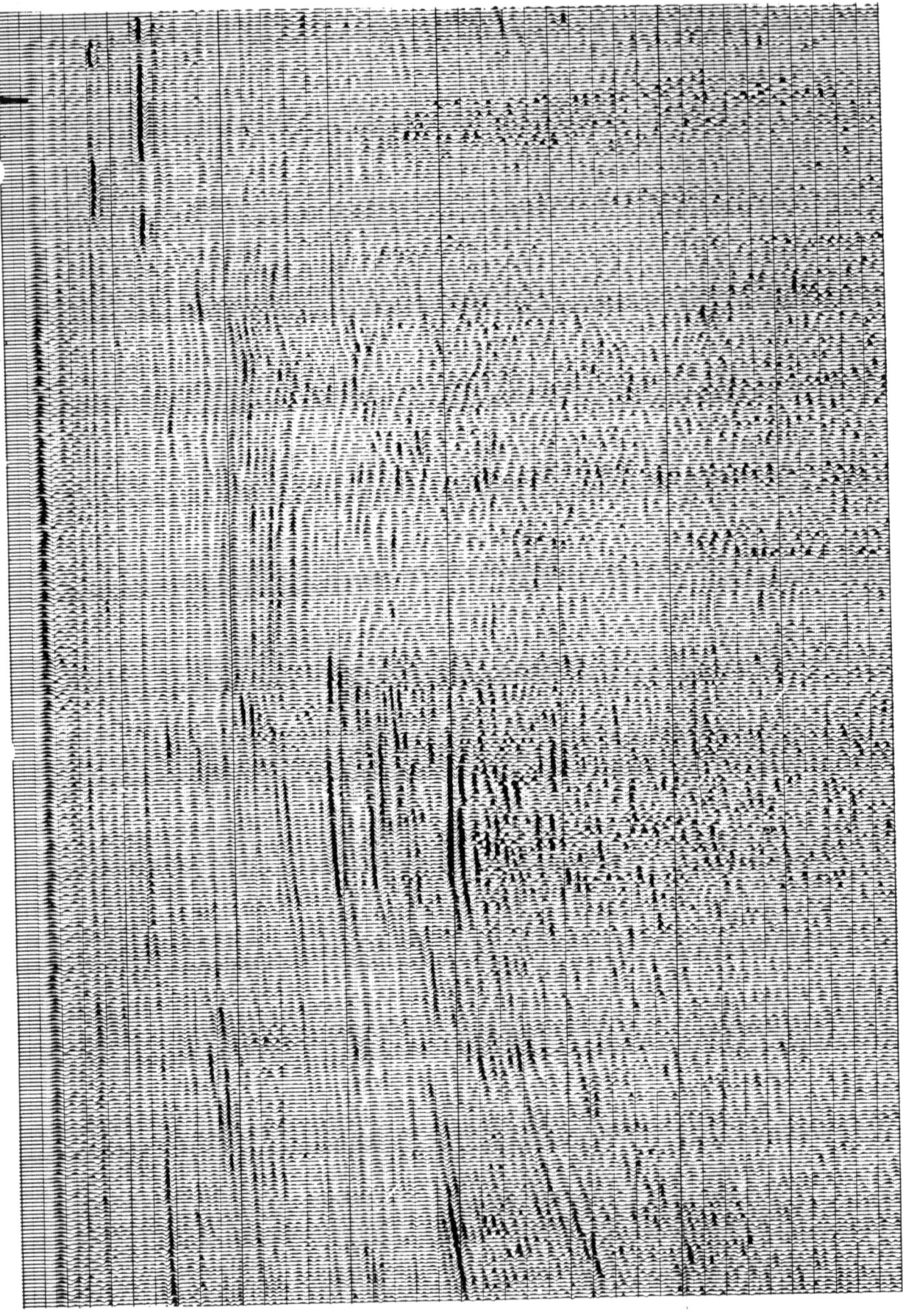

Courtesy of D.S. Paige and of Dresser Olympic, 1973

4-13

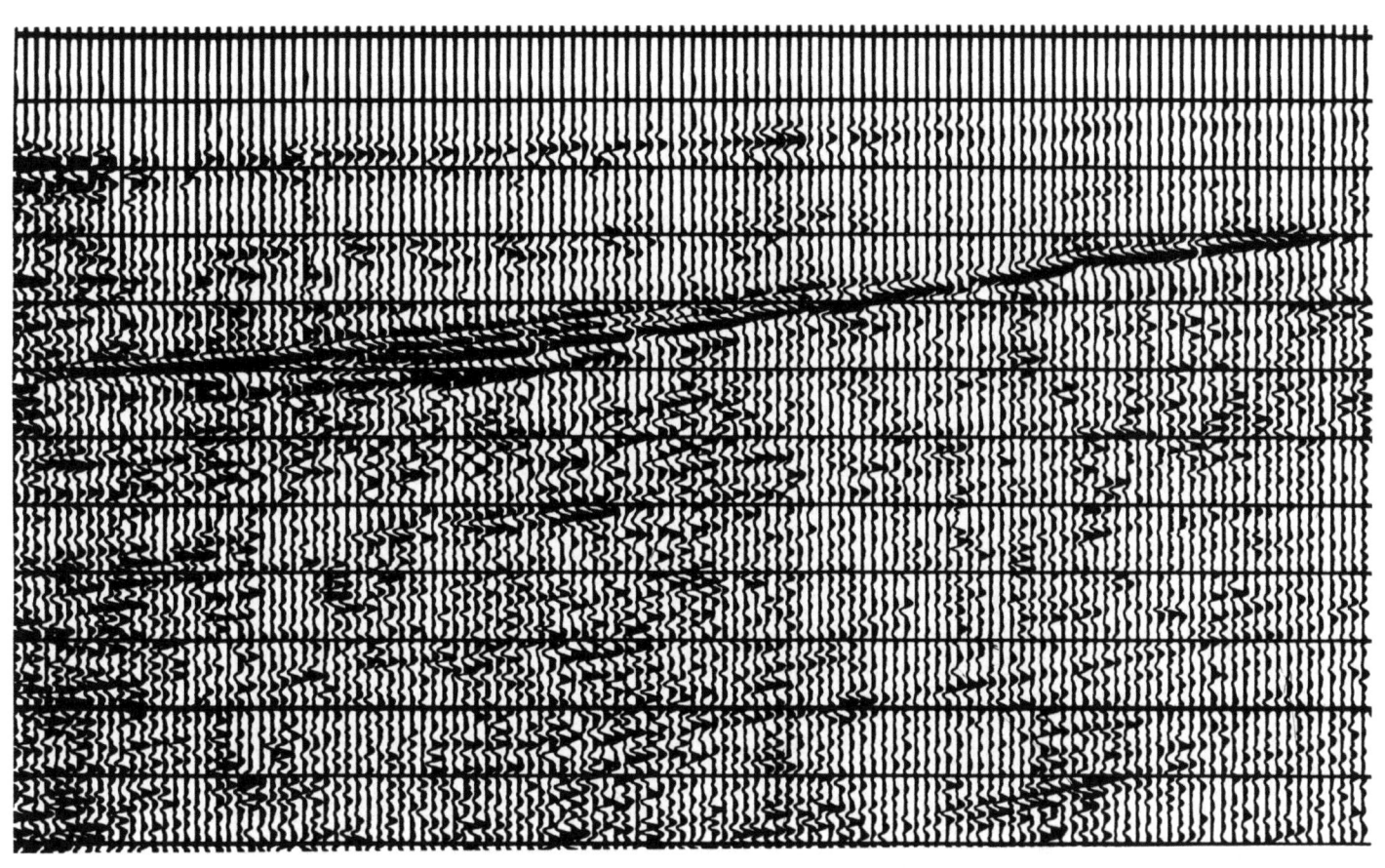

Courtesy of D.S. Paige and of Dresser Olympic, 1973

4-14

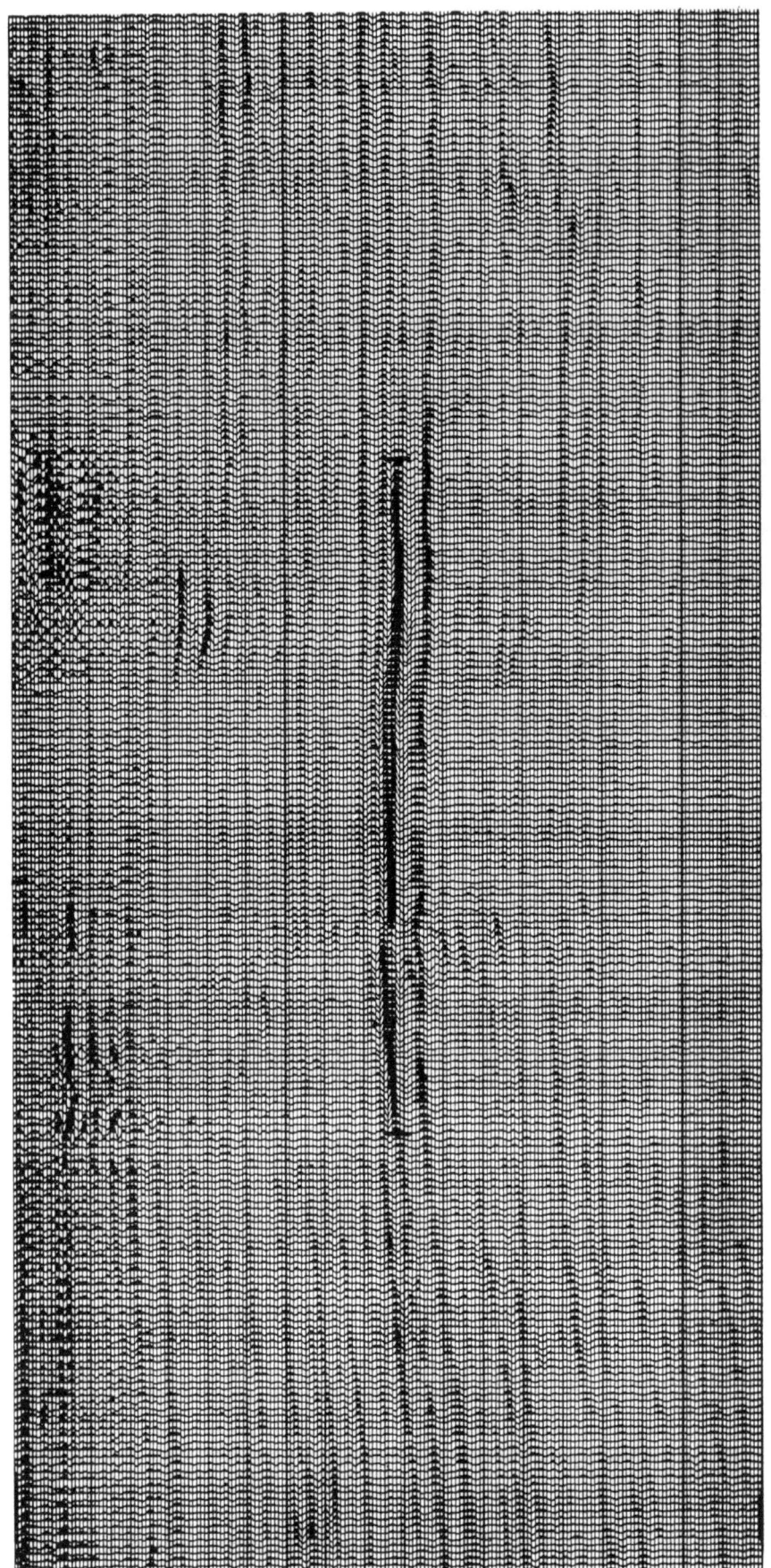

Courtesy of Larner, Mateker and Wu, Western Geophysical Company, 1973   4-15

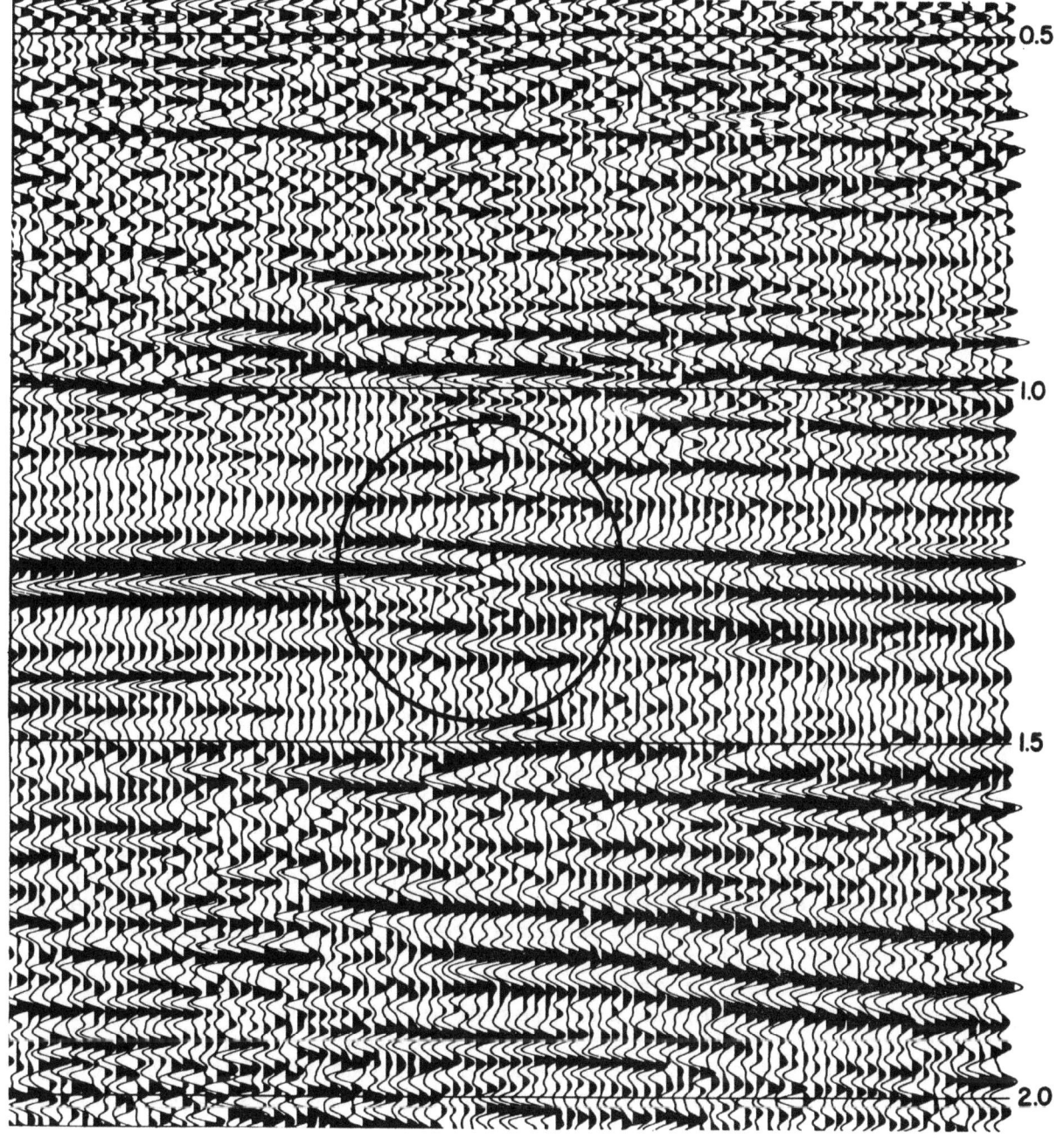

Courtesy of Barry and Shugart, Teledyne Exploration Company, 1973

4-16

orientations; since this constitutes a major additional **inter**pretational burden, it is better to use a suitable **symmetrical** or near-symmetrical display (for example, variable-area without significant bias) for bright-spot evaluation. This comment is restricted to situations where the subleties of the waveform are important; it does not apply to situations where the problem is signal-to-noise or statics, for which variable-area-wiggle or strongly-biased variable-area is preferred by many interpreters (see also pp.8-29, 30).

The dangers of variable-area-wiggle in the context of fluid contacts are exemplified by comparison of the sections on pp.4-19, 20. The circled event on the first of these sections is interpreted as a fluid contact; it is present, but might well have been missed, on the reversed display of the second. Many similar illustrations of this danger have been oberserved; it is disturbing to realize the extent to which our interpretations depend on the section display.

4. Polarity inversions are observed occasionally; as discussed in section 4.2.4 above, we expect no more than this. The circled feature on p.4-16 is interpreted as a polarity inversion, as are some of the features on p.4-12.

5. There is a major hazard associated with the identification of a "top" and a "bottom" of a gas accumulation. Sometimes, of course, the reservoir is so thin (say less than 20 or 30ms) that any such identification is obviously impossible. On the other hand, sometimes the reservoir is thick (perhaps even with a clear fluid contact), so that the identification can be done with assurance. But in the more usual intermediate cases, where there still remains a major exploration incentive to identify the two reflections separately if possible, such identification is hazardous. This separation is, of course, a matter of increasing the vertical resolution (section 8.4.1); in the present context it is achieved by source signature correction, by muting the stack to avoid significant nmo stretch, and by extreme statistical deconvolution (though, as we have learned, the last process would be used

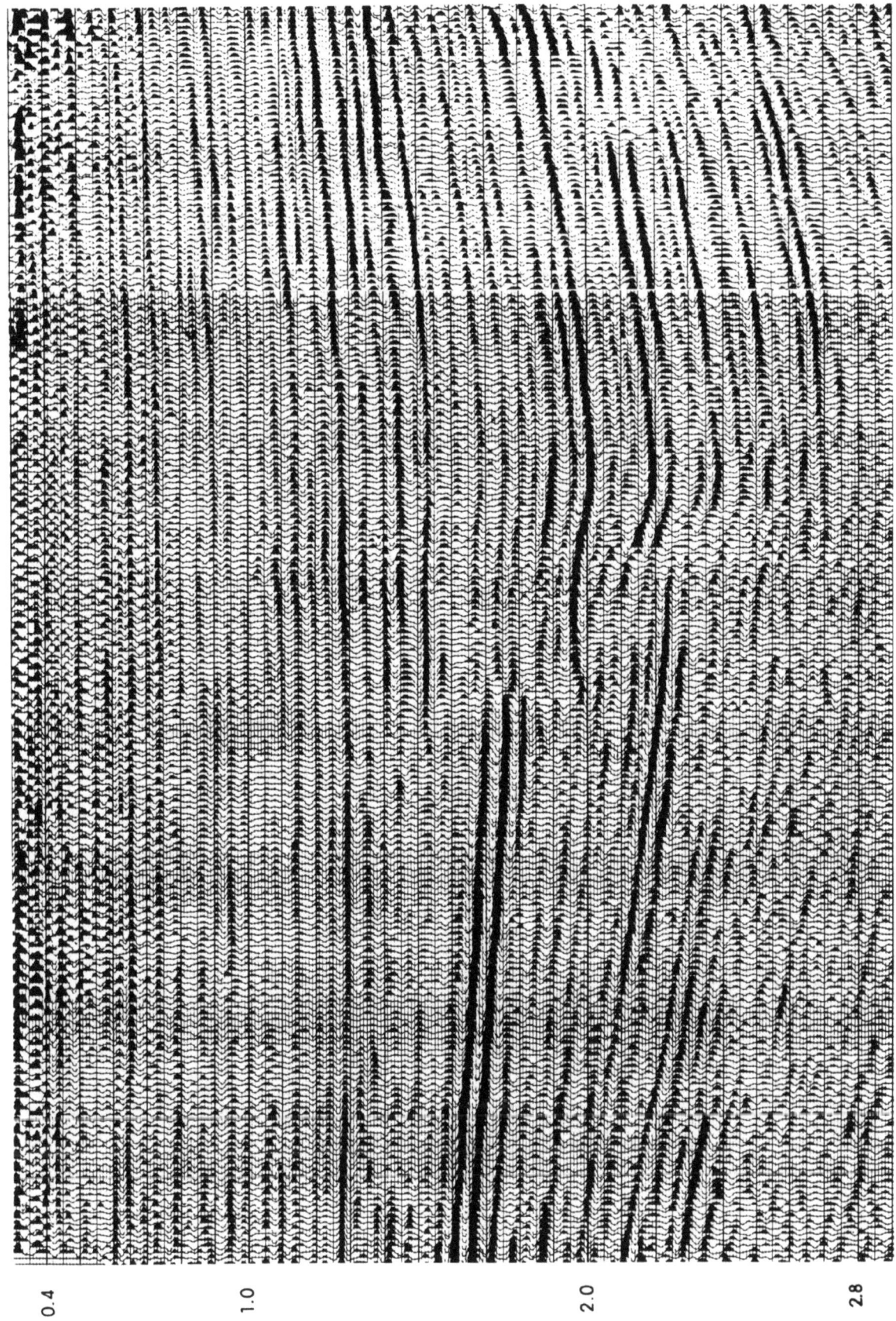

Courtesy of Larner, Mateker and Wu, Western Geophysical Company, 1973

4-18

325

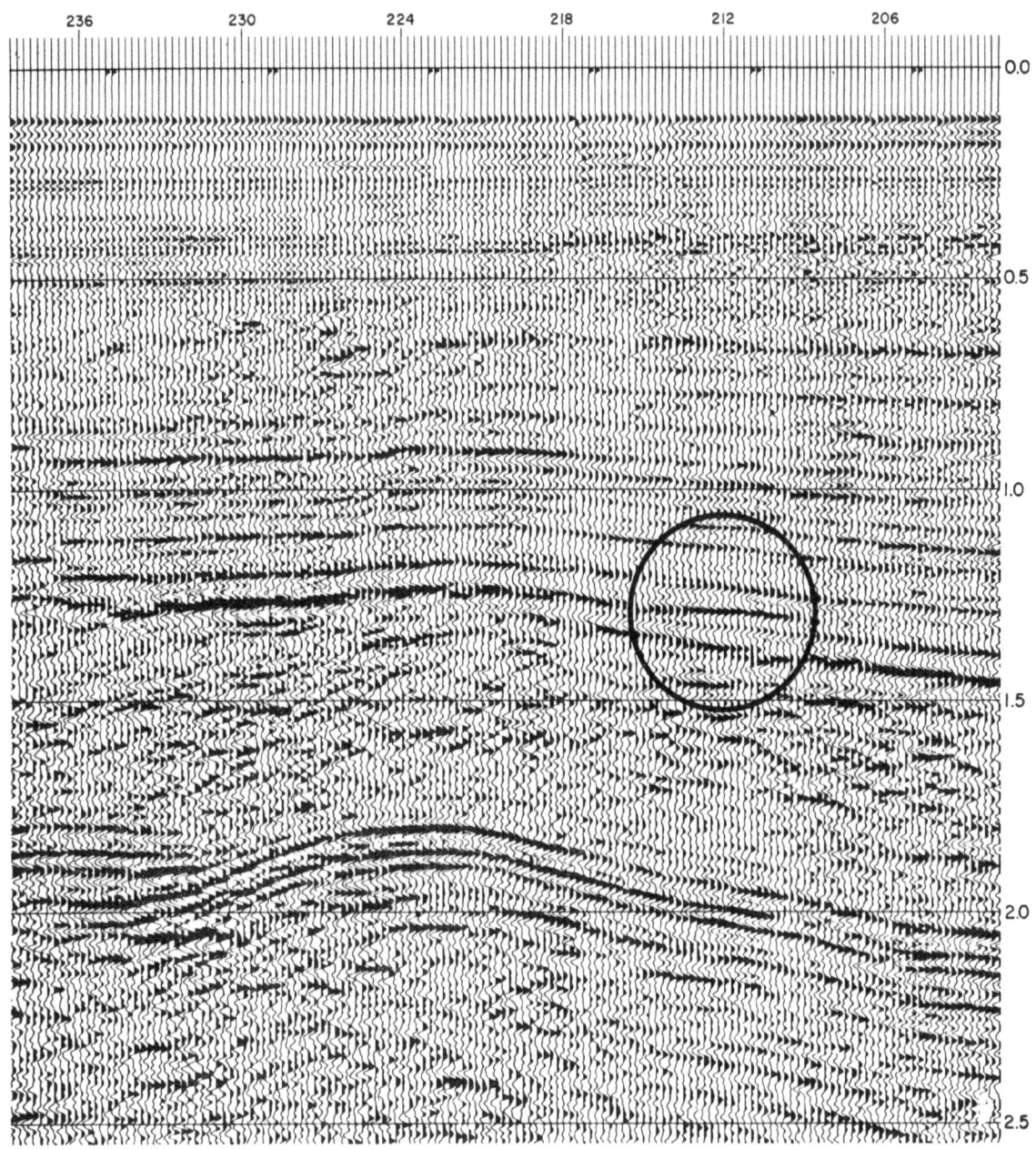

Courtesy of Barry and Shugart, Teledyne Exploration
Company, 1973

4-19

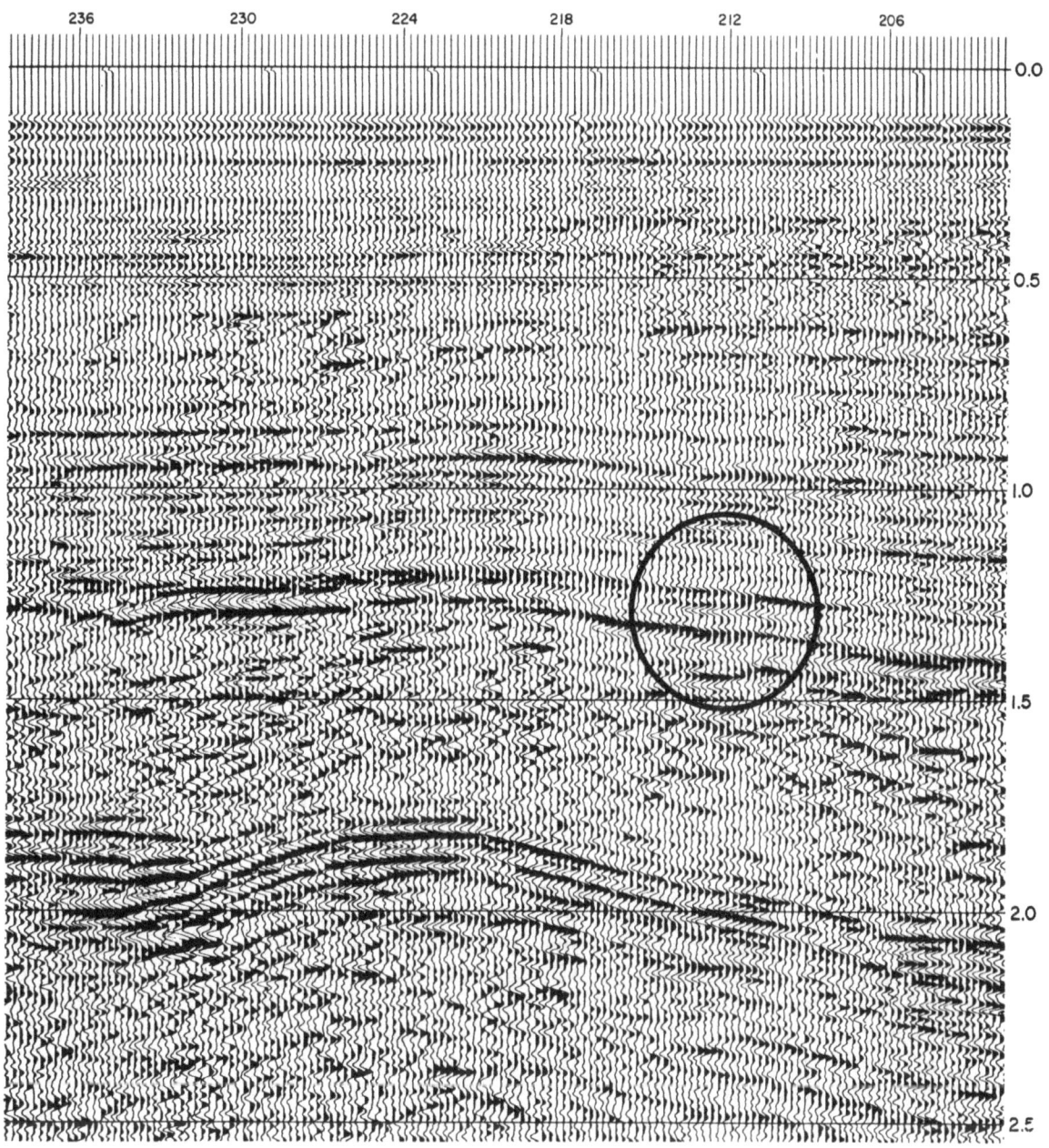

Courtesy of Barry and Shugart, Teledyne Exploration Company, 1973

4-20

recognizing the reservoir boundaries and understanding the reflection complex — not for purposes of accurate amplitude measurement).

6.   The section on p.4-22 is an example of a dim spot associated with the presence of commercial gas in a carbonate reservoir encased in a sand-shale sequence. In this example, there may also be a polarity inversion at the gas-liquid contact (about nine traces to the right of SP 420).

7.   Sections 4-10 and 4-13 are illustrations of the situation where the generation of stronger multiples far outweighs any amplitude "shadow" below the bright spots. Discarding also those sections which show clear evidence of base-levelling or other data-dependent amplitude manipulations, we come to the conclusion that a shadow is visible only on section 4-12 (of which we do not have sufficient to be completely sure). It is possible that both the simple amplitude effect (transmission coefficients) and the high-frequency-cut effect (absorption) are present; the conclusion — that there exist multiple gas zones of considerable total thickness — also appears feasible. We notice that in this case there is no suggestion of amplitude weakness above the bright spots. In this type of analysis we must be on our guard against the more sophisticated types of agc, which may not be detectable on the section; the only real solution is to ask penetrating questions of the processors.

8.   Section 4-12 also suggests a velocity pull-down, observable on reflections below the postulated gas sands.

9.   No obvious diffractions are associated with any of the fluid contacts illustrated. Further, many of the bright spots classify themselves as lenticular sands by their freedom from diffractions (section 4.2.5 above). Such diffractions as are visible (for example, that on p.4-9) are compatible with fault interpretations. Section 4-18 is included as a reminder that even very strong (and probably gas-indicative) reflection complexes can be faulted without obvious diffractions; probably, cementation from the fault conduit has introduced a gradual decrease of porosity toward the fault. Reefs, as we have noted, may have diffractions, or none, or a diffraction on one side only.

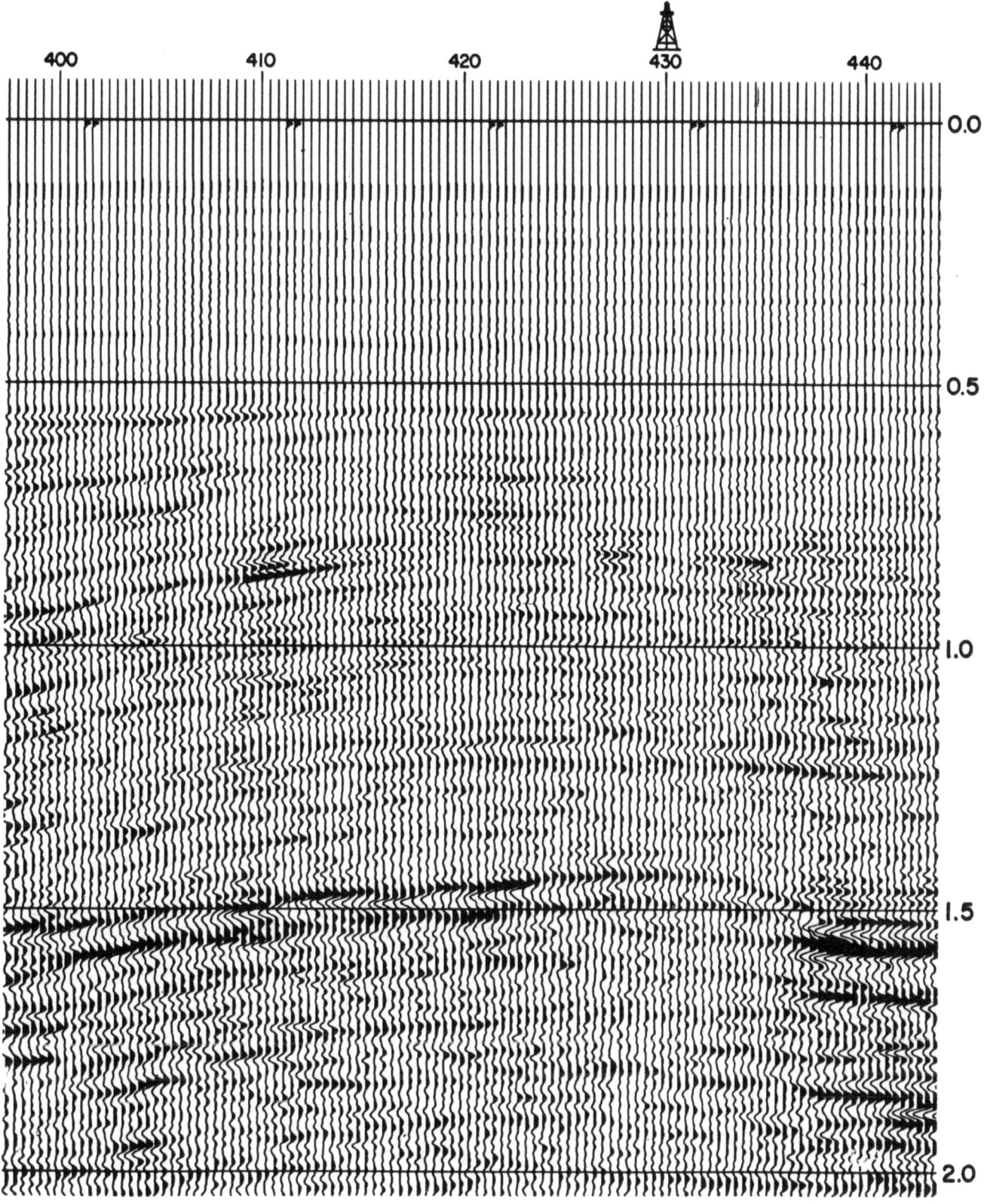

Courtesy of Barry and Shugart, Teledyne Exploration Company, 1973

4-22

10. In section 4.2.4 above, we made an important distinction between a frequency-selective effect observable on reflections 236(B) below the bright spot and a "low-frequency tail" on the bright spot itself. Of the illustrations we have just been considering, only section 4-12 has any suggestion of the first effect. Many of them, however, show a low-frequency tail. Since there have been efforts to equate the low-frequency tail with hydrocarbons, it seems wise to explore this further.

The first explanation for a low-frequency appearance to the later cycles of a bright-spot complex has to be improper stacking. Standard practice for the picking of stacking velocities for a reconnaissance or semi-detail survey often prohibits the correct pick for the bottom of the reservoir if the interval velocity within the reservoir is anomalously low. If the correct pick is not made, only the low-frequency components stack well.

The second explanation has to be inappropriate deconvolution. Spiking operators of inadequate length are known to produce low-frequency tails; they do this on all reflections, of course, but the effect becomes visually obvious only on the very strong reflections such as those associated with the bright spots. The old practice of preserving the decon operators on the bottom of the section had merit in this respect.

The third explanation notes that, in a structural gas accumulation, the strong reflector at the top of the gas and the strong reflector at the gas-liquid contact have a maximum separation at the crest. Toward the edge of the gas the complex must have a high-frequency appearance (section 2.7.1); at the crest the separation may be such as to give a low-frequency appearance. This explanation, then, does require a gas-liquid contact, and therefore gas. However, the low-frequency appearance is not really the indicator; it appears fortuitously, as a function of the gas column and the pulse bandwidth. The real indicator is the gas-liquid contact.

4-23

The fourth explanation is concerned with the particular overpressure phenomenon discussed on p.2-22. When a porous and highly permeable reservoir material is deposited on a thick shale, the shale compacts not only in the usual manner, from the bottom up, but also from the top down; water excluded from the upper zone of the shale passes through the permeable material, and thus allows the upper zone of the shale to become compacted and impermeable. In many cases this results in the sealing of the middle zone of the shale into an overpressured condition. Consequently, the log of acoustic impedance as a function of depth has the form indicated in the sketch; in this it is assumed that the overlying reservoir material is a gas-saturated porous sand.

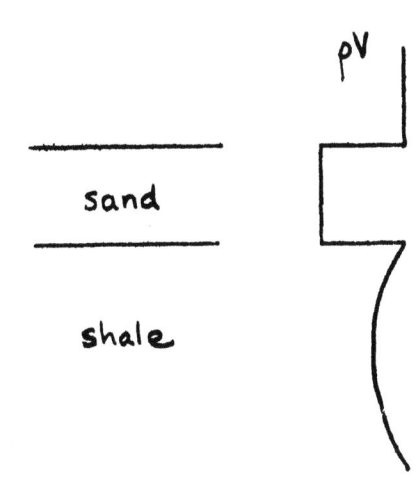

The seismic effect of the transition between the top zone of the shale (normally-pressured) and the middle zone (overpressured) depends on the degree of overpressure and the depth range over which it occurs. In practice, transition zones extending over as much as 100m are encountered. Such transition zones generate what is in reality a new (third) reflection, which is seen only at the low frequencies. The actual form of the bright-spot complex then depends on the mutual interference between the top-sand reflection, the bottom-sand reflection and the transitional third reflection; in general, clearly, the complex must show a low-frequency tail associated with the transition zone.

237

238(B)   The low-frequency tail, therefore, may be an indicator of an overpressured shale under the reservoir material. In a sense, then, the low-frequency tail is a good sign, because it means that the reservoir material has provided a path for primary migration; however, in itself it does not prove that hydrocarbons were present in the excluded water, nor whether any hydrocarbons were (or are) trapped in the reservoir material.

In summary, we have to say that a low-frequency tail is not in itself a hydrocarbon indicator. In particular, it is not a proof of anomalous absorption. Absorption requires that we see the high-frequency loss on a succession of reflections (of depth range compatible with wavefront healing) below a bright-spot complex.

4.4 POROSITY ESTIMATION IN GAS RESERVOIRS

First we remember the clear conclusion of section 2.3.6: that the porosity of a gas reservoir <u>cannot be estimated from velocity alone</u>. This is so for two independent reasons — the absence of a formula uniquely relating velocity and porosity in the gas-saturated rock, and the problem of water in the gas (p.2-62).

Therefore porosity estimation is best tackled through <u>density</u>.

4.4.1 Porosity estimation using the top-and-bottom method

The basis of the method is set out in pages 2-96B to 2-98B; the computational steps and the assumptions they imply are listed in pages 4-27 to 4-30.

We remember from the curves on p.2-98A that the top-and-bottom method is not suitable when the acoustic impedances of the materials above and below the gas accumulation are approximately equal. Let us take as our example, then, a marginal case which is also a very important one — that of a reservoir containing both gas and liquid, and overlain by shale. The "top" is the interface between the shale and the gas-saturated reservoir, and the "bottom" is the fluid contact. The success of the method then depends on the ratio of the acoustic impedances in the overlying shale and the liquid-saturated reservoir.

Let us consider the example of p.4-26, after Albright. We will do this, and accept the interpretation of the lower diagrams, without being concerned at this stage whether that interpretation is actually correct; all we need for present purposes is an illustration to fix the reservoir situation in the mind.

Since the depth of this example is very small (about 600m) we have a chance that the velocity and the acoustic impedance

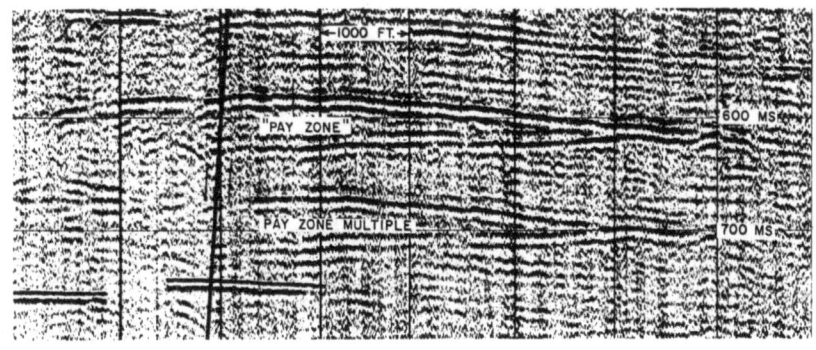

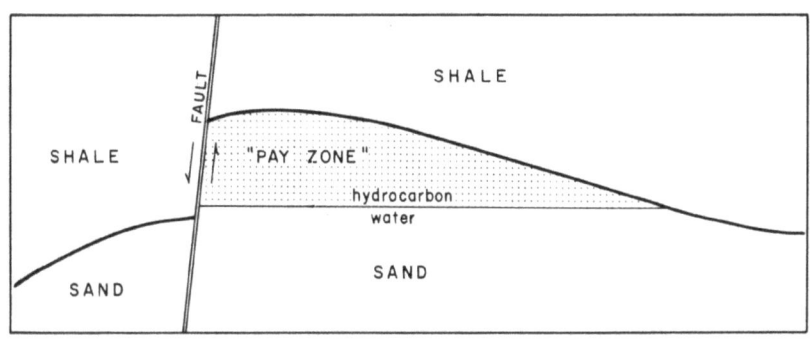

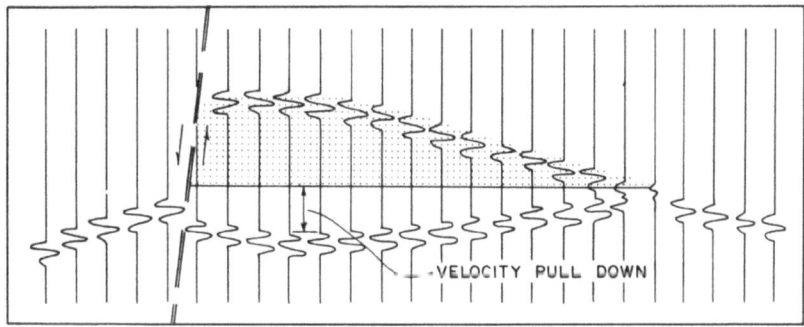

4-26

TOP - AND - BOTTOM METHOD

METHOD STEPS	ASSUMPTIONS
1. Identify top and bottom reflections.	That a gas-saturated reservoir is plausible in the location postulated.
	That the depth is such that the acoustic consequences of gas-saturation can be expected to be significant.
	That the top and bottom reflections are discrete and correctly identified.
	That the top-and bottom method is suitable ($r_1 \neq r_3$).
2. Measure the interval velocities above the top reflection and below the bottom reflection.	That reflections exist above and below, and that their character, quality and separation allow meaningful velocity computations.
	That the Dix interval velocity can be corrected if necessary.
3. If these intervals are not thin or deep, use the compaction law to derive local velocities just above top and below bottom reflections.	That a local compaction law is known, or that one of the published laws is appropriate (for example, pp. 2-36, 2-44, 2-60).
	That the zone is not markedly overpressured.

4. Use velocity values to obtain density values.

That a local velocity-density law is known, or that one of the published laws is appropriate if the lithology can be estimated (pp. 2-50, 2-52), or that the equation of p. 2-51 is appropriate if it cannot.

That the materials above the top and below the bottom are not markedly microfractured.

5. Compute acoustic impedances above and below, and their ratio.

6. Estimate magnitude and sign of ratio of the amplitudes of top and bottom reflections.

That a meaningful measure of amplitude can be made.

That the true amplitude relationships have been preserved through the recording, processing and display of the data.

7. Solve the top-and-bottom equations of the gas-saturated reservoir material.

That the Fresnel-zone requirements are satisfied, that there are no reflectors within the gas zone, and that the effect of absorption in the reservoir can be ignored.

8. Estimate the interval velocity between the top and bottom reflections, by conventional velocity analysis or by the degree of velocity pull-down.

That the gas zone is thick enough, and the top and bottom reflections good enough, to allow a meaningful velocity computation.

That the necessary time and other corrections can be made.

That there is not a marked vertical gradation of velocity within the gas zone.

If pull-down is used, that the true attitude of the bottom reflector can be assumed.

9. Compute the density of the gas-saturated reservoir material.

10. Estimate the density of the gas.

That the density of the gas at STP can be accepted.

That the pore pressure and temperature in the reservoir can be estimated.

Preferably, that a Z-factor is available for the reservoir conditions.

11. Compute the porosity on the assumption of a clean reservoir and zero water-saturation.

That the density of the rock grains is known.

4-29

12. Establish a range of effective porosities for likely contaminants and water-saturation.   That the density of the other components is known.

13. Make all possible checks.   See later sections.

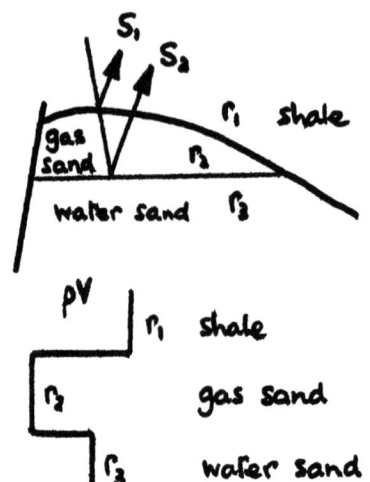

of the water sand are less than those of the overlying shale (p.2-60). We are therefore hoping for an acoustic impedance log of the type shown on the right, where the distinction between r_1 and r_3 is our justification for using the top-and-bottom method.

Let us assume that, since reflections exist both above and below those which delineate the pay-zone, we are able to compute interval velocities for the materials above and below (that is, for the shale and the water sand). These interval velocities, measured seismically, are what we have called "gross" interval velocities (p.2-54). If the intervals are thin we can accept them as representing the actual velocities just above and just below the gas; if not we must apply a compaction law to obtain these local velocities. An illustration of this calculation is given later, in the example of section 4.4.3; for the moment we assume that, by one means or the other, we satisfy ourselves that the shale velocity just above the gas is 2400m/s (7870 ft/s) and the water-sand velocity just below the gas is 2300m/s (7550 ft/s).

Then, using our lithologic assumptions of shale and water sand, we enter the velocity-density curves of p.2-50 and deduce corresponding values for the density: 2.21 for the shale and 2.07 for the water sand.

At this stage we must stop to remember the importance of using a coherent system of units. The only sensible system to use for calculations today is the SI system, in which densities are measured not in grams per cubic centimetre but in kilograms per cubic metre. The above figures then become 2.21×10^3 kg/m^3 and 2.07×10^3 kg/m^3.

Multiplying velocity and density, we now have values for the acoustic impedance above and below the gas:

$r_1 = 5300 \times 10$ kg/m^{2}s and $r_3 = 4770 \times 10$ kg/m^{2}s.

Whence their ratio $r_3/r_1 = 0.9$.

4-31

At this stage we must estimate the ratio of the amplitudes or strengths of the top and bottom reflections. This, of course, we must do very carefully, making sure that we are immune from processing errors and distortions, and permitting ourselves as much averaging as is allowed by the geology -- but no more. Let us suppose that we judge the average strength of the bottom reflection to be three-quarters of that of the top reflection. We must also satisfy ourselves as to sign; in the present case the polarity of the two reflections does appear to be generally opposite (as it must do if the interpreted picture is to be correct, at these depths), and so we arrive at the value:

$$S_2/S_1 = -3/4.$$

The sign, of course, is very important.

Now we can solve the top-and-bottom equations (p.2-98) for r_2, the acoustic impedance of the gas sand. This is conveniently done using the graphs on p.2-98A. The graphs drive home to us the insensitivity to error of the top-and-bottom method; the values we have used are clearly marginal for a satisfactory solution, since the graph for $S_2/S_1 = -3/4$ is substantially vertical at an r_3/r_1 value of 0.9. Therefore the value of r_2/r_1 could be anything from 0.65 down to less than 0.5.

Remembering this weakness, let us continue the illustration by accepting a middle conservative value of 0.57 for r_2/r_1. This yields a value for the acoustic impedance of the gas sand: $r_2 = 3020 \times 10^3$ kg/m^{2}s.

In passing, we may use the computed acoustic impedances to evaluate the reflection coefficients; they are -0.27 for the top of the gas zone and +0.22 for the fluid contact.

The low value for the acoustic impedance of the gas sand (3020) must be good, of course, but it is not yet interpretable in terms of porosity. To do that, we must obtain a velocity measurement within the gas sand.

Since the velocity in the overlying shale is known, the velocity in the gas sand can be computed from the time relief

of the top-sand reflection and the velocity pull-down of the
fluid contact, on the assumption that the latter is horizontal.
In the present case, the time relief is estimated to be 35%
more in the gas sand than in the shale, and so this approach
produces a velocity value of about 1770m/s (5800 ft/s).
Such a low value is feasible at this shallow depth, though of
course we would not expect it at normal commercial depths.

The velocity in the gas zone may also be directly measureable by conventional velocity-analysis techniques if (as in
this case) the top and bottom reflections are well separated.
Both in this approach and in the pull-down approach, it is
very important to make the time and/or velocity picks at the
appropriate part of the waveform (sections 3.2.7 and 3.5.3).

If then we accept the gas-sand velocity as 1770m/s, and
its acoustic impedance as 3020.10^3 kg/m^{2}s, it follows that
the density of the gas sand $\rho r = 1.71.10^3$ kg/m^3.

We now need the density of the gas, in its reservoir
condition. From p.2-17 we know that the density of methane at
a temperature of 0°C (273°K) and a pressure of 1 atmosphere
is 0.0007×10^3 kg/m^3. This is negligible, but of course this
is not the density of the gas in the reservoir. Manipulating the
gas laws, we have that the density is increased by the ratio
of pressures and decreased by the ratio of the absolute
temperatures. We can compute the reservoir pressure if we
assume a pore pressure gradient; in fact the reservoir
illustrated on p.4-26 is likely to be slightly overpressured,
but for the moment let us assume a normal pressure gradient
of 10.5kN/m^2/m (0.465 psi/ft). Then the reservoir pressure is
approximately 64 atmospheres,* and the ratio of pressures is 64.
The temperature ratio is less important; the temperature
gradient may be taken as 36°C/km, the surface temperature as
5°C, and the temperature ratio is therefore likely to be about
300°K/273°K, or 1.1. It follows that the density of the gas
in the reservoir is approximately $0.041.10^3$ kg/m^3, which may
not be negligible. (In a mature area, the reservoir engineers
could improve on this simple ideal-gas calculation, by the use
of their Z-factor (section 2.3.1).)

* 1 atmosphere = 101.3 kN/m^2.

At this stage we assume that the gas sand is composed of clean quartz sand grains of density $\rho_m = 2.65 \times 10^3$ kg/m^3, and that methane of density $\rho_f = 0.041 \cdot 10^3$ kg/m^3 occupies all the pore space. Then we can use the last equation of p.2-17 to tell us that

$$\emptyset = \frac{2.65 - 1.71}{2.65 - 0.04}$$

or $\quad \underline{\emptyset = 36\%}$.

The first observation to be stressed, in connection with this result, is the measure of protection which it affords against significant water-saturation of the gas. For, although we have no means for making a measure of the water-saturation, the approach through <u>density</u> allows us to compute a revised estimate of the porosity for any postulated water-saturation. If the water-saturation is 10%, for example, we can use the equation of p.2-18 to deduce that, to retain a density of 1.71, the porosity must now be 40%. If experience suggests that this is the largest porosity we can expect in this sand at this depth, then we have at least established a <u>range</u> of possible water-saturation, and so a range for the gas reserves (in this case, from 0.32 to 0.36 of the gas-reservoir volume).

Our second observation, of course, must concern the likely errors. There are five clear sources of error (apart from errors of arithmetic, which geophysicists never make):

- errors of interpretational judgement,

- errors of geological assumptions,

- errors in the recording, processing or correction of the data,

- errors in the picking of the data, and

- statistical errors representing noise in the total system.

We have already discussed the third and fourth of these (in sections 3.2 and 3.5), and we shall return to them in later discussions. For the moment, let us just say that we understand the considerations involved, we know what to do, and we can recognize the situations where we are at risk. The last of

the above sources of error — noise — we can assess by
observing the degree of scatter betweeen the results computed
for different traces, different groups of traces or different
lines judged to represent the same geological circumstances.
But we cannot pass so quickly across the first two of the
above sources of error — our interpretational judgement
and our geological assumptions. It is obvious that an error
in these can make the entire calculation meaningless. Supposing,
for example, that the lower reflection is not a fluid contact
but an unconformity — how Mother Earther would be chuckling!
Or perhaps there is a gradation of porosity within the sand
body, undetected by our velocity measurement, so that the ratio
of top-and-bottom reflection coefficients is changed without
our knowing. Or perhaps it is not a sand at all, and our
assumption of a grain density of $2.65.10^3 kg/m^3$ is consequently
erroneous. Such cautions make us appreciate that before we
make serious estimations of porosity we must identify the
limitations of our method, and list all the cross-checks we
can make to validate each step of the procedure This we shall
do, in some detail, in sections 4.5.1. amd 4.5.2.

But, while the present example is fresh in our minds, let
us just note the value of the test suggested in our initial
discussion of the top-and-bottom method (pp.2-97, 2-98B).
There we said that, in addition to using the ratio of the
r_1/r_2 and r_2/r_3 reflections, we would seek corroboration by
using the r_1/r_3 reflection (the shale/sand reflection in the
water-saturated zone). The assumptions we have used, and the
calulations of acoustic impedance that we have made, imply that
the reflection coefficient of the r_1/r_3 interface is - 0.05.
The shale/water-sand reflection should therefore have the
same sign as the shale/gas-sand reflection, but only one-fifth
of the amplitude. That the sign is correct is indisputable;
whether the amplitude is correct we cannot say definitely
from the illustration, because it has probably been base-
levelled or equalized. So we learn once more the importance
of avoiding data-dependent amplitude manipulations — not only
so that amplitude relationships are preserved vertically but
also horizontally. If (as seems very likely, from the appearances)
it is quite impossible that the amplitude of the shale/water-
sand reflection is one-fifth of that of the shale/gas-sand
reflection, then not only are all the above calculations in error
but the basic interpretation of p.4-26 is in error also. The
calculations are in error because the interpretation is in error.

4.4.2 Porosity estimation from velocity, in the presence of both liquid and gas

Earlier, we agreed quite positively that the porosity of a gas reservoir cannot be estimated from velocity alone. This was for two reasons — the lack of an explicit relation, and the problem of water-saturation. Consequently both the last section and the next section concentrate on the approach through density. The present section is inserted, however, to acknowledge the fact that there is a velocity equation which may be used in limited circumstances (though still without solving the water-saturation problem, which remains approachable only through density).

The equation is that of p.2-58, and the limiting circumstances are those of constant porosity in the liquid-saturated and gas-saturated parts of a reservoir body.

At the very beginning, we must remember the cautions of section 2.9 — that it is not necessary (perhaps even not likely) that the porosity in the gas-saturated zone should be identical to the porosity of the same original rock in the water-saturated zone. This is because there has been continued passage of mineral-rich water through the latter zone, but no such passage through the gas zone.

Nevertheless it is worth computing an example, because the method is sometimes useful as a check. The technique is based on the second paragraph of p.2-57, and the method steps are set out in p.4-37 and 4-38.

The measurements necessary to the method are just the two velocities — in the gas-saturated zone and in the liquid-saturated zone. However, we must be very conscious of the assumptions we make in using the equation, as regards both the lithology and the physical conditions of the reservoir.

Let us take as our example the case we have just computed by the top-and-bottom method. In that illustration we accepted that we had a gas-saturated sand of velocity 1770 m/s (5800 ft/s) and a water-saturated sand of velocity 2300 m/s (7550 ft/s).

VELOCITY METHOD

METHOD STEPS	ASSUMPTIONS
1. Identify the reflections bounding the top and bottom of the gas zone and the top and bottom of the liquid-saturated zone. (In the case of a gas-liquid contact, of course, two of these are the same.)	That a reservoir containing gas and liquid is plausible in the location postulated. That the porosities in the two zones are identical (or of known ratio). That the depth is such that the acoustic consequences of gas-saturation can be expected to be significant.
2. Measure the interval velocities within the two zones.	That the reflection quality and discreteness allow meaningful velocity computations. That the necessary time and velocity corrections can be made.
3. Establish the densities and bulk moduli for the reservoir material and the liquid pore-filling.	That the guess of reservoir lithology is correct. That the guess whether the liquid is oil or water is correct, and that the oil gravity or water salinity can be assigned.
4. Compute the density and bulk modulus for the gas in the reservoir.	That it is safe to assume the nature (and preferably the chemical constitution) of the gas That the depth of the reservoir can be estimated. That the pore pressure and temperature in the reservoir can be estimated.

5. Set up two simultaneous equations in k_s and $\emptyset$, and solve for porosity $\emptyset$.

That the equation is valid at seismic frequencies.

That the approximation is acceptable to set the first bracketed term (properly M, the P-wave modulus for the porous rock skeleton) equal to $2k_s$.

6. Make all possible checks.

See later sections.

Now we assume that the porosity is the same in both cases, and set up two equations of the form of p.2-58 with the same $\emptyset$. Because the porosity is the same and the material of the reservoir skeleton is the same, the value of k_s (the bulk modulus for the porous rock skeleton) is also the same. We accept a value of 2.65×10^3 kg/m^3 for the density of the sand grains, a medium value of 1.05×10^3 kg/m^3 for the density of salt pore-water, and a value of 41 kg/m^3 (computed as in the last section) for the density of the gas at a depth of 600m. We further accept the values of 38×10^9 N/m^2 and 2.6×10^9 N/m^2 for the bulk moduli of sand grains and salt pore-water, respectively. Then we must compute the value of the bulk modulus for the gas, under reservoir conditions; this is equal to the pressure of the gas times a factor (the ratio of the specific heats) which depends on the constitution of the gas. For methane this factor is 1.31, and so the bulk modulus is $64 \times 101.3 \times 1.31$ kN/m^2, or 8.39 MN/m^2.

If now we set up the equations for the velocity in the water sand and for the velocity in the gas sand, we have two equations with just two unknowns: k_s and $\emptyset$. This allows us to solve for the porosity $\emptyset$. The arithmetic is cumbersome, and is not reproduced here; it yields the answer:

$$\underline{\emptyset = 36\%}$$

Although we have said that we assume the porosity to be the same in the gas-saturated and water-saturated zones, local knowledge might possibly allow us to estimate a porosity ratio other than unity; we could then use this in the equations. It is also possible, in principle, to compute the effect of water-saturation; however, as we have said several times, this approach through velocity is unlikely to resolve the matter of water-saturation.

TAPE 25   4.4.3 <u>Porosity estimation by the top-only method</u>

A geophysicist producing porosity values which checked as well as those of the last two sections would feel the same kind of uneasy gratification that he experiences when his depth prognosis proves precise. So, to restore an air of reality to the discussion, and to show that the careful listing and consideration of our assumptions is a very healthy exercise, the illustration of our third porosity method will be the case history of a mistake.

The point to be made, of course, is not that the top-only method is unsound (because it is not), but that optimism must not blind us to the highly critical nature of our reflection identifications.

The top-only method is based on the discussion of multiple reflections which we struggled through in section 2.6. Some stages of the method are the same as those of the top-and-bottom method; the fundamental difference is that one of the reflection coefficients is determined by multiple-reflection techniques, rather than from a ratio of two primary reflections. The method steps and assumptions are set out in pp.4-41 to 4-43.

241 L
242
243

The example we shall use as an illustration of the method is that of page 4-44. As indicated, the line is 50km in length. Our attention is drawn immediately to the amplitude anomaly at about 1.65s.

As always, our first step is to assess the geological plausibility of a hydrocarbon accumulation. In this case, we note:

- The prospect is at the edge of a basin bounded by an ancient platform, visible at the extreme left of the section. The line is approximately perpendicular to the platform edge.

- The platform composition includes copious Paleozoic sandstone, as a source for coarse-grained sediments.

- The feature extends a considerable distance perpendicular to the line of profile.

- The circumstances are therefore compatible with coarse-grained deposition from a high-energy environment, possibly of sand beach or bar type.

- The region just below the deep unconformity (2.7-3s) is known to be a target level for oil accumulations, so that source and primary-migration circumstances are established.

- Secondary or tertiary migration toward and at the platform edge is plausible, as is the accumulation of such migrating hydrocarbons in the prospect reservoir. Evidence of some deep faulting supports this.

- There is also potential for favourable source and migration conditions within the Tertiary basin itself.

- The overlying material is likely to be impermeable shale.

A gas reservoir at the bright spot is therefore plausible. It is certainly not <u>established</u>, at this stage. The brightness could be due to Tertiary volcanics. Or the feature could be a shallow-water carbonate build-up, in which case the brightness may well mean extreme hardness and lack of porosity — the absence rather than the presence of gas. But, with these possibilities kept in mind, it is clear that the hypothesis of a gas-saturated sand body must be further explored.

The aspect of this target which leads us to adopt the top-only method is that there are no certain primary reflectors (except an unsuitable faulted horizon) below the feature. Therefore we have no means of measuring the interval velocity in the formation <u>below</u>. In this case the top-and-bottom equations are <u>insoluble</u>, however clear the actual top and bottom

TOP-ONLY METHOD

METHOD STEPS	ASSUMPTIONS
1. Identify the reflection corresponding to the top of the gas reservoir.	That a gas-saturated reservoir is plausible in the location postulated.
	That the depth is such that the acoustic consequences of gas-saturation can be expected to be significant.
	That the top reflection is discrete and correctly identified.
2. Determine, by all methods allowed by the data, the magnitude of the apparent reflection coefficient of this reflector as viewed from the surface.	That curvature and refraction effects, if significant, can be compensated.
	That the reflection coefficient of the surface is known, or can be accepted as -1.
	That the reflector is discrete, or that nearby reflectors exist which are discrete.
	That the correction for geometrical divergence can be properly determined and applied, and that other amplitude manipulations or distortions applied in processing can be compensated.
	That methods of determining the apparent reflection coefficient using stacked data are substantiated by clear picks on the velocity analyses.
	That the sea-floor reflection, if used in the determination, is properly recorded, and that it is sufficiently abrupt.
	That the target reflector is sufficiently extensive, having regard to the dimensions of the Fresnel zones.

4-41

| | That the target reflection occurs at small angle of incidence, and that the plane-wave approximation is justified. |

3. Establish the sign of the reflection coefficient.

If this is done by relating to a reference reflection, that the sign of the latter is known.

That pulse-shape changes caused by earth, instrumental and processing agencies do not mask the polarity indications.

4. Assess the transmission-coefficient losses at all <u>major</u> interfaces between the top-reservoir reflection (primarily, for the sea floor).

That the reflection coefficients for all treated interfaces are correctly computed.

5. Assess the frequency-dependent losses between the surface and the top-reservoir reflection (or, failing that, between the surface and a nearby discrete reflection).

That the amount of these losses is not so large that inevitable errors in measuring it are comparable with the amplitude variations it is desired to measure. (This means, in practice, that the lossy interval includes neither a great extent of highly-contrasty thin-bed stratification nor a superabundance of scattering inhomogeneities.)

<u>Either</u> that the velocity analyses are sufficiently clear to allow positive identification and stacking of multiple reflections generated from the topside and underside of the top-reservoir or a nearby discrete reflection, <u>or</u> that we know the downgoing pulse (from a discrete sea-floor reflection or otherwise) and can measure the frequency content of the target reflection

	(or a nearby discrete reflection) in isolation, <u>or</u> that experience in the area allows us to assume a value for the frequency-selective losses in dB/wavelength.
6. Use steps 2-5 to compute the actual reflection coefficient at the top of the reservoir.	That the result, which is biased low by the incompleteness of step 4, is still useful.
7. Measure the interval velocity above the top reflection.	That corrections can be made, as necessary, for zero-time errors and the Dix assumptions.
8. If the interval above the top reflection is not thin or deep, use the compaction law to derive the local velocity just above the sealing interface.	That a local compaction law is known, or that one of the published laws is appropriate. That the zone above the reservoir is not anomalously overpressured.
9. Use the velocity value to obtain a density value just above the interface.	That a local velocity-density law is known, or that one of the published laws is appropriate if the lithology can be estimated, or that the equation of p.2-51 is appropriate if it cannot. That the material above the interface is not markedly micro-fractured.
10. Compute the acoustic impedance above the reservoir.	
11. Use steps 6 and 10 to compute the acoustic impedance within the reservoir.	The top reflection is discrete.
12. Continue with steps 8-13 of the top-and-bottom method.	

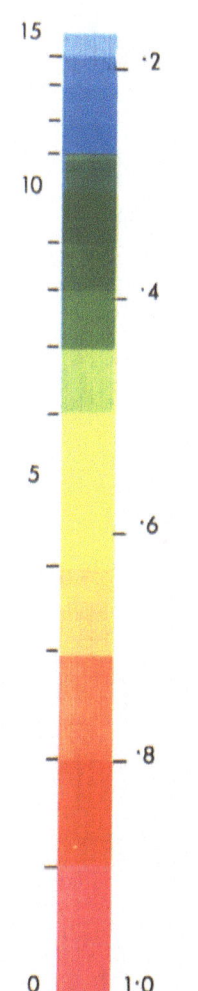

3-58B   Courtesy Seiscom

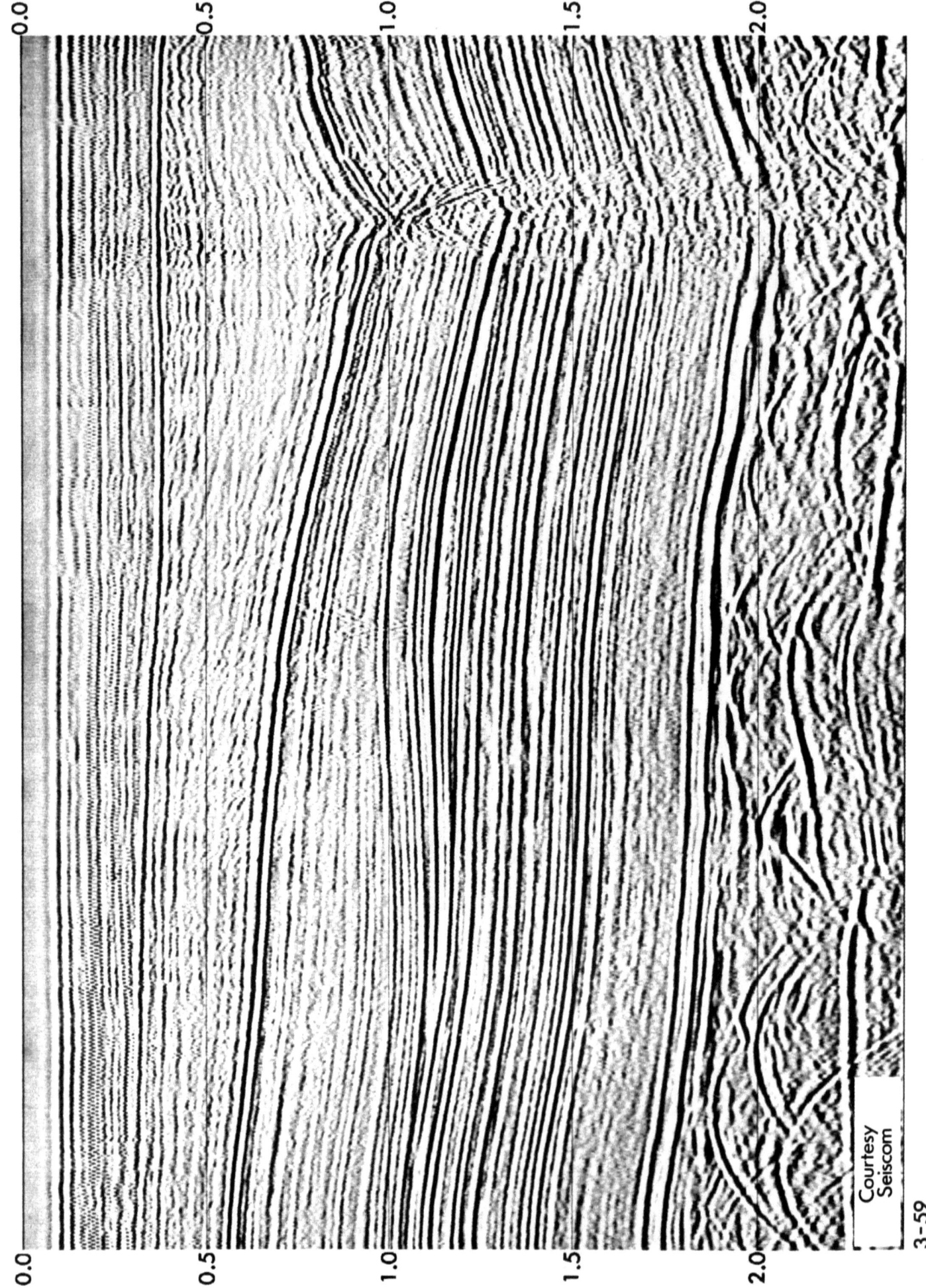

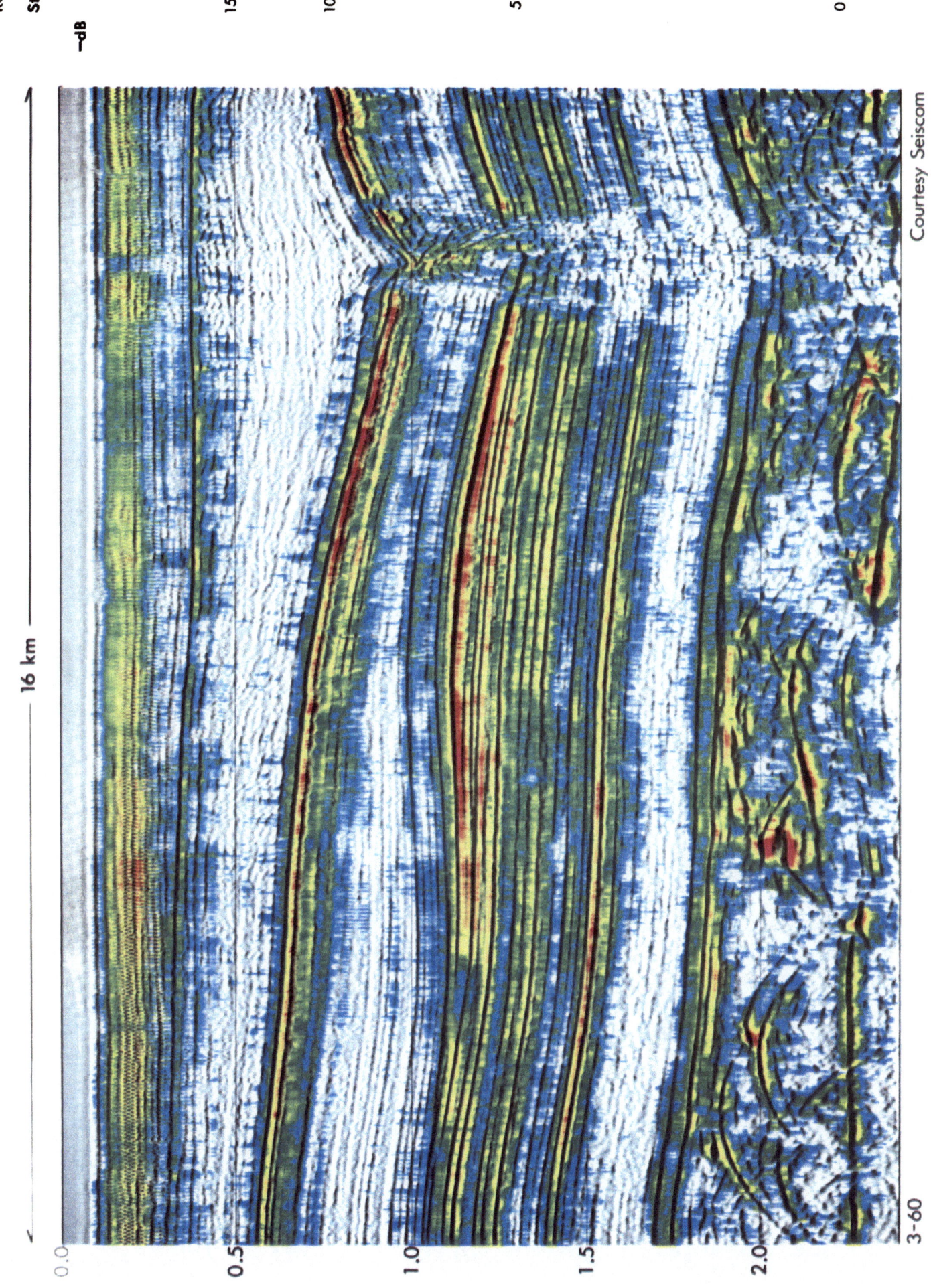

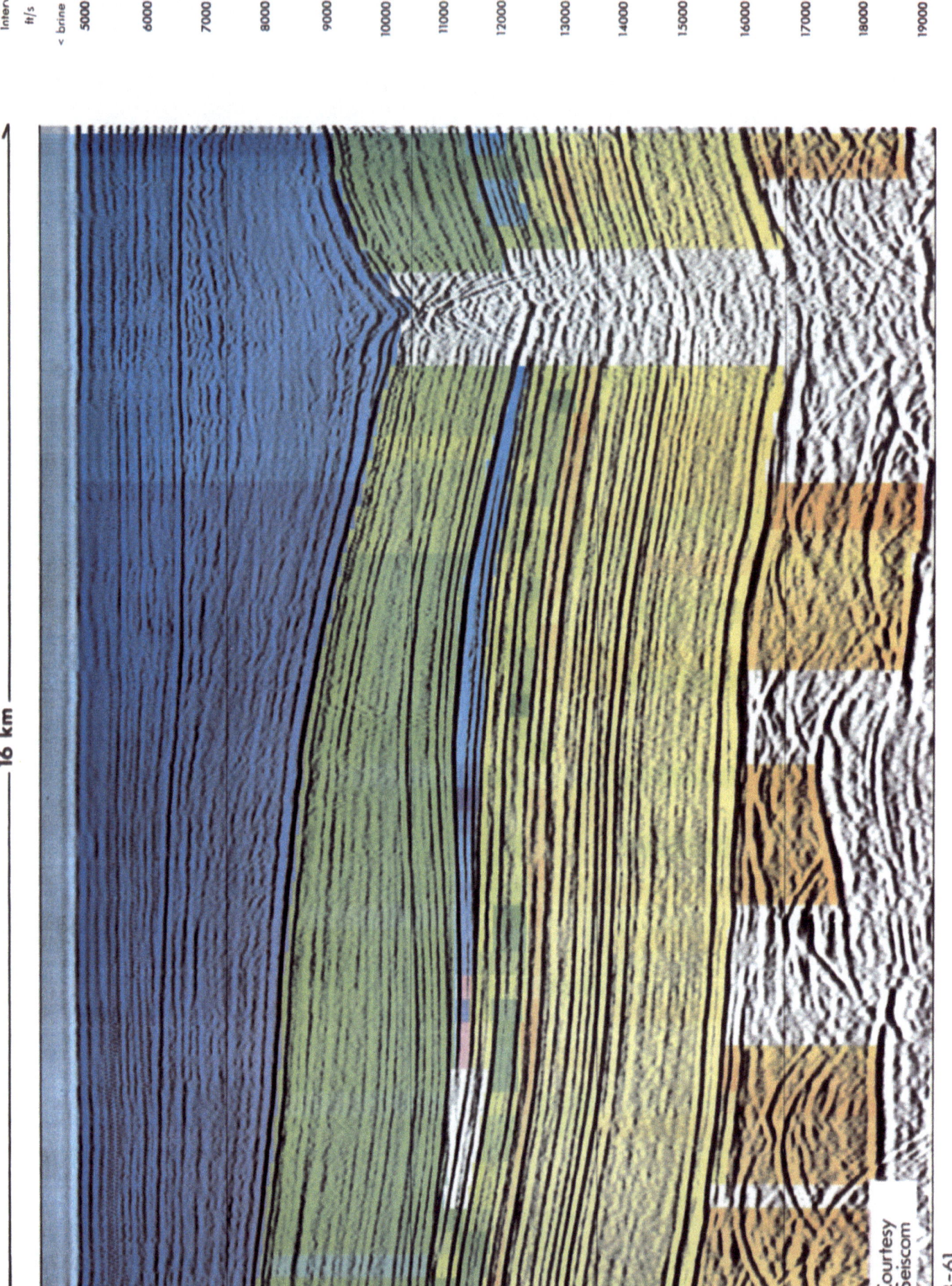

3-61 Courtesy Seiscom

16 km

3-63

Courtesy Seiscom

+   −

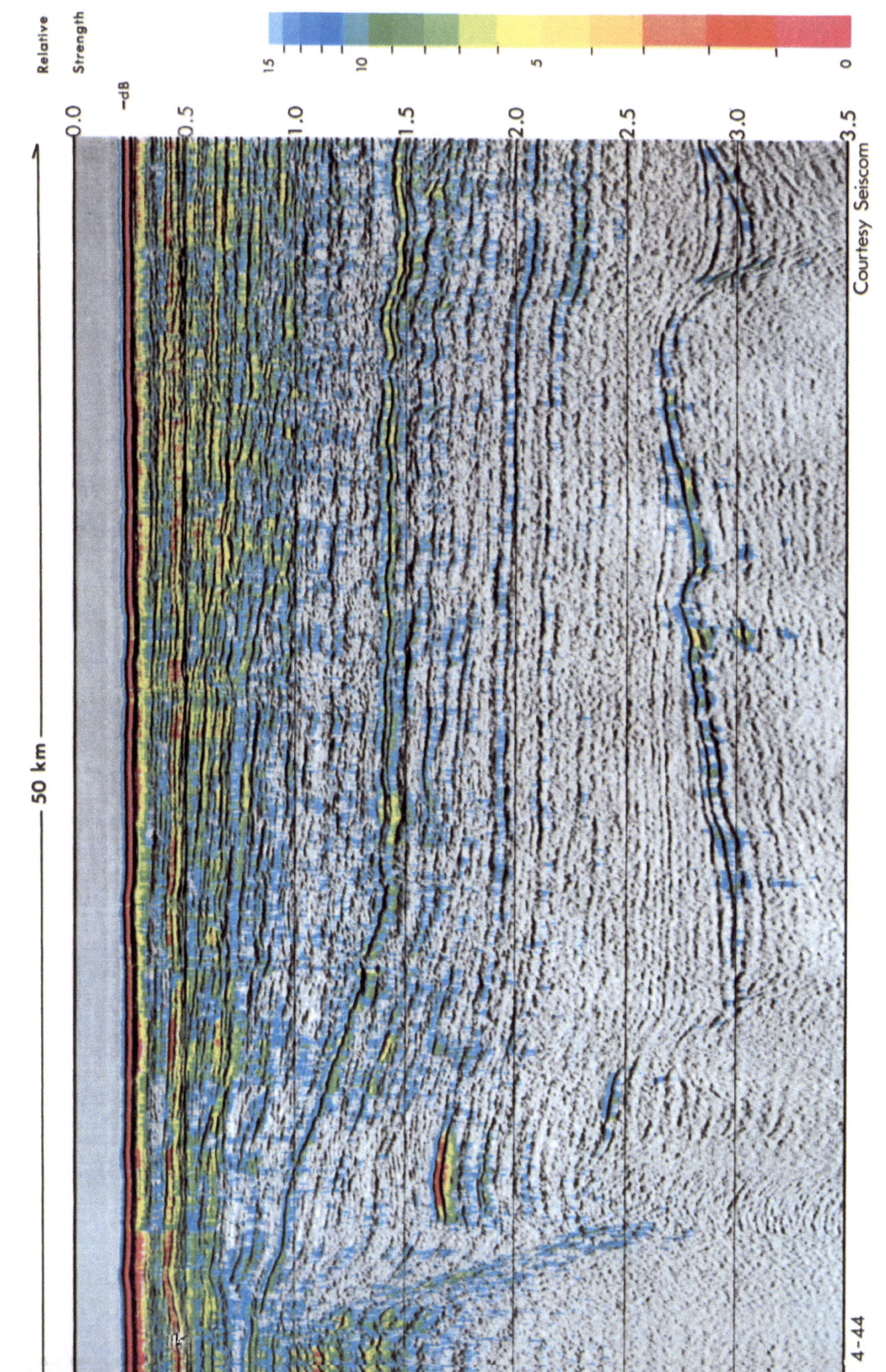

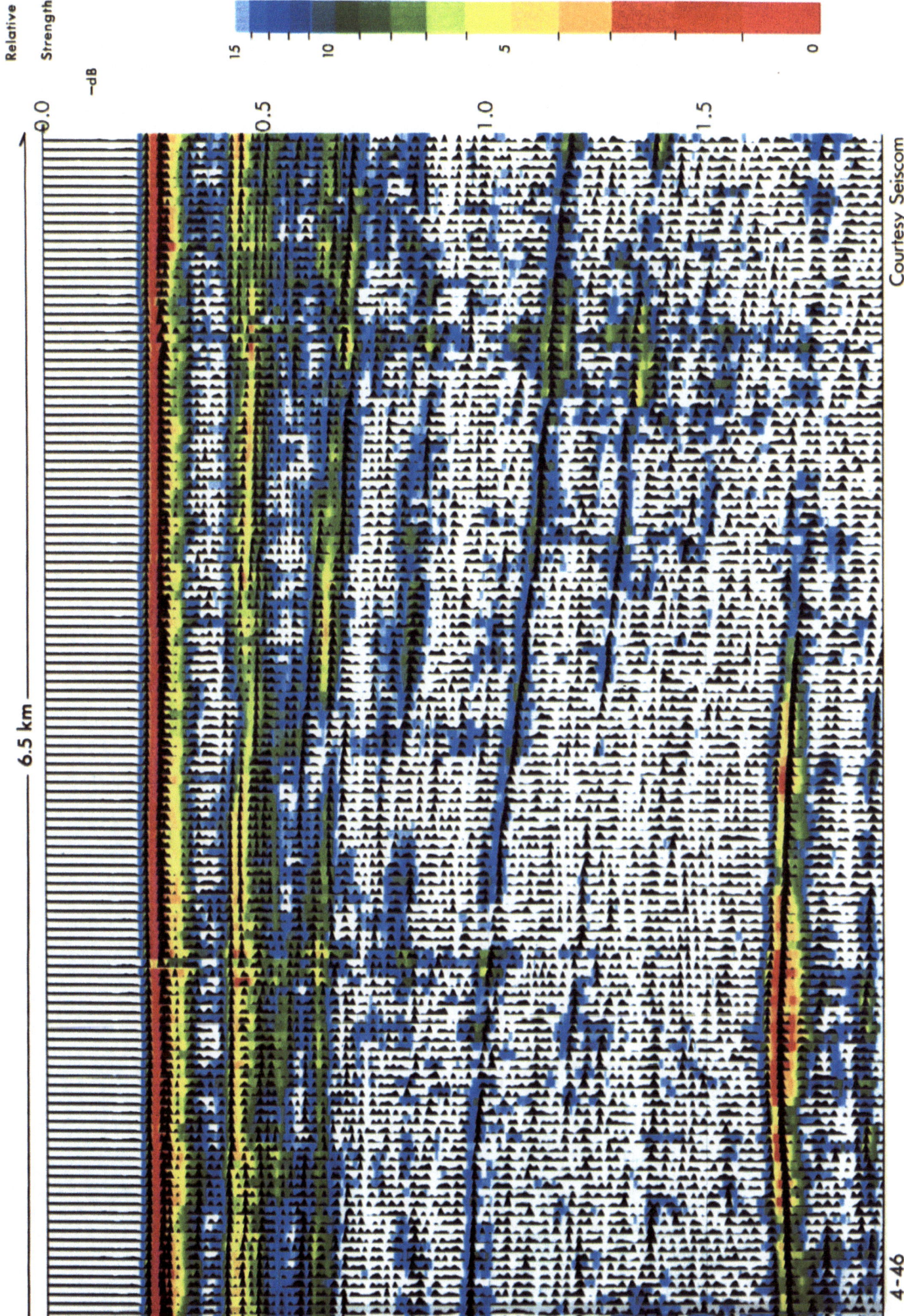

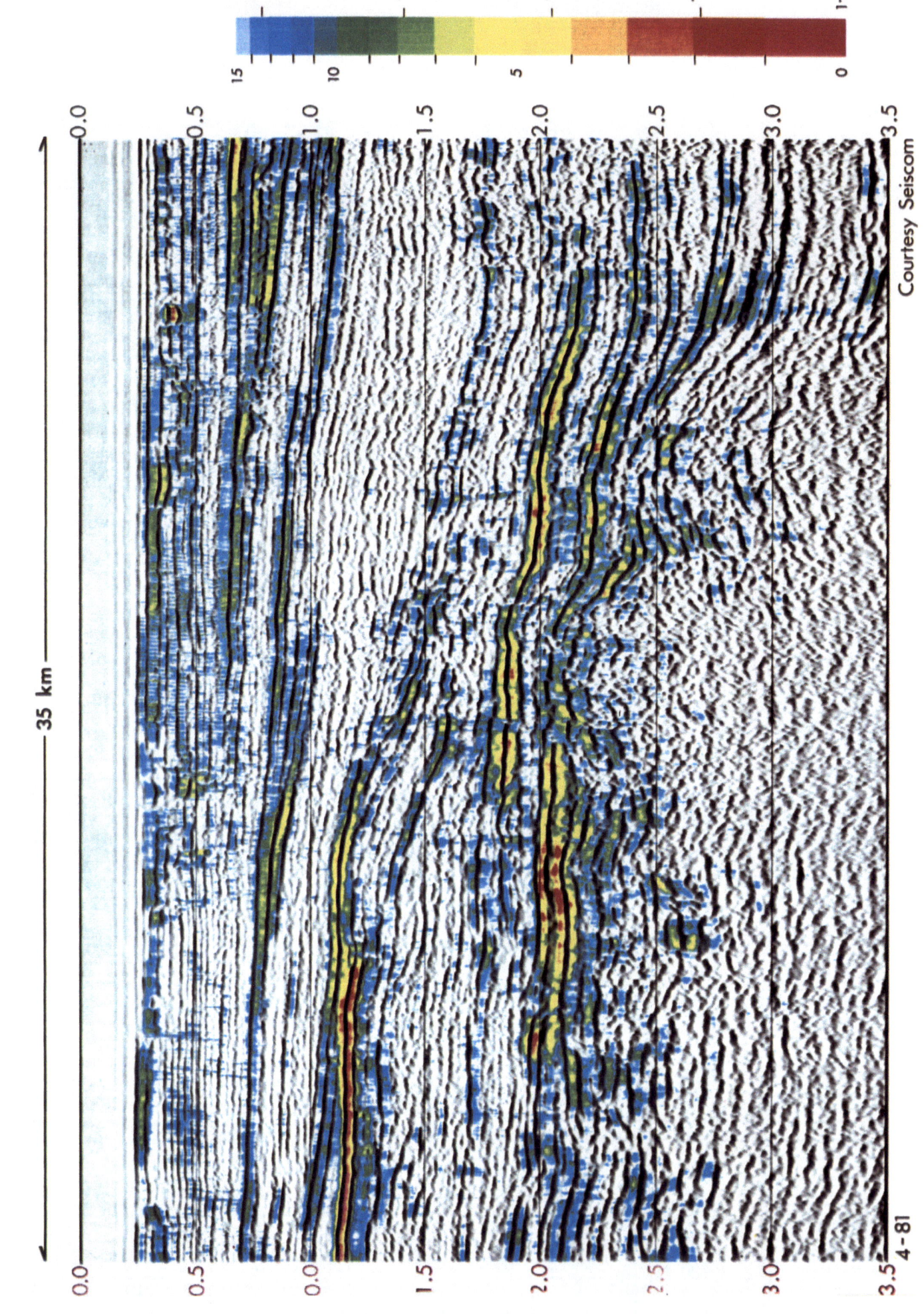

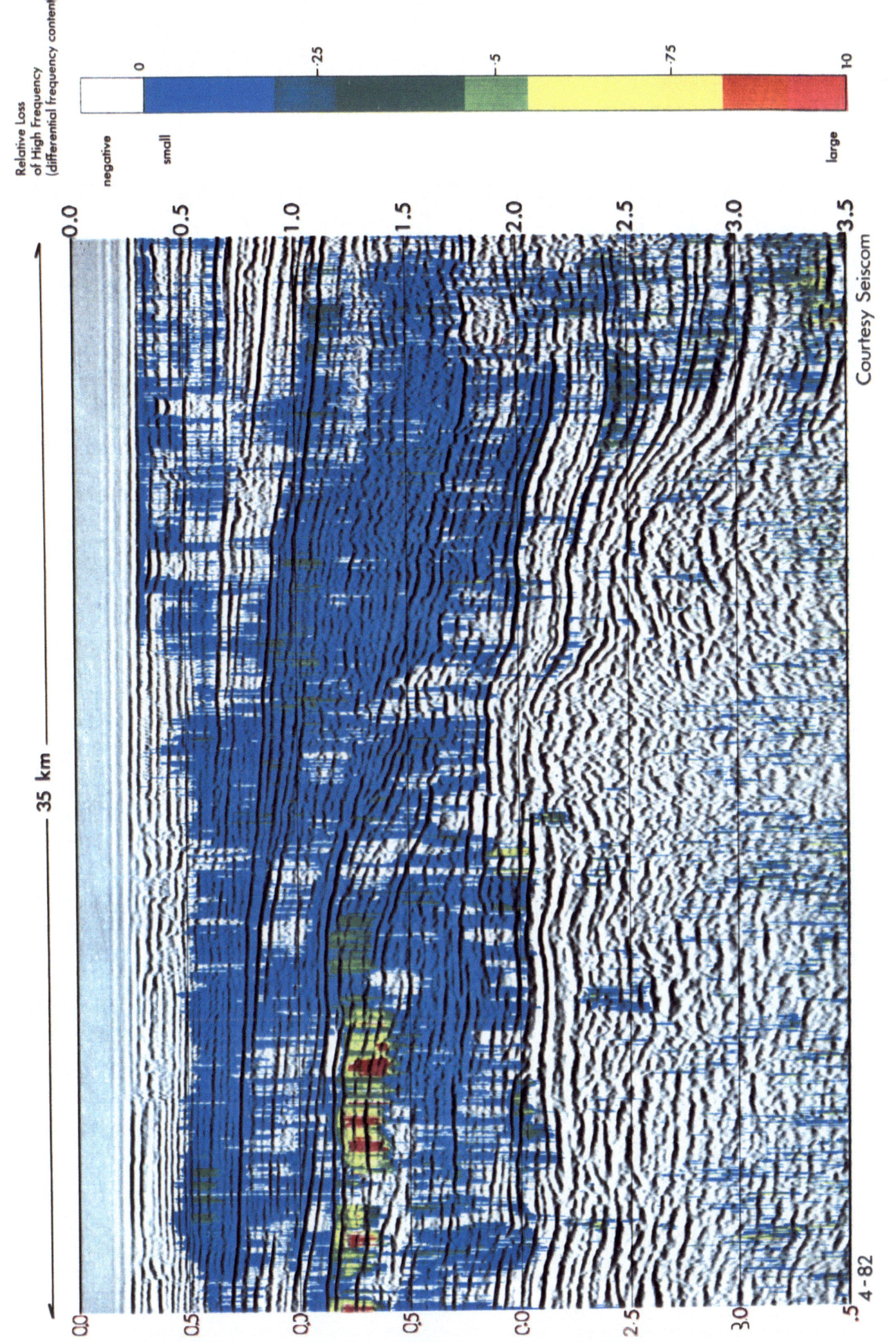

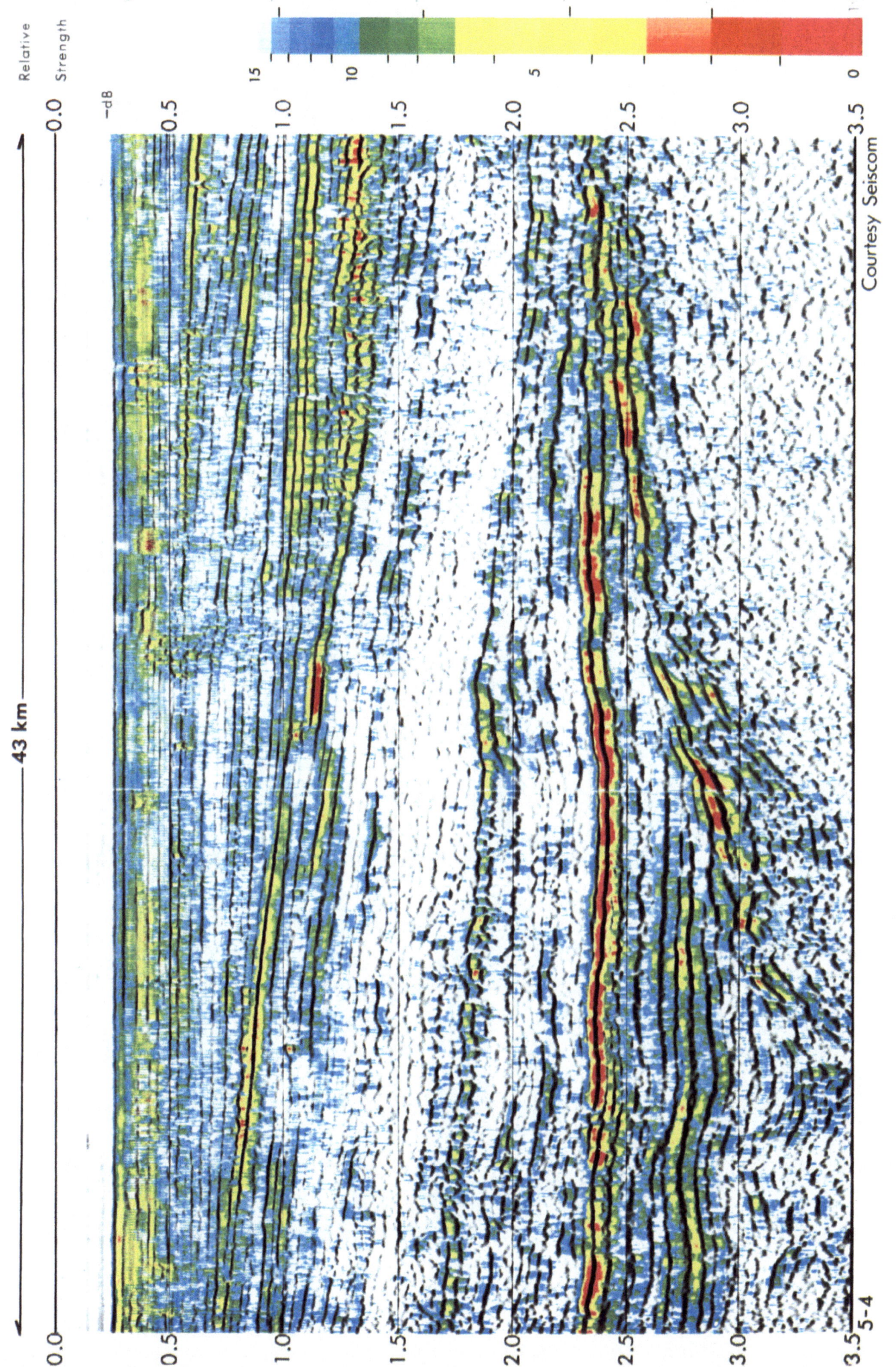

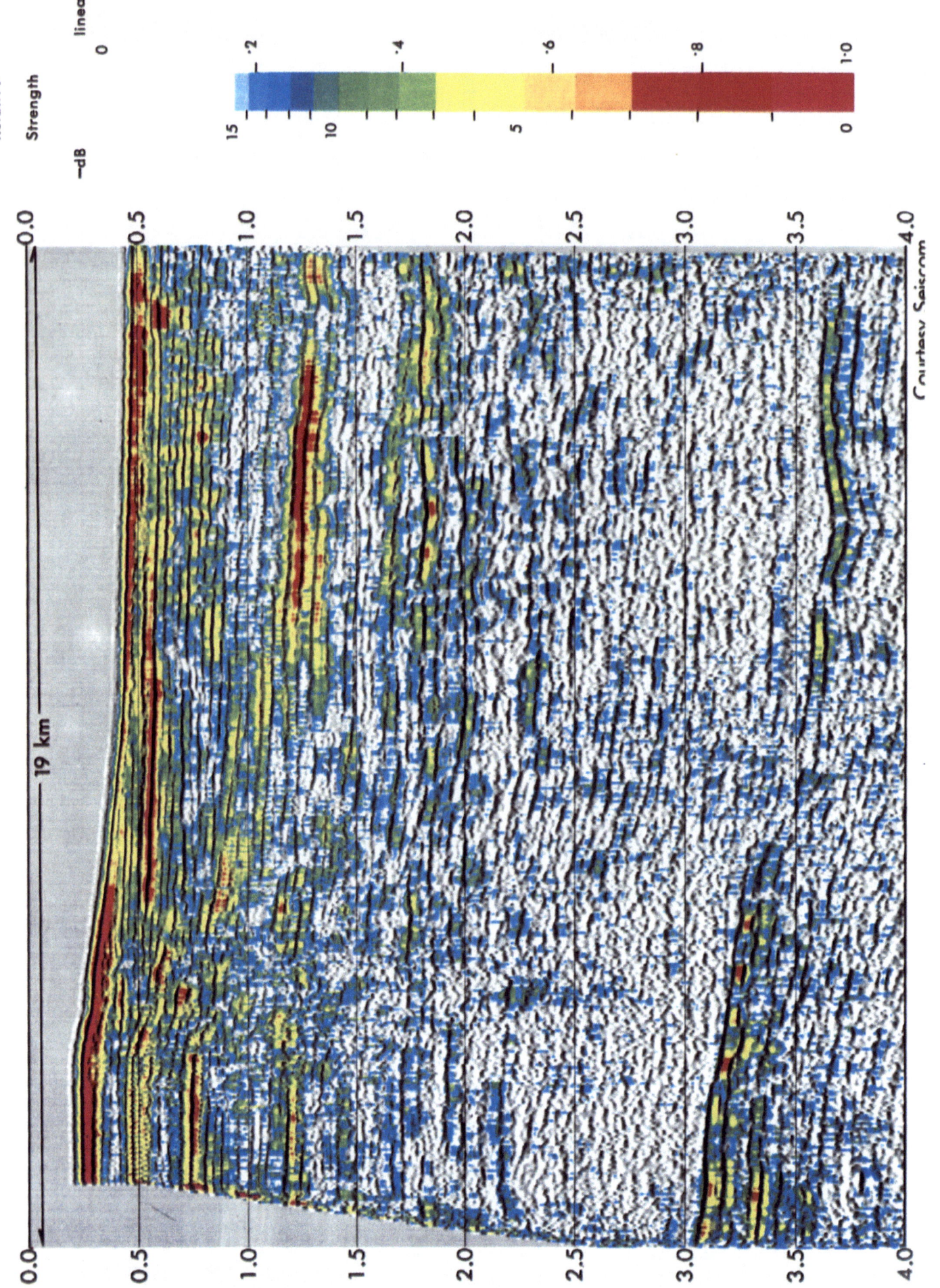

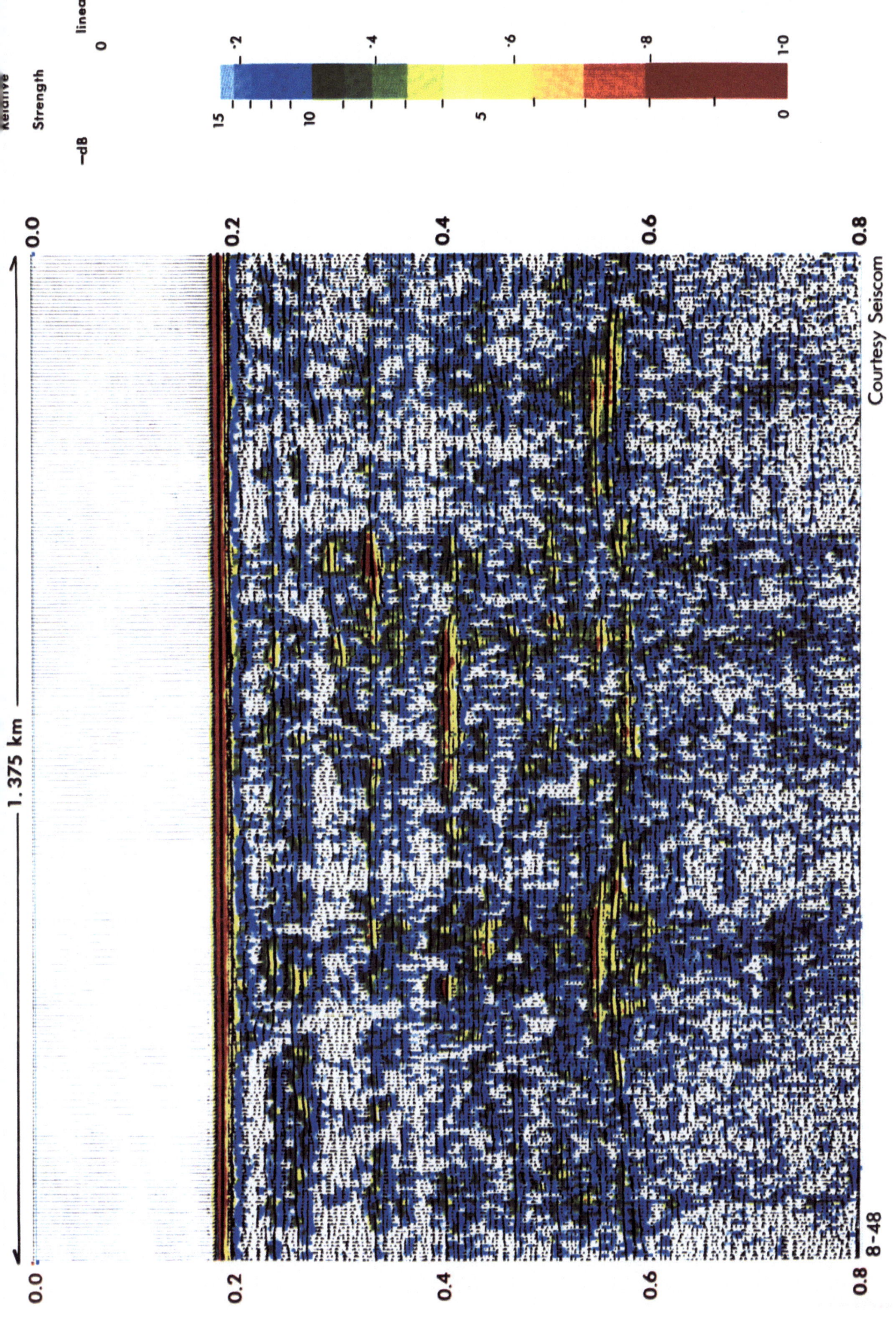

reflections may be.

In fact, there is also a problem with the top and bottom reflections themselves. In the top-only method we do not need the bottom reflection for strength measurement, but we still need it to give us a velocity within the reservoir (and so allow us to convert an acoustic impedance into a density). Further, it is still true that the general appearance of the bottom reflection must be compatible with our geological interpretation of the reservoir. This leads us to look very closely at the feature, which is shown at natural scale on p.4-46. We observe that the general configuration is compatible with interpretation as a lenticular sand. We observe also that there is a clear saddle in the envelope of the trace waveform in the thickest part of the sand, suggesting that the top and bottom reflections are separable over this part at least. Further, we observe that the bottom reflection is generally 2 or 3dB (2 or 3 colour steps) weaker than the top reflection. This might give us pause. It was reasonable enough in our example of the top-and-bottom method, for in that case the material under the gas accumulation was a water sand of acoustic impedance less than that of the overlying shale. But in the present case our first hope is a lenticular sand full of gas and bounded top and bottom by shale; this would oblige us to explain why the lower reflection was weaker, when our first expectation would be that it would be rather stronger. Our concern would grow when we realized that two of the alternative explanations for the feature — volcanics and a carbonate build-up — would both yield a top reflection stronger than the bottom reflection in the likely case that the lower shale was somewhat harder than the upper shale. And we would also note that interference between the top and bottom of a thin gas accumulation at the top of a lenticular water sand (or oil sand) might produce a similar effect. But it is also true that a close inspection of the waveforms reveals that the bottom reflection has a low-frequency appearance; this allows the optimist to invoke a locally over-pressured zone under the sand (item 10 of section 4.3), and so to dismiss worries about the slight weakness of the bottom reflection.

So let us stay with the optimist for the present, and follow him as he computes the porosity of his sand. As we do so, however, we make a mental note to ourselves to be sure to check all information concerned with reflection polarities and reservoir velocities, lest they indicate his basic interpretation to be wrong in its very first identifications.

The initial step in the top-only method is the determination of the apparent reflection coefficient of the top of the gas zone, by direct measurement.

The best way of doing this is usually the technique sketched as item 1 of section 2.6.5 — the direct comparison of primary and first multiple. This can be done either on unstacked data or on data stacked for the first multiple.

Although this is the preferred method, we have agreed repeatedly that where there are several methods of making a measurement they should all be checked. Accordingly the interpreter (even our optimist) would certainly confirm his measurement by using the methods identified as items 3 and 4 of the same section. It is helpful tutorially, in the present case, to set out the procedure for item 4, since it illustrates many of the methods described in the whole of section 2.6. We stress, however, that this is for tutorial reasons only; the method to be used as first choice in a case like this is the comparison of the primary and first surface multiple from the top of the target feature.

It is visually apparent from p.4-46 that the reflection strength of the top of the gas sand, at its thickest, is about 1-2dB (one or two colour steps) below that of the sea floor. So the next step is to compute the reflection coefficient of the sea floor.

The method which (though the most complex) gives the best tutorial illustration of the principles involved is that of p.2-103. A typical velocity analysis in the bright-spot zone is shown on p.4-48; on this the train of water reverberations is very clear. Then the section is stacked, and a divergence correction is applied, using a velocity distribution passing through the top of the postulated gas sand and through all the water reverberations;

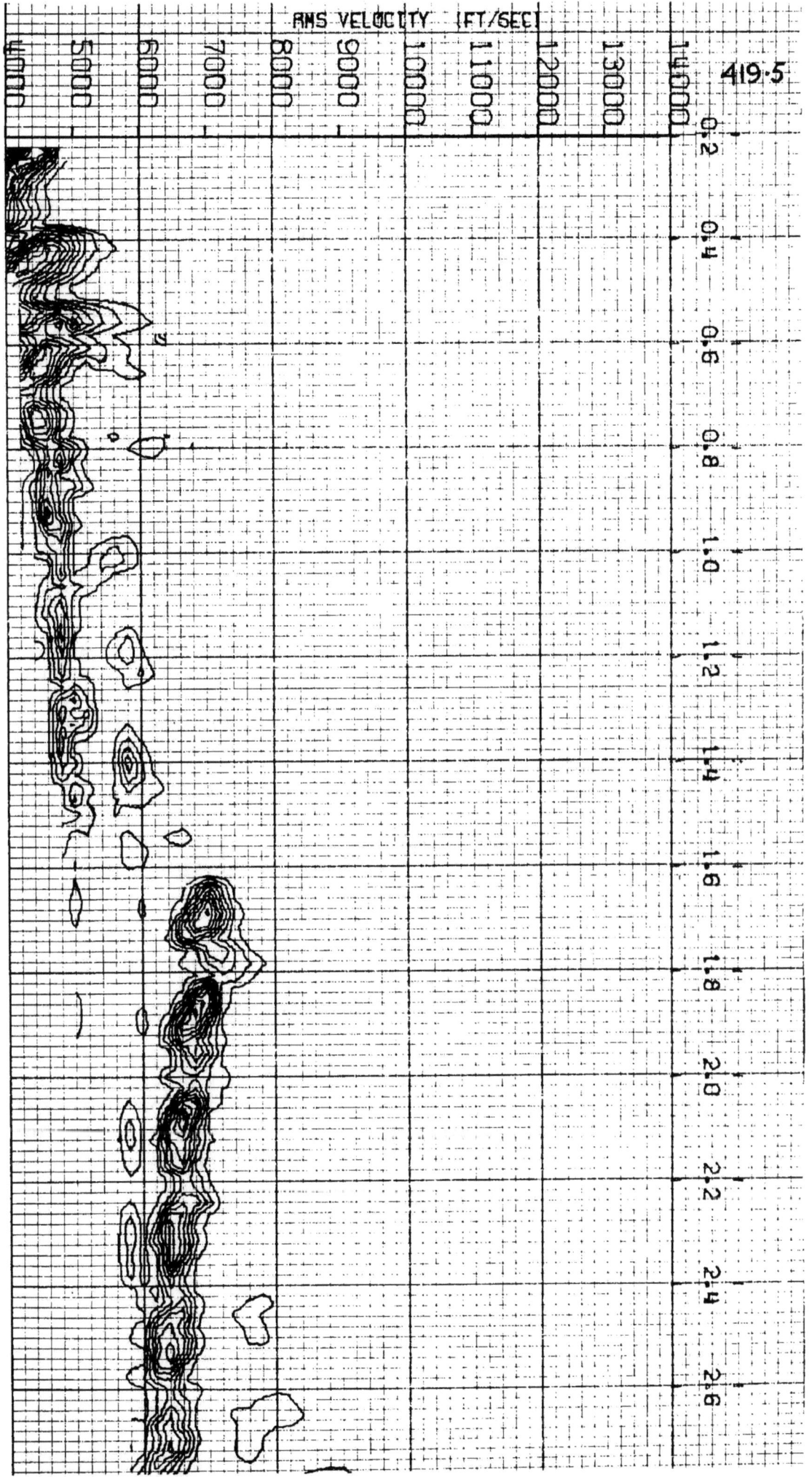

248 the result is shown in the upper illustration of p.4-50. Then, after a check that the multiple train does not include any significant primary reflections, and that the first multiple is a tolerably good (inverted) version of the primary, autocorrelation functions can be computed over the window shown. Typical of the resulting autocorrelation functions are those
249 shown in the lower illustration of p.4-50; obviously these are of classic form, highly suitable for use in deriving the sea-floor reflection coefficient. Since the reverberations represented are <u>reflected</u> reverberations, the upper curve of p.1-22 is appropriate; the average ratio A_1/A_0 is -0.67, so the reflection coefficient is $+0.38$.

Before the application of the necessary corrections, therefore, the computed value for the magnitude of the reflection coefficient of the top of the gas sand is 1-2dB (say 1½dB) less than 0.38, which is 0.32.

The first correction to be applied becomes necessary if relentless probing in the dark recesses of the processing programs reveals that amplitude manipulations other than or supplementary to the geometrical divergence correction have been applied at early times. In the present case, we are assured (Why does it have to be so difficult to find out with certainty?) that an amplitude expansion of 5½dB has been applied, substantially all of it between the sea-floor reflection and 1 second. Our best value for the apparent magnitude of the reflection coefficient of the top of the postulated gas sand is therefore 5½dB down on 0.32, which is 0.17.

The second correction concerns the losses which must occur between the sea floor and the gas sand; the value of 0.17 is, of course, only the <u>apparent</u> magnitude as viewed from the surface.

We remember that these losses fall into two categories: the frequency-independent effect of transmission coefficients at major interfaces, and the frequency-dependent effects of absorption, scattering and short-path multiples.

0.0
0.2
0.4
0.6
0.8
1.0
1.2
1.4
1.6
1.8
2.0
2.2
2.4
2.6
2.8
3.0

0.0
0.2

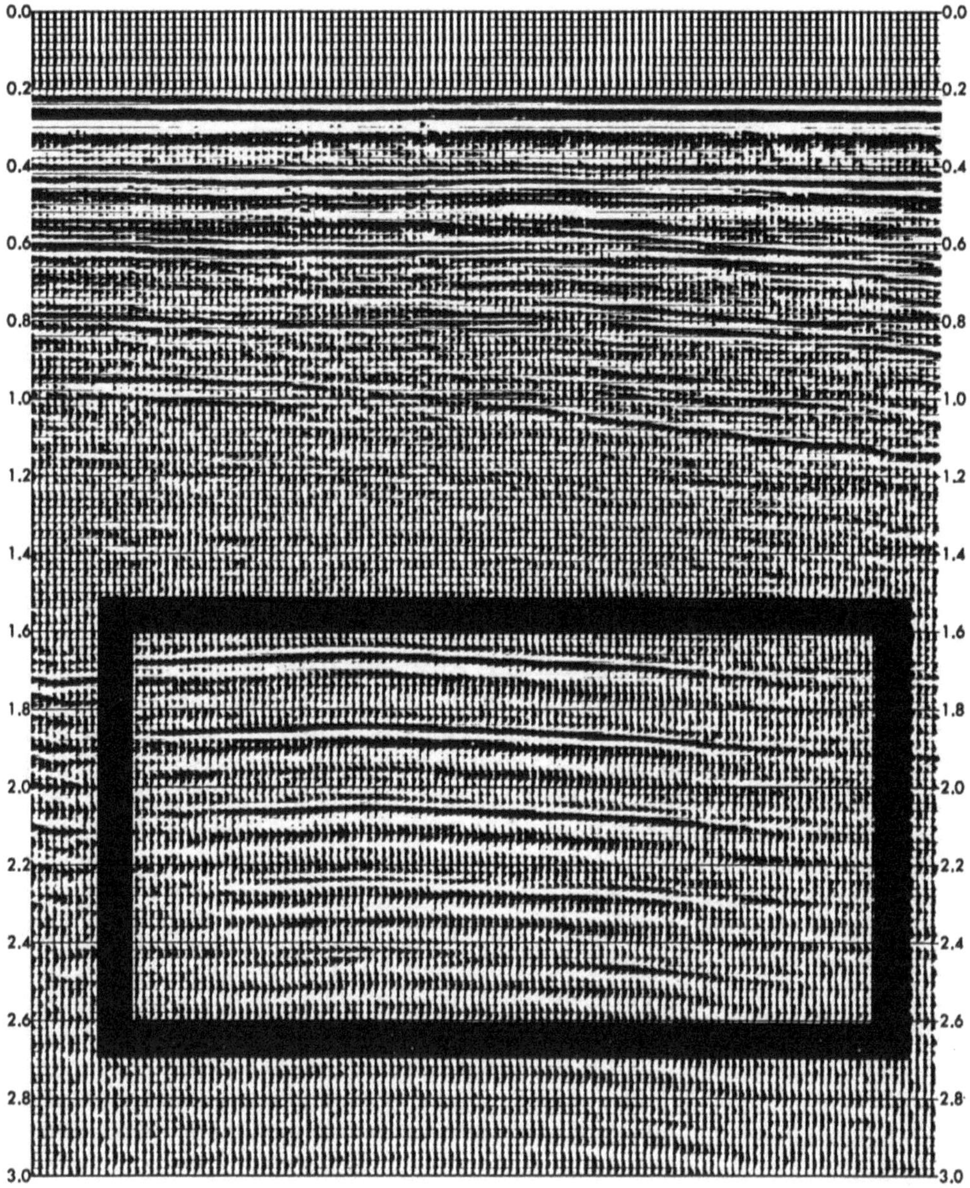

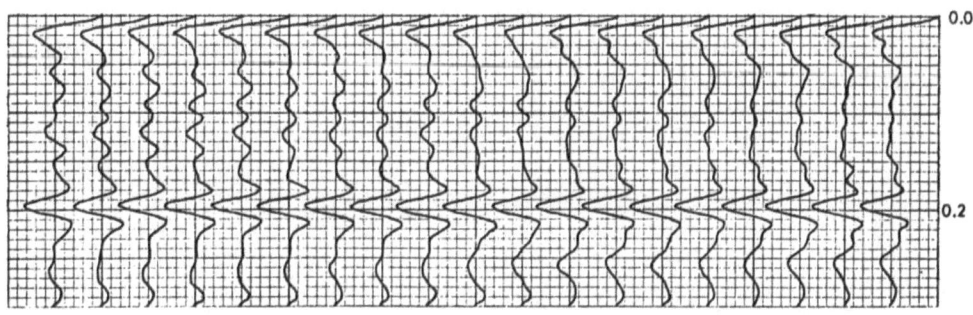

In the first category there are two interfaces — the sea floor, and a reflector at about 1 second. The sea floor has a reflection coefficient of + 0.38, and its two-way transmission coefficient is $1-R^2$ or 0.86. The effect of transmission through the sea floor is therefore compensated by dividing the observed reflection coefficient of the top of the gas sand by 0.86 (section 3.2.6); this yields a corrected reflection-coefficient magnitude of nearly 0.2. The reflection at 1 second has a strength visually at least 12dB below that of the bright spot, with comparable processing; therefore the magnitude of its reflection coefficient is about 0.04, and the correction for its two-way transmission coefficient is 0.998 — which is immaterial.

In the second category, we have to decide what to do about the frequency-dependent losses.

There are several judgements we might make before formulating this decision. First we could scrutinize the section to see whether it contained any evidence of the likely causes of significant high-frequency loss. Certainly we would expect some absorption in the shallow unconsolidated materials. Certainly we could not discount the possibility of local high-frequency loss associated with inhomogeneities; the bright spot itself is sufficient proof of the likelihood of inhomogeneities in that depositional environment. However, we shall learn in Part 5 to recognize the consolidated formations between about 0.5s and the bright spot as deep-water shales, in which we do not expect major absorption (section 2.5.1). Further, this recognition also precludes the conditions leading to a major transmission-coefficient loss and short-path-multiple effect. The same conclusion is supported by the scarcely-visible reflection "grain" in this interval; most of what is visible (except for the reflection at 1s, of course) can be seen to be due to recurring sea-floor multiples. Of course, it is theoretically possible that a highly-contrasty sequence could involve layers all so thin that no significant reflection was produced anywhere within it, but this is unusual; certainly the first conclusion from this scarcely-visible grain is that no major short-path effect is likely. So we shall be expecting some high-frequency loss at normal frequencies, but not of excessive proportions.

When we come to assess the magnitude of the loss, we find that the material of Parts 2 and 3 has really left us with only three methods.

The first of these is that which allows the separation of reflection coefficients and losses by the use of the upward and downward reflections from the same interface (section 2.6.4). Since there is no suitable deep reflector below the bright spot, this method cannot be used to obtain the losses in the whole of the section above the bright spot; however, it might well have been used to obtain the losses in the material down to the reflector at 1s, and (since these losses in the less-consolidated upper material are likely to represent the greater part of the total loss down to the bright spot) it is unfortunate that the special stacking necessary for this approach was not attempted in the present case. Somebody goofed.

The second and third methods (section 2.6.4, 3.4.7 and 3.4.8) are both concerned with measuring the change in frequency content between the reference reflection (in this case, the sea floor) and the target reflection, on the assumption that all operative agencies are high-cut. The second method attempts to measure this change by computing an amplitude spectrum over an interval straddling each of the two reflections, and the third attempts it by envelope amplitude measurement in several frequency bands (the "filter panel").

Because frequency spectra do not "live" to the interpreter in the way that wiggles do, we should always be careful not to draw an important conclusion from frequency spectra without studying the wiggles also. In fact, the present case illustrates well the cautions that are present in the time waveforms. On page 4-54 we see the individual stacked traces over the central part of the bright spot, and we can compare visually the detailed wave shapes of the sea-floor and bright-spot reflections. On the sea-floor reflection, for example, we see a low-frequency tail and an indication of interference, both of which must distort the spectrum obtained with any window of practical length. These effects may be due to nmo stretch and/or a likely transition of density just below the sea floor (p.2-16) and/or an actual depositional discontinuity of acoustic impedance at very shallow depth. The first two, and possibly the third, make a spectrum unduly rich in low frequencies. At the bright-spot level we again see two complications: the low-frequency appearance of what our optimist has defined as the bottom-sand reflection, and the generally skew-symmetric appearance of what he has defined as the top-sand reflection. The first of these must make the spectrum unduly rich in the low frequencies, but the second — taken in relation to the more symmetrical form of the sea-floor reflection — is one of the hallmarks of a dipole reflection doublet (page 2-110) which would generally favour the high frequencies. All of this demonstrates very well the problem set out in item 5 of section 3.2.8 — that measured changes of frequency content are suspect unless the reflectors involved are spaced sufficiently well to be discrete. Only if every consideration (including those of geological likelihood as well as those of reflection waveform and frequency spectrum) allows us to believe that we are seeing <u>only</u> the effect of the high-cut agencies we wish to compensate — only then dare we go through the arithmetic of computing the ratios of the spectra at the reference and target levels, and so of computing the amplitude loss brought about by these agencies. In the present case, clearly we dare not.

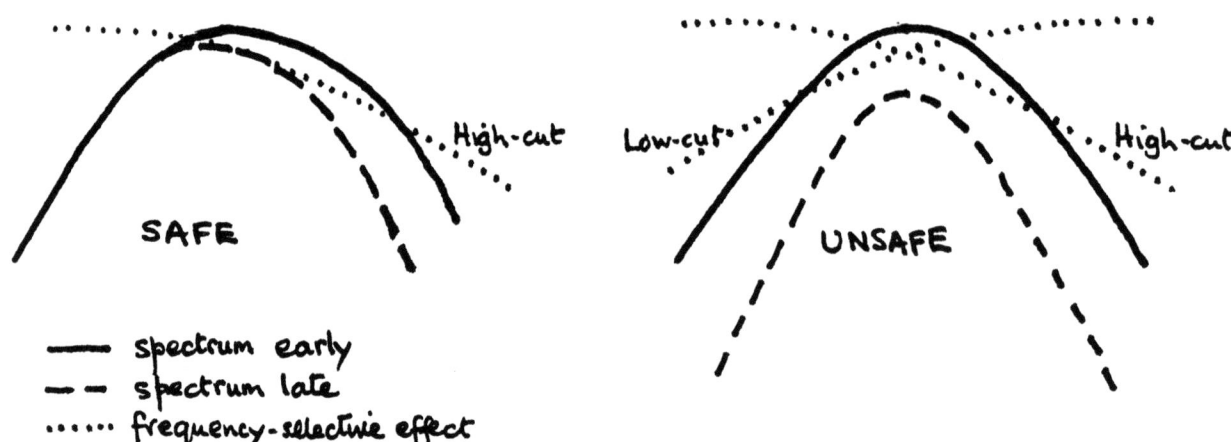

——— spectrum early
– – spectrum late
······ frequency-selective effect

The third method — the filter-panel approach of sections 2.6.4 and 3.4.7 — has some advantage over the directed frequency analysis just discussed, in that its nature eliminates the risk of interpretation of the frequency domain in isolation from the time domain. In the present context, we wish to apply

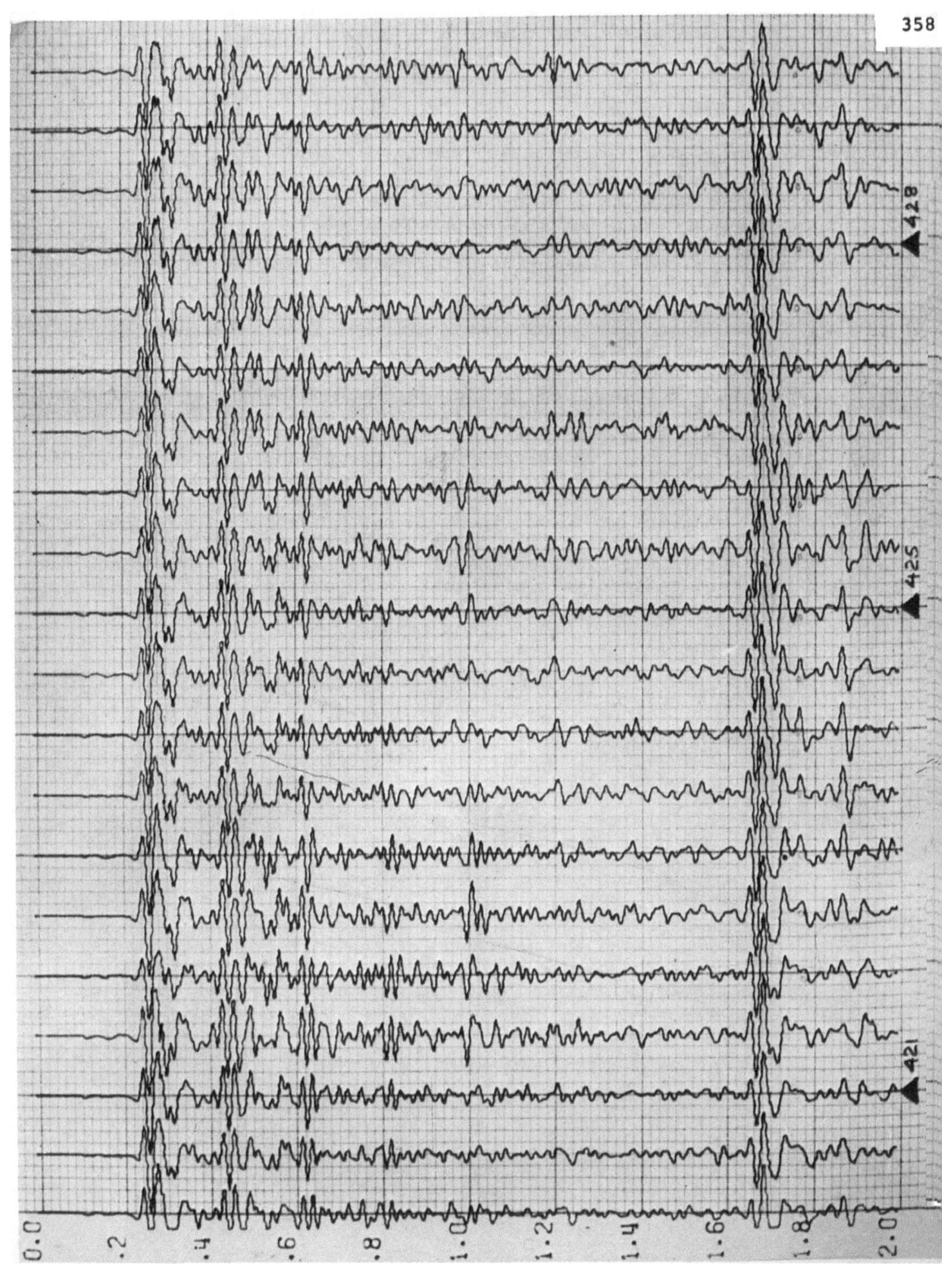

successively lower high-cut filters, until the ratio of the strength of the target reflection to the strength to the reference reflection no longer increases significantly; this ratio is then substantially immune from absorption and other high-cut effects.

However, we remember the limitations of this method: the fact that the different filter operators create different possibilities for interference between closely-spaced reflections, and the fact that, as in the last method, errors arise if effects exist other than the high-cut effects we are trying to compensate. The conclusion therefore has to be that, while the method makes a good check in those cases where target and reference reflections are discrete, it fails in the present instance.

What are we to do, then, in our present example? How can we compensate the high-cut frequency-selective effects, and turn our apparent reflection coefficient viewed from the surface into the real reflection coefficient as it exists in the earth? In this case, we cannot. The failure to try the first loss-estimating method was a minor tragedy.

All we can do is to ignore the frequency-selective losses, and take the apparent reflection coefficient as the true reflection coefficient. This must be unreal, of course; it is inconceivable that the losses do not exist at all. However, two observations can be made. The first is that neglecting the losses leads us to accept a reflection coefficient which is numerically too small; if the target is indeed a gas-saturated sand this yields a porosity estimate which is <u>conservative</u>. Sometimes a minimum value for porosity is of positive value. The second observation is that, whatever the real magnitude of the frequency-selective losses, they certainly do not approach the 0.2 dB/wavelength used as an illustration on p.2-80. For the mid-band loss over a travel-time of 1.6s would be about 10dB, which would make the magnitude of the corrected reflection coefficient of the target reflector more than 0.6. This is unthinkable. The continual recurrence of examples like this adds more and more credibility to the belief aired at the end of section 2.5.1 — that experimenters have erred in ascribing to absorption alone which are due in part to other agencies, and that commonly-quoted absorption values of the order of 0.2 dB/wavelength are too high. In a case like the present one, where highly cyclic stratification is not expected, the combined loss due to absorption and scattering could scarcely be even as large as 0.1 dB/wavelength.

So let us continue with our optimist's attempt to calculate the porosity of his gas sand, accepting that what emerges will represent a <u>minimum</u> value for the porosity. To do this we proceed with our previous computed value of 0.2 for the magnitude of the reflection coefficient at the top of the sand — satisfied at least that it cannot be <u>less</u> than this.

The next step is to compute the acoustic impedance of the shale just above the postulated sand. First we need the velocity.

There is no layer above the bright spot which would be thin enough to allow its interval velocity to be applied without the use of a compaction law. In fact there is no significant reflector above the bright spot except the one at 1s, and that (as we have already noted) has a reflection coefficient which is rather small. We would still use it, of course, if it were associated with good consistent velocity picks. However, primarily for the sake of illustration, we shall ignore this reflection, and compute the velocity above the bright spot on the basis of a single compaction law from the sea floor down.

As has been stressed many times in the foregoing material, it is one of the concerns of the interpreter to assemble all borehole and other information available in a mature area, and to deduce the appropriate compaction laws for density and velocity wherever possible. If, as in the present case, we have no boreholes in the area, we can do no better than to use a general law applicable to broadly comparable geology; the outcome will be no better than this approximation.

Herring's compaction law (p.2-36) is derived from a deeper part of the basin which contains our present example. So our best approximation is to accept the general form of his law, and also his value for the gradient B. Thus, from p.2-35, we have that
$$T = Be^{-A/B} (e^{Z/B} - 1),$$

where B has the value -14000 if we are working in feet, or -4270 if we are working in metres.

The two-way time to the top of the postulated gas sand, at its thickest point, is 1.65 s, and the stacking velocity is 2135 m/s (7000 ft/s). The T_O correction, known from study of the auto-correlations (section 3.5.3), is 0.05s, so that the true two-way time is 1.6s and the one-way time is 0.8s. Properly, the stacking velocity should now be converted to average velocity; in the present case it will make very little difference, so we ignore this correction, and compute our depth z as 1700 m (5600 ft). Then, substituting into the above equation (remembering that T enters the equation in microseconds), we emerge with a value for A. It is 72200 if we are working in feet, or 27100 (note this not a simple conversion) if we are working in metres.

Then we use the equation for the local velocity V_z, from p.2-35:

$$V_z = e^{(A - z) / B}.$$

From this we compute that the velocity in the shale just above the bright spot is 2620 m/s (8600 ft/s). This is the value we need.

If we were apprehensive about the geological propriety of accepting the gradient B as equal in two parts of the basin, while expecting the near-surface velocity implicit in A to be different, we could easily check whether the assumption makes any difference. In this case, the V_z value obtained by keeping A constant and allowing B to change is substantially the same.

While we are doing these calculations, we should check the reasonableness of the near-surface velocity resulting from the values of A and B employed; for z=0 we compute the local velocity as 1760 m/s (5770 ft/s). Therefore we can draw, as shown here, a numerical version of our previous sketch 2-54. This drives home to us the importance of the compaction law in a shallow or thick layer; it would be quite wrong to accept the seismically-measured interval velocity of 2135 m/s (7000 ft/s) as the local velocity just above the target, since the latter is nearly 500 m/s higher.

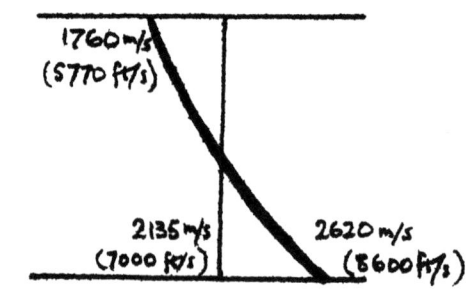

Now we need the density in the shale just above the target. Again, we would be using local borehole data for the velocity-density-lithology relation if the area is mature, and only in an area without boreholes would we use the curves of p.2-50 or the $V^{¼}$ relation. In the present example the best we can do is the shale curve of p.2-50, which yields a density of 2.28×10^3 kg/m^3.

The acoustic impedance of the shale above the postulated gas sand is therefore $2620 \times 2.27 \times 10^3$, or 5950×10^3 kg/m^{2}s.

We can now use our value of the magnitude of the reflection coefficient to compute the acoustic impedance below the bright reflection. The magnitude, we have said, is 0.2; if our optimist is right in postulating a highly porous gas sand, the reflection coefficient (at such a magnitude) could only be negative. Therefore we insert a reflection coefficient of -0.2 into the equation of section 2.4.1, and emerge with a value of 3970×10^3 kg/m^{2}s for the acoustic impedance of the postulated gas sand.

At this stage we are on common ground with the top-and-bottom method; we need the velocity within the sand in order to turn the measured acoustic impedance into a density, and so into a porosity.

But the time separation between the postulated "top" and the postulated "bottom" is only 40ms, in this case. Is it possible to obtain an interval velocity, with present techniques, over such a small interval?

TAPE 26   The question is of the greatest importance. If our optimist is right, this interval velocity leads to a minimum porosity estimate. If he is wrong, the interval velocity measurement contains the proof that he is wrong.

So we resolve to try. We know that if it is possible it is only just possible; therefore we act on the golden rule for such circumstances — use only the good data. First we scrutinize the stacked traces, and then the individual traces of each gather, to discard all input which is corrupted by noise or by local distortions of spectrum (poor shots, deep or shallow source or hydrophones, weathering anomalies, doubtful offsets, etc.). Then we decide where to locate the velocity analyses, and how many depth-points may safely be included at each location. These are critical decisions; once again we are reminded that painstaking selection of a few measurements is much better than blind averaging of many.

Over the thick part of the bright spot, only three velocity locations are acceptable on these criteria — one with 5 depth-points, and two with 7 depth-points. The locations are marked by arrows on p.4-54. We then compute these analyses using every means available to optimize the resolution: extreme spiking deconvolution, a short window (24ms), and small increments between successive positions of the window. At this level the interpreter is even interested in the details of the contouring program used in the velocity analysis.

251 L The three resulting analyses are shown on pp.4-60 to 4-62. Initially let us look at the analyses through the prejudiced eyes of our optimist. They light up when he sees the elongated appearance of the contour at and below the bright zone, 1.64 - 1.68s. The same elongation is visible on the water reverberation, 1.83 - 1.87s, and he notes (quite correctly) that there is no reason why reverberated arrivals should not be used for corroborative velocity measurement. The axis of the elongated zone is substantially at a constant rms velocity of 7000 ft/s (2135 m/s), which could imply an interval velocity within this zone of the same magnitude. By this stage our optimist is too excited to look very carefully at the details of the arrival times; he

252 L sees the picture he expects. He even sees (on the analysis of p.4-60) some evidence for two distinct arrivals which — again in an excess of enthusiasm — he quickly assumes to be his "top" and "bottom". So he accepts an interval velocity of 2135 m/s in the postulated gas sand, and uses this and the previously-

253 L computed acoustic impedance of 3970×10^3 kg/m^{2}s to calculate the density in the sand; this emerges as 1.86×10^3 kg/m^3. Then, working at white heat, and following the assumptions and technique set out in section 4.4.1 on the top-and bottom method, he computes the gas density at a depth of 1700 m as 0.1×10^3, enters the final equation, and concludes that total gas-saturation implies a porosity of 31%.

He can scarcely contain himself.

In fact, he is wrong.

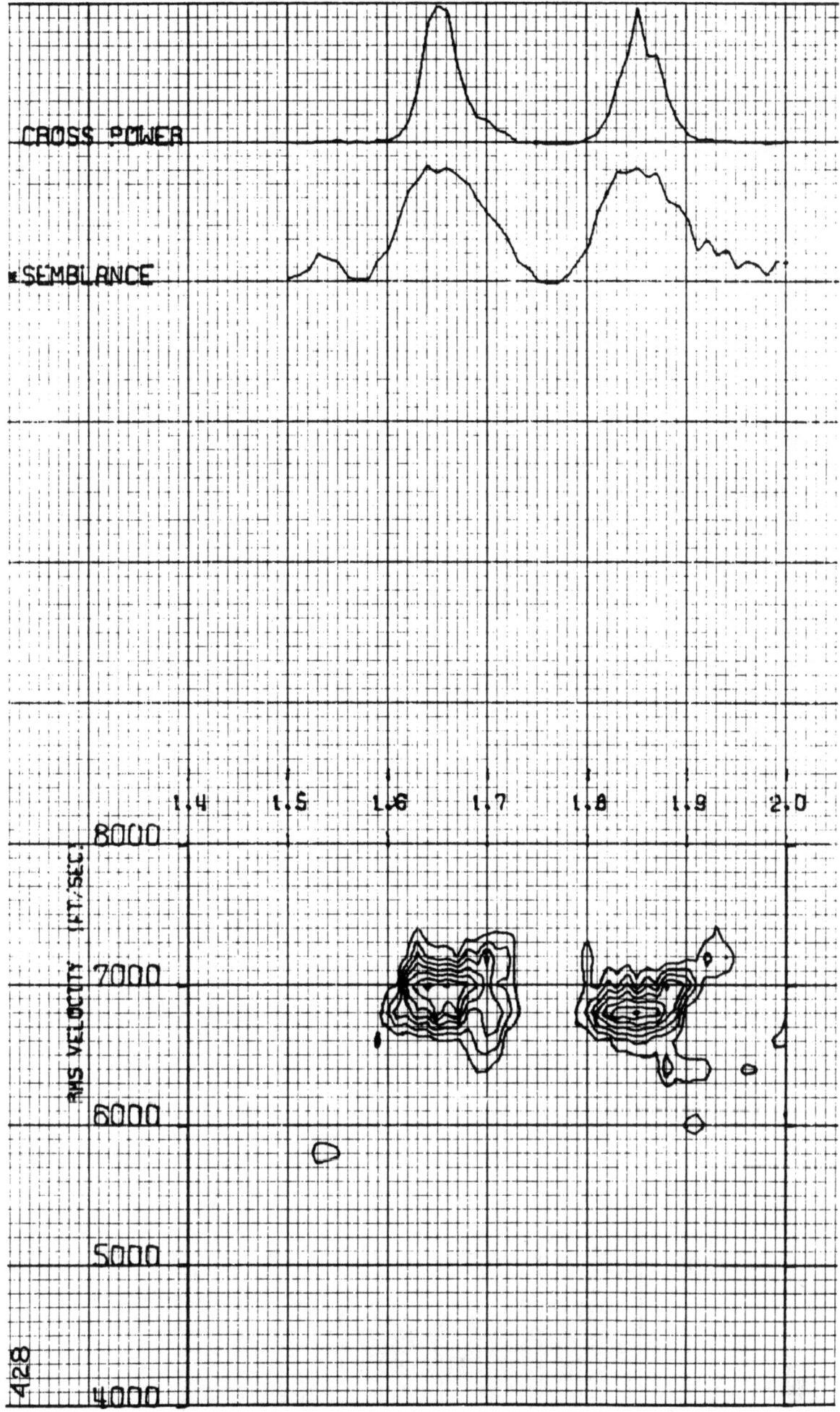

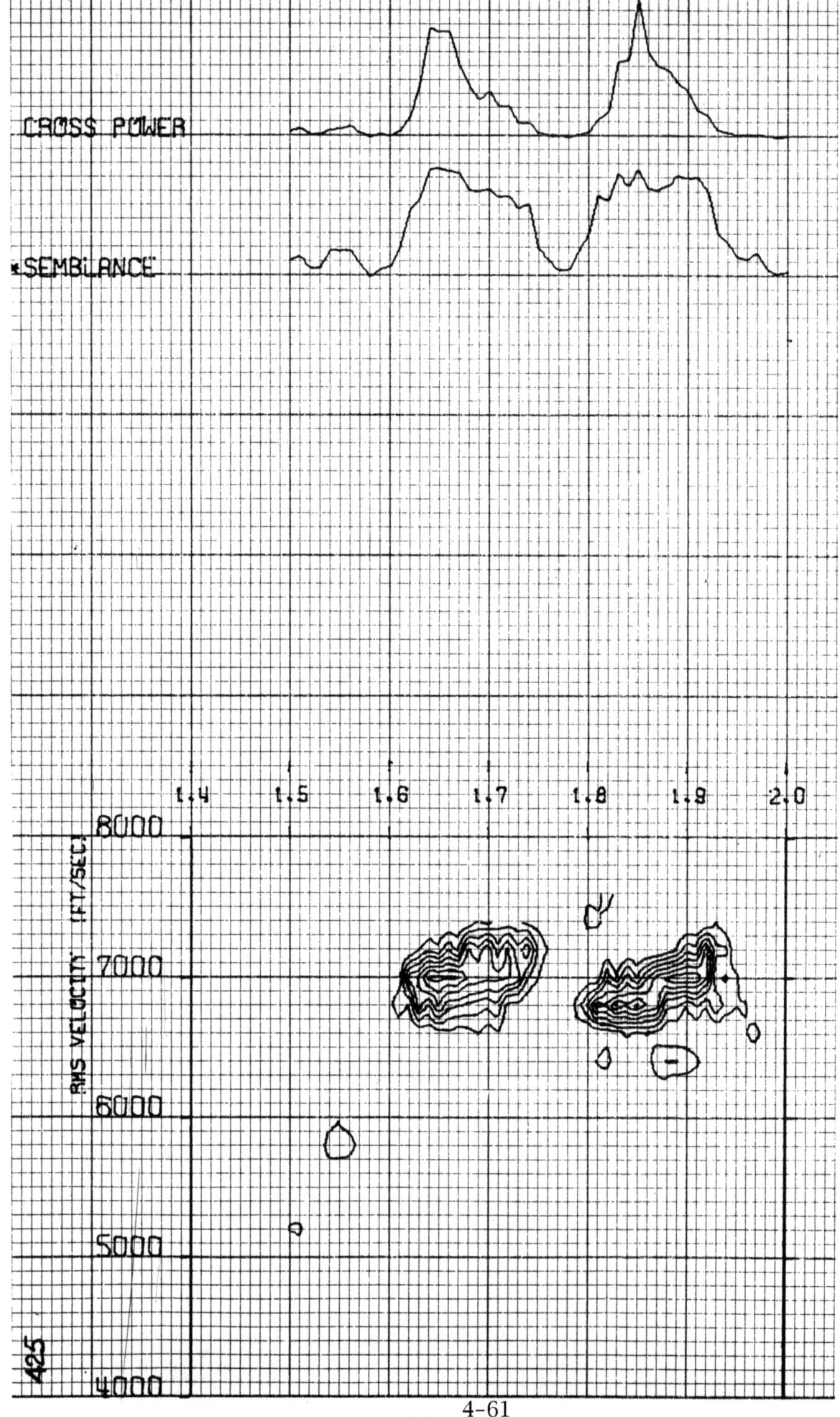

4-61

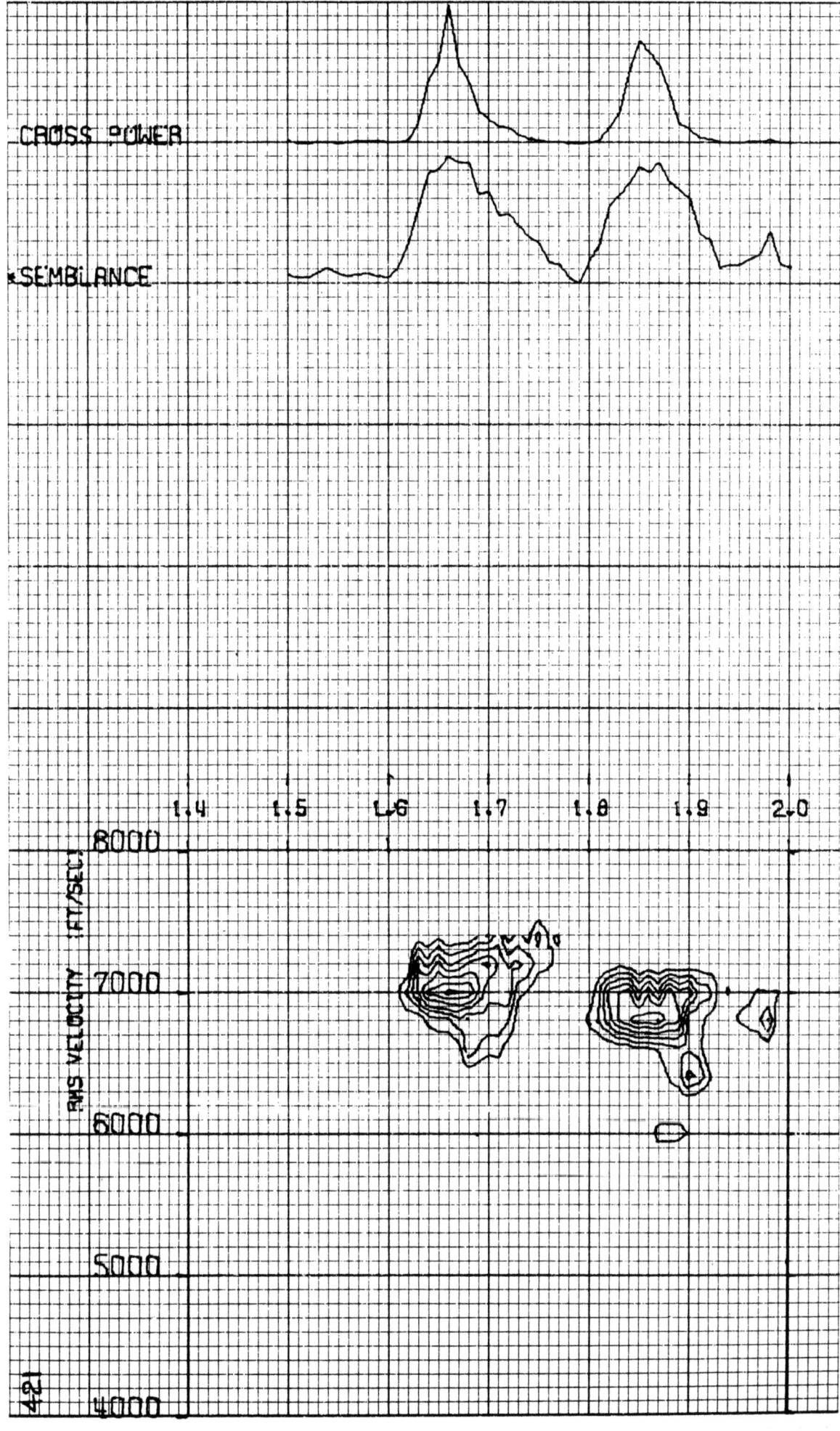

4-62

The first point which must be stressed about this is that his error does not lie in the top-only method as such, nor indeed in any of the physical or computational steps he followed. His error lies in his geological identifications, and in ignoring the subtle evidence in the seismic data which was telling him that those identifications were dubious.

A well was drilled on this bright spot, about 2km from the line displayed on p.4-46, and was announced as an oil discovery. The art of coy diffidence, once the prerogative of modest maidens, now survives only in oil-company announcements; we are not surprised to learn very little from the release. We are also unsurprised by the industry rumour that the oil is not at the bright-spot level, but below. How far below we do not know — perhaps it is associated with the deep fault.

At this juncture, let us return to our original discussion of the possible interpretations of the bright-spot feature. In the early part of this section, we considered the possibilities sketched on p.4-64:

- a lenticular sand full of gas,

- a lenticular sand partially gas-saturated and partially oil-and/or water-saturated,

- a totally non-prospective acoustically-hard anomaly (for example, volcanics or a carbonate build-up).

Our optimist selected the first of these, and went wrong. Let us review the seismic appearance of each of these interpretations, and see to what extent our optimist could have protected himself against his error.

All three interpretations are easily reconciled with the observed reflection-strength anomaly; in the first and third interpretations we have a major acoustic contrast, and in the second (particularly if the thickness of the gas zone is of the order of a quarter-wavelength) we have the additional possibility of constructive interference. However, the strong reflection of the first interpretation must be negative, that of the third must be positive, while that of the second is likely to be of indeterminate polarity but should show the evidence of interference. So we must check the polarity evidence very carefully.

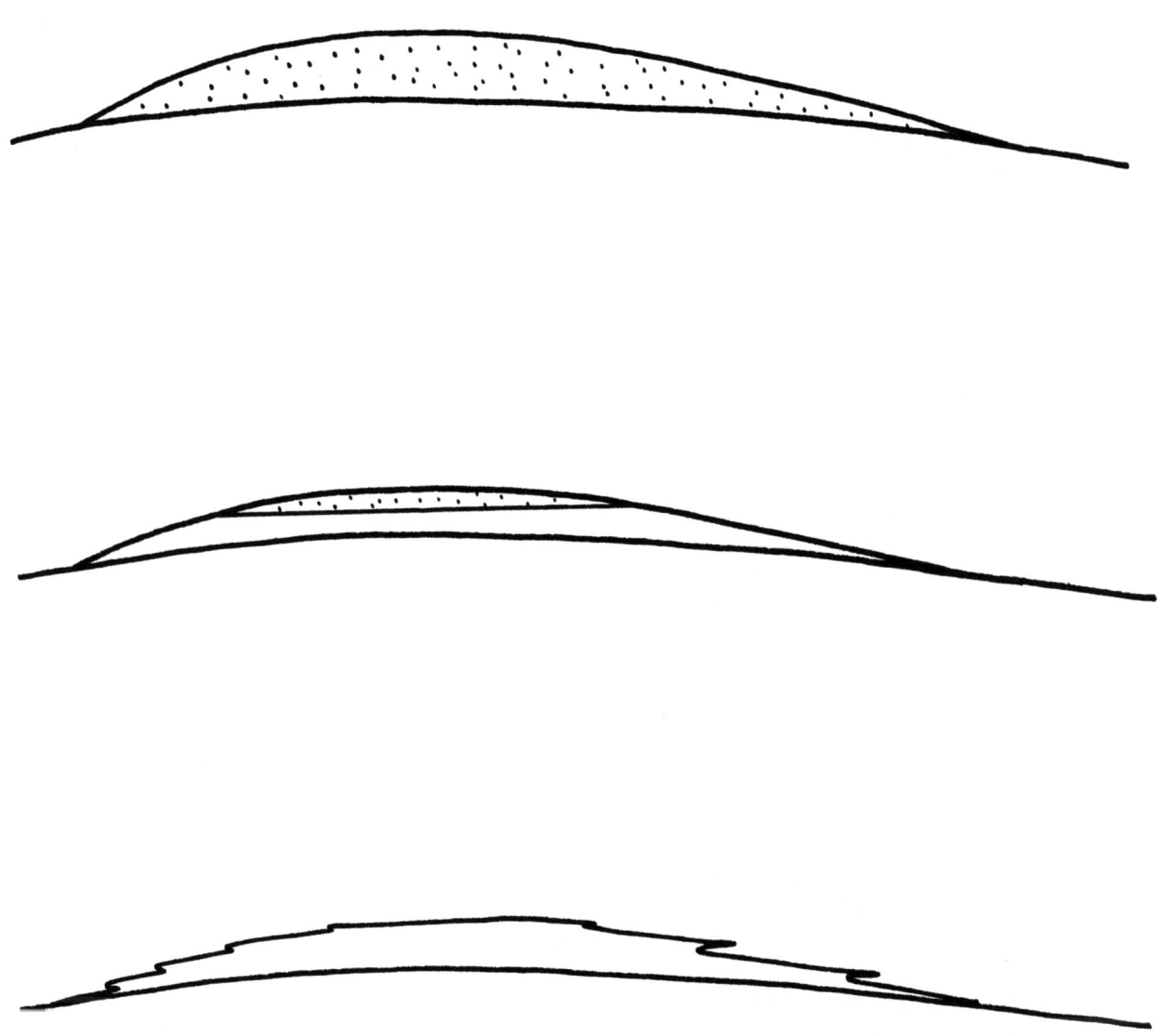

4-64

We have already discussed (p.4-45) the problem of the "bottom" reflection and its strength. The other key variable, of course, is the interval velocity; the first interpretation must be associated with a low velocity, the third with a high velocity, and the second with low velocity in the gas zone and medium velocity in the oil and/or water zone. Therefore we look again, and more carefully this time, at the representative individual traces of p.4-54 and the velocity analyses of pp.4-60 to 4-62.

From p.4-54, we see that the bright-spot event is clearly a complex. The general appearance of the event is of a small trough followed by a large peak and a large trough of about equal magnitude, followed by a hickey indicating an early leg of the "bottom" event, and then the bottom event itself. The first small trough and the large peak, when compared to the sea-floor reflection, suggest that the "top" reflection is positive. However, the large peak and large trough, when compared to the sea-floor reflection, suggest that the "top" reflection is itself a dipole complex. We are looking for a dipole complex if we favour the second of our possible interpretations, but for that the uppermost reflection of the complex must be negative. If the uppermost reflection is positive, therefore, we have a suggestion that we have a thin hard layer overlying the unit of interest. This is a new possibility which we did not consider on p.4-64.

However, in making comparisons to the shape of the sea-floor reflection, we must remember that the latter may itself be a complex. As we discussed earlier, there is some suggestion of a second reflection (also positive, and possibly the density increase of p.2-16) just below the sea floor. This drives home to us the importance of using as a pulse-shape reference not the sea-floor reflection but the recorded source pulse, processed just like the reflection data as recommended in sections 3.5.3 and 8.2.4. In the present case the source pulse was not recorded, and so we cannot be entirely sure about the polarities.

The other polarity approach would be to look solely for a polarity change along the reflector. Such a change is apparent to the right of the bright spot if we view it at very compressed horizontal scale. However, there still remains the possibility

4-65

that this is really just a change in the units or the spacing present in the complex, or the consequence of minor faulting.

So we have to say that polarity studies (with the limited data available) do not completely resolve the question. What they do say seems to favour a two-layer possibility — with a thin hard unit as the upper layer.

Therefore we turn to the velocity analyses again (pp.4-60, 61 and 62). And now we pay much more attention to the details.

Out hope is to separate the "top" and the "bottom", and to this end we have used windows of 24ms. This should be sufficiently small to separate them, but if the "top" is actually a dipole complex (whether a thin gas layer or a thin hard skin) it cannot hope to resolve this complex. In fact, support for the complex view comes from the elongated appearance of the velocity contours in the zone 1.64 - 1.68s. This appearance was too quickly assumed by our optimist to be diagnostic of a low-velocity gas zone; this interpretation is possible, but not necessary. Visualizing a 24ms window traversing across the pulse shapes of p.4-54, we can be sure that it cannot be expected to assess a moveout difference between the large peak and the large trough of the complex. So there is probably no interval-velocity inference to be drawn from the elongated contours in this case; the thin upper layer could be of high or low velocity. So our final hope is for a useful interval-velocity measurement between the upper complex and the "bottom". The "bottom" event is clear enough (though our optimist chose not to see it); it is at about 1.7s on all three analyses. Its stacking velocity is about 7180 ft/s (2190 m/s). The resulting interval velocity is very dependent on the time pick that we assign to the upper complex; a mean pick leads to an interval velocity range of 3660-3970 m/s (12-13000 ft/s) over the three analyses, while realistic upper and lower bounds on the possible time picks enlarge the range by 350 m/s (800 ft/s). What is clear, of course, is that we have a high-velocity unit.

Very tentatively, then, we emerge with the conclusion that the feature is a unit of fairly high velocity, capped by a thin layer which (since we think it represents a dipole complex) must be even harder.

If we now accept that the "top" reflection is a complex too thin to be resolved, we can see that it was a mistake to compute its reflection coefficient, as our optimist did, assuming it was discrete. The computed reflection coefficient represented the interference of reflections from the upper and lower boundaries of the thin upper layer. Consequently, it is entirely feasible that the upper layer is a small fraction of a pulse length in thickness, and therefore that the reflection coefficient at the top of this layer is actually larger than the 0.2 computed for the complex. The range of possibilities can be explored by the modelling techniques of section 6.2, varying the thickness, velocity and density of the top layer to the end that the resultant complex shows an effective reflection coefficient of 0.2.

What is very clear from all of this, of course, is that the bright spot does not signal gas, and that the careful interpreter could have established this fact from the seismic data.

Why was it good to take this example, and to pursue it so far?

- Because it was an illustration of circumstances — not too easy and not too difficult — in which the top-only method was the key to quantitative conclusions.

- Because it illustrated the application of a compaction law, and the calculation of reflection coefficients by auto-correlation methods.

- Because it was a cautionary tale, warning us against a too-facile association of bright-spot amplitudes with gas.

- Because it exemplified the tenet that the different seismic measurements (time, velocity, strength, frequency content, polarity) must not be viewed as unrelated measurements; they must be <u>mutually compatible</u>, or the conclusions from any one of them are erroneous.

- Because it illustrates well the boundaries of interval-velocity analysis; nowadays we can get useful velocities in a layer whose two-way thickness is a mere 40 ms if the record quality is good, but only an improvement in bandwidth will allow us to do better than this.

- Because it illustrates clearly that the basic data in the sequence of calculations, and all the data used in the oh-how-important checks, must be recorded and processed with an altogether new standard of detailed care and understanding which only the <u>interpreter</u> can dictate.

Finally, we see very clearly the importance of iterating around a processing-interpretation loop. It often happens that each new conclusion sends us back to reprocess the data, so that the processing shall be more specifically appropriate to that conclusion. Of the many examples present in the case just studied, let us identify just one; at an early stage we were trying to draw conclusions from the low-frequency appearance of the "bottom" reflection (overpressure? transitional reflection? absorption?), but at a later stage, when detail velocities became available, we found that the lack of detail on production (48 ms) velocity analyses had let us to mis-stack this reflection originally. A final conclusion would need to be based on restacked data.

4.5 THE LIMITATIONS AND VALIDATION OF DIRECT DETECTION

Repeatedly, in the foregoing material, we have agreed on the importance of recognizing the factors which limit the techniques of direct detection, and of utilizing all possible means of checking each conclusion. It is inexcusable to drill a bright spot without going through this exercise; as we have just seen, a not-too-onerous analysis of seismic details can save (or justify) the expenditure of millions of drilling dollars. Therefore, let us set out the limitations formally, and then do the same for the validation techniques. Safety may be boring, but it is very very important.

4.5.1 The limitations

1. <u>The theory</u> Always, in the backs of our minds, we should remember that there are several relevant features of the basic theory which are not yet fully solved. We do not have a complete theoretical solution for our prime concern — the reflection of spherical waves on the boundary between thin layers of porous granular material. We use simple ray theory for many highly important operations (including velocity analysis), and our techniques must be deficient in the presence of small faults, inhomogeneities and the phenomenon of wavefront healing. We know about these problems, but we have only rudimentary and non-quantitative ways of allowing for them in the course of everyday interpretation work.

2. <u>The geology</u> Reminding ourselves that we can measure only the <u>acoustic</u> consequences of geologic features, we must be very conscious of the ambiguities this allows. We have learned that there are fundamental acoustic ambiguities between basic lithology, degree and type of cementation, fluid content, tectonic stress, differential pressure, burial history, fracturing (including scale, type and dominant direction), shale content, and many other properties. Reflection-amplitude anomalies and velocity anomalies can be analysed in detail <u>only</u> when some of these variables are known or can be guessed in a geologically prudent manner. In particular, porosity estimates are impossible without

a knowledge of the matrix material and the saturating fluid. And we know of no seismic distinction between interconnected pores of good permeability and sealed pores of zero permeability.

Further, we must always be conscious of the poor vision of the seismic method where two very real and very important geological features are concerned — the thin bed and the slow vertical gradation. Our vision of the first is limited by the seismic bandwidth at the high-frequency end, and our vision of the second by the bandwidth at the low-frequency end. We have seen many times in the foregoing material that our numerical techniques are sound if the reflectors considered are separable and discrete; these techniques must be applied with understanding or not at all if the reflections are actually complexes or transitional. Fortunately the top of a gas-saturated reservoir is usually an abrupt discontinuity of acoustic impedance, but, as we have seen, the reservoir itself may evince a gradation of porosity, and the shale underneath may evince gradation associated with local overpressure.

Further again, we have to accept that realistic degrees of geological inhomogeneity must sometimes distort reflection amplitudes and spectra, locally, just as they distort reflection times. Numerical computations are suspect if they are made below faults, or overthrusts, or old shore lines, or deltaic zones, or intrusions, or shallow gas-charged sediments, or any other feature which in its nature must distort the seismic waves from the simplicity we are forced to assume for computational purposes.

3. The field work   Obviously a new stringency is required in all aspects of the field work which can affect the validity of our porosity computations. Among the quality-control criteria which acquire a new importance are:

- the performance of the source,

- the constancy of depth of the source,

- the constancy of depth (or depth pattern) of the streamer,

- the leakage on the high-impedance (hydrophone) side of the streamer transformers,

- the feathering of the streamer,

- the offset distance,
- the total avoidance of clipping in the detectors and instruments,
- the trim of the detectors and instruments,
- the scrupulous logging of the field work.

Even more important than these considerations, however, is that the field work should be designed for the problem. This is discussed in more detail in Part 9; for the present let us stress and stress again the importance of ensuring sufficient bandwidth and sufficient signal-to-noise. In the processing we can, to some extent, trade one for the other; only in the field can we ensure sufficient of both. And the meaning of "sufficient" can be decided only after definition of the geologic problem.

4. The processing We have already discussed, in section 3.2, the processing corrections which must be applied if the field data are to have genuine geological significance. We remember in particular the corrections for source and streamer depth, and the compensation for variations in the strength or frequency content of the source. We remember the importance of using a proper correction for geometrical divergence, and the dependence of this on good velocity analysis and geologically reasonable interpolations in both vertical and horizontal directions. We remember that while we may find it expedient to apply exponential expansions or other "curve-fitting" techniques at intermediate stages of the processing, we must certainly replace these by a geometrical divergence correction, in fact or with a calculator, before we make any numerical comparisons of reflection amplitude (either vertically or laterally). We remember from section 3.1.3 the problems of maintaining a valid amplitude ratio between shallow and deep events, when the former represent only a few traces in the stack and the latter represent many. We remember the importance of preserving valid changes of frequency content by judicious muting on a criterion of nmo stretch. We remember the importance of the interpreter's input to the eternal problem of finding the appropriate compromise between signal-to-noise ratio and resolution. We remember the dependence of the optimum velocity-analysis parameters on the nature of the geological problem,

and the special difficulties which arise when one or more of the reflectors analysed is complex or transitional.

We are also particularly conscious of the limitations introduced into nearly all of our processing techniques by our failure to treat the third dimension properly. It is often necessary to iterate around the processing-interpretation loop specifically for this.

At this stage we should also take note of a fundamental indeterminacy in seismic processing. This is discussed in greater detail in sections 8.3.2 and 8.3.3; for the present let us just observe that an autostatics program can optimize the stack on each stacked trace (and thereby make us reasonably confident of the stacked amplitudes) but it cannot distinguish between a dipping section and a slowly thickening weathering. In the same way we need to be cautious of automatic amplitude-adjusting programs, which are very good at compensating individual bad shots or individual bad geophones, but cannot distinguish between slowly brightening reflectors and slowly improving source efficiency. This remains a fundamental problem to which the only answer is the recording of the outgoing source signal; if this is not done the interpreter must remain conscious of the risk it represents.

4.5.2 Methods of internal validation

In all efforts to make interpretation quantitative, the geophysicist is constantly on the edge of what is possible. Therefore it is of the utmost importance that he school himself to make all feasible checks on his conclusions (or at least on all conclusions which commit considerable sums of exploration money).

The first fundamental criterion, obviously, is that all quantitative measurements and all geological interpretations should be mutually compatible. The standard test of this is the modelling of the interpretation, as we shall discuss in Part 6. However, as we shall find in that discussion, the modelling techniques are plagued with an embarrassment of options; therefore it is a good practice to explore several other specific validation techniques before setting up a model.

1. <u>Geological considerations</u> It is essential that all inferences from the physical measurements should be geologically reasonable, both with regard to the structural picture and with regard to the depositional history.

Often the geophysicist cannot resolve, from the seismic measurements alone, whether an observed lateral variation of interval velocity is real or not; on some of these occasions the geologist can identify features of the depositional or tectonic history which are likely to have led to a facies change in the zone in question. Often the geophysicist is at risk by reason of slow vertical gradations of acoustic impedance, too slow to yield a visible reflection with the usual low-frequency limit of bandwidth; on some of these occasions the geologist knows where in the geology such gradations are likely. Often the geophysicist cannot tell the reason for an observed loss of high frequencies, when the geologist would know of circumstances likely to produce highly contrasty cyclic sedimentation.

We have remarked earlier on the problem of distinguishing polarity inversion from small faults. We might make our first approach to this by studying the reflection pulse shapes. But we must also adopt the geological approach — to map the questionable feature, and to see whether its areal configuration makes more geological sense as a fault pattern or as the limit of a gas-liquid contact.

If geological and physical methods agree, the conclusion is strengthened; if they do not, we must determine why.

2. <u>The sea floor</u> In retrospect, it is remarkable that it took us so long to realize the value of recording the sea-floor reflection. Of course, in the early days of marine work the water was not sufficiently deep, but most of our current work is in water-depths sufficient to permit the practical recording of a near trace at a reasonable angle of incidence (20-25°).

The first benefit we obtain from the sea-floor reflection is the ability to determine, easily and with authority, one numerical reflection coefficient. Then, if we use this to "calibrate" our section, we have (subject to proper processing techniques, as described earlier) a first validation of real amplitude anomalies.

The second benefit is known polarity, determinable by autocorrelation. In the open sea this is always positive. Off major deltas and in inland seas, where considerable vegetation decays on the sea floor, a negative coefficient may be produced by trapped gas bubbles; this is a warning, of course, that the reflection is complex, because the actual sea floor below the vegetation must be positive. In general, however, this knowledge of sea-floor polarity gives us a polarity calibration for the whole section; even if the source pulse has been recorded directly, the sea-floor reflection gives us proof that it has been recorded and processed properly.

A further benefit is that, if the source pulse has not been recorded, it yields supplementary evidence for the necessary T_0 correction (which is the correction that yields the proper value of about 1500 m/s for the interval velocity in the water).

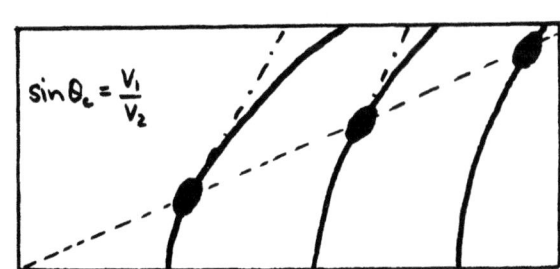

A further benefit is that it is usually possible to find the alignment of high-amplitude sea-floor multiples which indicates the critical angle. From this, since we know the water velocity, we can deduce the effective velocity in the material just below the sea-floor. This supplements the usual refraction measurements in providing a useful validation of compaction-law calculations.

Further benefits accrue in the delineation of near-surface anomalies; these have direct exploration usefulness which we shall discuss in later sections.

4-74

3. <u>Other reflection coefficients</u> An apparent reflection coefficient can be derived for all significant reflections by studying their multiples (section 2.6.2). As soon as one reflection coefficient (which may be the sea-floor) is known, then the apparent reflection coefficient of all other reflectors follows by simple comparison of amplitudes. Therefore there are always two methods of determining the apparent reflection coefficients; these two methods must check, or the reason why not must be understood. <u>The whole system of reflections and their various inter-multiples must all be coherent numerically</u>. Sometimes the general system makes good sense, but observations involving the topside and underside of one particular reflector cannot be reconciled with the others; this, as we have noted earlier, ordinarily indicates a complex.

In section 2.6.1 the top-and-bottom method was described primarily in terms of two reflections, though in the discussion and in section 4.4.1 we noted that the down-flank reflection provides a useful check. In fact, as soon as any one reflection coefficient has been determined by the top-and-bottom method, it is possible to work out from this one to every other reflection (provided, of course, we are confident that the processing has maintained the true amplitude relationships).

The coherent picture can be built up still further by careful attention to the interval velocities. Whenever the intervals are thick or shallow, this attention must include a compaction law. Our insistence on <u>full compatibility between the velocity measurements and the amplitude measurements</u> leads to the preparation of synthetic displays of reflection strength (section 3.5, p.3-98). In these, we remember, the interval-velocity measurements (as modified by the compaction law if necessary) are combined with the density estimates (from p.2-50 or from local knowledge) to yield a computed reflection coefficient at each velocity boundary, and the results are displayed in sectional form. Then the display of this synthetic reflection strength is compared with the display of observed reflection strength. Any one reflection computed by the top-and-bottom method (or estimated by any other method) may be used to set the display scales to be equal at that level.

Features of interest and value which emerge from synthetic strength displays are:

- The reflections at which we expect to see a good match are the discrete ones, between rather massive units permitting good interval-velocity measurements.

- No match is possible (of amplitude, and occasionally even of sign) if inadequate provision has been made for compaction (p.3-98).

- A local mis-match may indicate a failure to pick the velocity data at all significant boundaries of <u>acoustic impedance</u> (not velocity). In this case the mis-match may lead to identification of anomalous densities associated with lithology (salt, lignite, basalt, etc.) or erroneous densities introduced by ignoring overpressure or fracturing.

- Alternatively, a local mis-match may indicate a slow vertical gradation of major proportions, too slow to be seen as a reflection at normal frequencies. The test, if reflector spacing allows it, is to explore the very low frequencies.

- Alternatively again, a local mis-match may furnish proof of the complex nature of a particular reflection, and contribute to its decipherment.

- A local mis-match confined to the shallow section (say the first 1 second) is usually caused by errors of the compaction law, or possibly by absorption in the shallow sediments. The two are distinguishable, of course, by the frequency-selective effect of the latter.

- A mis-match confined to all reflectors below a particular level indicates anomalous absorption or other frequency-selective agency (which should be apparent also in the frequency content).

4. <u>Conversion to depth</u>   A very strong validation of the velocity picking follows from the fact that a reflection which is continuous on the time section should be continuous also after segment-by-segment three-dimensional inverse modelling into depth. We remember that in order to convert stacking velocities into velocities for depth conversion it is necessary (sections 3.5, 7.3) to model the ray-paths entering the gather — a process which requires the travel time, time gradient and stacking velocities of each event in turn. Normally this is done in segments perhaps 5-25 depth-points in width (depending on the reflector curvature). Such segments are suggested graphically on the lowest reflection of the time section in the upper half of p.4-78. Then, when the modelling is complete, we emerge with a fully migrated depth section on a segment-by-segment basis. If the segments do not butt together to form a continuous event, then we know that some part of the input is not correct; assuming that the gradients have been correctly measured and the segment width chosen with due regard to the reflector curvature and extent, this failure to butt together is an excellent indicator of:

- inadequate attention to the third dimension,

- slow vertical gradations of velocity not detected as reflections,

- reflections due to the density component of acoustic impedance,

- failure to incorporate all the velocity breaks in the picking,

- lateral gradation of interval velocities.

These problems must be unravelled before the velocities are used for depth conversion (Paturet, 1971); when this is done the validation of the synthetic strength section becomes major. These matters are taken up again in Part 7.

5. <u>The frequency domain</u>   Part of the material of previous sections leads us to a general feeling of disappointment with the frequency domain. Certainly none of our discussions or examples has shown frequency measurement as the kind of prime exploration tool which is furnished by amplitude or polarity.

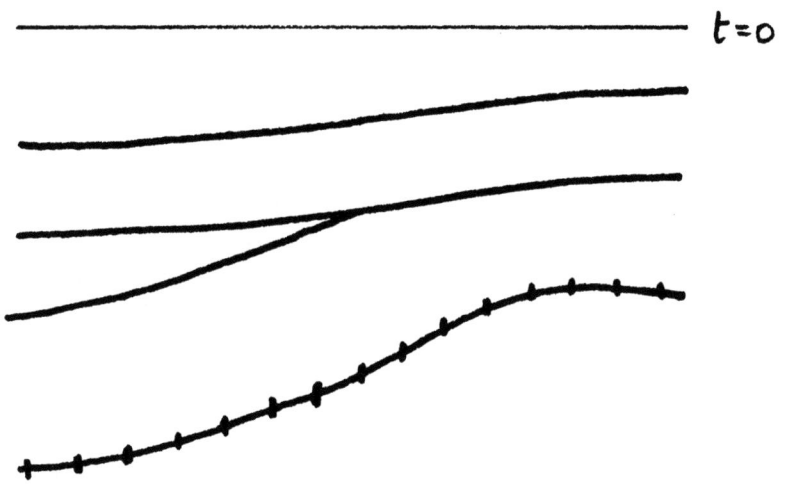

$t=0$

$z=0$

4-78

We have seen that notches in the frequency spectrum appear at about the same reflector spacing as that which allows us to see the presence of two interfering events directly, in the time domain. We have seen that the frequency changes produced by reservoir materials are small, and that they have often been confused with frequency effects due to incorrect stacking, or poor deconvolution, or simple thickening, or transition into an overpressured zone.

Nevertheless, with all that said, it is still true that measurements in the frequency domain have a corroborative value. All our other conclusions must be compatible with what we observe in the frequency domain. And we remember that the top-only method relies on frequency-domain measurements for the estimation of amplitude loss due to frequency-selective agencies, and possibly for an allocation of loss to each such agency.

Further, although we have said (ever since we computed the answer to question 2 on p.2-87) that the degree of frequency change introduced by absorption in a reservoir is unlikely to be visible, we have never excluded the possibility that it might be detectable by machine. This would need a program designed to be blind to low-frequency tails as such, but still able to detect a loss of high frequencies over a window of length appropriate to the expected degree of wavefront healing. It would also require correct stacking, and no deconvolution.

Page 4-81 illustrates the reflection-strength display associated with a believed gas accumulation at 1.1s, and p.4-82 shows the corresponding machine-generated measurement of the change of frequency content. The maximum (red) values on the latter display represent a shift of the order of 1Hz in the median spectral frequency. So we have a very sensitive machine measurement of changing frequency content, which measurement appears to correlate with a probable absorptive reservoir material. But the old problem (of section 3.2.8.5) remains: Instead of a depression of frequencies below, due to absorption, are we seeing a richness of high frequencies above, due to the dipole complex from the top and bottom of a thin reservoir?

6. <u>Velocity-lithology checks</u> Since it is necessary to make lithologic assumptions before computing porosities by any method, we take every opportunity to check the consequences of the computed porosity for velocity. For example, in our

illustration of the top-and-bottom method we had a velocity below the fluid contact of 2300 m/s (7500 ft/s), and, on the assumption that the material was a water-saturated sand, we computed a gas porosity above the fluid contact of 0.36. If the fluid contact is horizontal in depth, it is reasonable to accept that the porosity in the postulated water-saturated sand would not be more than that in the hydrocarbon zone. According to the time-average equation, a porosity of 0.36 in a consolidated water-sand would lead us to expect a velocity of 2950 m/s (9700 ft/s). We remember that this example is very shallow, and we are therefore not surprised to find the measured velocity so far below the time-average velocity. The result means that our lithologic identification is at least possible; a time-average answer less than 2300 m/s would have suggested an error.

Color plates 4-81 through 4-83
and Color plates 5-4 and 5-6 in the next chapter
follow page number 349.

TAPE 27   4.6   DIRECT HYDROCARBON DETECTION IN OLDER ROCKS

271L
Most of the preparatory material of Part 2 prepares us to accept that the major application of direct detection techniques is concerned with gas in Tertiary sand-shale sequences. This is true, of course. However, just one glance at p.4-84 is sufficient to convince us that other applications exist; the illustration is of a proven gas accumulation at 1.15 seconds in a Triassic sandstone.

272

PROJ.
273L
On a deconvolved display the top and bottom of the gas are clearly identifiable, with obviously opposite polarities (p.4-86). Over the most favourable region the velocities too are clear — above, within and below the reservoir. Further, we note that it would probably be unnecessary to consider a compaction law over the intervals and with the lithology represented here, and that there is an acceptable S_2/S_1 ratio for the top and bottom amplitudes. Accordingly we would have no hesitation in using the top-and-bottom method, as illustrated in section 4.4.1.

Since this is an example where the porosity might conceivably be the same in the gas-saturated and liquid-saturated conditions, we might also try the velocity method, as we did in section 4.4.2; before we trusted the results, however, we would have to observe that a measurement of the liquid-saturated velocity down the flank would need to be properly compensated for the dips, and that fracturing induced by the folding is likely to have introduced its own differences in velocity between crest and flank.

Of the two methods, then, we would almost certainly prefer the top-and-bottom method. However, if the answers proved discordant, we would at least try to establish the reasons for the discordance. Possibly we would find, in these studies, support for a warning that this structure should be drilled with some care; in general, direct-detection indications which are more obvious than we expect them to be (either because they are very deep or because they involve rocks not ordinarily likely to have very large porosity)

4-83

alert us to to the likelihood that pores or fractures are being held open by abnormal pressure in the reservoir. This particular structure produced a dramatic blow-out.

274
275L
276
The region from which this illustration comes is replete with examples of direct gas indications in Triassic sandstones. However, they are all at depths less than 2000 m. Where major gas-induced amplitude anomalies occur at greater depth, then the gas should certainly be assumed to be at abnormal pressure.

Several examples of direct gas indication have been found in Upper Cretaceous chalk. On a first consideration this would appear unlikely, since the matrix velocity of the chalk is quite high. However, as we have agreed many times, the geologic age and chemical composition of a reservoir are often less significant (acoustically speaking) than the conditions of fracture and pore pressure.

Since several of the world's large fields (and at least one of its giants) have been found in chalk reservoirs, it is worth developing this point a little. The chalk is highly variable in properties, being sometimes a reservoir and sometimes a seal. The exploration problem is not only to establish porosity but to indicate permeability; the porosity may well be 40% in a zone where the permeability is effectively zero. Now nothing in the foregoing material, or in the published literature, gives much hope of a seismic measure of permeability as such (at least while we are confined to compressional waves in the usual seismic frequency band). The best that the seismic method can do is to indicate the presence of extensive fracturing; beyond that, we must take our chance as to whether the fractures are inter-connected. We must also take our chance as to whether the porosity provided by the fracture system is large or small, since the velocity is not very dependent on the porosity as such. For an estimate of porosity we must work through density; this problem can be approached by standard application of the top-and-bottom or top-only methods. In practice, of course, we also introduce many other considerations. We always demand that the structural position of the postulated reservoir should be geologically plausible, and the conditions

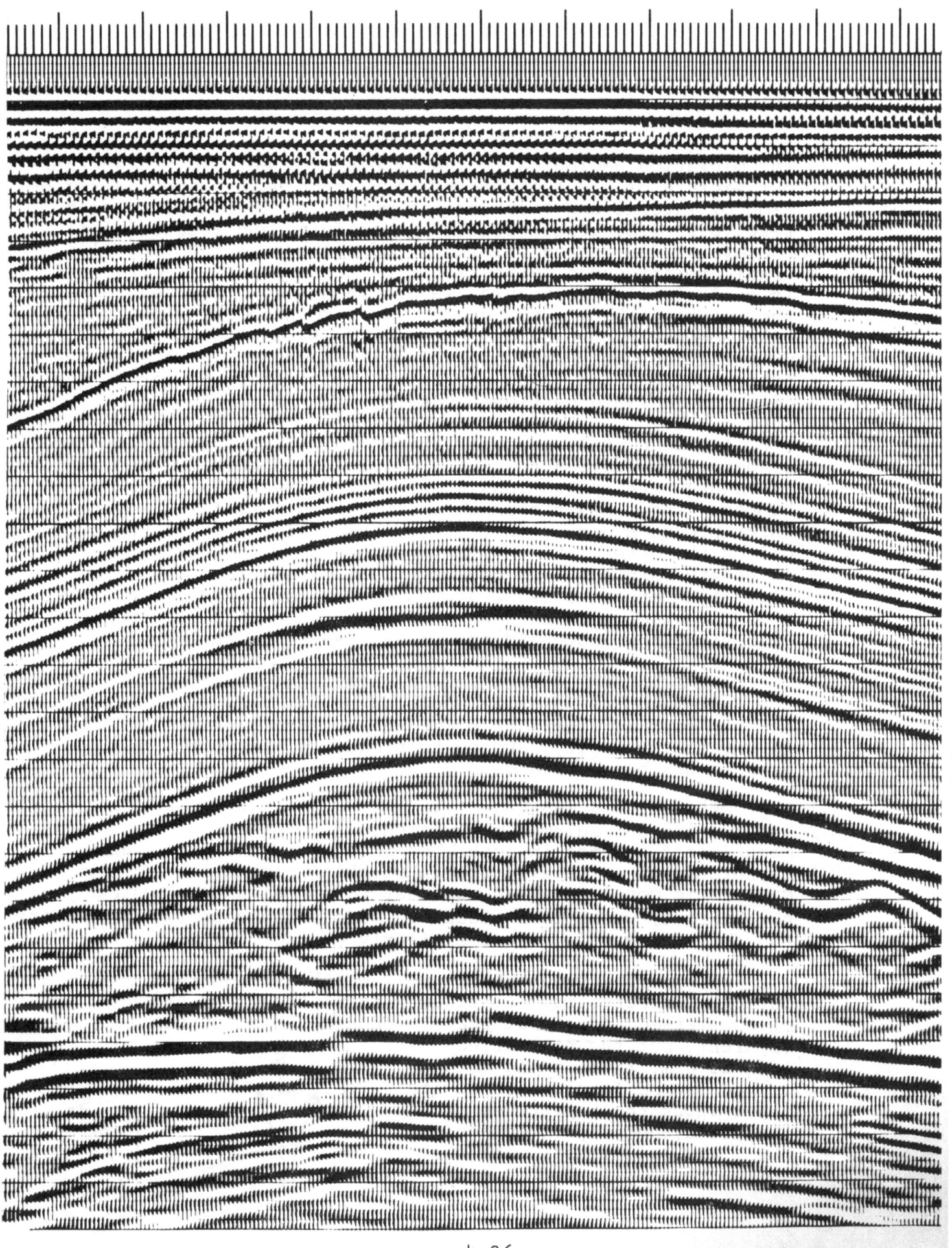

for migration geologically favourable; now we also ask whether our picture of the tectonic history leads us to expect a reduction of lateral and overburden compression in the postulated reservoir — by reason of folding activity and relaxation of the overburden during uplift, or of changes in the thermal gradient introduced by the local tectonics. And, if fracture is geologically reasonable, we are then much concerned to know whether the fractures are likely to have relaxed, or to have been held open by over-pressure. Later, in section 8.6.4, we shall see how positively the seismic method identifies the overpressure which is an essential contributory factor to the giant chalk field mentioned earlier.

Although it is unusual for gas accumulations in carbonates of normal porosities to have acoustic impedances low enough to yield negative reflections when overlain by shale, this can happen. The classic example is a faulted anticlinal reservoir which produces gas on the upthrown side of the fault but only oil on the other; the reflection from the gas-saturated carbonate is bright and negative, while that from the oil-saturated carbonate is so dim as to be scarcely visible. For this to happen in a carbonate, however, probably requires both fairly shallow depth and some overpressure in the reservoir.

In carbonate reservoirs generally, as we decided earlier, the depression of velocity produced by gas (or, to a lesser extent, oil) is likely to bring the reservoir velocity closer to the seal velocity, and so to produce a dim spot. One example has been given on p.4-22.

Obviously, dimness is less desirable than brightness if we are to make our seismic measurements quantitative. The measurement itself is more in doubt, on account of the inevitable background of noise. Errors in stacking give a falsely <u>optimistic</u> conclusion (in contrast to the situation with a bright spot, where poor stacking leads to a conservative answer). And defocusing effects are similarly dangerous.

Fluid contacts are still well worth looking for in carbonate reservoirs. As always, the indications are best seen at compressed horizontal scale. When a fluid contact is apparent, but the top reflection is dim, the top-and-bottom method is modified to use as its major input the amplitudes of the fluid-contact reflection and the flank reflection, with the top reflection used now as the corroboration. In this case the transmission coefficient of the top reflection is virtually 1, so that the governing equation reduces to a quadratic. One additional assumption is involved: that the velocity in the overlying material is the same above the hydrocarbons and off to the side. We remember the caution of section 2.9, about differential mineralization.

In more complicated structural situations, it sometimes happens (more often, in fact, that anyone but Murphy would expect) that the postulated fluid contact could be a multiple. The upper sketch of page 4-90 illustrates one case; the fluid contact could be a peg-leg multiple between the second and third horizontal reflections (after due allowance for the T_o correction, of course). We would hope to eliminate this explanation by observing whether the multiple persists beyond the possible limits of a fluid contact, but we have to remember that sometimes multiples are visible where there are no other reflections but not visible where there are. Further, it is possible that the third reflector is locally strong over the top of the anticline, because different materials are brought into contact across the truncation; this would make the multiple between the second and third reflectors locally strong over the region of the postulated fluid contact. So we are interested in all possible methods of removing the doubt.

A first approach would be by study of the velocity analyses. It is important to consider them all; to the right of the upper sketch is shown a velocity analysis over the high, which shows that in this particular location it would be possible to interpret the multiple velocity as yielding a very low interval velocity within the postulated gas zone — obviously a dangerous situation.

Another analysis away from the high would very quickly show whether the velocity of the questionable event moved (to keep the interval velocity in the gas zone constant) or stayed stationary (indicating a multiple). If the fluid contact is confirmed, it is obviously very important to stack it correctly.

A second approach is by study of the autocorrelations. For example, the autocorrelation function illustrated at the left of the upper sketch would be regarded as favouring the fluid contact; it shows the effect of each of the first three reflections sliding across the others of the three, but it gives no indication of multiples generated between the second and third. (For this type of analysis, of course, the proper divergence correction and appropriately long autocorrelation windows are most important.)

So much for the direct techniques in older rocks; eventually, as we consider reservoir materials of high matrix velocity deeper and deeper in the section, our expectations of direct hydrocarbon indicators must evaporate. Under these considerations we cannot ask what is acoustically impossible, but we still retail a hope of using our quantitative techniques as a supplement to the traditional approach. For example, we established in Question 5 on p.2-71 that there do exist circumstances — reasonable circumstances — in which a distinction between oil saturation and water saturation can be expected; however, although the ratio of reflection amplitudes is satisfactory the actual magnitudes are small. Therefore it is not possible to make this distinction on an undrilled target. But as soon as the prospect has been drilled and the oil-water contact is known in one place, then the seismic data — and in particular the amplitudes — must be reviewed for evidence of a prospect-wide indication of the oil-water contact.

Another case where amplitude studies are of value even at great depth arises when the reservoir is locally contaminated by shale or other solid filling. Thus the acoustic distinction between gas and water as a pore filling becomes unusable below perhaps 2000 or 2500 m (at normal pressures), but the distinction between gas and solid filling remains useful to greater depths. This is of value in the siting of wells into a gas reservoir which is locally shaled out; in one important case the zones in which the reservoir pores are full of clay may be distinguished from those in which the reservoir pores are full of gas by clear variations in the amplitude of the top-reservoir reflection.

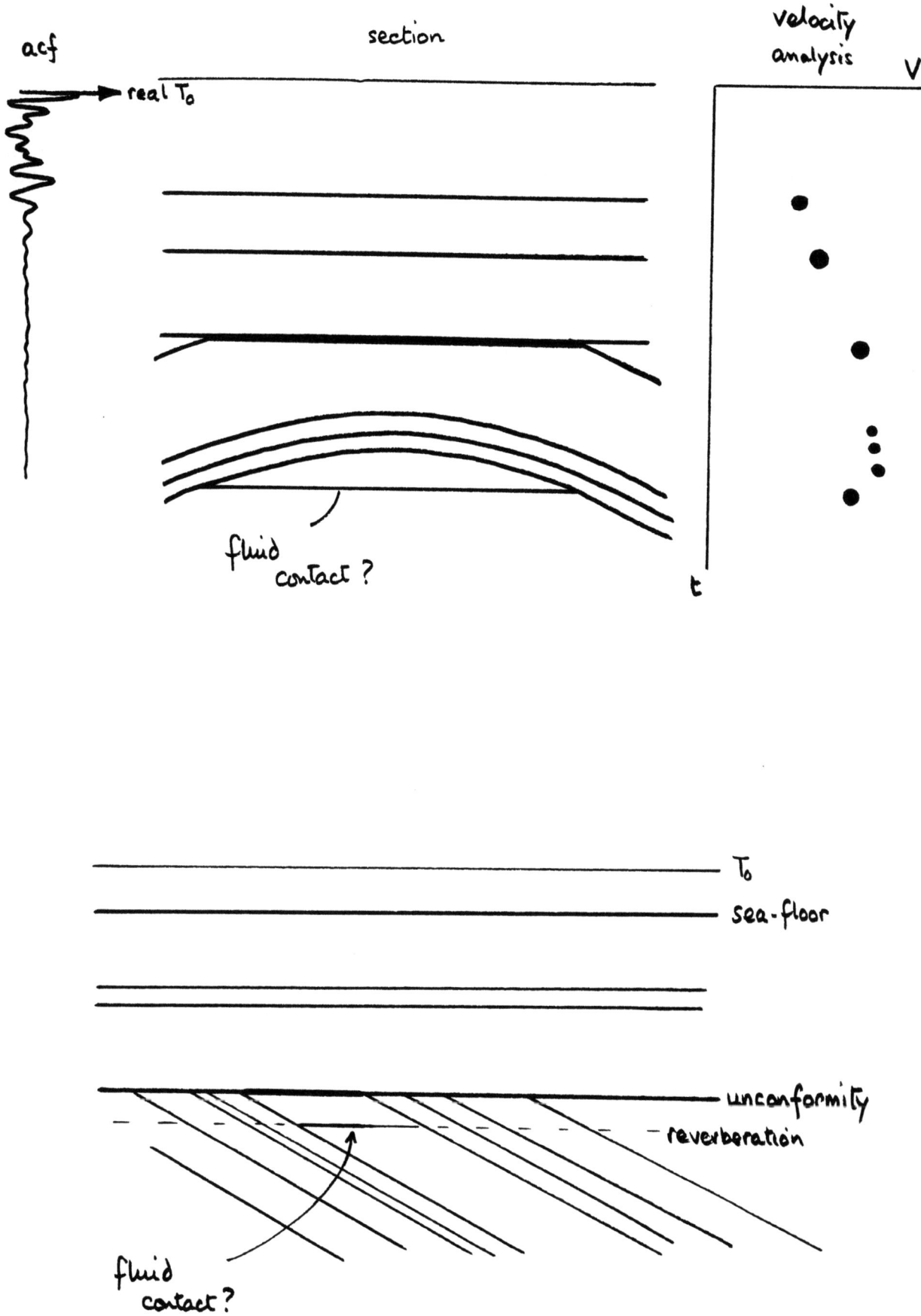

4-90

Such cases are good examples of the situation where the study of reflection amplitudes is quite hopeless until we have some borehole data — but should become mandatory afterwards. We <u>must</u> remember to go back to the seismic data as the logs come in and the geology becomes established, to see whether those amplitudes have a message concerning the limits of the reservoir. Amplitudes must mean something — we cannot just shrug.

PART 5

INDIRECT

HYDROCARBON DETECTION

TAPE 28   The traditional methods of seismic interpretation are, of course, indirect. The basic techniques by which seismic sections are transformed into structural contour maps and profiles are well known; they are set out in Fitch's book and summarized in the pre-course notes, and they will not be repeated here. The classical discussion of the risks in these techniques is given by Tucker and Yorsten in "Pitfalls in Seismic Interpretation"; it is assumed that all of us are familiar with their monograph.

Our concern in Part 5, therefore, is to set out those features of modern practice which supplement the traditional indirect techniques. In fact this is also done to some extent in other Parts — structural modelling in Part 6, migration in Part 7, and contouring in Part 9. Here we concentrate on two main features: indirect applications of direct detection techniques, and seismic stratigraphy.

5.1 INDIRECT APPLICATIONS OF DIRECT TECHNIQUES

We look at these applications first, while the discussion of the direct techniques is still fresh in our minds.

5.1.1 Shallow gas as an indicator of deep hydrocarbons

The basic thesis here is that of Pirson — that no caprocks provide a perfect reservoir seal. Indeed we now realize that in many stratigraphic accumulations the concept of an impermeable caprock is outdated; the "seal" is one of porosity and capillary action rather than of permeability. Only a negative pressure gradient provides a perfect seal. Above most deep accumulations of hydrocarbons, therefore, there tends to be a greater concentration of volatile gas than is normal for the section. If minor reservoir and trapping systems occur at shallow depth, it becomes possible that the presence of the deep hydrocarbons may be inferred from a local richness of bright-spot activity in the shallow section.

In this it is assumed that the deep hydrocarbons are invisible to direct detection techniques (either by reason of their depth or their liquid form); it is only when the volatile gas reaches shallow depth that the acoustic properties are changed sufficiently to produce bright spots.

Obviously some cautions must be stated immediately. The shallow section often contains indigenous gas pockets which owe absolutely nothing to deep hydrocarbons. Even in the presence of upward-percolating gas, bright spots can appear only if suitable reservoirs and traps have been provided by the later geology. And, as we well know by now, not all bright spots indicate gas. So it is very clear that the most we are hoping for is a corroboration — an anomalous degree of bright-spot activity over some but not all deep hydrocarbons.

We must also note that the shallow gas may be all that is left of deep hydrocarbons which escaped.

However, with those reservations fairly stated, it must be said that some petroleum provinces show a very clear correlation between shallow-bright-spot activity and deep hydrocarbons. Probably the distinguishing feature of these provinces is the existence of a young sand-shale sequence at shallow depth, to provide the necessary reservoirs and traps.

In this context, of course, there is no concern with the bright spots as commercial targets in their own right; we are interested in them — however small — solely as indicators of what is (or was) below.

The section on page 5-4 is typical of bright spots found above (in this case almost vertically above) a deep oil trap at the edge of a basin. In such a case we might well be encouraged to drill the deep trap. However, we must first satisfy ourselves that the gas is not locally generated (for example, in the sediments to the right of the bright spot). To do this we map bright spots areally, on a grid of lines, to see whether there is a general correlation between the limit of the deep trap and the configuration of the bright spots.

Sometimes the path by which the upward gas migration occurred can be identified. The classic case of this, of course, is the fault plane with bright spots branching off it. An illustration can be seen at the right-hand side of p.4-13. Further, the presence of the shallow bright spots in the sketch would add to our hopes of finding hydrocarbons in the anticline, even though in this case no direct hydrocarbon indications are given by the latter.

The risk in this is that local anomalies of reflection strength branching off faults can be expected for a variety of other reasons — differential compaction, relaxation and mineralization among them. These other explanations would also be in our minds when we observed a concentration of bright spots over a prospective deep uplift.

Having located a deep prospect, then, we scan the shallow section for a local concentration of bright spots, paying little attention to their individual dimensions. If we do not find them, we ask first whether we have a young sand-shale sequence in the shallow section, or whether other depositional conditions in the shallow section are likely to have left an abundance of small reservoirs and traps. If not, then no evidence is expected one way or the other. But if the reservoirs and traps are there, but the gas is not, this has to be taken as mildly downgrading the deep prospect. On the other hand, if a local concentration of bright spots is present over the deep prospect, and they are not obviously associated with locally generated gas or local tectonics, then this has to be taken as encouraging.

5.1.2 Sea-floor seeps

The evidence for hydrocarbons seeps through the sea floor is now well established. The seep areas — and even the columns of rising gas bubbles — can be seen on standard sonar depth-sounders such as those used on seismic vessels. The gas vents in the sea floor can be seen on side-scan sonar. Sparkers take the indications one stage further, since gas-charged zones can sometimes be seen in and under the sea-floor sediments. The full seismic method is better still (provided it has a short offset to the near trace), since it shows not only the sea-floor seeps and the shallow gas-charged sediments but the deep mechanisim which gave rise to the seep. Thus the seismic method alone can distinguish between uninteresting seeps due to shallow indigenous gas and highly significant seeps indicative of deep hydrocarbons.

On a seismic section a sea-floor seep shows as a <u>weakening</u> of the sea-floor reflection. This is because the presence of the gas decreases both the density and the velocity of the sea-floor sediments, and so decreases the acoustic contrast with the water.

Page 5-6 illustrates several of these features. The flexure or fault affecting the deep reflector (some 8km from the left-hand side) is seen to have reprercussions all the way to the sea floor, and to represent a possible migration path for gas. The bright spot at 1.25 s to the right of the section is a good gas sand, responding well to the top-only method previously discussed. The bright spot at 0.55 s in the central part of the section is another gas sand which, under the poorly compacted clays overlying it, must be leaking. The effect of the seeping gas is seen clearly in the sea-floor reflection, which loses a full 6dB of amplitude.

(The interpreter's first reaction, of course, would be to question whether agc or base-levelling had been used in the processing. In the present case, he would be comforted to observe that the loss of sea-floor strength can be seen clearly on the depth-sounder charts — scruffy though they always are.)

In favourable conditions, therefore, the combination of this section and the last one means that gas migration paths can often be tracked by amplitude anomalies — all the way from their origin in a deep trap to their termination in a sea-floor seep. Of course, this must be done in three dimensions; like the detailed delineation of faults, it can be done only on a grid of lines spaced more closely than would be necessary for structure alone.

5.1.3 Recognition of a permeable path to the basin margin

In a full investigation of a new basin, it is desirable to recognize any permeable path by which hydrocarbon-bearing water from deep source rocks may have been excluded towards the basin margin. Then prospective targets exist wherever traps may have formed along this rising permeable path — in folds or reefs or faults. The appeal of the new basin is obviously enhanced if any one of such traps shows evidence of direct gas indicators. In the situation depicted in the sketch, for example, the presence of gas-induced brightness at A would encourage us to hope for oil at B and/or C. It is unfortunate that the "natural cracker" phenomenon has the effect of filling the _deepest_ trap with the gas; we would have preferred the opposite, which would have given a greater chance for the direct detection of the gas.

5.1.4 Differential mineralization

In principle at least, the fact that rocks above hydrocarbon accumulations tend to be differently cemented (section 2.9) might be used as an indirect exploration tool. Primarily, this would show as a difference in the velocity to the target level, as measured on and off the accumulation. There is also the possibility of consequent variations in reflection amplitude.

Both of these features represent complications to our direct-detection techniques. Apart from the references given in section 2.9, we do not have much to guide us on the seriousness of these complications. However, at least one explained amplitude anomaly has been observed, of dimensions corresponding to an oil accumulation but considerably above it, where it was established that the anomaly was not caused by gas. Perhaps this anomaly is merely a consequence of the thinning of a shale over the structure, but it does not appear so. And a reflection from an oil-water contact has been observed which was not due to the acoustic contrast between oil and water but to the much greater acoustic contrast between different degrees of cementation above and below the contact.

5.2 SEISMIC STRATIGRAPHY

Many of the techniques of what is now called seismic stratigraphy have been practised by interpreters for decades. However, there has recently been a healthy movement to formalize these techniques, and to associate them with several new developments; the results warrant the discussion of seismic stratigraphy as a special study in its own right.

5.2.1 Review of the objectives

Earlier, in section 1.2.2 of the pre-course notes, we summarized in the briefest fashion the interpretation objectives as the petroleum geologist sees them. Let us set them out here in a little more detail. In so doing, let us accept that there are many objectives to which the seismic method cannot hope to contribute; let us concentrate on those where we have at least half a chance.

1. We wish to recognize likely source rocks, and to estimate their volume. We know these are probably organic-rich fine-grain clay-shales or carbonates, and as such laid down some distance from shore. We accept that we have no way of telling whether the rock is rich in organic material; failing this, we wish to know whether burial was sufficiently rapid to guarantee anaerobic conditions soon after deposition, and whether geothermal conditions after burial were favourable to the generation of petroleum from any organic matter which may have been present. If we have experience in another basin in the same geological province, we also wish to be able to specify the geologic age of the postulated source rock.

2. We wish to establish the direction of hydrocarbon migration out of the source rocks. In the simplest case this is nothing more than determining which direction now is equivalent to "up" at the time of migration. In more complex cases it involves recognizing impermeable barriers above the source rock, and recognizing permeable paths to the basin margin; these latter paths are likely to be coarse-grained or otherwise open material laid down under shallow-water or non-marine conditions, and so are likely to be above the source rock in a regressive sequence and below it in a transgressive sequence. Therefore we are concerned to recognize the seismic signatures of regressive and transgressive situations.

3. We wish to recognize the depositional environment likely to produce favourable reservoirs. Typically this involves the recognition of high-energy depositional environments (where the velocity of river flow or waves or currents is such that only the coarsest range of sediment sizes can accumulate), or continental environments, or reef-building environments such as transgressive situations having an appropriate rate of rise of relative sea level, or some of the fractured conditions we have discussed previously.

4. We wish to recognize the path by which hydrocarbons may have migrated from source to reservoir; failing this, we wish to recognize any evidence concerning the fluid potential field in past time which might affect a postulated path. Both this investigation and item 2 above may require us to establish successive positions of the axis of a basin, for example, or to explore successive conditions of a sequence which has been uplifted from its maximum depth of burial.

5. We wish to recognize sealing and trapping situations associated with the aforesaid reservoirs. These involve recognition of depositional environment, tectonic sealing situations and distortions of the fluid potential field. The depositional history is the essential key to the recognition of environments likely to yield stratigraphic traps.

6. In more general terms, we wish to recognize all indications of the type of lithology and the successive directions of sediment transport.

7. We wish to recognize growth features, and to relate the time of their growth to other features such as primary and secondary migration.

8. We wish to recognize any disturbance which could have caused the loss of trapped hydrocarbons, and to establish the relative time at which it occurred.

9. In a new area, we wish to establish the geologic age of the major units, and to determine which units in the geologic column are not represented.

So let us see to what extent these objectives can now be addressed by seismic stratigraphy.

5.2.2 Reflection amplitude and continuity
(ref. Sangree & Widmier)

The basic assumption of seismic stratigraphy is that a reflection alignment corresponds to what the geologist

calls a time-stratigraphic horizon — a representation
of the surface of the solid earth at a particular time
(such as the end of a particular geologic period).

  Apart from the obvious exception of a fluid contact,
and the less obvious exception of false dip produced
locally by interference (section 2.7.1), this assumption
is a fair one.

  In geological terms, there is no assurance that a
time-stratigraphic horizon is also a lithologic boundary.
In seismic terms, there is no assurance that a time-
stratigraphic horizon is an acoustic contrast; neither
is there an assurance that a lithologic horizon is an
acoustic contrast, though this is more likely. Thus at
one place a time-stratigraphic horizon (say the top
Paleocene) may be a lithologic boundary (clay to sand)
and a good seismic reflector (reflection coefficient + 0.1);
further basinward the quieter depositional conditions may
cause the sand to be replaced by clay, the same time-
stratigraphic horizon to cease to be a lithologic boundary,
and the reflection coefficient to fall to zero. Notionally
there is still a reflection alignment representing the
time-stratigraphic horizon, but its amplitude has fallen
to zero. The amplitude is <u>variable</u> and the reflection
is <u>discontinuous</u>. Clearly, <u>this is</u> telling us something
<u>about the depositional environment</u>; our plan then is to
formalize the connection betweeen the reflection indications and the depositional message.

  We cannot do better than use the Sangree and Widmier
illustrations, starting with that on page 5-14. The initial
division of reflection patterns is:

- reflection-free (zero amplitude, zero continuity),
  indicating unstratified material; for example
  basement, or salt, or the interior of some types
  of reef,

- chaotic (variable amplitude, poor continuity),
  indicating depositional or tectonic complications,

324L • layered (amplitude variations slow or absent, continuity fair or good), indicating a degree of calm in the depositional environment.

325L As shown in the diagram, the layered division may be further sub-divided — into parallel and divergent simple layering, and into at least two forms of complex layering.

326 Simple parallel layering is illustrated very well in the Jurassic-Triassic interval (1.1-1.7 s at the left edge) of our much-used section on p.3-60. Apart from a few minor accidentals, the continuity is extremely good; further, the amplitude of most reflections stays remarkably constant, often within 2 or 3 dB over many kilometres.

327L The frequency content is also rather uniform, suggesting beds of constant thickness. Except for the limited times and places of the accidentals, therefore, the depositional environment for the whole of this sequence was very stable despite the fact that a considerable range of lithologies is represented.

The range of lithologies is evident, of course, from the actual magnitude of the reflections in the time zone 1.1 - 1.4 s (left edge). The interval from 1.4 to 1.7 s, however, is very different; it has a "grain" of quasi-continuous reflections, but virtually nothing with an amplitude more than -10 dB (one-third) relative to the 1.1 - 1.4 s zone. Here, then, we have a major unit (about 500 m thick) devoid of significant contrast in an acoustic — and probably lithologic — sense. The absence of significant reflections makes it difficult to assess the continuity, but the grain makes it clear that this is not a reflection-free unit; as such the grain is an important distinction. The inference is that the unit was deposited in an extremely low-energy environment — probably on the deeper part of the shelf — and as such very likely to be a fine-grained clay/shale or uniform carbonate.

The same conclusion could be drawn for a Tertiary interval (0.3 - 0.55 s at left edge) and for a Cretaceous

TYPICAL REFLECTION CONFIGURATIONS

- REFLECTION - FREE
- LAYERED
  - SIMPLE LAYERED
    - PARALLEL
    - DIVERGENT
  - COMPLEX LAYERED
    - OBLIQUE
    - SIGMOID
- CHAOTIC

Courtesy of Sangree and Widmier (Exxon) 1974

interval below 0.55 s. In fact the Tertiary interval and the Triassic interval are both clay/shale, while the Cretaceous interval is carbonate. The lithologic distinction cannot be made on these considerations (though of course it can be made on considerations of velocity, coupled with a compaction law); here we are concerned with depositional environment alone.

We therefore carry forward a very important conclusion: that low reflection amplitudes within a unit, combined with at least a grain of continuity, indicate a quiet environment which is certainly marine, and a facies which is probably fine-grained. If in addition to these characteristics we observe a tendency for the grain to be maintained concordant over underlying erosional topography, then we can add that the deposition was in deep water (Figure 1.2.2 C, and top left of p.5-20). The depositional conditions were such that the sediment was deposited from suspension (in contradistinction to the general situation on the continental shelves, where a measure of bottom transport is usual).

At this point let us inject three asides. The first concerns the odd appearance of the interval between 1.75 and 2.25 s at the left edge of p.3-60. The profusion of diffraction patterns is believed* due to blocks of limestone locally sinking through salt; only by looking in zones clear of diffractions can the reflection-free character of the salt be recognized. The second concerns the broken appearance of the conformable sequence about 2½ km from the right side of the section; this is almost certainly due to the distortion of the ray paths in the velocity lens at the top-Cretaceous level, and must not be used to make inferences about the local depositional environment at greater depth. Third, the interpreter must be very conscious of the risk of being tricked by the processing:

- If a grain 10 dB down on the preceding reflection is to be taken as diagnostic, it is very important that the dereverberation is well done.

5-15

*More sophisticated explanations have recently emerged, but the conclusion in this context is unchanged.

- Base-levelling is undesirable, since it may obscure or create lateral variations of reflection strength interpreted as changes of depositional environment.

- A section with conventional agc is unacceptable, because it may obscure or create vertical variations of reflection strength interpreted as low energy depositional environments.

- A section without agc is also less than desirable, because the grain cannot be seen and there is a risk of mistaking low-energy depositional environments for salt or other reflection-free units.

One solution to the last two problems is exactly that illustrated in the section on p.3-60: agc on the black-and-white background (so that the grain can be seen) but the true reflection strength superimposed in colour (so that the low-energy environment is correctly — even quantitatively — recognized). We see how clearly the low-energy depositional zones stand out on the section; perhaps we never before realized the degree of this effect.

But back to the mainstream, and the next figure (p.5-18) from Sangree & Widmier. In the middle of the top row we see the case we have just been discussing — the low-amplitude grain. The authors are careful to associate this with <u>uniform</u> depositional energy, and indeed to state that it could be a sand; our working assumption that the uniform depositional energy is actually low, and the material fine-grained, reflects the usual situation if the low-amplitude grain extends over a considerable area and for a considerable depth. Once again we are reminded that all geophysical inferences must be measured against general geological likelihoods.

The same observation applies to the top left-hand diagram. Here the authors associate strong amplitudes and good continuity with interbedded high-energy and low-energy deposits. There are, of course, geological

5-16

conditions which can produce strong continuous reflections from low-energy environments; again the interpreter looks at the extent of the sediments (and perhaps also at the frequency content) to see whether he has the areally-stable conditions of the 1.1 - 1.4 s zone we discussed just now, or whether he has the locally-constrained conditions of a transgressive or regressive shoreline zone. In the latter case, he will undoubtedly see the strong amplitudes and locally good continuity as an indicator of interbedded high-energy and low-energy environments, and perhaps hope for an interfingered series of clay source rocks and sand reservoirs.

The top-right diagram depicts an interval of poor continuity and variable amplitude, suggesting variable depositional energy. Frequently this is diagnostic of nonmarine sediments.

The authors illustrate these three situations with the diagram on p.5-19. Many features of the geological picture are evident from the seismic section — the marine shale, the non-marine clastics, and the important interfingering sand-shale members. With these identified, the pattern of transgression and regression also becomes clear.

Another section which responds to the same type of analysis is that on p.5-4 (left half, 0.7-1.0 s).

The extreme case of poor continuity and variable amplitude corresponds to the "chaotic" situation of p.5-14. This condition is often associated with mounding and fill features, which are themselves associated with a local sediment source (for example, a river, a submarine canyon) or a local sediment sink (for example, a glacial channel). Typical cases are illustrated at the bottom of p.5-18 and on all except the top-left sketch of p.5-20.

SHELF SEISMIC FACIES UNITS

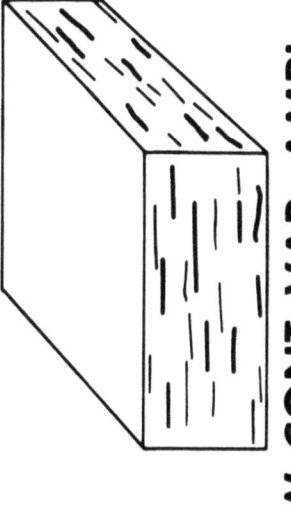

LOW CONT., VAR. AMPL.
(VARIABLE ENERGY)

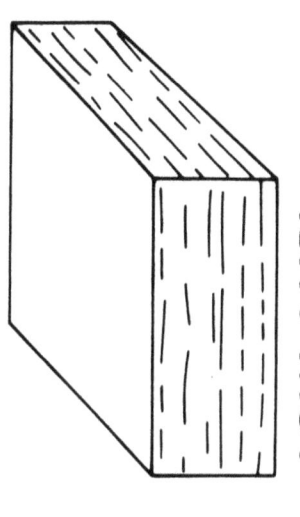

LOW AMPL.
(UNIFORM ENERGY)

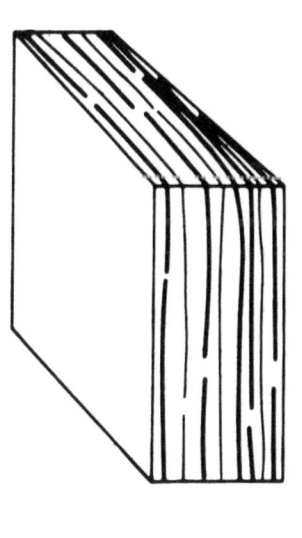

HIGH AMPL. AND CONT.
(INTBD HIGH AND LOW ENERGY)

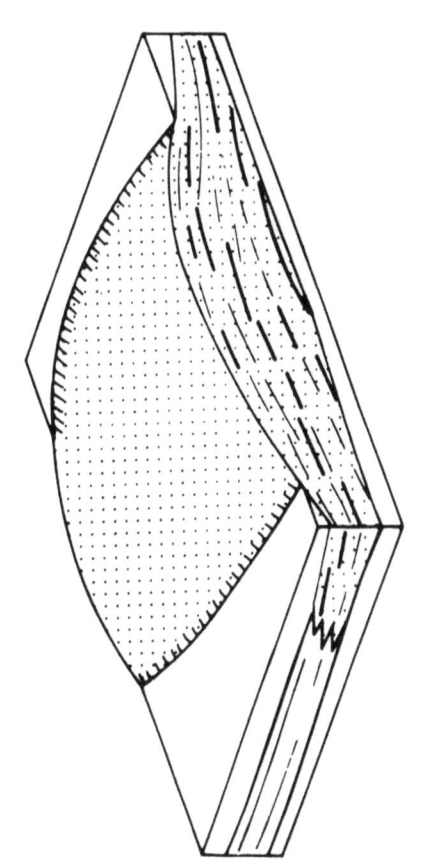

LOBATE MOUND
VAR. AMPL. AND CONT.
(VARIABLE ENERGY)

Courtesy of Sangree and Widmier (Exxon) 1974

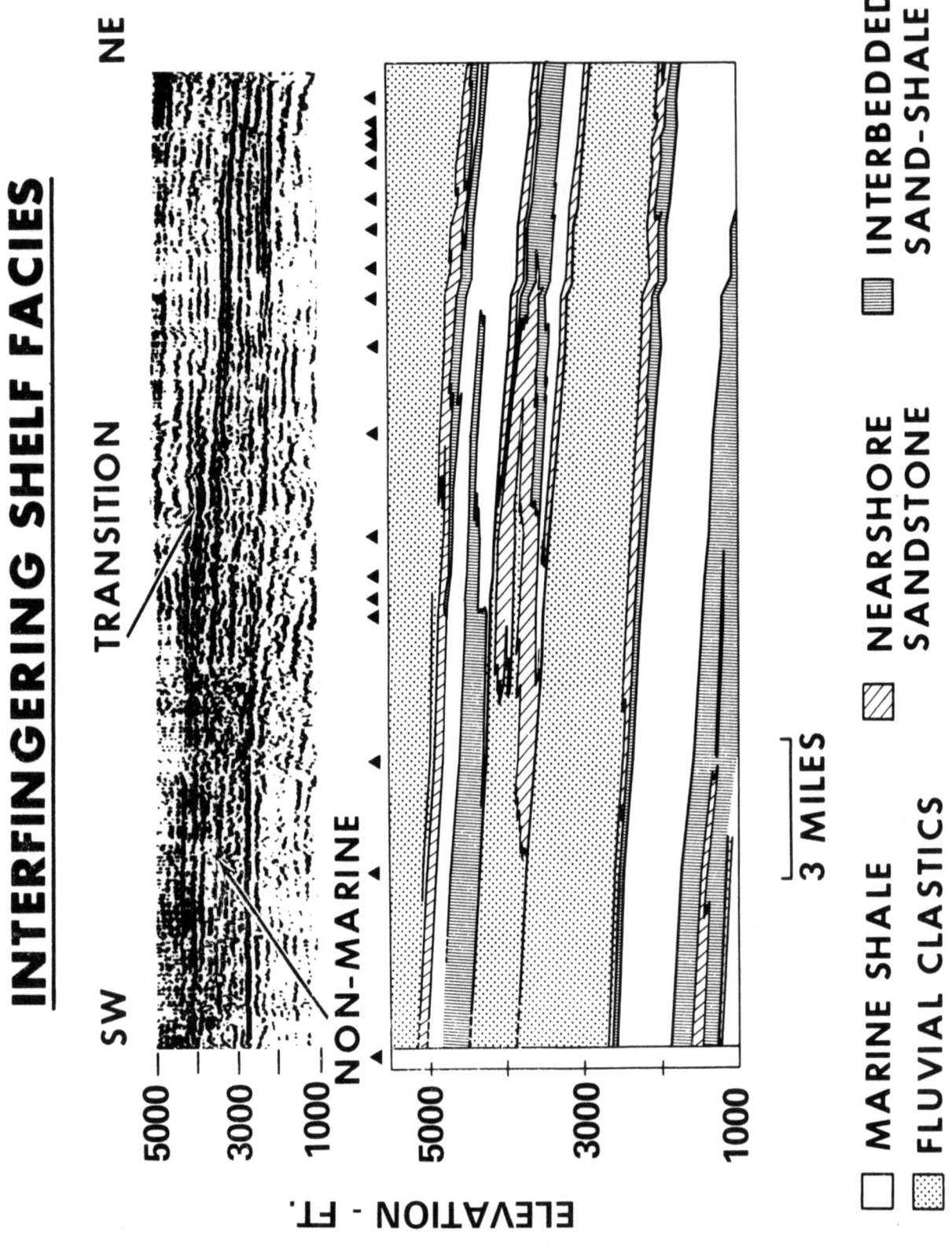

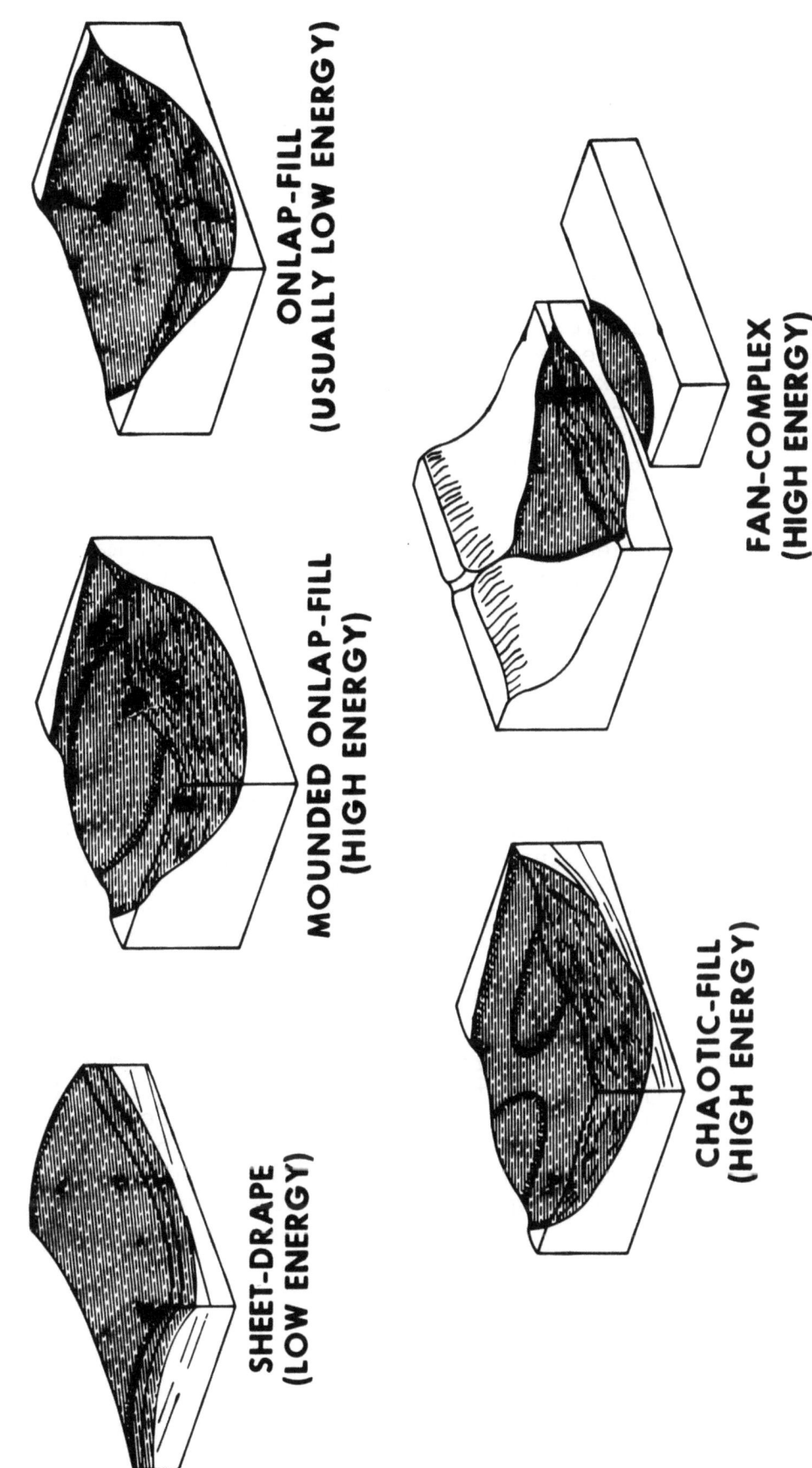

Courtesy of Sangree and Widmier (Exxon) 1974

Within these features, as is true generally, the presence of chaotic alignments and very variable amplitudes indicates high-energy depositional conditions, while horizontal alignments and better continuity indicate a low-energy environment.

The recognition of the presence and type of these features is useful not only in that the high energy varieties are potential reservoirs, but also in that they serve to define the position of old sediment effluxes, shelf edges, and paleotopographic surfaces generally. We shall return to this latter point in a moment. But, before we proceed, let us just insert a proviso about our association of high-energy depositional conditions with potential reservoirs: the high-energy conditions lead to the selective accumulation of coarse-grained material provided that such material is available. In the earth sciences we have little need of Heisenberg to tell us about uncertainty.

5.2.3 Angular relationships and discontinuities
(Chapman, 1972; Dobrin, 1975)

We have already considered the plane-parellel relationship indicative of calm depositional conditions, and the chaotic condition indicative of highly erratic depositional conditions or subsequent tectonic disturbance. (Perhaps we should have made specific mention of slumping and sliding in this context.) Now we look at specific depositional features and their consequences for the angular relationships between reflections.

1. The prograding shelf A standard exploration objective is the prograding shale shelf (where sand reservoirs — often multiple — are likely at and near the top edge of the shelf) and, in particular, the portion associated with a major delta. The basic forms to be expected are shown on p.5-24. The left-hand sketch illustrates

332L the usual situation, where a major outbuilding is
333 associated with a significant upbuilding to give a
clear regressive situation. The bottom sets show a
characteristic offlap pattern thinning toward the
334 basin, and the transition from top sets to foresets
is smooth. An example occurs in the right half of
the section on p.5-4, between 0.5 and 1.0 s. The
position of potential reservoirs on such a sigmoid
configuration depends on episodic high-energy periods
335 disturbing a generally low-energy environment.

Continuing high-energy conditions, on the other
hand, are likely to yield the oblique configuration
at the right of p.5-24, in which the major growth is
an outbuilding. Here high-velocity flow sweeps the
sediment off the truncated top surface, and volumes
336 of coarse-grained material are therefore to be expected
just over the edge. Obviously, this is a very attractive
situation; porous and permeable reservoir rocks are
accumulated on top of potential source rocks, with the
natural paths for fluid flow leading directly to the
337 reservoirs. An example occurs in the left-centre part
of the section on page 4-81, at about 1.2 - 1.4 s.

Both types of prograding sequence are likely to
be cut by small normal faults, and it may be necessary
to put these faults together again before the pattern
is clearly seen.

338(B) 2. <u>Divergent layers</u> This is the second category of
L(B) simple layering introduced on p.5-14. The important
cases are thickening into a basin, thinning into a
basin, and thinning over an uplift. As illustrated in
the first example of Tucker and Yorsten's monograph,
our first concern in all of these cases is to prove by
interval-velocity studies that the divergence is real.
A normal compaction law, obviously, implies that parallel
layering appears to thin into a basin, and that divergent
layering may appear parallel; the only caution here is
that, as we shall see in a moment, the compaction law in
a rapidly-deposited basin may not be normal. However,
the point stands — our concern is with <u>real</u> thickening
or thinning.

Thickening into a basin indicates the combination of subsidence and an abundance of sediment, with the rate of subsidence being greater in the centre of the basin and the sediment influx in that zone being sufficient to allow the thicker deposition. Lateral changes in either the rate of subsidence or the availability of sediment must affect the pattern.

It often happens that a basin has the seismic expression shown in the sketch — virtual absence of any reflections (sometimes even a grain) in the central part of the basin, but clear 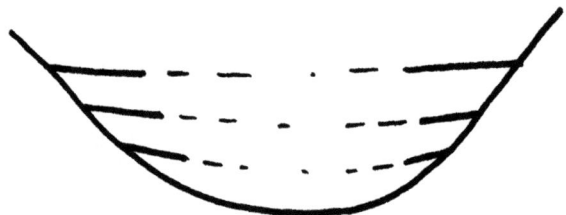 divergence evident toward the edges. These edge reflections may be due to lithologic changes associated with changes in the nature of the sediments closer to shore, or they may represent simple breaks in deposition. The absence of reflections in the deep basin, however, suggests continuous deposition. If the unit is thick, this in turn suggests rapid subsidence and rapid deposition. This is an interesting situation, because it improves the chances of rapid burial of organic matter to anaerobic conditions — believed to be a prerequisite for hydrocarbon generation. Further, the rapid deposition increases the likelihood of overpressured conditions; because this reduces the geothermal heat loss by convection, it has the effect of keeping the hydrocarbon-forming materials at high temperature. The potential therefore exists for good source conditions, and the interpreter must then analyse the chances for migration created by subsequent geological processes.

Thinning into a basin may indicate an inadequacy of sedimentary material; in this there is an analogy with the offlap thinning of a prograding sequence just considered. Or it may be the effect of greater compaction of the deep

SHELF-MARGIN AND PROGRADED-SLOPE SEISMIC FACIES UNITS

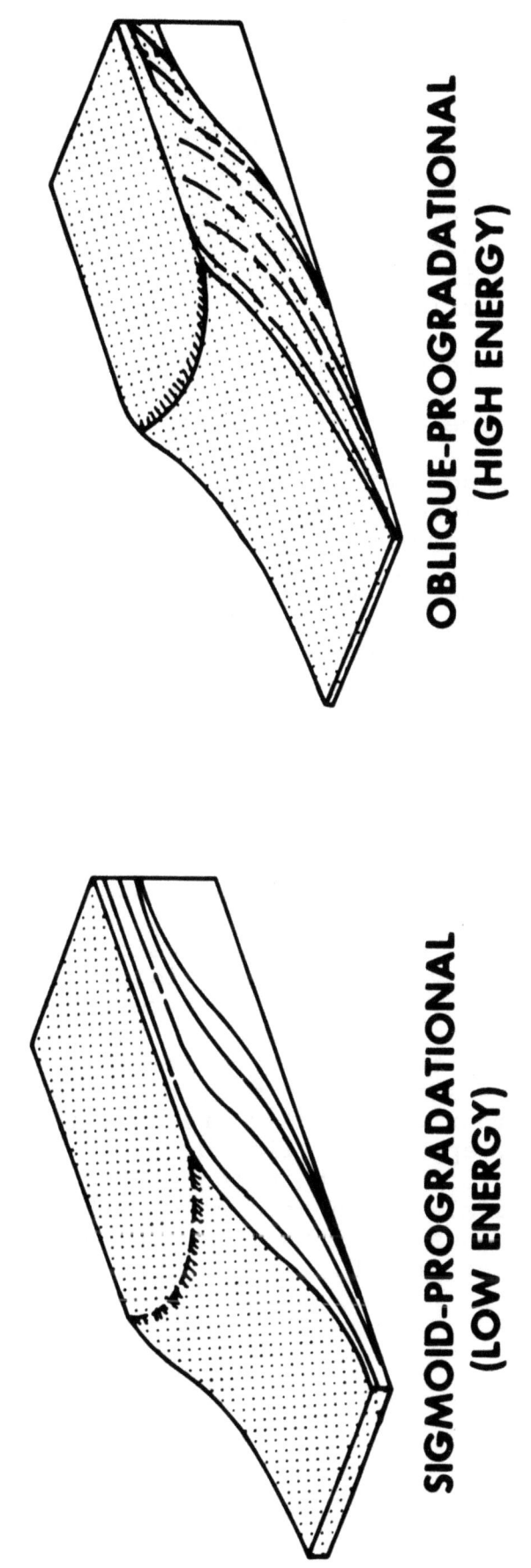

Courtesy of Sangree and Widmier (Exxon) 1974

layers by reason of the greater overburden; this situation suggests a compactible lithology. Or it may be entirely illusory — regional tectonics may have rotated a basin flank and confused the basinward direction. This last point, obviously, constitutes a caution for all stratigraphic studies involving thickening and thinning, onlap and offlap.

339   Where thinning is due to starvation of sedimentary material, it may be interpretable to show the direction of the source of that material (Sheriff, 1976).

Thinning over an anticline is an important situation. This is because it may indicate a growth structure, and because we remember that one of our exploration objectives is to identify situations where the migration and trapping of hydrocarbons occurred simultaneously, in such structures. The normal inference from thinning over an anticline is that although both the structure and the region around it were subsiding, the structure was subsiding less (which, in this context, means that it was growing). We need the usual caution about sediment supply, of course. Subject to the availability of sediment, the duration of the growth periods can be estimated from those layers exhibit-
340 ing the thinning. The likely cause of the growth is shale or salt diapirism, and so the interpreter is alert for all other evidences of these often-rewarding situations.

Although differential subsidence is the main cause of thinning over a growth anticline, there can be other contributions. These complicate the issue. For example, shales tend to slide downhill, and so both thin and stretch over the high; but before we identify as shales the members showing the most thinning over the high, we must remember that shales are also the most compactible, and so may appear parallel or even thin down-flank. Therefore it is the thinning of the non-shale units which is most safely diagnostic of growth. In some structural contexts, these factors limit our ability to recognize intermittent growth.

Some help may be obtained from the study of amplitudes; reflections which change strength in a manner which is systematic with structure may suggest growth **during** deposition.

Thinning over a growth anticline is to be distinguished from thinning in a drape over a reef or other rigid feature. In this case the thinning may be completely explained by differential compaction alone.

3. <u>Faults</u> In a lithologic context, the behaviour of rocks near faults sometimes allows us to distinguish between plastic and brittle rocks, and it is common practice to search for the termination of a fault as indicating a mobile material. The strength of techniques like this is increased enormously by modern migration methods (provided, of course, that the line direction has been selected to be normal to the fault plane).

In a stratigraphic context, major interest attaches to growth faults, in which correlating stratigraphic units show greater thickness on the downthrown side. The throw generally increases with depth, and the fault plane is usually curved — concave basinward and concave up. If the correlation across the fault is good, this suggests a superfluity of sediment and hence a regressive situation; it also suggests that both sides of the fault — including the "up"-thrown side — were subsiding, but at different rates. Clearly these conditions exclude deposition from suspension (that is, very deep water); bottom transport of sediment "over the edge" is necessary. Some sand is likely. A sand on the upthrown side tends to be well-sorted but of good permeability; it is more free of contaminating clay than its counterpart on the downthrown side.

In principle, a compactible clay might be recognized in that it **deforms** the curved fault plane, locally making it less steep; in practice this is difficult.

4. <u>Diapirs</u>   The importance of diapirs in the exploration scene is major. Salt diapirs form anticlinal traps above them, stratigraphic and fault traps on their flanks, and structural traps between their rim synclines. Shale diapirs combine the important features of a potential source rock, retained and "cooked" fluids, and the mechanical instability capable of deforming adjacent materials into traps. In the terms of seismic stratigraphy, diapirs are recognized by:

- thickening and thinning which suggests plastic movement of the material within one unit,

- thinning above, as discussed earlier,

- movement of the axis of the adjacent syncline as it develops into the peripheral sink,

- in the case of salt, a reflection-free appearance within the material,

- in the case of shale, very poorly defined reflections before movement; chaotic reflections (often with local brightness associated with gas pockets) after movement.

In the case of shale diapirs, additional seismic evidence is afforded by the anomalously low velocities characteristic of uncompacted clay materials (section 2.3.4).

We conclude this section with the illustration on p.5-28. The details of this are left, as the real textbooks say, "as an exercise for the reader". However, just a casual inspection reveals some low-energy deposition, considerable interbedded high-energy and low-energy deposition, some littoral complexities, thinning to the left, thinning to the right, offlap, salt, thinning over the uplift, materials of different compactibility, a peripheral syncline, movement of the axis of the peripheral syncline, and an anticline formed by several peripheral synclines (and a caution about multiple reflections falling within the salt!).

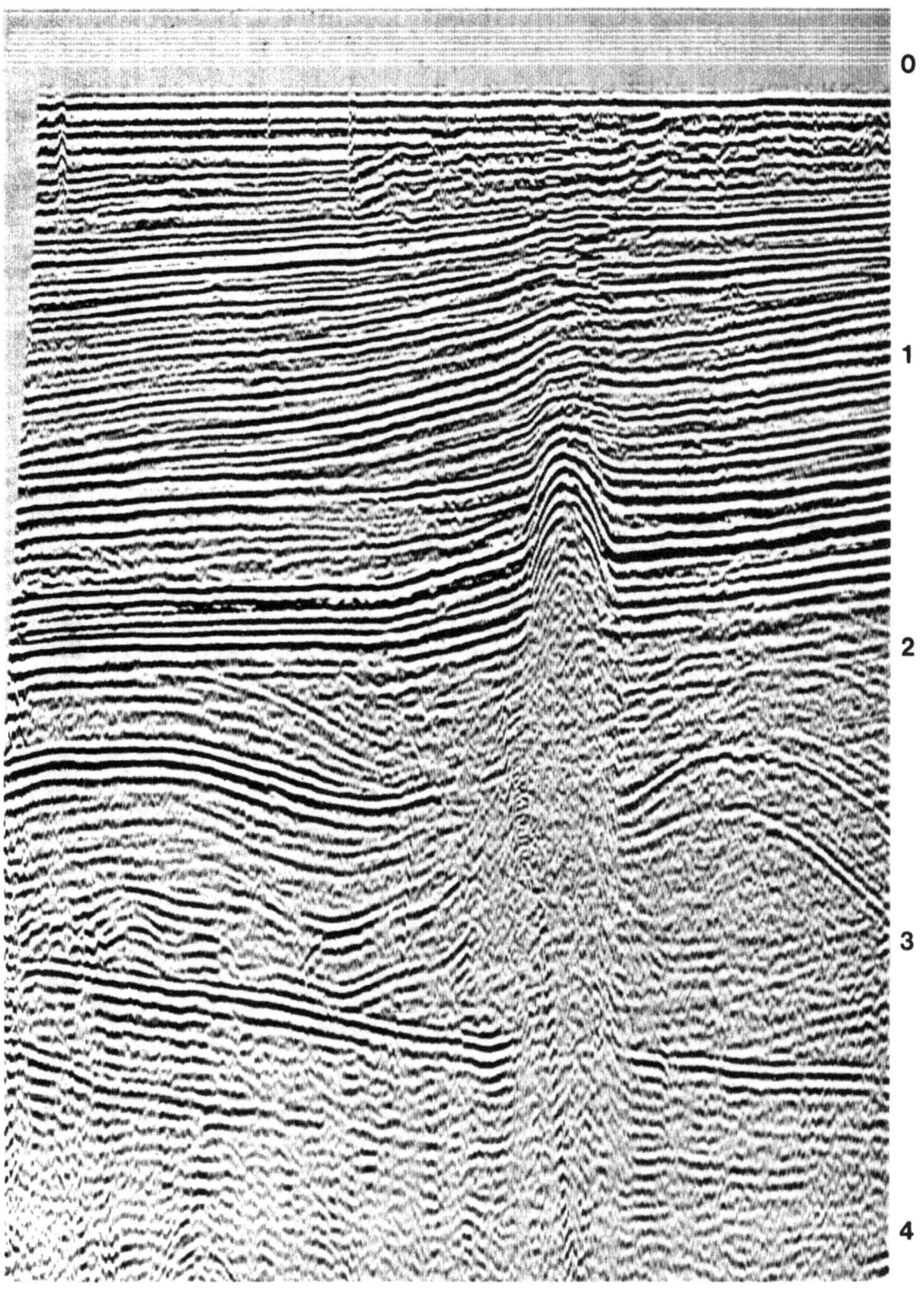

5-28

343 5. <u>Unconformities</u> Stratigraphic studies must embrace not only deposition but non-deposition and erosion. Our seismic tool for studying erosion is that most fundamental example of an angular relationship — the unconformity. The type of interpretation exercise involved is illustrated very well by the central part of the unconformity which meets the left edge of p.3-60 at about 1.05 s.

We also wish, as far as is possible, to include two seismic manifestations which are more difficult to recognize — the case of non-deposition in the marine environment, and the case of locally-derived non-marine sediments above the erosional surface. The first requires areal studies to the limits of the unconformity; the second depends on whether the degree and type of cementation in the products of erosion is sufficient to allow acoustic distinction between them and the mother rock, and often becomes a study in diffraction patterns.

344 Unconformity surfaces are also of interest to us in that they sometimes represent paths for the migration of hydrocarbons from source rocks in the deep basin to reservoir rocks at the basin margin.

5.2.4 <u>Sequences</u>   (Vail et al., 1975; Zaaza et al., 1975)

345 A seismic sequence is a stratigraphic unit bounded at its top and base by unconformities or their correlative conformities. This concept is of great importance, for sequences are the units which provide the framework
346 for the depositional interpretation of a basin. Thus the
347 type of local deposition is established by studying the configuration of the reflections <u>within</u> the unit, as we have discussed above, but the broad picture of basin development is established by studying the <u>areal</u> configuration of the <u>sequence</u>.

As an example, a deltaic sequence can sometimes be internally distinguished from a regular prograding-shelf sequence by the presence of the "oblique" characteristic at its upper limit; usually it can also be distinguished by the <u>areal</u> configuration of the <u>unit</u>, showing a

basinward bulge. As we observed before, it is the areal
configuration of a contrasty reflection sequence which
allows us to distinguish a local interbedded high-and-
low-energy series from a widespread change of sediment
type which is not primarily due to a change of energy.

However, an even greater significance attaches to
sequences by reason of the fact that sequences associated
with coastal onlap are found to be correlatable from
basin to basin. This we develop in the next section.

Since the areal configuration of a sequence is
diagnostic, it now becomes standard practice to <u>map</u>
such configuration. Since the vertical configuration is
also diagnostic, there is merit in making <u>three-dimensional</u>
displays of each sequence of interest. In doing this,
it is sometimes advantageous to distort the seismic
display so that a particular surface known to have been
substantially horizontal at the time of deposition (such
as a deltaic plain) is restored to the horizontal. A
type of display which lends itself to these three-
dimensional representations (including the datumized
ones) is decribed in section 9.1.3.

5.2.5 <u>Eustatic cycles and basin dating</u>

Eustatic cycles are concerned with worldwide rises
and falls of relative sea level. It is well established
that such rises and falls, of comparatively minor proportions, are associated with the periods of glaciation.
Now it emerges that periods of rapid sea-floor spreading,
when uncommonly large volumes of basalt exist in the
oceans, tend to be associated with high relative sea
levels, and periods of slow spreading with widespread
regressive situations. The interest in this, of course,
is as the basis for a worldwide scheme of basin dating.
Perhaps we should note, before we proceed, that there is
not universal agreement on these matters. But universal
agreement might make us anxious too.

For the purposes of global correlation, the concept of relative changes of sea level is more useful than the concepts of transgression and regression. Thus although Figure 1.2.2A associates regression with falling sea level, regression can also occur when the relative sea level rises. A rise of relative sea level can produce a transgression, a regression, or a balanced stillstand — it all depends on the volume of erosional material available.

Let us illustrate the consequences of oscillating relative sea level with the hypothetical (and rather artificial) situations depicted on p.5-32. In the sketch (a) we allow ourselves the ultimate geological luxury — a start button. At the time we push it, we have no sediments — merely a monoclinal surface and a defined sea level. In sketch (b) we see the picture after the situation of the first sketch has remained stable for a period. Accepting that everything here is oversimplified, we have generally coarse-grained sediments (full-line) deposited close to shore, grading into fine-grained sediments (dashed line) in deep water.

Then we let the relative sea level rise slowly, while we maintain a suitable volume of sediment influx. After a period characterized by successive coastal onlap, we obtain the upbuilt situation of sketch (c). In the real world, this situation would be accompanied by isostatic adjustment; for present purposes we ignore the adjustment, which affects the details of the picture but not our conclusions from it.

Sketch (d) suggests the effect we might expect to observe if we maintained the sediment influx but allowed the relative sea level to fall slowly. The upbuilding of the transgressive sequence is separated from the out-building of the regression, with a clear sequence boundary between them. In reality, this situation would involve erosion of some or most of the newly emergent land surface above the fallen sea level, but we would still hope to be able to establish the sequence boundary by the offlap distinction at the original monocline surface.

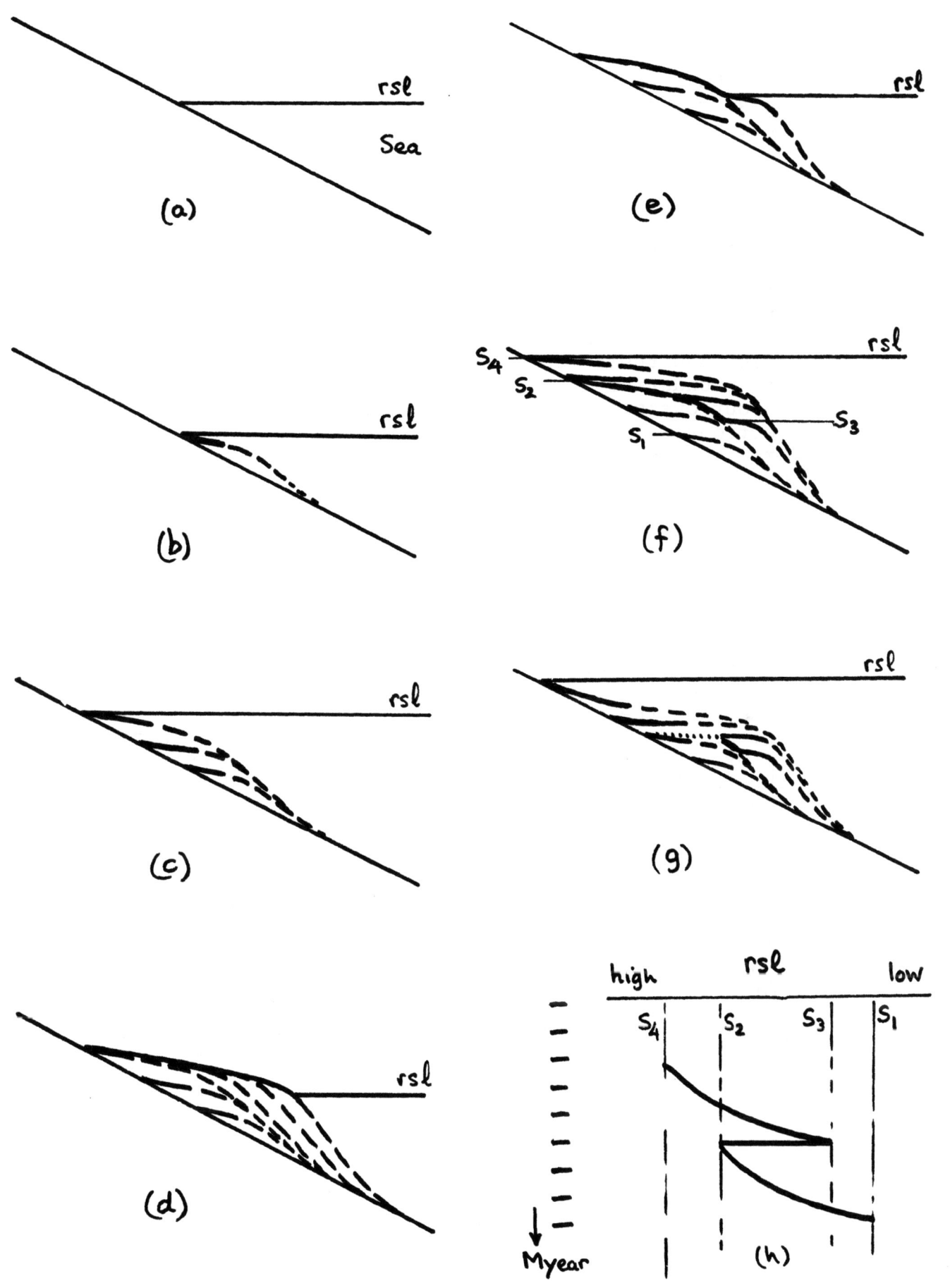

5-32

However, this type of pattern is seldom seen on seismic sections, and this has led to the assertion that <u>falls of relative sea level are generally abrupt.</u>

An abrupt fall of relative sea level, after the transgressive stage of sketch (c), yields the pattern of sketch (e). If this is followed by a gradual rise of relative sea level without significant erosion during the period of lowstand, we expect the pattern of sketch (f). If there is significant erosion, we expect the pattern of sketch (g). The important thing in both cases is that we recognize the sequence boundaries and their message. On real sections, of course, the boundaries and their message are less clear than in our oversimplified sketches; however, on real sections we have the help not only of the high-amplitude evidences of coastal interfingering, but also of mappable zones of fill in which the reflection continuity and amplitude tell us about the prevailing water depth and depositional conditions.

In sketch (f) we note in particular the evidence for four critical values of relative sea level. The value S_1 corresponds to the start of the first rise. The value S_2 corresponds to the end of the first rise and the start of the fall. The value S_3 corresponds to the end of the fall and the start of the second rise, and the value S_4 corresponds to the end of the period sketched. If we are doing this in an area of extensive borehole control, we can now construct the sea-level chart (h), which displays the relative sea level (horizontal scale, in metres or feet) as a function of geologic time (vertical scale, in millions of years). The asymmetrical signature of a sea-level cycle is evident.

If we imagine a basin in which the entire column is preserved, we could construct such a sea-level chart from the present back into the Paleozoic. The general form of this master-chart would be as suggested at the top of p.5-36; we see a number of asymmetrical cycles each characterized by a gradual rise and a rapid fall, and we see that some cycles also include an extended period of stillstand. Of course no such perfect basis for the chart exists, but it begins to emerge that it can be pieced together by assembling the evidence from many basins.

To the degree that this can be done (and publication of such a master chart is expected soon), we have a seismic tool for the age-dating of a new basin. This tool relies on:

- The observation that the major sea-level falls are global,

- the master chart,

- our ability to distinguish between the global pattern and the "noise" of local effects,

- Our ability to obtain a quantitative measure of the rises and falls, so that we can locate our new basin on the master chart, and

- our ability to cope with the problem of erosion.

The problem of measuring the rises quantitatively is attacked by assessing the coastal aggradation associated with each visible step in the onlap series within one sequence. It is necessary to proceed in this manner to obtain some immunity from problems of isostatic adjustment and thickening into the basin; the error involved by making mid-basin rather than coastal measurements has been reported as high as a factor of 5. If we can do this for perhaps 5 cycles, we might emerge with figures of (say) 250, 350, 150, 500 and

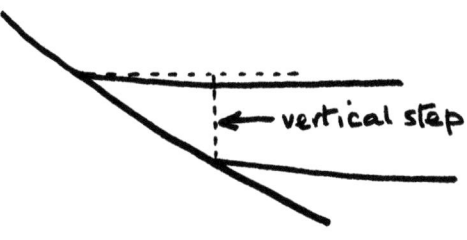

200 m; these might be sufficient to locate us on the master chart. But in general, and to add a quantitative measure of the falls, we must tackle the problem of erosion.

353(B)

There seems no solution to this problem other than the usual type of geological detective work. First we identify all the indications of erosion (that is, truncations at upper unconformable sequence boundaries). Then we search the basin for zones showing the least erosion at each level of interest. The message from these is then combined with that from making plausible upward projections from within each sequence, and coupled with the supplementary information from identified deltas, zones of fill, shallow-water reefs and other seismically-recognizable indicators.

To the extent that we are successful in unravelling the evidence, we can construct chronographic sections. A chronographic section is illustrated in the lower sketch of p.5-36. The vertical scale is that of the master sea-level chart — geologic time in millions of years. The horizontal axis represents the present-day landward limit of individual stratigraphic units. In this type of display all reflection (being deemed time-stratigraphic surfaces) are horizontal, and the vertical extent of a unit represents its duration of deposition — not its thickness. The parts of the graph which have the same general form as the master sea-level chart are the parts representing still-visible coastal onlap; the parts curving away to the top represent periods of erosion, as a result of which the unit is found only at progressively more-basinward locations. One part of the total stratigraphic sequence is shown as missing.

5.2.6 The consequences for indirect hydrocarbon location

We now consider what sections 5.2.2 - 5.2.5 have given us, in answer to the problems posed in section 5.2.1.

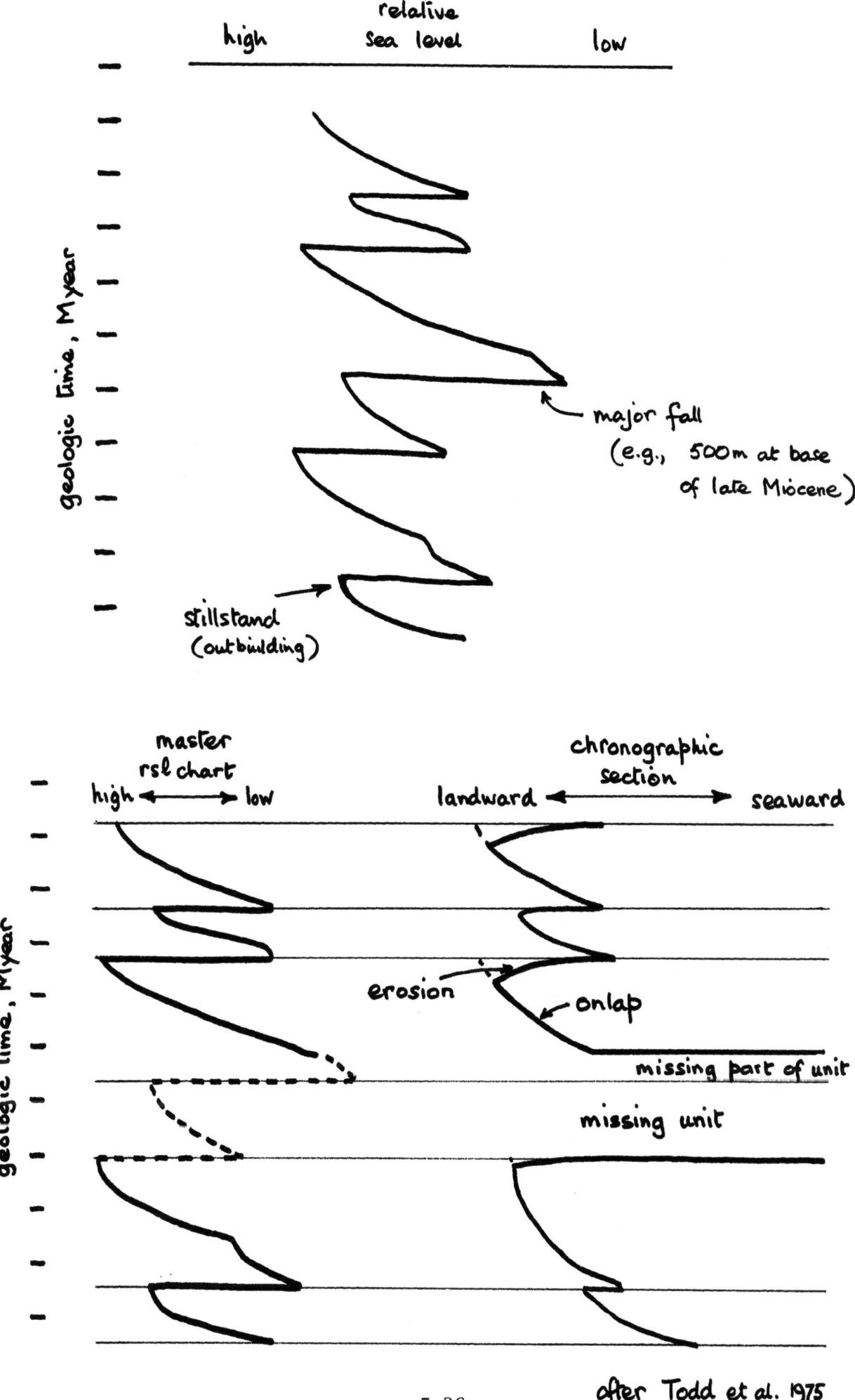

5-36 after Todd et al. 1975

Before we become specific, let us just note how seismic stratigraphy has changed our view of the ultimate product of the seismic method. In former years we would have said that this ultimate product is an acoustic impedance log on every trace. This remains, of course, an important objective. Now, however, we would add to it a new objective: the making of a motion picture in which successive frames represent the depositional and erosional conditions existing at intervals of perhaps one million years. In some areas this second objective is both more feasible and more useful to the petroleum geologist.

Now let us list our specific conclusions with regard to source rocks, reservoir rocks, and the chances of hydrocarbons.

1. Source rocks   Seismically, we have the following tools to assist us in the recognition of source rocks:

- When the source rocks are fine-grained, we have the seismic indications of a low-energy depositional environment (low amplitudes, a grain of continuity, sheet drape; pp.5-13 to 5-16).

- When the source rock is a shale, we have the additional evidence of seismic velocity according to a shale compaction curve (pp.2-35 to 2-40), plasticity as revealed by the behaviour near and at the termination of faults (p.5-26) and compactability as revealed by the behaviour near growth structures (p.5-25, 26).

- Anaerobic conditions are likely if the reflection indications suggest uninterrupted deposition into the basin at its edges (p.5-23).

- Favourable cooking conditions are likely if the shale is shown to be, or to have been, overpressured (p.5-23); geothermal methods may confirm this. Subduction zones are also of interest in this context, in that they provide rapid deep burial to high temperatures, under conditions which are likely to foster migration.

- Extreme compactability suggests a significant gas content.

- Where a knowledge of geologic age helps us to assess source-rock potential, some help may be obtained from the thickness of the unit (which at least establishes a minimum depositional period) and more specific help from age-dating by eustatic-cycle techniques (p.5-33 to 5-36).

Where we can identify a source rock and measure its interval velocity, we can compute its volume. If the geology is not disturbed, we might hope to recover oil equivalent to 0.0001 of this volume; if the geology is complicated by faults and other lossy features we cannot expect more than a small fraction of this.

2. <u>Reservoir rocks</u> Our seismic tools for identifying potential reservoir rocks are as follows:

- We can recognize a near-shore high-energy depositional environment by locally strong and fairly continuous reflections, provided that the indications occur in an appropriate areal configuration (pp.5-16, 17; 5-21, 22; 5-26).

- We can often recognize continental deposits by the variations of amplitude and continuity of reflections within them (p.5-17).

- We can recognize massive carbonates by their low-energy appearance and by their velocity in zones which are not uplifted; we can then assess the possibility of reservoir characteristics by the change of velocity over the highs (pp.2-45 to 2-51; 4-85 to 4-87).

- To identify reefs we use their velocity, their rigidity (recognized by differential compaction in draping sediments), their reflection-free interior, and their areal configuration. We also demand agreement with our picture of the depositional environment and of the rate of rise of sea level at the time of growth.

- We are always interested to know the source of sediments (p.5-25).

- Where we identify a transitional reflection at the boundary of an overpressured zone, we know that permeable beds are near (pp.2-22; 4-23, 24).

- Reservoir materials which are frictional in nature may produce a measurable loss of high frequencies by absorption (section 2.5.1).

We make a seismic contribution to the assessment of the chances of hydrocarbon accumulations in these reservoirs, by means of the following tools:

- We search for shallow gas indications above the postulated reservoir (pp.5-2 to 5-4).

- We search for shallow gas indications or sea-floor seeps which, in conjunction with a defined fault plane or other permeable path, indicate a hydrocarbon path into the reservoir (pp.5-3; 5-5 to 5-7).

- Our attention is attracted by seismic evidences of growth in anticlines or faults, occurring simultaneously with the compaction of source rocks (pp.5-25, 26). Our interest is increased further if the cause of growth is diapirism, which we recognize by the configuration of reflectors above and below, and the reflection-free or chaotic appearance within the diapir (pp.5-26, 27).

- If the source and reservoir rocks are not juxtaposed, we search for permeable connections between them. These may be fault planes, fracture zones or the porous weathered zone below an unconformity surface (p.5-29). In this we may have to make a judgement whether an observed fault plane is likely to be a conduit or a barrier. Transcurrent faults are said to be the most likely to maintain a seal. And it may be that the actual fault plane remains a conduit long after the formations near the fault have become totally cemented by the minerals passing up the fault plane.

- We establish the spatial relationship of key source and reservoir features in past geologic time by datumizing our sections appropriately (p.5-30). This may involve making subcrop displays, or worm's-eye displays, or displays from which regional dip is eliminated.

- We study all evidences of overpressure in relation to the picture of transgression/regression, in particular to resolve whether hydrocarbons are likely to have migrated upwards or downwards.

3. <u>The fluid potential field</u> In general we do not expect anomalies of fluid pressure before the Mesozoic. The estimation of formation age is therefore relevant. Where overpressure is feasible, our tools for recognizing it are:

- Overpressure always produces anomalously low velocities and densities (pp.2-23, 24; 2-33; 2-39; 2-49 to 2-52).

- Overpressure in a shale is likely if the deposition has been rapid; this can be assessed by the thickness of the unit and the correlation of its associated onlap sequence with a sea-level chart (pp.2-21; 5-23; 5-35).

- Overpressure is unlikely in thin beds.

- Transitional (low-frequency) reflections are likely at both the top and bottom of massive overpressured units (pp.4-23, 24).

- Faults do not traverse an overpressured shale, except by removing the overpressure locally; faults may occur, and indeed be caused, within nearby units of other lithology.

- Overpressure in a shale is transmitted to a permeable sand or carbonate in contact with the shale if the permeable bed is itself sealed; this is the basis of production from overpressured fractured chalk reservoirs (pp.4-85; 8-52 to 8-60).

- An active overpressured shale diapir may evince chaotic internal reflections coupled with bright spots due to gas pockets (p.5-27).

4. <u>Lithology generally</u> The seismic tools for studying lithology are velocity, acoustic impedance, absorption, depositional environment, and deformation under tectonic stress. We have to accept that although these tools are occasionally unequivocal, they usually require assistance from geological considerations before they speak clearly. In fact neither the geophysical evidence nor the geological evidence is self-sufficient, but, as in so many other things, we seek the correct interpretation by bringing the two together.

Yet another tedious subscription to virtue.

PART 6

MODELLING

INTRODUCTION

Modelling is the process by which a postulated depth section is converted into the corresponding time section.

In a structural context, this process is inverse to the process of migration into depth. It is used to demonstrate the validity of a postulated interpretation in depth, by showing that the postulated interpretation does indeed yield the observed time section.

In a stratigraphic context, the modelling process is the final validation of an interpretation involving a bright spot or some other related anomaly.

The value of the modelling process is that it takes us from a postulated geological situation to a seismic time section with a certainty much greater than that associated with the reverse process. The weakness of the modelling process, as we shall see, is that the specification of the postulated geological situation requires the correct choices for an embarrassingly large number of variables.

6.1 STRUCTURAL MODELLING

Structural modelling is available at many levels, from the very simple to the mathematically complex. The interpreter is not concerned with the details of these solutions, but he is concerned to know what level of complexity (and cost) is appropriate to his problem. He is also interested in knowing how to monitor the performance of the modelling program, and in particular how to establish the mechanism by which a particular event on the modelled time section comes to appear as it does.

The first level of structural modelling is illustrated on p.6-4. At the right hand side, normal-incidence raypaths have been constructed from equally-spaced points on the reflector (c) specified in _depth_. The length and emergence point of each ray to the surface define a normal-incidence travel time observed by a closely-spaced source and geophone at the surface. The plotting of these times under the surface points allows the construction of the corresponding _time_ section (lower sketch). To conform to the normal arrangement of equal spacing of surface points (rather than points on the reflector), the patterns may be modified by simple interpolation or iteration.

Obviously, the travel times and emergence points of the rays are incorrect if refraction occurs at high levels. For example, if a compaction law is know for the material above reflector (c), then the corresponding modifications to the time section must be computed. Or if a velocity break is known to occur at reflectors (a) and (b), the corresponding refraction is again required; these computations are simple applications of Snell's law.

Even at this elementary stage, we can see some of the features and problems of structural modelling:

- The third dimension is critically important, and we ignore it at our peril.

- If the reflectors (a) and (b) are not actually the velocity breaks in the upper section, or the compaction law is not apt, then the modelled time section must be incorrect.

- On a simple ray-path construction, there are points on the surface from which a certain reflector may be seen at normal incidence twice, or not at all; this is the phenomenon of velocity lensing discussed briefly in section 2.8.4.

- Sometimes it is very difficult to see which reflection on the time section comes from which reflector on the depth section. The tool to resolve these difficulties is the ray-path diagram, which shows the connection between the two.

The last point is emphasized by studying the time sections corresponding to a variety of geological models. A selection of the slides illustrated (by courtesy of M.T.Taner) is reproduced in simplified line form on pages 6 - 5, 6, 7, 8, 10.

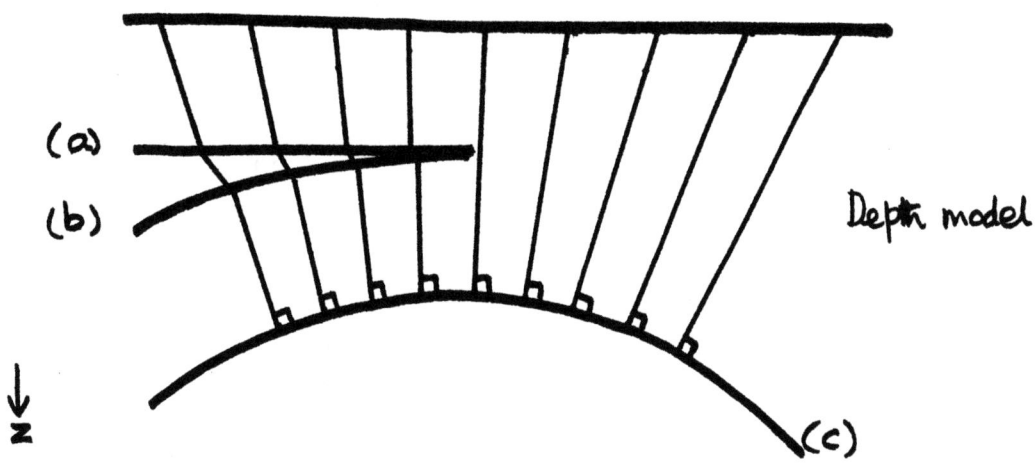

(a)
(b)
Depth model
↓ z
(c)

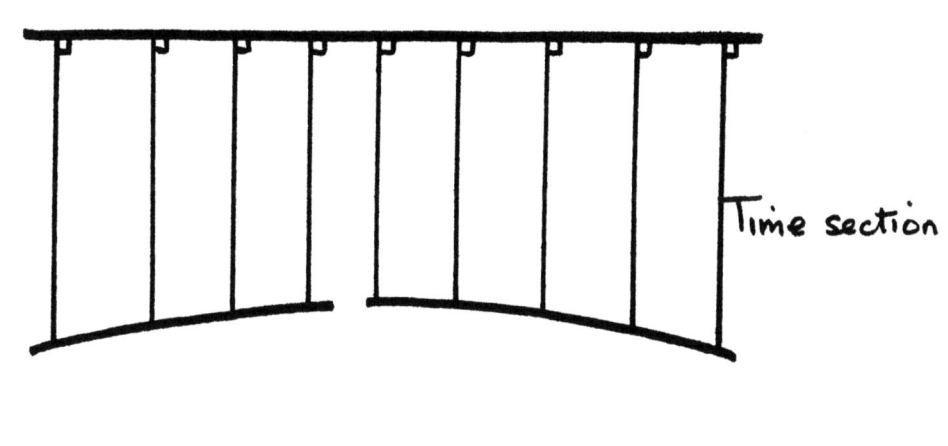

Time section
↓ t

6-4

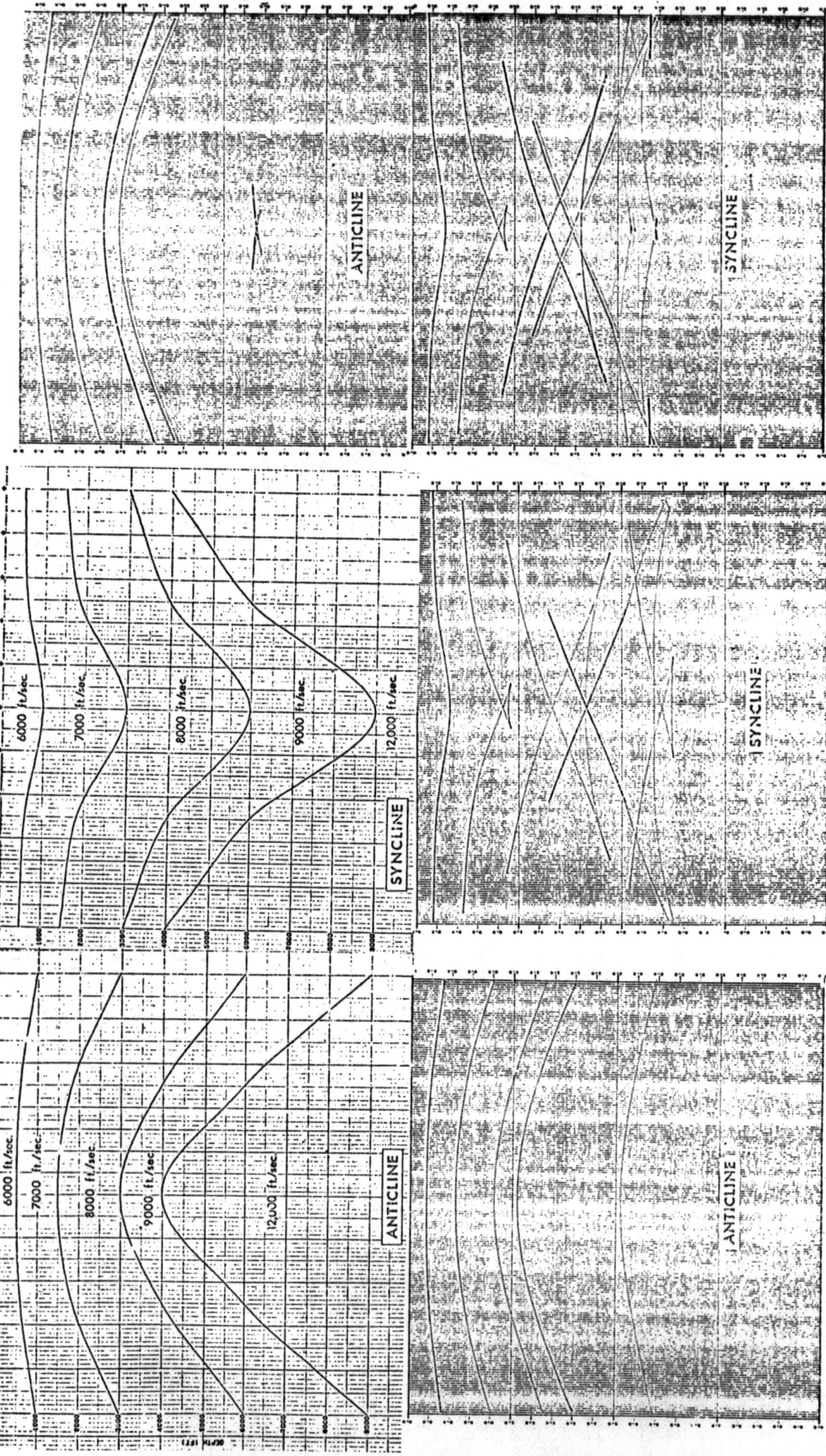

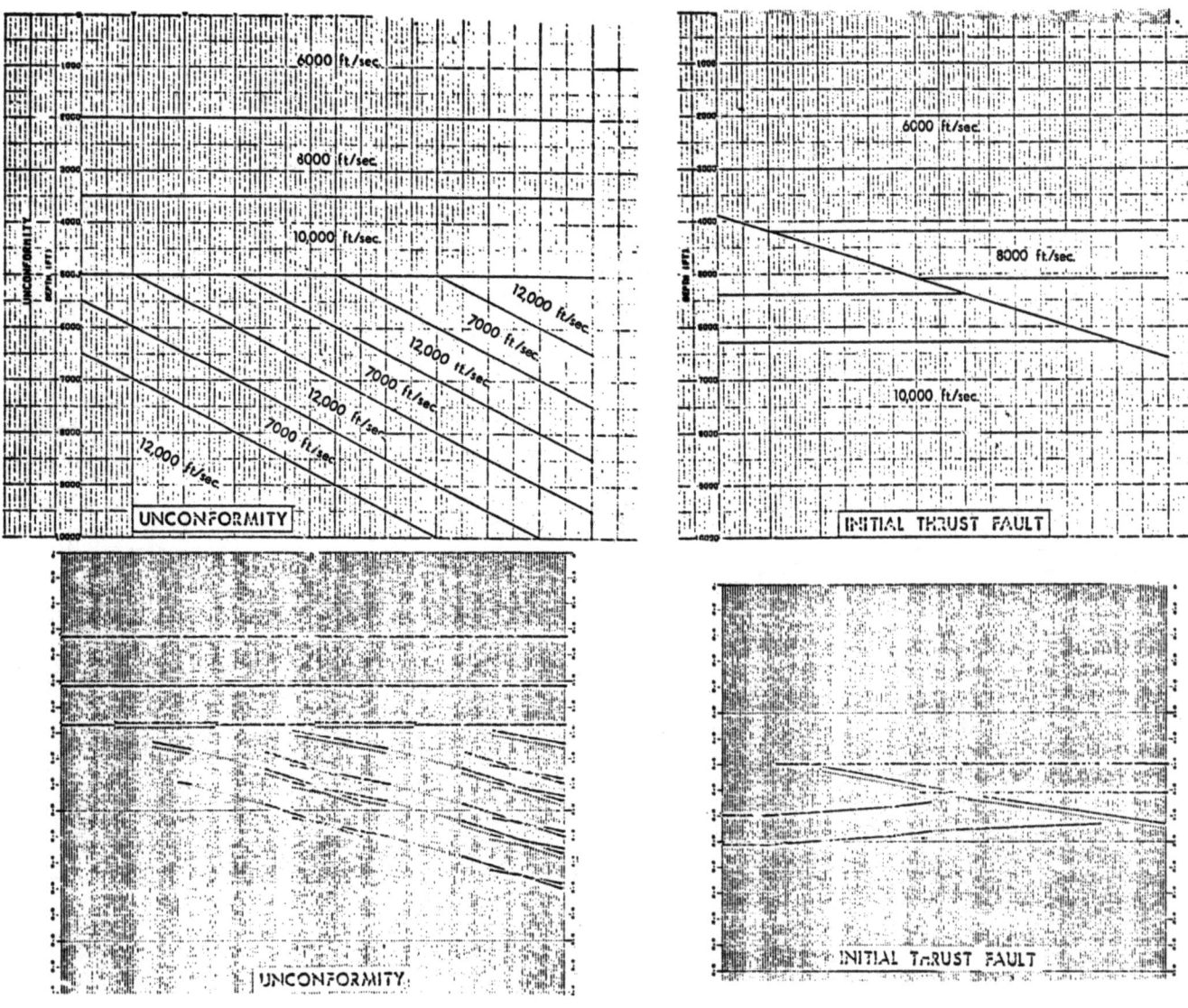

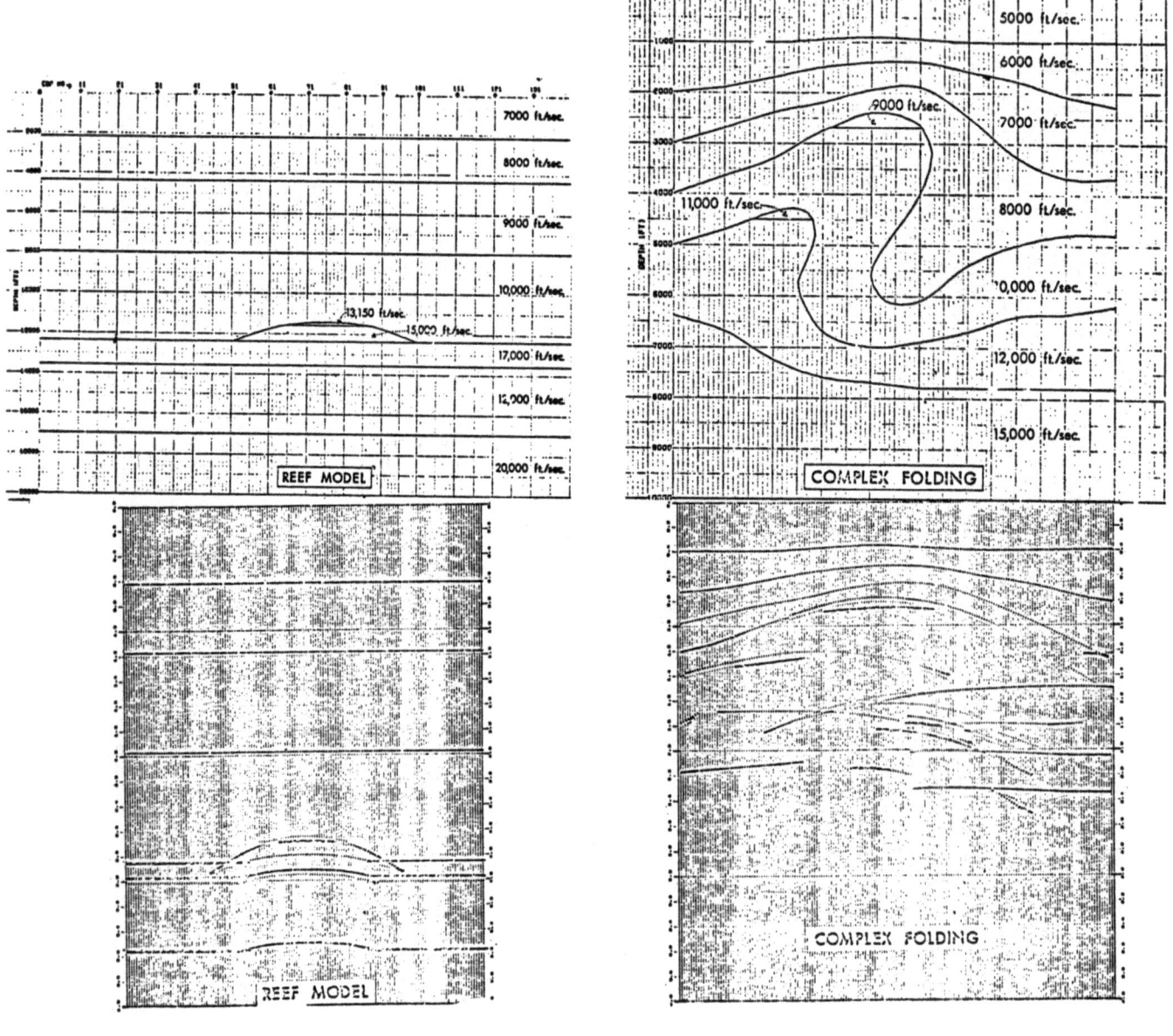

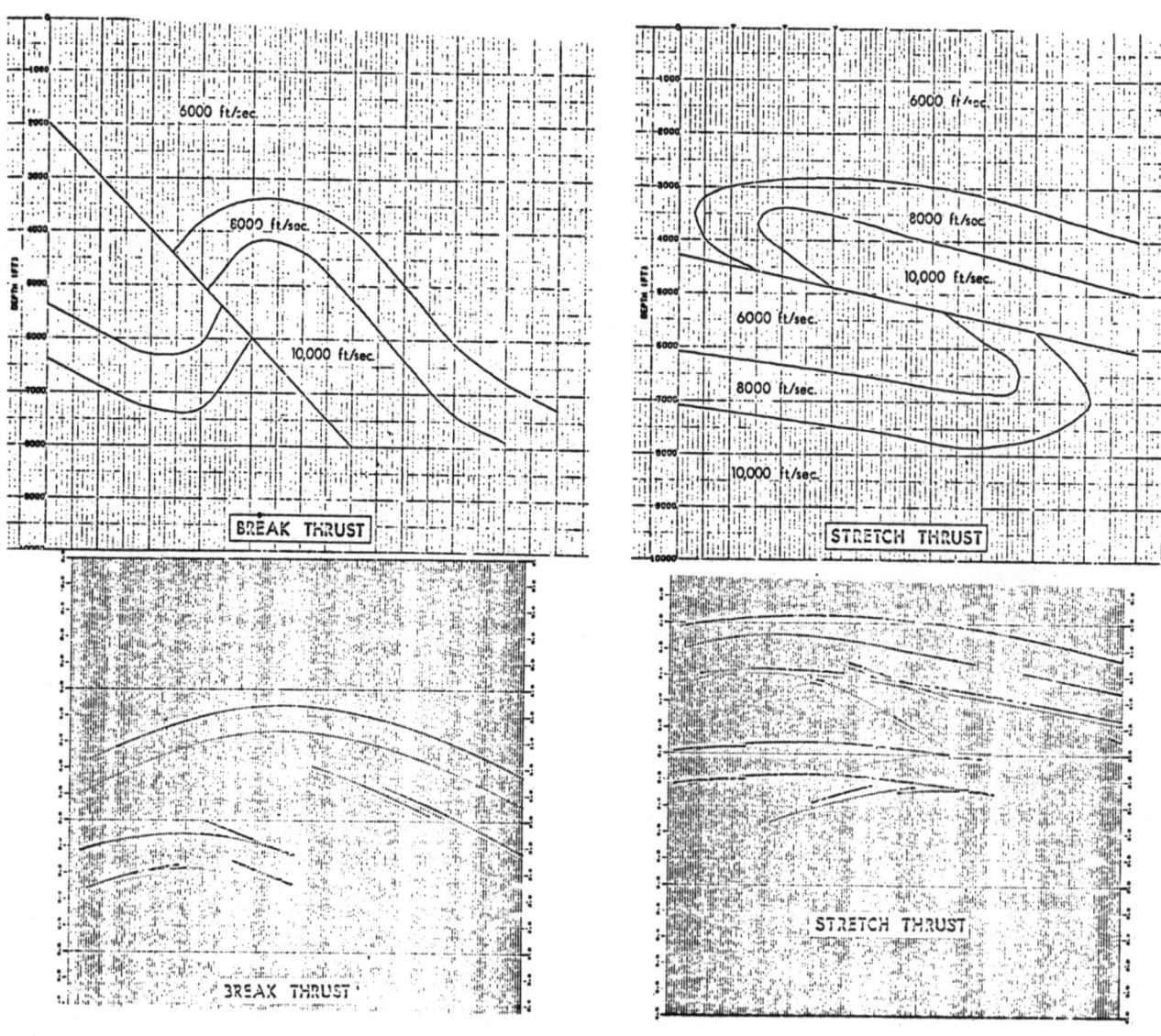

TAPE 31   Although some of the slides illustrated obviously
370L represent extreme geology, they serve to remind us that
371   there is a level of geological complexity beyond which
372L interpretation is impossible (certainly unless the con-
373   ditions allow precise migration in the necessary number
374L of dimensions). Under such circumstances the good inter-
375   preter resists the internal and external pressures imply-
376   ing that it is his duty to make an interpretation, and
377L affirms that the section is <u>not interpretable</u> in its
378   existing form.

379   Even at the present simple level, there is merit in
380L adding a pulse shape to the modelled time section. This
381   may be done by extracting the primary pulse from the
382L comparison time section, using the techniques of White
383   and O'Brien discussed in section 3.1; these techniques
384L are very stable if the whole of the comparison time
385   section is used as input for the estimation, although
386L some uncertainty usually remains in the details of the
387   timing. Alternatively, the pulse shape used for the
388L model may be that recorded from the source in the field,
389   and processed in the same manner as the comparison
390L section (as set out later, in section 8.2); the only
391   uncertainty remaining is then the effect of the earth
392L filtering, which is not usually of first concern in
393   structural modelling.
394L
395
396L
397
398L
399
400L
401
402L
403

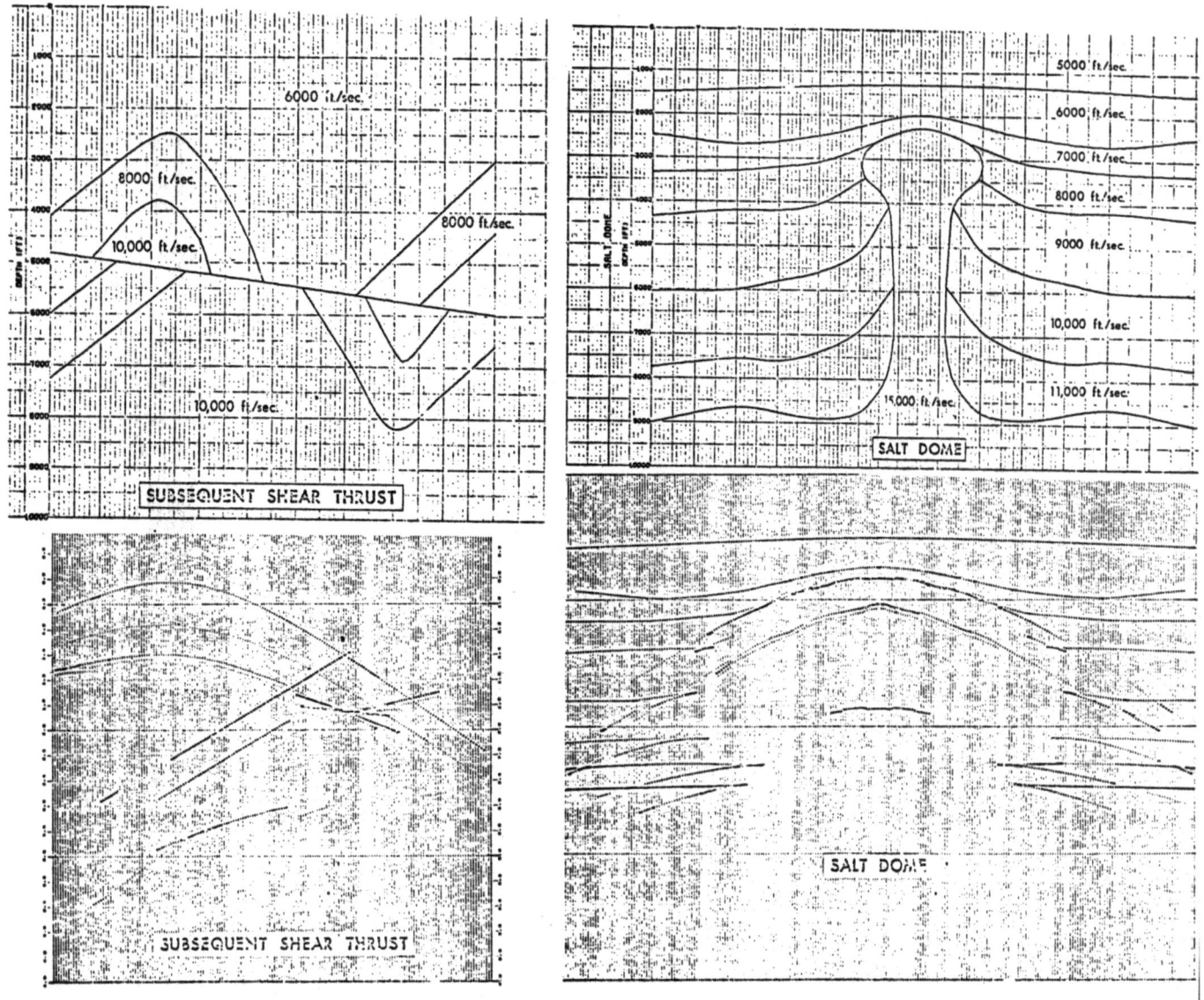

6-10

Some additional conclusions are now apparent from the modelling.

- Velocity lensing can be a major effect. It can itself create the "bow-tie" pattern more usually associated with tight curvature, and it can cause important fluid-contact reflections to appear in quite unexpected locations far from other reflections associated with the reservoir.

- The same refraction effects can give rise to quite different time sections from the same depth model, depending on whether a particular interface refracts one way or the other. A simple anticline may appear as a simple anticline with one velocity distribution, but as a buried focus with another.

- Therefore the interpreter should not regard a buried-focus bow-tie as <u>necessarily</u> indicating a tight concave curvature.

- Simple models do not properly represent the mélange zone of a fault, and so tend to produce modelled fault-plane reflections far stronger than those normally observed on real sections.

- Refracting (velocity-lens) features in the shallower section can cause a deep reflector to appear faulted, discontinuous or irregular — entirely spuriously.

- Simple ray-path modelling at normal incidence is adequate to show the expected broadening effect on tight convex structure, but of course the truly "diffracted" signal is missing.

A first approach to the introduction of diffractions is the Huygens technique of section 2.8, in which each point on each reflector is viewed as generating a symmetrical Huygens hyperbola. Computationally this is simple (being closely inverse to a classical migration program). Page 6-13 shows the transition from a depth model (of one horizontal and three dipping reflectors) through an intermediate situation in which the spacing of individual Huygens hyperbolas is still such that they can be seen, to the final physical situation in which the hyperbolas are spaced infinitesimally. This last situation is, of course, the one approximating the real effects in the earth (though at the ends of the reflector we remember the caution, from section 2.8.5, that a terminating reflector does not correspond to a fault).

Page 6-14 shows the same transition for anticlines and synclines of progressively increasing tightness. We notice, as we expect, that the increasing tightness of the tighter anticlines is not apparent in the breadth of the features on the time sections; it is apparent in the amplitudes (as discussed in section 2.8.4), and it may also be apparent, in its effect on the bow-tie, in the concave-upwards curvature at the limits of the anticline.

The Huygens technique can be extended easily to any complex reflector model. It yields a useful (though in detail approximate) delineation of reflection configuration and diffraction patterns.

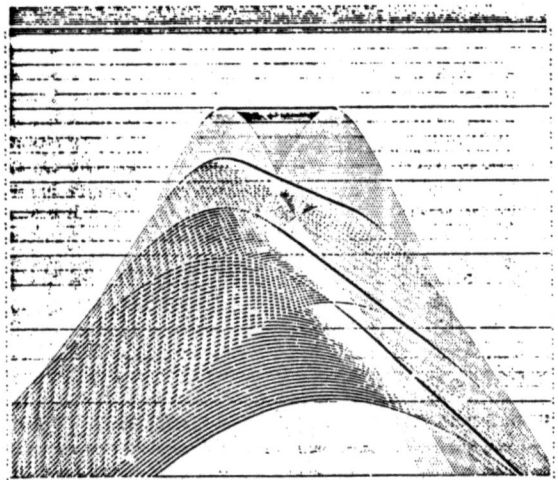
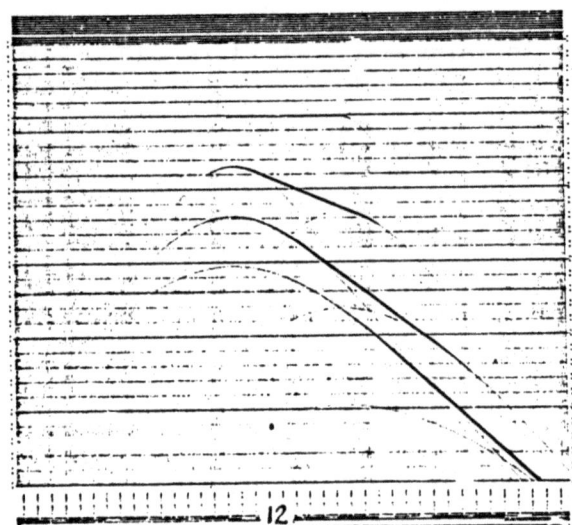

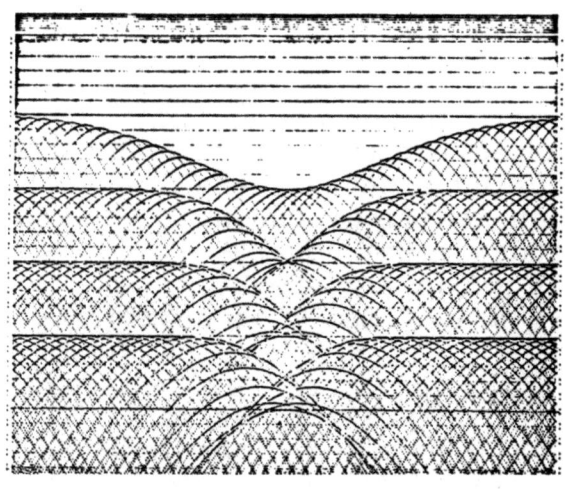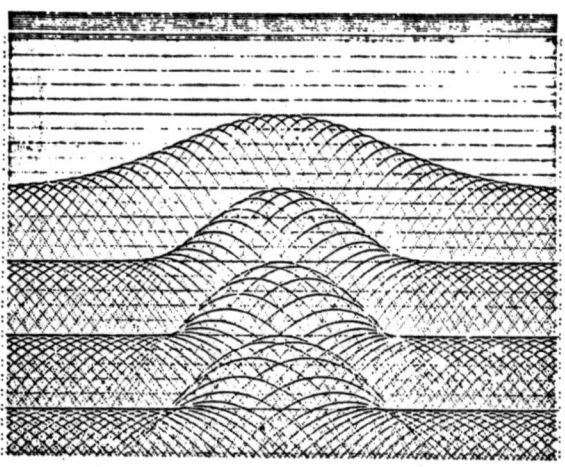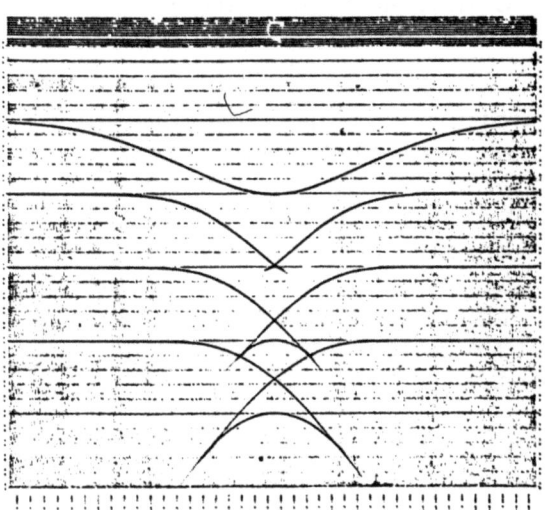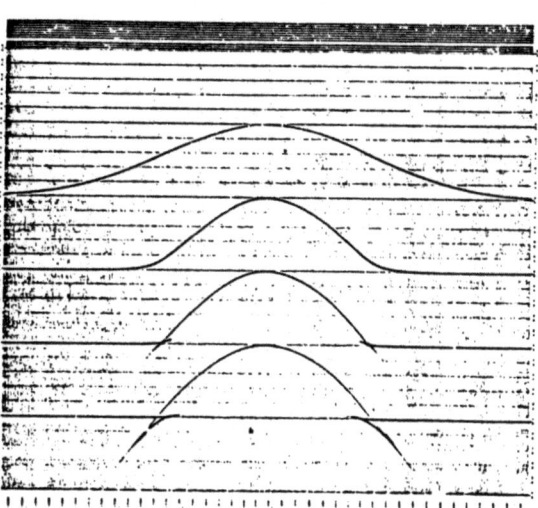

There are two situations which (added to a niggling disquiet about Huygens, which we will discuss in Part 7) may lead us to be dissatisfied with this technique.

- If the technique assumes Huygens sources of <u>equal</u> amplitude and equal spacing, it requires modification to accommodate those variations in the Huygens sources brought about by the downward path from the surface. The modification often decreases the degree of focusing associated with a particular geological configuration, in that the focusing constructed for the return path is offset by a defocusing effect on the downward path. Unless such modifications are done, the relative amplitude of the upward-travelling diffractions is subject to doubt, and the initially-downward-travelling ("phantom") diffractions are omitted.

- If lateral variations of velocity exist (as is inevitable in any geologically-complex situation) the surface response to a single Huygens source is not generally hyperbolic and not generally symmetrical.

To some extent these objections can be overcome by additional ray-tracing. For example, the geometrical divergence and velocity-lens focusing introduced by the downward path can be calculated as in section 2.8.1, and an amplitude assigned to each secondary Huygens source accordingly. Further, the appropriate modification to the hyperbolic form of the surface arrivals from each Huygens source can also be calculated by ray tracing upwards from such source.

As these additional refinements are incorporated, the amount of computation increases until the cost is no longer trivial (particularly if, as is sometimes necessary, the modelling is of a two-dimensional time section from a three-dimensional geologic model). If the required degree of detailed accuracy is extreme, our minds turn to a full wave-equation solution by the method of finite differences (Alford et al., 1974); the computing time then becomes formidable. Accordingly, the search is for a method which uses the ray-tracing of the normal-incidence path to define a tube, within which the Huygens solution or the complete wave solution is effected (May and Hron, 1975). Another approach is outlined by Dedman and Lindsey (1975).

But we have said that our concern is not with the details of the modelling calculations, but in asking, "As interpreters, how do we know what degree of sophistication to ask for in the modelling, how do we check its adequacy, and how do we understand its message?" The answers, for any particular application, lie in a balance between the following considerations.

- Normal-incidence ray tracing is very simple and very cheap; even in the most complex cases it is wise to use a ray-tracing method first, to identify the gross effects.

- Where the problem is strictly one of validating a structural interpretation in complex geology, at constrained cost, a simple normal-incidence ray-tracing solution performed in three dimensions may be a better buy than a wave solution in two dimensions; incorporating the third dimension may be more important than incorporating the diffractions and other wave subtleties.

- Normal-incidence ray-tracing yields all the specular reflections (including such effects as buried-focus-bow-ties, broadened anticlines, etc.); it also yields all the distortions of reflection position produced by refraction.

- Therefore a time section produced by simple ray-tracing presents the main reflection alignments in approximately the correct places. However, the reflection alignments terminate and flex in an "unnatural" way, being generally more abrupt than

425   is observed on real time sections; the section also evinces "holes", blind zones" and breaks-in-continuity which again create an unreal appearance. On the real section these holes are filled partly by diffractions and partly by unmodelled "round-the-edge" paths associated with the usual multiple-coverage cdp technique. These two contributions can sometimes be separated by comparing the stacked section with the near-trace single-cover section.

(In passing, let us note the significance of this with regard to cdp stacking. Probably we have all wondered, from time to time, why stacking is as good as it is. The improvement often seems more than a $\sqrt{N}$ enhancement of signal-to-noise ratio, often seems more than that to be expected from attenuation of the multiples. Of course we are lucky in that so many real situations preserve, by good chance, the hyperbolic reflection alignment assumed in ordinary stacking. But the unexpected benefit probably comes from the "round-the-edge" blank-filling effect of the triangle of cdp paths.)

- No modelling method can be better than the velocities it uses. The position and attitude of reflection segments is critically subject to the correct specification of the interval velocities, the breaks between the velocities, the compaction law(s) and the transitional reflectors. The neglect of the compaction law(s) and other _gradual_ transitions of velocity over-emphasizes refraction effects. Velocities which are in error to the degree of replacing a high-low contrast by a low-high contrast (or the reverse) can produce major errors in the position of reflection segments, since the rays are refracted in the wrong direction. As a consequence of all these facts, <u>it is futile to adopt a very sophisticated modelling program unless the velocity field is well known</u>; since this velocity field is unlikely to be well known in cases of extremely complicated geology, there is a practical limit to what can be expected from structural modelling in these cases.

426 • As a monitor of the correctness of the modelling, and as a means of understanding the position of events on the modelled section, the ray-path diagram is invaluable to the interpreter. It (or its equivalent) should be requested whatever the type of modelling program.

427B
L(B) • The incompleteness of a time section obtained by normal-incidence ray-tracing becomes evident whenever the radius of curvature of the structure becomes less than the radius of curvature of the wavefront (section 2.8.4), since this is the condition leading to marked diffractions. This situation existing in the depth model can be flagged very easily by the machine.

• The chances of precise modelling of the diffractions from faults are poor with any program, however sophisticated. This is because the basic requirement for visible diffractions is not just structural curvature but <u>volume rate-of-change</u> of acoustic properties. Many of the features of real faults (the melange zone, curvature and tearing behaviour near the fault, differential mineralization, straightness of the fault, etc.) militate against a sufficiently precise specification of the model. Further, if the fault persists to levels higher than that of interest, the chances are poor of correct specification of the disturbed conditions governing the wave propagation down to the level of interest and back. It is very easy to become obsessed with the computational possibilities to the exclusion of the geological realities.

- The seriousness of the complications above the level of interest (which complicate and distort the Huygens hyperbola to be used in modelling the diffractions) can be assessed to some extent by studying the <u>gathers</u> and their deviation from the hyperbolic form. However, the usefulness of this is obviously limited by the fact that the common-depth-point represented by the gather is unlikely to be truly common under the complex conditions postulated. Clearly, this sort of sophistication in modelling is reserved for special cases concerned with defined reservoirs. In such cases it may indeed be true that it is helpful to model the diffractions at the level of interest before attempting to compress them with an appropriately-sophisticated migration program (sections 7.3, 7.4).

- When complex geology is modelled by simple normal-incidence ray-tracing techniques, modelled amplitudes are unlikely to bear much relation to observed amplitudes. Neither is simple Huygens modelling sufficient; it is necessary to adopt one of the sophisticated "tube" or wave-equation methods. Again the modelling can be no better than the velocities used — and now we have the additional complications of the likely variations in real reflection coefficients brought about by the tectonic complexities. Amplitude studies in geologically contorted areas are therefore reserved for the very finest degree of detail in the resolution of a reservoir situation.

In summary, it is clear that the first and greatest usefulness of structural modelling is in recognizing the origin of reflection segments, and thereby validating structural picks, in areas of tectonic complexity. Modelling can be extended progressively, to the generation of diffractions and the synthesis of reflection forms and amplitudes; the limit of usefulness is set not by the physics but our ability to specify all the relevant details of the model.

Structural modelling would be unnecessary if we could perform migration perfectly. However, the noisy records which ordinarily partner tectonic complexity mean that perfect migration is not possible when we need it most. It is often more sound to match the model to the unmigrated section than to attempt migration, and cheaper to perturb the model specification (in particular, the velocities) than to repeat the migration.

6-2 BRIGHT-SPOT AND STRATIGRAPHIC MODELLING

Some modelling situations can ignore refraction and migration considerations altogether, and treat all normal-incidence paths as vertical; a simple channel-sand reservoir would be an example. This removes the doubts and uncertainties of the velocity-dependent refraction above the zone of interest; we are no longer concerned with the <u>positions</u> of problematic reflection alignments, and this makes it realistic to consider more subtle features such as amplitudes, polarities, pulse shapes, absorption, transitional reflections, and interference patterns.

We may still need to consider faults, where they represent reservoir terminations, but these are more likely to be adequately modelled (for example, by the simple Huygens approach) when the overlying velocity field no longer contains the problems of gross tectonic deformation.

<u>TAPE 32</u>

428

429

When all reflection paths are considered vertical, the modelling of a simple reservoir situation is elementary. One illustration has already been presented in the top-and-bottom example of section 4.4.1 (p.4-26). A more detailed illustration is given opposite; the feature has been interpreted on the seismic section (top diagram) as a single gas sand, picks have been made for its upper and lower surfaces, these have been converted into depth using estimated velocities (bottom diagram), and the seismic response of the sand has been modelled using assumed or estimated pulse shapes (middle diagram). The relative amplitudes of the reflections (and hence the shape of the inevitable interference complexes) also require the calculation of the reflection coefficients, which is based on assumed or estimated values for velocity and density. The final comparison, between the field section and the modelled section, is reasonably satisfactory; the seismic interpretation is therefore acceptable (though in general there will be others which would be equally acceptably — the answer illustrated is possible, but not necessarily right).

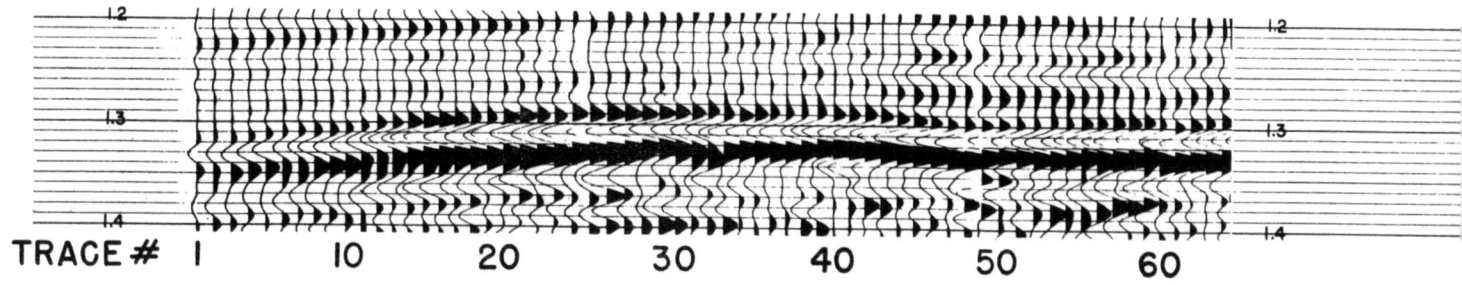

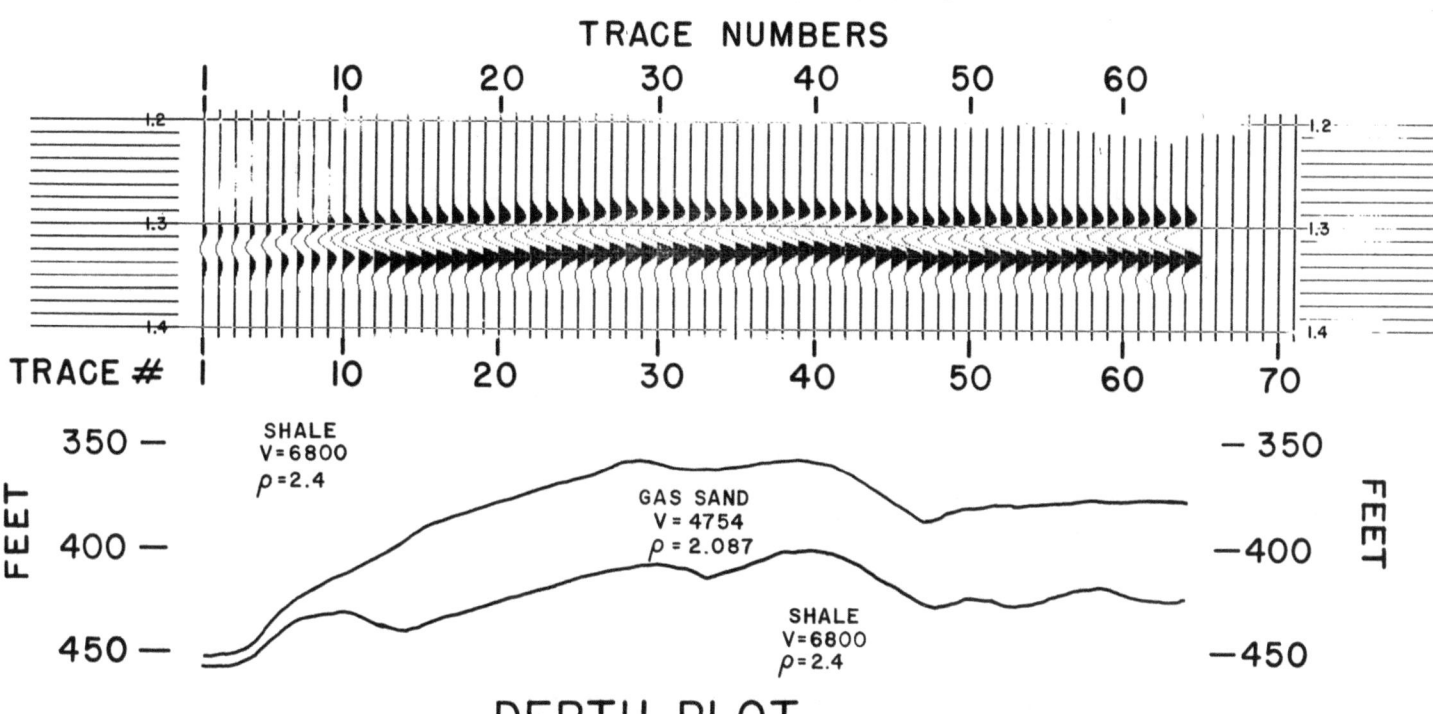

MODEL OF A SINGLE GAS SAND

430 A further illustration is shown opposite. Here, although we all salute the interpreter responsible, there exist several zones where we are left feeling uneasy. Clearly the evidence for the presence or absence of some undesirable unit within the sand hinges critically on our having the right pulse shape in the model. Further, we are immediately led to check the curvature at the two uppermost points of the sand — is it fair to use single vertical ray-paths, or are the conditions such that we can model the amplitude weakness at these points only by using Huygens hyperbolas? We are worried, therefore, about the means of deciding on the pulse shape and the aptness of our modelling technique, but we are still fairly satisfied with the corroboration of the outline interpretation.

431 It is when we come to consider a model of the complexity of p.6 - 26 that we realize fully the problem of setting up the original outline interpretation. Here we must not only salute the interpreter, but give him a cheer as well.

There is an embarrassing profusion of options in any such model. This profusion is such that it is clearly desirable to perform the modelling at an interactive terminal. However, even at a terminal it is still scarcely feasible to consider all the possibilities. So let us list the profusion of options, and see what guidance can be given for their selection.

The variables which must be specified in setting up the model are:

- the time configuration of all relevant horizons, and in particular the thickness of the reservoir(s),

- the pulse shape,

- the acoustic impedances (that is, the interval velocities and densities) within, above and below the reservoir(s),

SEISMIC SECTION

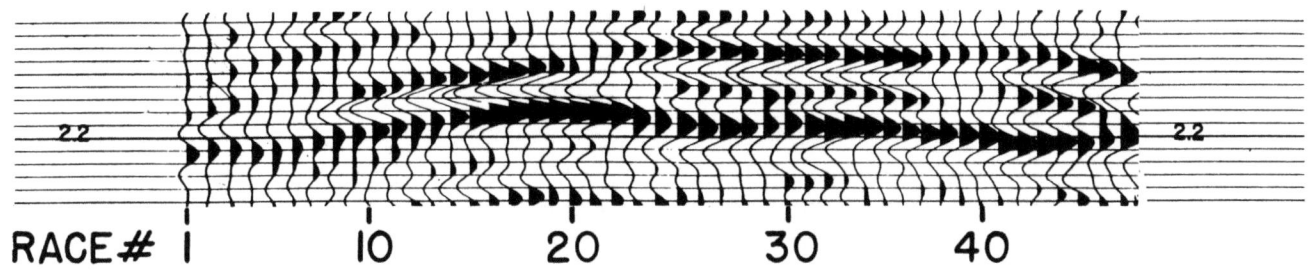

SYNTHETIC SECTION

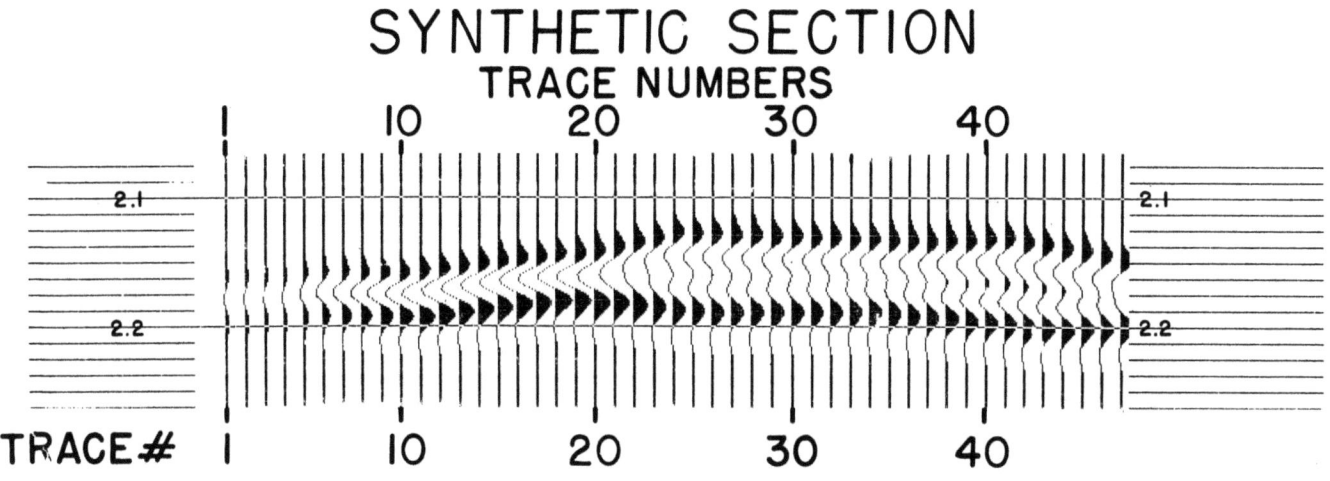

DEPTH PLOT

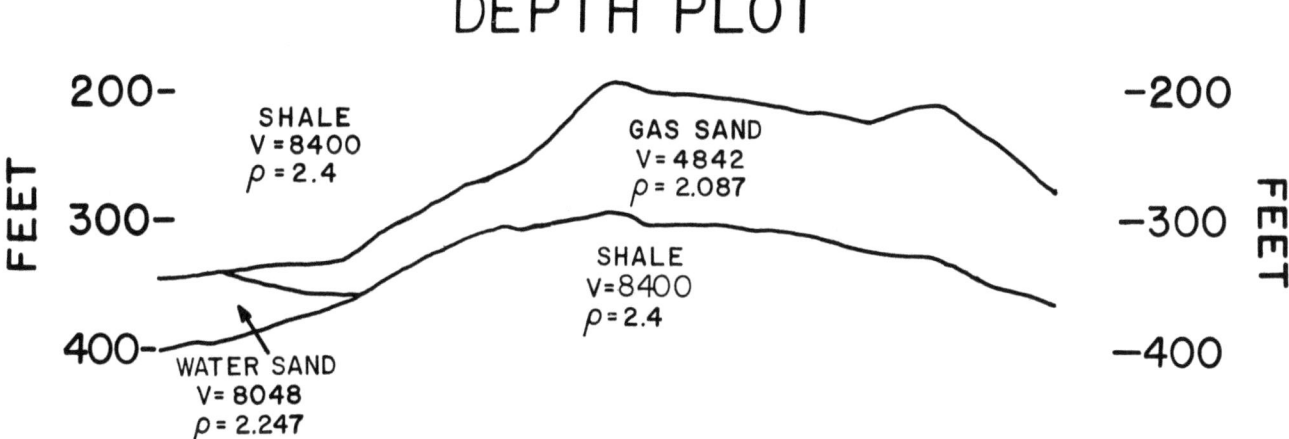

MODEL OF THICK GAS SAND WITH WATER CONTACT

After Patch (Geocom) 1973

- the existence and extent of transitional reflections, and transitions of acoustic impedance within the reservoir(s),

- the absorption (if significant) within the reservoir(s),

- the migration velocities appropriate to the reservoir time(s), the areal configuration of the reservoir(s), and the direction and configuration of any faults or abrupt curvatures (to permit modelling of the diffractions).

The guidelines for the selection of these variables are as follows.

- The initial picking of the interfaces is, of course, the most crucial step. This picking is different from normal picking for structure, since it cannot be done without making judgements about the acoustic properties of the rocks forming the interface. For example, the reflection from the top of a shallow gas sand is often strong and negative; at the gas-liquid contact the same reflection may become very weak and positive. The correct pick, therefore, may be on a peak of the strong reflection waveform over the gas, changing to a nondescript position near the contact, and continuing on a very weak trough over the water; this, we remember, is for <u>one single continuous interface</u>. Minor variations of rock properties could turn the last weak trough into a weak peak. It is clear, therefore, that the basic picking to establish the interfaces to be modelled requires a complete interpretation (not to mention courage), and that this picking must be consistent with all the velocity information, and the petroleum likelihoods. Where alternative picks exist both must later be modelled.

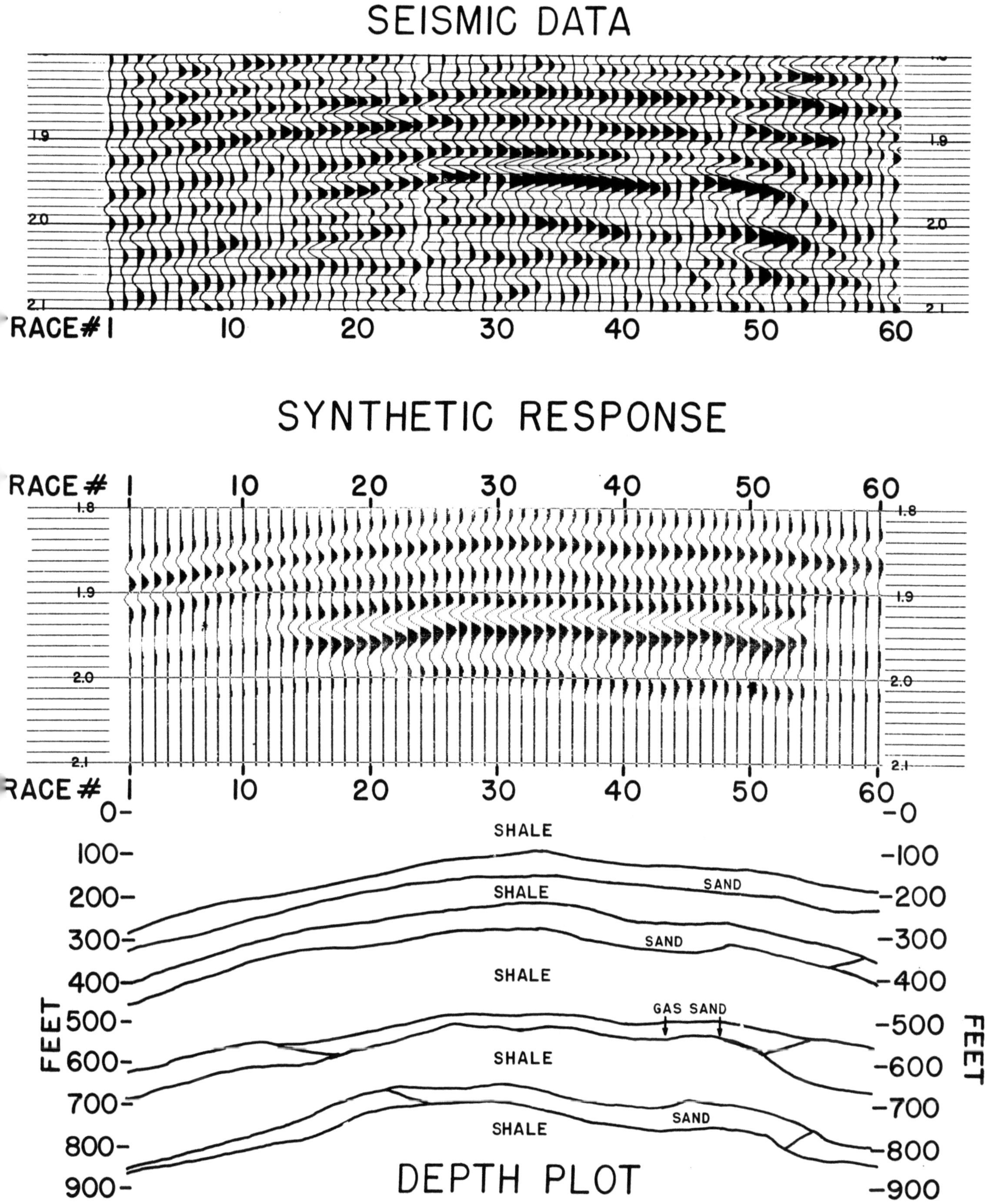

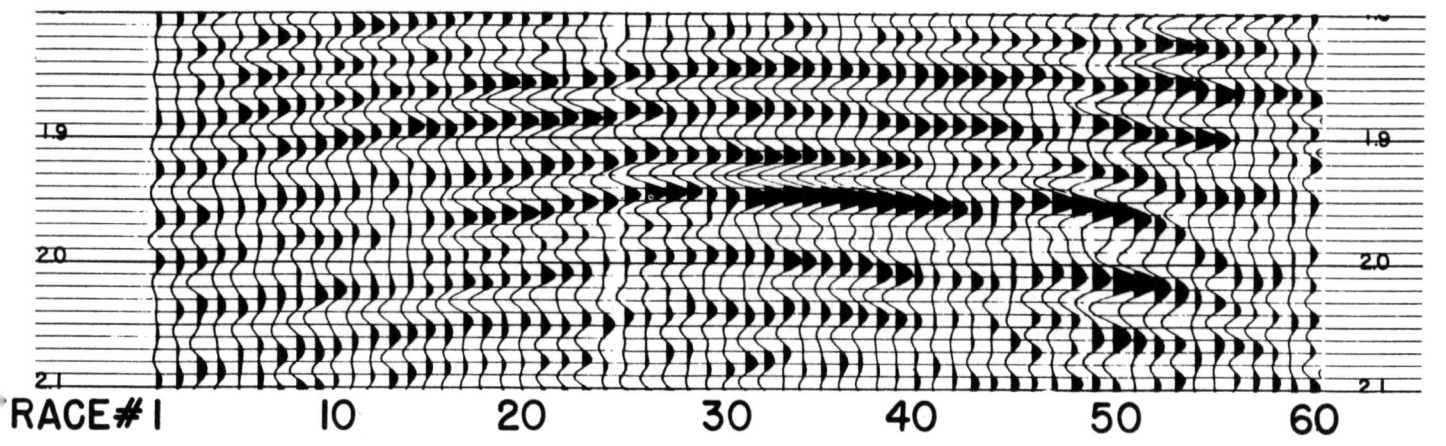

FOUR ZONE MODEL FINAL SOLUTION

6-26

- The thickness ascribed to the reservoir is a critically-important feature of the initial picking. If the top and bottom reflections are well separated, the only major worry is the correct identification of the polarities (since incorrect identification creates a half-period error in the time thickness, and a need for care in the calculation of the interval velocity). However, if the top and bottom reflections interfere (even after spiking deconvolution, which should always be used in this context — section 3.4.3), then the estimation of reservoir thickness is manifestly hazardous. This thickness is obviously one of the main variables on which we shall iterate in the final stages of modelling, but there is a strong case for using every tool available at picking time to minimize the range. The essential tool in this context is <u>knowledge of the pulse shape</u>, derived either from the section (in the form used) or from the recorded source pulse (after appropriate processing). Then we search for the characteristic high-frequency skewed (differentiated) appearance associated with closely-spaced reflectors of opposite polarity, or the lower-frequency skewed appearance associated with like polarities, or the presence of "hickies" or other indicators of the nature of the interfering elements. Thus the estimation of reservoir thickness at picking time may require the testing of different thicknesses by constructing interference complexes; this may be done either at the terminal or by template addition at the desk (not as tedious as it sounds) using the derived pulse shape. There is great value in this pre-modelling as a means of reducing the profusions of options at the final modelling stage, and of establishing the intervals within which the velocities must be estimated.

- It follows that another critical variable is the pulse shape. Here we invoke again a question which recurs repeatedly during this course — Why guess at something which can be measured? It is surprising that so many of the modelling illustrations in the literature are in terms of assumed pulses (even Ricker pulses, which are most unlikely to be correct). So the conclusion is clear: we use the estimation techniques of section 3.1.1, or we record the source pulse and process it appropriately (as discussed later, in section 8.2.4). If we are using estimation techniques from the section itself, we are careful to use a wide window of medium length (section 3.1.2). The pulse shape should not be something on which we have to iterate in the modelling.

- The point on the pulse which is taken as the reference time should be the point which yields the maximum of the velocity pick on velocity analyses made with the same processing. This matter must therefore be absorbed into the initial picking discussed above; it is important that all timing (whether on pulses judged positive or negative) should be consistent on section and on velocity analysis (section 3.5.3).

433
434
- The additional processing step to bring the final pulse to the symmetrical zero-phase condition (sections 3.1.1, 3.2.8.3; p.3-56 is of significant benefit in all problems requiring maximum resolution and identification of polarity (see also Dedman et al., 1975b; Barry, 1975). In particular, the skewed property associated with closely-spaced reflectors of opposite polarity becomes more easily identified.

- The interval velocities above and below the reservoir(s) can usually be obtained (with a confidence dependent on the number of analysis points available) by the detailed techniques set out in section 3.5.3. We recall that there may be a necessity for a compaction law if the intervals used are thick (p.3-102). The densities above and below are estimated from the curves of p.2-58 (provided microfracturing is not expected), or from local equivalents derived from borehole data.

- If the reservoir is more than 40 or 50 ms in two-way thickness, and the spread is of appropriate length, its interval velocity may be determinable by the techniques of section 3.5.3 as illustrated in section 4.4.3. Otherwise it must be one of the iterated variables in the model.

- The density of the reservoir, of course, if the variable which we ordinarily wish to use as the final objective of the model iterations (in accordance with section 4.4.1 and 4.4.3).

- It is trivial to include the absorptive Q of the reservoir if it is known. It is also trivial to include as a convolution operator any transition of acoustic impedance for which the limiting values, depth range and "shape" of the transition are known. But in general, of course, these quantities are not known, and so must be included as trial variables. We decide whether to include absorption or transitions (or both) according to the criteria of section 3.2.8.5, 4.2.5, 4.3.9 and 4.5.2.5; the major check is whether a low-frequency appearance is visible only at the bottom of the reservoir, or consistently over a significant zone below. We

6-29

remember also the caution that the comparison section must be properly stacked and properly deconvolved (section 4.3.9).

439L • We know that it is necessary to include diffractions*
440 if the reservoir is faulted, or has any rapid spatial change of acoustic properties, or is smaller in any horizontal direction than the Fresnel zones of section 2.8.3. In all of these, the third dimension is critically important; the angle between the fault and the profile must be known, and the direction of tightest curvature or greatest rate-of-change of properties must be known, and the minimum dimension
441 of the reservoir (in plan) must be known. It is important to remember that a profile following a channel sand, for example, shows a reflection amplitude which depends on the width of the channel sand if this width is small (less than a few hundred metres, depending on bandwidth, velocities and depth). These conditions must enter the model if they are relevant.

This discussion of the profusion of variables drives home to us the importance of narrowing the range of as many variables as possible before commencing the trial-and-error mating. Even with an interactive capability the number of permutations of variables is too great to allow an exhaustive test by a "try-everything" approach.

---

* The simplest Huygens method discussed in sections 2.8.5 and 6.1 is usually adequate here, since in the absence of structural complexity and marked velocity-lensing it is reasonable to assume that equally-spaced Huygens sources on the reflector are equally strong.

A satisfactory match with an appropriate model is, of course, the ultimate validation sought in section 4.5.2 -- the ultimate demonstration that the interpretation is self-consistent and therefore probably correct, and the most complete way of determining reservoir densities and porosities.

6.3 COMBINED STRUCTURAL AND STRATIGRAPHIC MODELLING

The above division between stratigraphic and structural modelling is artificial, of course. Just as, year by year, it becomes increasingly clear that a great number of so-called structural fields actually include a significant stratigraphic element, so it becomes increasingly important to model the combined picture.

Thus modelling programs are now asked to tackle such problems as a thinning reservoir sand, containing gas, oil and water, within a tilted fault block affected by additional minor faulting; or an uplifted and fractured reservoir showing a pronounced change of interval velocity and occurring below a section complicated by volcanics; or a curved reservoir on the overhung flank of a salt dome.

It is clear from the preceding material that such situations cannot be modelled with any confidence unless some of the enormous profusion of variables can be specified from borehole data. The problem then is to bring together all that is known from seismic results and all that is known from the boreholes, and to decide on an appropriate level of modelling sophistication. This is done having regard to the discussion in the last two sections, with particular emphasis on the principle that everything which can be measured should be measured — not regarded as a wide-ranging variable. Thus pulse shapes can be derived, near-trace sections can be compared with cdp sections, gathers can be scrutinized for deviations from a hyperbola, the important transitions of acoustic impedance can be identified from the borehole logs, velocity-density-porosity relations can be studied quantitatively, faults can be mapped in three dimensions, near-surface anomalies can be located, and synthetic seismograms can be constructed (section 8.1). Using these and other obvious tools, the interpreter can establish what sophistication and cost is justified for the modelling program. In this, of course, he must never err on the side of inadequate sophistication; however, there is no point in wasting money either, particularly as the most sophisticated techniques tend to produce models which the interpreter cannot monitor for correctness.

In the case of these complex problems, there is merit in extending the modelling to reflectors <u>below</u> the zone of major interest; there should be a satisfactory match not only down to and including this zone, but also at levels which represent paths passing through the zone. This may require a full modelling of both the downgoing and the upcoming diffractions (or, in other words, of wavefront healing).

442
443L
444
445

6.4 MODELLING THE CDP GATHER

The process of using ray-tracing to model diffractions in the presence of laterally changing overburden (as discussed in section 6.1) can be simply extended to allow modelling of the "common"-depth-point gather in these same circumstances. This gives us an additional tool for establishing the correctness of a complicated model. For we may demand not only that the position of the reflection amplitudes and polarities should check, and that the interference complexes should check, <u>but also that the velocities should check.</u>

The simplest case we might consider is that of an overburden consisting of layers of varying thickness, but in each of which the interval velocity is laterally constant. The tutorial conclusions are well known (Taner, Cook and Neidell, 1970; Levin, 1971; Neidell and Lindsey, 1974; Dedman and Lindsey, 1975a); major swings of stacking velocity are introduced, Dix interval velocities computed between reflectors of different dip are sometimes wildly in error, and conventional methods of horizontal interpolation between spaced velocity analyses yield a locally poor stack. (In connection with the last point, we recall the distinction, stressed in section 3.5.1, between the picking and interpolation of velocity analyses for purposes of genuine velocity measurement.)

The more general case we might consider is that in which both the layer thicknesses and the interval velocities are allowed to change. In terms of real geology this is clearly the only acceptable model; the above restriction to constant interval velocity is convenient for tutorial purposes, but cannot apply in practice. The derivations become complicated (Levin, 1973); the gather modelling must incorporate not only the <u>time</u> differences introduced into the various paths of the gather by the different thicknesses of layers of different velocity, but also the different <u>refraction</u> behaviour imposed on the various paths <u>both at the interfaces and within the layers</u>.

Thus the interpreter is faced once more with a decision between levels of modelling sophistication. Part of the evidence for the choice must be the degree to which the lateral variations of interval velocity are really known. And part is concerned with the rate-of-change of interval velocities laterally; if the change of interval velocity across the horizontal extent encompassed by a single gather is insignificant, there is no reason to use the more complicated modelling technique. This can be tested at the place where the rate-of-change is reckoned to be most severe; the comparison of the modelling techniques over this one zone allows the interpreter to assess the need for a complicated solution.

The result of modelling the cdp gather under real conditions is almost always sobering. Although the time and appearance of the reflection obtained by stacking the gather are not critically sensitive to the overburden complications, the derived velocities are <u>extremely sensitive</u>. In one context this is beneficial, <u>in that it increases the validation power of velocity modelling</u>; in another it reminds us that good validation cannot be expected unless the velocity field and the structural picture are very well defined.

The interpreter can usually visualize the general effect of a specified disruption of the velocity field fairly easily, by estimating the likely time shifts between different paths of the gather, and then by thinking in terms of the effect of statics on velocity (as discussed in section 3.5.1). Thus it is generally true that:

- A local anomaly of the velocity field (for example, a low-velocity gas sand, or a reef) produces a statics-type variation of stacking velocity, similar to those of pp.3-84, but present only on horizons <u>below</u> the anomaly.

465 466 467	• The largest variations of stacking velocity occur in the zones below the ends of the anomaly.
468(B) L(B)	• The variations grow in magnitude and horizontal extent with depth below the anomaly. These three basic effects are illustrated on p.6-38, following Pollet (1974).

- If dip develops at depth, without lateral change of interval velocity in any layer, then the "depth-point" for the far traces of the gather moves up-dip, the ΔT decreases accordingly, and the stacking velocity increases from its "correct" value. Rather interestingly, this effect is less marked if the shallow overburden shows a generally progressive increase of velocity with depth, as it usually does (Neidell and Lindsey, 1974).

- If there is no dip, but a smooth lateral change of interval velocity in any layer, the far-trace "depth-point" on deeper layers moves in the direction of the lower velocity, the ΔT decreases, and the stacking velocity increases from its "correct" value.

- In the general case involving both dip and lateral velocity change, the final effect on stacking velocities is an amalgam of the above effects. The two component effects act in the same direction (to produce spuriously high stacking velocities) in the usual situation in which interval velocity increases down-dip.

- Whenever a more abrupt change of interval velocity occurs (within the triangle of the cdp geometry) very large swings of stacking velocity occur, positive and negative, after the manner of p.6-38.

At last we understand!

In the early days of velocity analysis, hopes were high for a new lithologic tool. These hopes were quickly dashed when interpreters commissioned sequences of closely-spaced velocity analyses, and found that the resultant velocities (even from one common-depth-point to the next) showed wild swings. These were regarded as observational error, of a quite unacceptable magnitude, and velocity analysis fell into disrepute as a lithologic tool.

Year by year, our understanding of the causes of these wild swings has increased. Much of this increase of understanding has come from the published studies of velocity modelling. We now know that the swings are not generally the fault of our velocity-analysis techniques, but are primarily due to:

- statics-causing near-surface anomalies (section 3.5.1),

- lateral changes in interval velocity,

- complication of the gather ray-paths by passage through faulted zones and velocity lenses, and

- the forcing of velocity picks on reflections subject to interference between component reflections or between reflection and diffraction (for example, near faults on the picked reflector).

When these agencies are taken into account, and spurious picks rejected, velocity analysis re-emerges as an accurate and powerful tool. Many interpreters have been slow to realize this.

Additional references:

Larner (1974)
Lucas, Al-Chalabi and Shaw (1975)

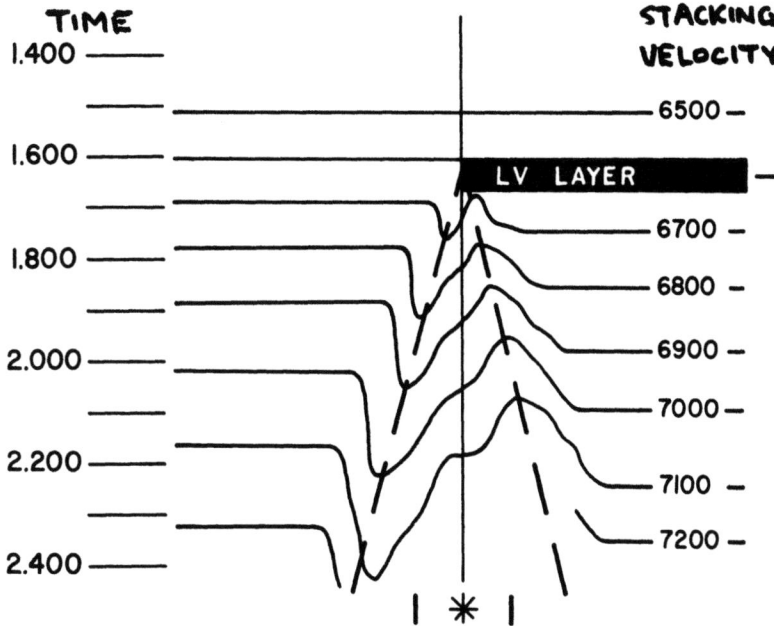

PART 7

MIGRATION

7.1 INTRODUCTION

Basically we are all familiar with migration viewed as a final processing stage in areas of steep dips, faults and general structural complexity. In applying the process, we seek to collapse the diffraction patterns generated by faults, to reduce the breadth of anticlines to their correct values, to increase the breadth of synclines, and finally to achieve a true picture of the geology in depth.

The subject is a fascinating one (particularly as expounded by Peterson and Walter, 1974). However, for present purposes we must abstract and concentrate on those facets of direct concern to the interpreter.

In the early literature, the process of migration was regarded as the direct transformation of a reflection segment from its position on a time section to its revised position on a depth section. Nowadays, the process is more likely to be regarded in two stages:

- Time migration, in which the reflection segment is moved to its revised position horizontally, but in which the vertical axis is still reflection time.

- Depth conversion, in which the vertical scale is changed to depth.

These stages are illustrated on p.7-4. The central diagram represents, let us say, the real configuration of reflectors plotted as a function of depth. Then, for the depths, dips, layer thicknesses and velocities shown, the normal-incidence time section (taken here as adequately approximated by the cdp stacked section) is as shown at the left. In the depth model, the vertical line represents the vertical through a particular surface point; in the cdp time section, it represents a trace plotted in a position corresponding to that surface point.

Thus the reflection from the left-dipping reflector of the depth model follows the dashed normal-incidence path, and occurs at a "point" some 360m to the right of the vertical through the surface point. This reflection appears on the corresponding trace of the cdp time section at a time of 0.915s and with a negative time gradient of 5.6ms per unit distance. Similarly the reflection from the right-dipping reflector of the depth model follows the illustrated normal-incidence path; it then appears on the time section at a time of 2.158s and with a positive time gradient of 6.5ms per unit distance. The fourth reflector of the depth model, which is actually horizontal, appears on the time section dipping to the right.

The process of time migration transforms the left diagram into the right diagram. The reflection pulse formerly plotted at 0.915s on the trace through the surface point now appears at a time of 0.846s on a _different_ trace corresponding to a horizontal offset of 360m to the right. Similarly the pulse formerly plotted at 2.158s on the trace through the surface point now appears on a different trace corresponding to a horizontal offset of 1079m to the left. Therefore the horizontal positions of the time-migrated reflection indications are now correct. The time at which each of these indications occurs represents the two-way vertical time from the surface to the reflection point; it is important to appreciate that this time — the reflection time on a time-migrated section — _does not correspond to any realizable seismic path in the earth_.

The process of depth conversion transforms the right diagram to the central diagram. This is a simple conversion from time to depth, using the appropriate relation between _vertical_ travel time and _average_ velocity.

In the present illustration, it is this **last** process which restores the fourth reflection to the horizontal; the dipping event on the time-migrated section at this level is a consequence of the lateral gradient of average velocity introduced by the dipping interface between the two upper layers of contrasting velocity.

<u>TAPE 33</u>   Why is migration generally split into the two stages of time migration and depth conversion? Because the first process — which is intrinsically useful in that it accomplishes all the objectives concerned with the collapse of diffractions and the general clarification of the structure — is not overly sensitive to the correct choice of migration velocity. A fairly small error of depth-conversion velocity, however, can be a disaster for a structural prospect — turning something into nothing, or nothing into something.

Time migration

.160 .493 .994 ← .846
360 m
1.762 ← 1.751
2.360
1079 m

Depth model

Surface
122 m
305 m
610 m
762 m
1524 m/s
1829 m/s
2438 m/s
−26°
360 m
1981 m/s
3048 m/s
Layer thickness at this cdp position
1079 m
26°
Vertical depth at this cdp position
Depth of normal-incidence point
Horizontal displacement of normal-incidence point

Cdp time section

Surface
.160
.493
g = −5.6
.915
g = 1.5
1.757
g = 6.5
2.158

7-4

7.2 THE CLASSICAL TECHNIQUE FOR TIME MIGRATION

There are several ways in which the classical technique may be described. Sometimes it is associated with the names of Huygens, or Kirchoff, or Fresnel; sometimes it is called the common-tangent method, sometimes the hyperbolic-integration or hyperbolic-summation method. These descriptions represent the same method, which follows from Huygens' principle.

We consider first the dashed ray to the left of the upper diagram opposite. It represents the normal-incidence path from the surface to the reflection "point". A coincident source and receiver at the surface point record a reflection pulse at a certain time, and the time section displays this pulse, at that time, on a vertical trace (full line) passing through the surface point. All that we know, from this one observation, is that the reflection "point" must be somewhere on a wavefront (curved full line) representing the one-way travel time in the earth.

Similarly we may record normal-incidence reflections from other surface points, as suggested in the sketch. In each case we know only that the reflection point lies somewhere on a wavefront; however, as we repeat the operation many times, it becomes evident that the only possible position of the reflector is the common tangent to these wavefronts.

Computationally, therefore, we may perform time migration by taking each sample value on each trace of the unmigrated section, distributing it to all possible positions around the wavefront corresponding to its time, and adding together the results of doing this to all samples of all traces. This is migration by the common-tangent method.

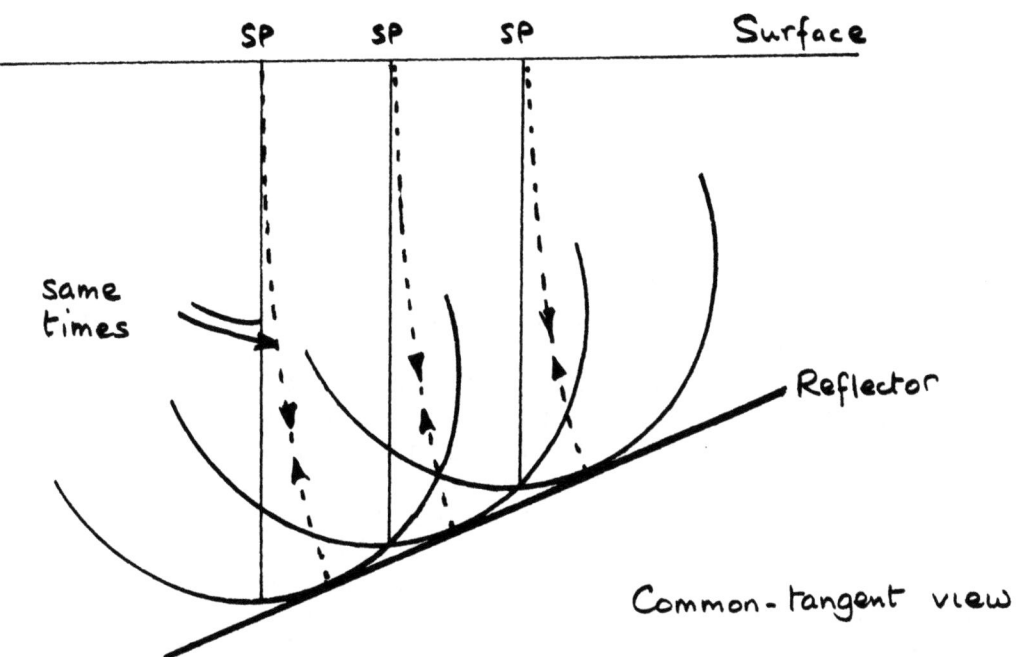

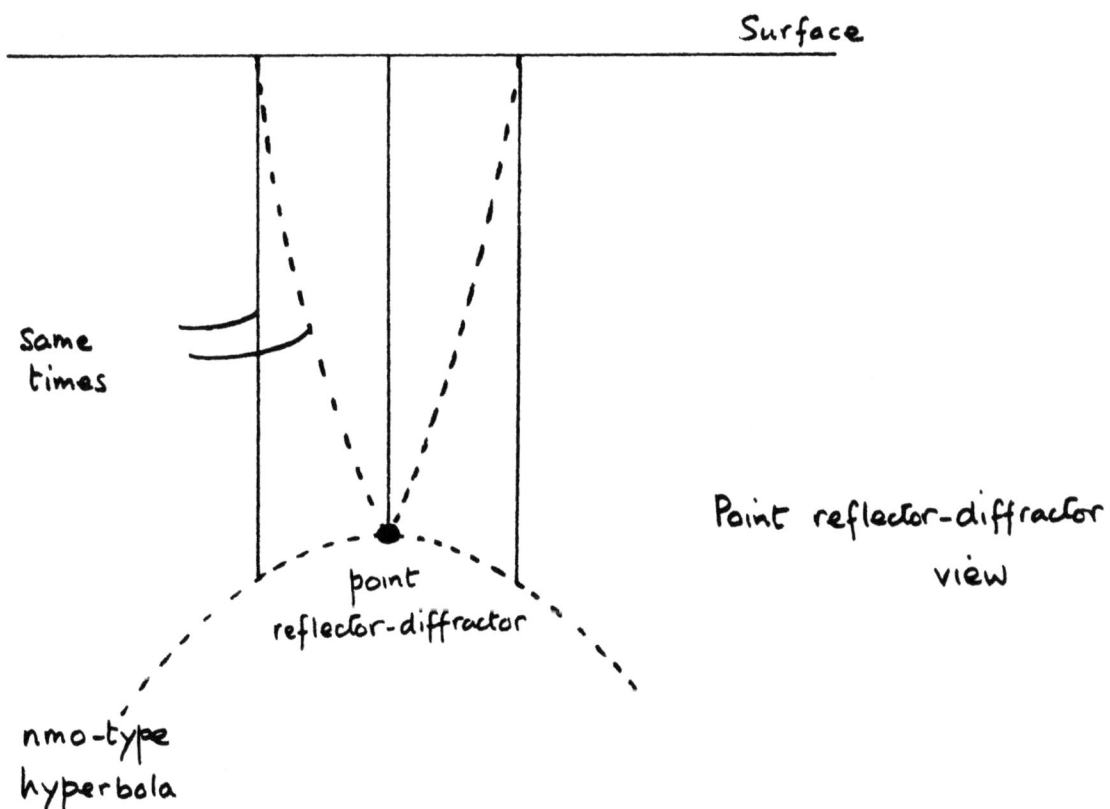

7-6

Alternatively the operation may be performed by the hyperbolic-integration method, which is illustrated in the lower sketch. This relies on the observation (section 2.8.5) that any point which returns energy to the surface (whether we call the process reflection, or diffraction, or scattering) may be regarded as a secondary Huygens source. The signal from this source, as received by geophones spaced along the surface and displayed on traces spaced similarly, appears as a hyperbola. The totality of signal radiating generally upwards from the source may be obtained by adding, for each sample value on each trace of the unmigrated section, all the sample values on the appropriate hyperbola having that sample location as apex.

Thus the <u>common-tangent</u> method is implemented by adding, for each point on the section, the results of distributing every sample value over its corresponding wavefront; the <u>hyperbolic-integration</u> method is implemented by adding, for each point on the section, the sample values along a "diffraction" hyperbola whose apex lies on that point. Although the implementations appear to be different, the methods are the same (both being expressions of Huygens' principle) On the diagram on p.7-8 the common-tangent method of time migration moves the reflection pulse from its time-section position O to its time-migrated position P along the wavefront (dashed); the hyperbolic-integration method moves it between the same points along the hyperbola (full curve).

If the source and receiver are not coincident, the details (though not the principle) become more complicated. These complications must be taken into account for a migration before stack; for migration after stack, however, the stack is usually regarded as a sufficient approximation to a normal-incidence section recorded with coincident source and receiver.

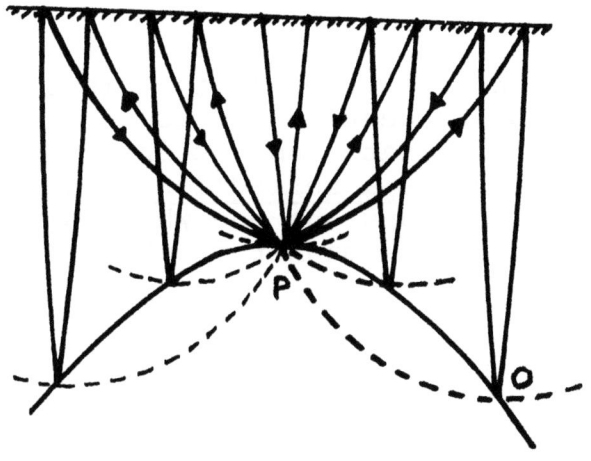

after Peterson (UGC) 1974

Of the two views of the classical migration process, the common-tangent method is perhaps the one which gives the most immediate insight into the manner in which migration <u>moves a reflection</u>; the hyperbolic-integration method is perhaps the one which gives the most immediate insight into the manner in which migration <u>collapses a diffraction</u>.

The interpreter's first reaction to the techniques of time migration is <u>horror</u>. Here his sacred target reflection — resolved and localized on the time section — is going to be spewed all over the reflections above it, and added to the results of doing the same for all other reflections. Or, his sacred target reflection is going to have added to it portions of all the reflections (and noise) below it. Either way, he is apprehensive that all this distribution and adding can only degrade his genuine reflections, introduce spurious alignments, and generally increase the noise.

In fact, these worries are groundless. After all, it is Huygens' message that where the secondary wavelets are coherent (along the common tangent) they add to give the full undistorted wavefront, and that <u>everywhere else they cancel</u>.

But let us retain a vestige of wonder, just the same.

The degree to which the cancellation occurs is illustrated on p.7-12, which also has a very important practical message for the interpreter. Let us suppose that we are performing the migration by the hyperbolic-integration method. Then one trace of the migrated section is obtained by adding many traces of the unmigrated section,

485   along Huygens hyperbolas. This is conveniently done by applying time shifts to the traces to bring the hyperbolas to a straight line, and then by simple adding of the shifted
486   traces; clearly this can use programs developed for nmo correction and stacking, since computationally the operations
487   are the same. The illustration of p.7-12 represents the effect of applying the time shifts to the many traces; it is the counterpart of a corrected gather in the stacking operation. Then the final single migrated trace is obtained by horizontal addition, on a sample-by-sample basis, of the traces illustrated; this is the counterpart

L(B)   of the actual stacking.

In a migration gather, however, the reflections are not <u>horizontally</u> aligned — they curve upwards as in the figure. The zones of each curved reflection alignment which contribute to the final reflection on the migrated section are those <u>where the reflection is horizontal</u>; the upward-curving parts <u>cancel</u> when added.

The important <u>horizontal</u> zones which provide the final migrated <u>reflection</u> can be seen to occur at a horizontal offset from the centre of the migration gather (which centre represents the position of the final summed trace, and in the present illustration is somewhat to the left of the centre of the page). This horizontal offset is to the right if the dip is to the right. In the case illustrated, therefore, the dip is substantially zero at shallow depth; it then steepens to the right, and finally becomes nearly horizontal again at the bottom of the page.

The value of this type of display is that it shows very clearly and positively <u>how many traces (what "aperture") must be used for the migration.</u> The number of traces must be such as to include (and preferably to extend somewhat beyond) the horizontal zones which provide the final migrated reflection.

The importance of this to the interpreter cannot be overstated. The choice of aperture is not a matter of taste, or of what was done last time, or what is the contractor's standard practice, or what the program can handle, or what is included in the price — either we use the proper aperture, or we do not migrate.

Of course the necessary aperture, which is related to the horizontal displacement of the reflection point (p.7-4), can be calculated from the time gradient on the unmigrated section and the velocity distribution. However, if this is the means used by the processors, the interpreter is well advised to call for a migration gather like that opposite, to be performed in the zone of steepest dip. This gives an immediate visual check, which is not subject to the assumptions necessary for the calculation. It also gives a meaningful indication of the extent to which the aperture should extend beyond the point of horizontality given by the calculation; as in all such operations, it is unwise to truncate at or near a maximum, and the aperture should preferably extend to the region in which the waveforms of the upward-curving reflection start to cancel.

A related matter, on which the interpreter is also entitled to be informed, concerns the use of tapering or weighting across the migration gather. In the illustration it is clear that the ideal aperture would widen progresively with time, would drift across to the right and then abruptly back again, and would be tapered symmetrically about its (time-variant) centre. Such adaptive tapered-aperture schemes are widely available, but, since the aperture and tapering system varies for each sample and from trace to trace, they are necessarily expensive. More rudimentary schemes maintain a symmetrical aperture and symmetrical tapering about the central trace; clearly this is far from ideal in the case illustrated, since it must weaken the steeply-dipping reflection. To maintain

7-11

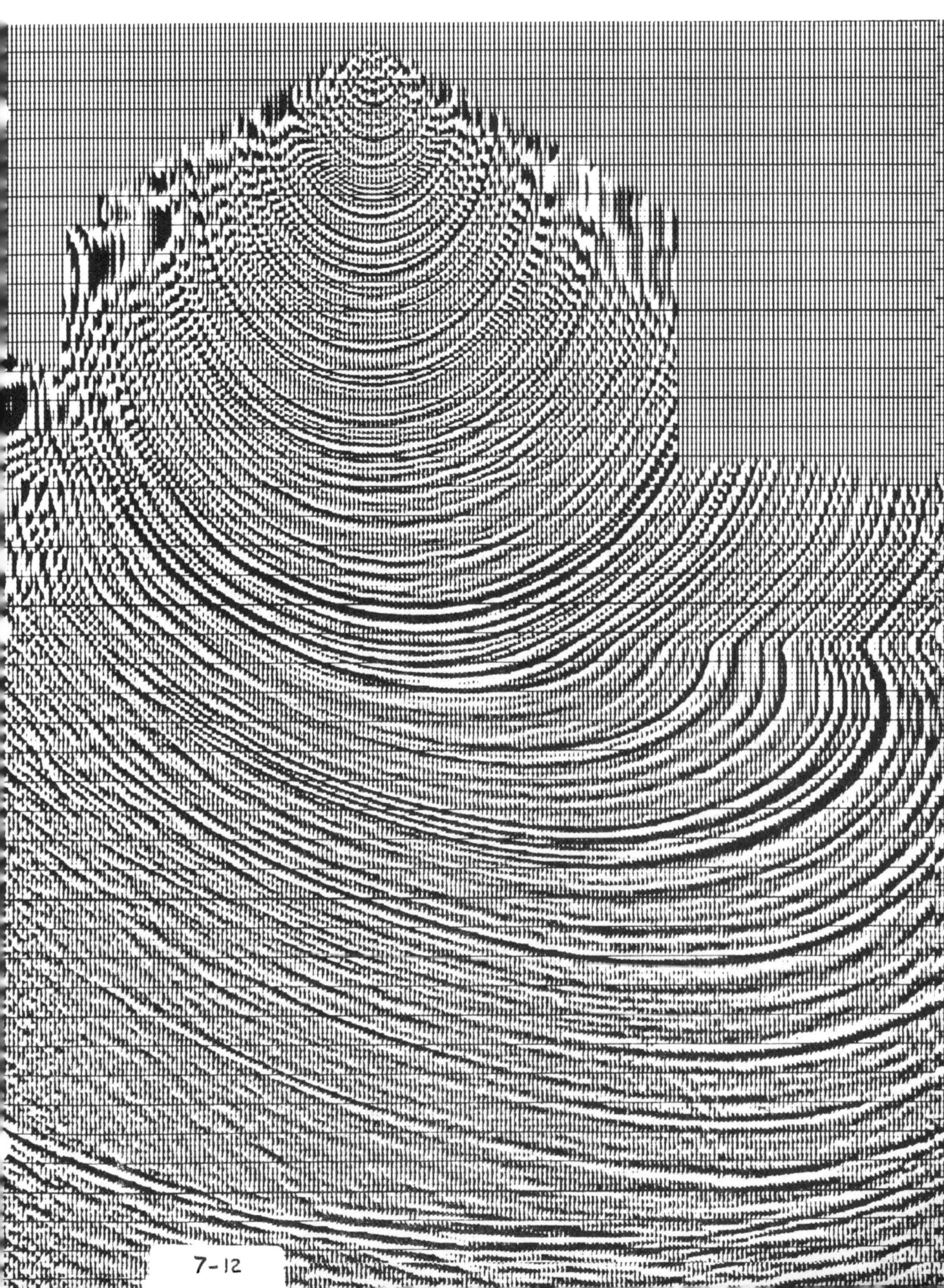

7-12

proper amplitudes on steeply-dipping reflections, with the classical migration method, we must use a sufficient aperture; then to minimize the "migration noise" we must taper symmetrically about the zone of horizontality.

A more detailed discussion of many of these matters is given by Gardner, French and Matzuk (1975) and by Larner et al., (1976).

Another concern of the interpreter is that the migration has not involved any aliasing. Aliasing conditions can be seen, for example, on the shallow reflections of p.7-12 near the edges of the mute pattern; horizontal "events" are fabricated when the step-out from trace to trace approximates to a pulse period. It is essential, of course, that any such aliasing should be excluded from the addition. In the case illustrated it can be done (by suitable time-variant choice of the aperture) without any sacrifice; this means that the cdp interval was appropriate to the geology. However, the warning is there: unless the choice of aperture and tapering system is attended to carefully, migration imposes a limit on the horizontal distance between traces. Beware the migration of alternate traces in a misguided attempt to save money! Even with appropriate choice of aperture and tapering, it remains true that the migrated section appears smoother and more akin to real geology if the trace spacing is small.

Continuing our concentration on those matters which the interpreter needs to know, we should note that, while a reflection appears concave-upwards on a migration gather, a fault diffraction appears as an alignment which is horizontal or near-horizontal; the time-shift corrections turn the fault near-hyperbola into a straight line.

Further, we should discuss what processing is appropriate before migration. Clearly, we do not wish to migrate multiple reflections and noise all over the section, so it is wise to use all legitimate weapons for their

minimization. For example, widely-available programs which
modulate the stack by the coherence of the gather are often
beneficial before migration; ordinarily such programs are
not much used because they contain an element of risk if
the stacking velocity is not well known, but by the stage
of detailed migration we should certainly know the stacking velocity rather well.

Then we have to decide what amplitude manipulations
may be applied to the data before migration.

We can see an immediate risk: if base-levelling is
applied anywhere in the processing before migration, the
trace-to-trace amplitudes along the Huygens hyperbolas may
not represent the real effect of the earth. At a first
level of concern this probably does not matter very much;
after all, the diffractions are unlikely to be correctly
stacked anyway, in general.

More serious, at an everyday level, is the effect of
agc before migration. For example, let us consider the
reflector at the top of the upper diagram on p.7-16, and
let us say that for some good geological reason the
reflection coefficient increases markedly and abruptly
over the central portion. Then the resulting reflection
on our unmigrated time section shows a gradual diminution
of amplitude at the edges of the strong zone, and
diffractions curving away as shown generally at the
second level of the upper diagram. The process of migration
collapses the diffractions and restores the abruptness of
the amplitude change. Fine.

Now let us consider a second reflection close to the
first, as in the lower sketch, and let us say for present
purposes that its true amplitude is constant. Then the
action of a conventional agc or equalization program is

to reduce the amplitude of the second reflection, as shown, on those traces where the first reflection is strong. In so doing, it obviously does not generate the diffractions which would be present on the second reflection if this amplitude reduction was genuine. Consequently, the effect of migration is to generate events having wavefront-type curvature upwards from the edges of the affected zone of the second reflection (bottom sketch).

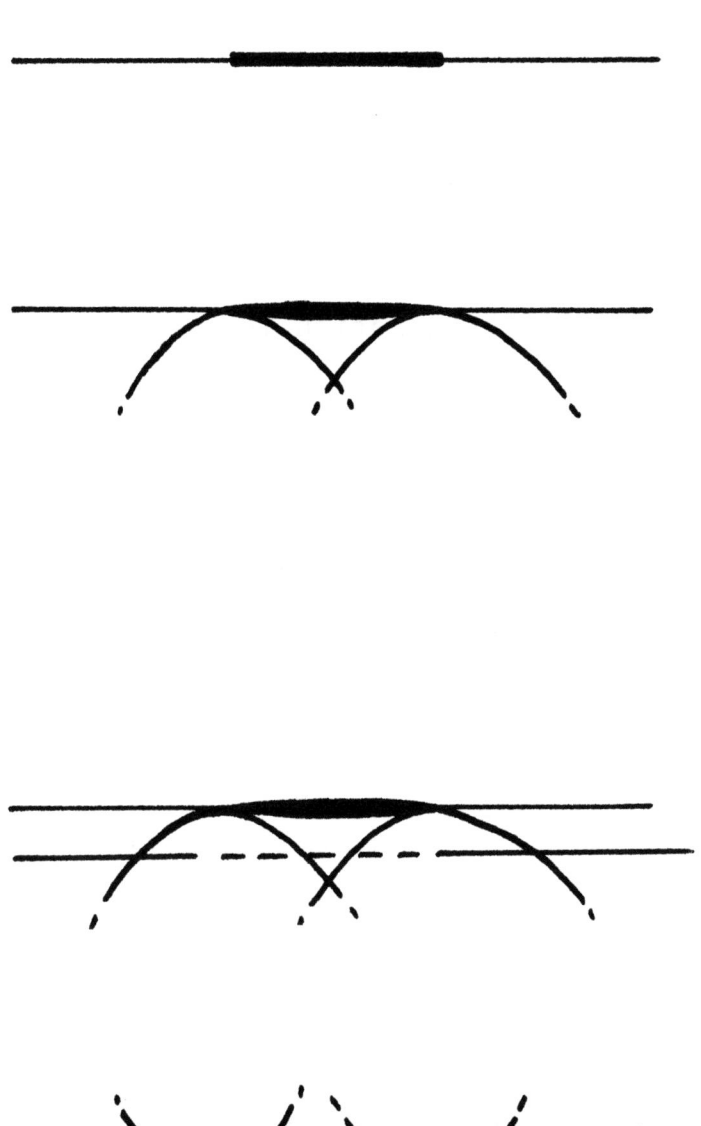

7-16

7.3 THE CHOICE OF MIGRATION VELOCITY

Although migration "velocities" have rather more association with real velocities than do stacking "velocities", we must remember that the migration velocity is most safely viewed just as the defining variable for a hyperbola (along which the summation is to be made).

Occasionally, migration is performed with velocities known to be "wrong" (for example, in a test to see whether an apparent diffraction from depth is actually an off-line mound in the sea floor, or for the collapse of diffractions from an important fault known to be not perpendicular to the line). More usually, however, the problem is to find the migration velocity which defines the effective Huygens hyperbola on the assumption that all the Huygens sources are in the plane of the section.

We must remember that the diffractions evident on the section do not themselves represent migration hyperbolas; they are interference patterns of migration hyperbolas, and their detailed form depends on the size and shape of the diffraction-generating feature. So direct measurements on diffractions do not accurately define migration velocities.

Neither, of course, do stacking velocities (except in the absence of dip and lateral velocity change, when the common depth point is really common). In practice, therefore, we have the choice of performing migration with stacking velocities which are known to be (usually) too high, or to attempt the correction from stacking velocity to migration velocity. Industry opinion varies on this matter. Clearly the correction is desirable in principle; the problem is to balance the theoretical benefit against the risk of mis-correction, which varies from case to case.

The correction is done by considering each reflection in turn, from the first downwards. Measurement of the reflection time, time gradient and stacking velocity for the first reflection allows us to deduce the normal-incidence reflection point, and then to arrive at the migration velocity for that point by modification of the modelling techniques of section 6.4. With the true position of the first reflector established in space, we can now proceed to the second reflection and repeat the operation (taking proper account of the refraction at the first reflector). Thereafter the operation proceeds downwards, allowing the computation of migration velocities, entry and emergence angles, interval velocities, depth-conversion velocities and layer thicknesses for successive reflectors.

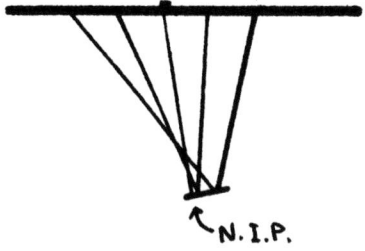

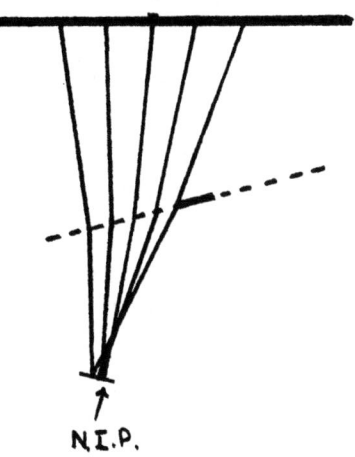

This is the process, mentioned in section 3.5.3, by which a correction is applied to interval velocities for dip and for refraction in the layers above. In that context we agreed that such a correction may be unnecessary for interval-velocity measurement in thin conformable layers (though it may become necessary if we need to know exactly where the interval velocity measurement was made). It is essential for interval velocities between uncomformable boundaries; it is also generally desirable, as we have said, for all migration and depth-conversion purposes — provided that the conditions allow it to be done well.

But what do we mean by conditions which allow it to be done well?

The assumptions made by such a correction technique, in its simplest form, are as follows:

- There is no cross-dip, and no diffractors out of the plane of section.

- All reflectors are plane, and extend back <u>either</u> to the vertical through the surface analysis <u>point</u> (where layer thicknesses are defined, according to p.7-4), <u>or</u> to the zone where refraction occurs in the path to deeper layers (whichever is greater).

- The picked reflections represent all the velocity breaks, and represent them at the correct time.

- There is no lateral change of interval velocity across the dimensions of a gather.

More sophisticated versions of the technique can extend its use to situations where the cross-dip is known, where the reflectors are not plane (but are continuously specified in terms of time gradient) and where vertical and horizontal transitions of velocity occur. The interpreter must satisfy himself that a technique appropriate to his circumstances has been used, and that the accuracy of his input gradient and stacking velocities is sufficient for the correction to be meaningful; the error of neglecting cross-dip, or compaction in the first layer, may exceed the errors he wishes to correct.

Here we should remind ourselves of the diagram of p.4-78 (reproduced as p.7-20), and its message that one of the best tests for the adequacy of the derived corrections is the continuity of a reflection after segment-by-segment migration into depth by means of the correction program.

506
507L
508
509(B)
 L(B)

Additional reference: Sattlegger et al., 1974, 1975.

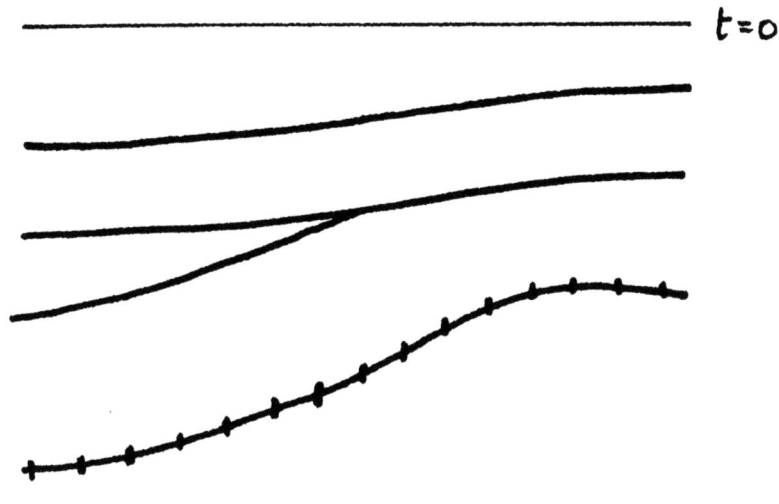

7-20

7.4 HOW MUCH SOPHISTICATION IS WORTH WHILE?

Just as in the modelling situation the interpreter has to make a decision concerning the degree of sophistication and cost which is appropriate to each of his problems.

The first point to make, of course, is that there is no snap answer; the degree must depend on the geological nature of the prospect and the quality of the data. The decision is one of informed interpretive judgement.

It is probably true that if all reflections of interest can be picked with certainty on the unmigrated sections, then migration is not justified; it is safer and better to migrate the contours (Hagedoorn, 1954; Haas and Viallix, 1974). Migration becomes essential only when the section cannot be picked without the clarification which migration affords (particularly, the clarification of faults and tight anticlines).

At the reconnaissance level, the migration afforded by the simple Huygens technique, using stacking velocities, is often adequate; the major caution is the one already stressed — that the aperture must be sufficiently large. The simple technique is also indicated when the problem is one of fault resolution in layers of small dip (less than 10-15) under an uncomplicated overburden, and when the fault orientation relative to the line is near perpendicular and known. It is essential, of course, that the stacking velocities should be reviewed very carefully (using the principles set out in section 3.5) before the migration is done; the term "stacking velocities" in this context means the velocities obtained from a conventional velocity analysis, and not necessarily those actually used at the time of stacking.

Some of our concern with degrees of sophistication must involve obvious questions like "What happens to the amplitudes during migration? And the pulse shapes?"

On a first view, accepting the validity of Huygens at the level discussed in section 2.8.5, the treatment of amplitudes by Huygens migration is quite good, provided that the aperture is appropriate and the tapering correct; the compensation of such effects as structural focusing and defocusing, for example, is at least approximately correct.

- We remember the difficulty of being sure of the ratio of the amplitudes of two reflections on a stacked trace, when the two reflections represent different numbers of traces in the gather (section 3.1.3); on a migration gather we have the same problem, and in this case it is just not possible to perform a special migration using the same traces at all times.

- Physically we can see that the migration process is likely to reduce the high-frequency content of the reflection pulse, and probably to lag its phase. Modifications of the migration technique to correct this have been proposed by Newman (1975) and by Michon and Tariel (1975) — though the latter approach then requires separate compensation for amplitude effects due to focusing and defocusing.

- We remember from sections 6.1 and 6.4 that in the presence of dipping layers and/or laterally changing interval velocity the Huygens hyperbola is no longer symmetrical (page 7-24). Conventional migration programs ignore this. The result must be to degrade the amplitudes and pulse shapes locally. (Incidentally, stacking programs ignore this effect also, even though they may be using laterally changing velocities.)

More fundamental is our concern about Huygens' principle itself. It is easily demonstrated (Peterson, 1974) that the simple form of the principle, as we have used it so far, cannot be correct; it involves a loss of energy from the forward-travelling wave, and it turns an initial spike into a long-tailed pulse. The corrections which make Huygens mathematically satisfactory have been known since Kirchoff; they involve a directivity function and a differentiation to be applied to each secondary source, and they apply only in an odd number of dimensions. In our own time, these corrections have come to be regarded as physically unsatisfactory (John, 1975). Accordingly, Huygens' principle is today regarded as a very convenient way of looking at wave phenomena qualitatively, but not as a physical law.

All of these considerations have provided the impetus for the development of a method of migration which can be viewed in isolation from Huygens' principle. This is called wave-equation migration (or finite-difference migration, or the Stanford or Claerbout method). At the present stage most of the information on this method (except for Claerbout, 1972; Loewenthal et al., 1974) is restricted to participants in the Stanford project, and cannot be published here. It has undoubted advantages — not so much that the final product looks vastly different, but because of the greater assurance of correct amplitudes and wave shapes. It is also more likely to cope correctly with a complex geological situation in which part of a reflection path actually depends on diffraction above the reflector.

However, Larner (1976) has pointed out that the classical method and the Stanford method are both solutions of the wave equation — but with different approximations. There is therefore no question of one method's being right and the other's being wrong; the only question is whether one set of approximations proves to be more desirable than another in terms of aptness to the general situation, or of practicality, or of cost.

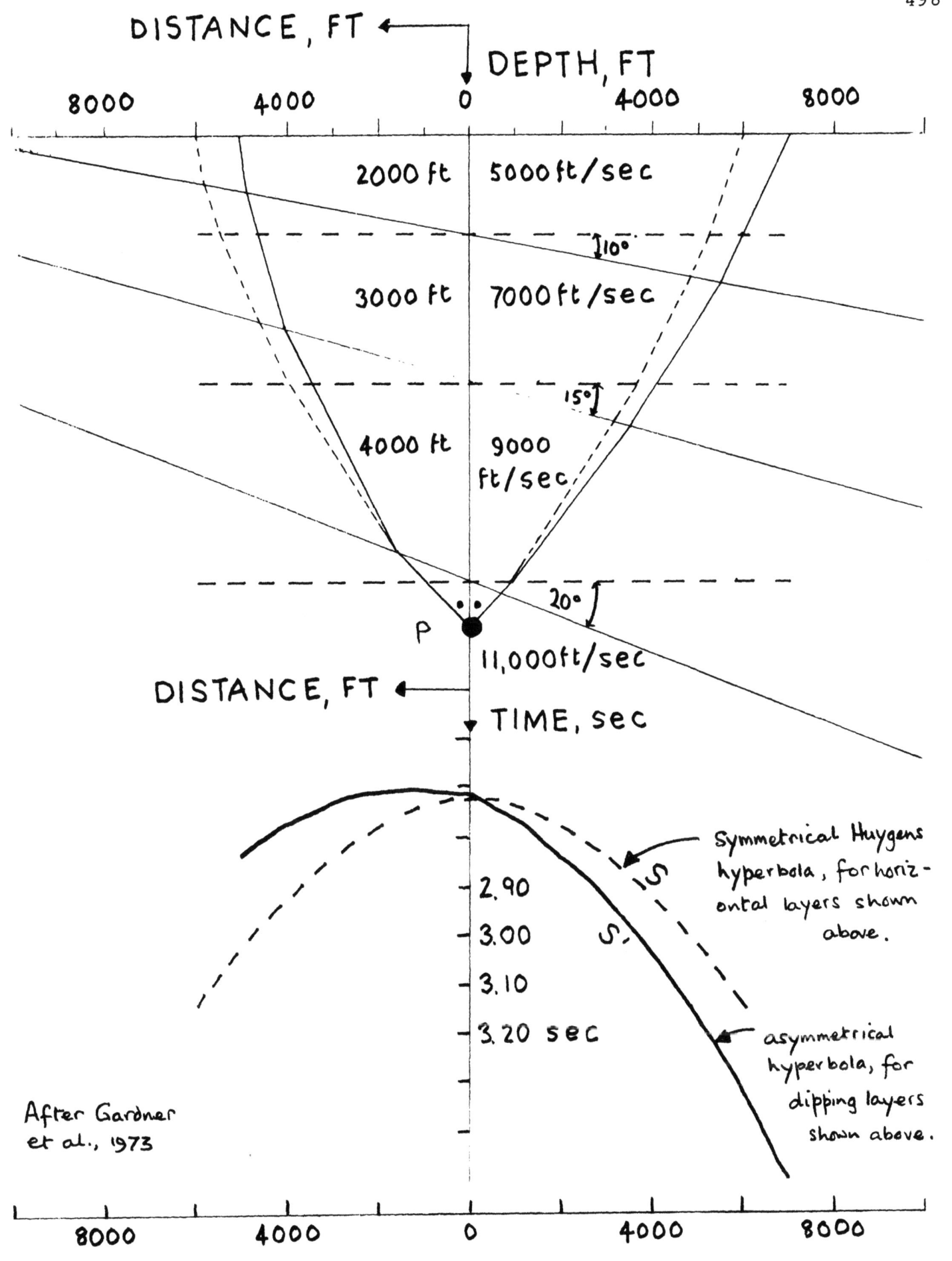

After Gardner et al., 1973

7-24

Both methods assume that the earth is liquid and only weakly inhomogeneous (which may sound like a poor start). The classical method — as noted above — does not usually make proper provision for lateral changes of velocity, and often not for the necessary pulse-shape modifications. It also generates "migration noise" associated with the incompleteness or unsuitability of the aliasing, muting and tapering features discussed in connection with the migration gather of p.7-12 (though it is a favourable point that this is readily understood and checked by the interpreter). However, the performance is good on steep dips. The Stanford method, on the other hand, fails at steep dips, has some difficulty in migrating different frequencies to the same place, and generates a different kind of migration noise; these problems are being overcome, but it probably still remains true that the non-mathematical interpreter is less readily convinced that the method is treating his data correctly.

7.5 APPROACHES TO THREE-DIMENSIONAL MIGRATION

1. In Part 9 we shall stress that the correct time to be concerned about the third dimension is when the detail surveys are planned -- <u>not after they have been shot with an inappropriate grid</u>.

2. The best possible positioning of lines cannot completely avoid the problem which arises when there is cross-dip at some levels but not at others. Thus we may choose to accept the results of normal two-dimensional migration at some levels on some lines and at other levels on other lines.

3. The techniques of two-dimensional section migration can be used in the restricted three-dimensional case of parallel and horizontal structural ridges, when the seismic line is not at right angles to the ridges. The migration velocity must be increased by division by $\cos \theta$ where $90-\theta$ is the angle between the line and the direction of the ridges. It is important to increase the aperture in the same proportion. The lateral distance over which the ridges must be horizontal is defined by the circular Fresnel zones of section 2.8.3, and is time-variant. The division by $\cos \theta$ is also used locally when it is essential to collapse diffractions from a particular fault not perpendicular to the line.

4. Another approach which gives some protection against the third dimension is that which searches for the optimum migration velocity by the migration process itself, and takes due account of dip (Sattlegger et al., 1975).

5. The process of correcting stacking velocities to migration velocities (section 7.3) can be performed in three dimensions at all the line intersections.

6. Some critically-important prospects have been shot with a line spacing equal to half the group interval, so that a complete three-dimensional matrix of data is obtained. This can then be migrated by a three-dimensional Huygens technique, in which each hyperbola becomes a hyperboloid. The storage requirement is large, but the concept is a simple extension of the classical one.

7. As we agreed earlier, the migration by reference to the time contours is generally the best course whenever the sections can be picked confidently without migration. Thus a problem of the estimation of reserves in a gentle anticlinal structure — Are there any left? — would probably be tackled by such migration. This is a migration of reflection segments, not of the full mass of reflection data.

8. The correct drawing of the contours is undoubtedly facilitated if one of the elegant broad-line field techniques is used (Michon, 1972).

9. The migration of contours is effected along lines of steepest gradient drawn everywhere at right angles to the contours; this migration is therefore immune to cross-dip problems. It is usually sufficient to define these lines (which may be curved, of course) by reference to the time contours, since rapid variation of velocities along strike is unlikely.

10. Alternatively, we can migrate the data points in three dimensions by reference to the contour map, and then contour the migrated time values.

11. The migration requires a velocity estimate at each migrated point. The general matter of contouring velocities and reflection times is discussed in section 9.1.1. The migration and the velocity estimation are both intertwined with the matter of depth conversion, discussed in the next section.

7.6 DEPTH CONVERSION

7.6.1 The importance of selecting the data

Although most of this Part 7 has been concerned with time migration, no interpreter needs reminding of the critical importance of the final stage — depth conversion. We all know (because it happened to someone else) that incorrect depth conversion can seriously distort reserves estimates, and even cause us to miss a prospect.

If there are no major lateral changes of the effective depth-conversion velocity (or "average" velocity), then we shall not suffer the ultimate embarrassment of missing a prospect; our velocities may be wrong, but our embarrassment will be just that of an error in our depth prognosis. Under circumstances where this can be tolerated, much work has been done with the simplest of techniques: an elementary conversion from stacking velocity (assumed to be rms) to average velocity, with perhaps a gross cos θ dip correction if dips are generally conformable down to the target horizon.

But the real problem arises in the type of geology for which lateral changes of depth-conversion velocity are likely. Then the proper approach to the selection, correction and smoothing of the velocities becomes extremely important.

Let us discuss the matter of <u>selection</u> first.

We may be deriving our velocities from fairly widely-spaced individual analyses (section 3.5.1; the reconnaissance survey), or from continuous velocity analysis (section 3.5.2; the semi-detail survey) or from closely-spaced individual analyses (section 3.5.3; the detail survey). In all of these cases, the moral from our previous discussions is that it is unwise merely to accept without review the picks made by the processors or by the machine, and then to force plausible velocity behaviour by massive smoothing. If we apply intelligent editing of the picks first, we obtain more valid velocities without so much smoothing; we thereby minimize the risk of missing a prospect.

Specifically, we discard those velocity analyses (or segments of continuous analysis) which we can see from the stacked section to be in the vicinity of near-surface anomalies, or geologically likely to be subject to abrupt lateral changes of interval velocity, or for which the ray-paths must pass through geologically disturbed zones. Further, we review all the picks on the remaining analyses (or segments of continuous analysis), discarding all levels associated with obvious interference, or pinch-outs, or discontinuous reflections, or faults, or even major abrupt variations of reflection strength.

We also discard from the velocity analyses those portions of line which are not straight (including particularly the run-in portion of marine lines), those portions where a streamer has changed its depth or become subject to extreme feathering in the presence of cross-dip, and all locations where the spread distances are in doubt. (Who thinks it never happens?)

Whatever magic may be claimed for statistical smoothing schemes, there is no such scheme which is not vastly improved by the prior elimination of the bad data. Once again we are reminded that if valid geological or geophysical considerations brand a data value as bad, it should be discarded — not smoothed out.

7.6.2 The need for smoothing

Even after the best that can be done by editing, it is usually necessary and desirable to apply a measure of smoothing. Immediately it is obvious that we have a decision to make: should we smooth stacking velocities, or average velocities, or interval velocities?

The answer depends on the cause of the variations from one analysis point to another. We have seen that the causes can be:

- A genuine change of velocity within one or more intervals.

- A genuine change of thickness of one or more layers (either of constant or varying interval velocity).

- Residual problems of the near-surface or of distortion of the ray-paths, which we have not been able to eliminate by editing.

- The need for changing corrections for refraction and dip.

From this the proper course is clear. We should smooth the stacking velocities if we recognize that the reason for the variations is residual near-surface or ray-path problems. However, stacking velocity is not a variable to which maximum genuine rates-of-change can be assigned on geological considerations. If we are using geological criteria for determining likely permissible variations, that operation must be done on interval velocities. And, in principle at least, the situation where the observed variation is due to changing refraction and dip above the target level should be tackled by smoothing the average velocities, after correction for these effects.

7.6.3 The smoothing of stacking velocities

1. This type of smoothing is appropriate when the problem is due to the near-surface.

2. The smoothing may be done along the profile or on an areal basis. Smoothing along the profile is easier to control according to the interpreter's judgement.

3. Along the profile, the smoothing operator is some kind of running average. The average need not be a simple arithmetic average: some interpreters use an arithmetic average of the central 50% of the values, or some similar median scheme which is not affected by the very wild values.

4. The horizontal extent of the operator is, of course, the most problematic decision for the interpreter. From p.3-84, it is clear that this should not be less than a spread length (actually, the effective spread length within the mute pattern of the gather).

5. The type of autostatics program which has been used is very relevant. This is discussed further in section 8.3.2. For the present, let us just note that the first type of program (which assumes that the statics over a spread length average to zero) requires an operator length chosen to minimize the spread-length pseudo-structure introduced by the program. The second type (the complete matrix solution along a line) is less likely to require major smoothing of stacking velocities, but requires judicious treatment of the line intersections; specifically, it may be necessary to adjust the statics with bulk shifts or linear gradients in order to optimize the line ties.

6. Areal smoothing of stacking velocities over very large extents is utilized by some processing centres; this is specifically directed at establishing gross regional trends, and is done in full knowledge of the fact that it must obscure local variations of velocity which may be highly important for a particular structure. Sophisticated versions of such programs attempt to combine the smoothing of stacking velocities with some measure of reasonableness in the interval velocities. Other advanced programs,

discussed more fully in section 9.1.1, adjust the operator extent in different directions according to an automated measure of the trends (which may, of course, be statics trends, like rivers or sand dunes).

7. Perhaps we should be explicit that the smoothing of stacking velocities here considered is an initial step in the depth-conversion process; there is no suggestion that such smoothing should be used for <u>stacking</u>.

7.6.4 The smoothing of interval velocities

1. This is the type of smoothing which is concerned with rates-of-change of genuine interval velocities, rather than with the surface phenomena. In effect, the interpreter has to make a judgement (based on borehole data and/or experience) as to how quickly an interval velocity can be expected to change — in m/s per km. To make this judgement locally, the interpreter needs a fairly good picture of the structural trends (since the rate-of-change is clearly expected to be less on strike) and the depositional history (since concentrated changes are to be expected near ancient shorelines).

2. Any smoothing operator applied to interval velocities must be reconciled to the observed stacking velocities in some fashion. A convenient way to achieve this is to do the smoothing in $V_{st}^2 t$, as discussed in section 3.5.2.

3. Again the averaging may be of median type as discussed in section 7.6.3.3 above.

4. The horizontal extent of the operator is now selected, and locally varied so that it does not obscure genuine changes of interval velocity having the maximum rate-of-change estimated for the local geology. Near a fault, for example, the interpreter is forced to select a small operator because a large rate-of-change is likely; the change may

represent a decrease of velocity (fracturing) or an increase (subsequent cementation). Near a structural high a large rate-of-change is again likely; the change may represent an increase (uplift compaction) or a decrease (fracturing). The rate-of-change associated with fracturing may be particularly large; we remember that the major change of velocity occurs with the very first occurrence of fracturing.

5. The boundaries between which the interval velocity is smoothed should be chosen with some care. Since the rate-of-change criterion is a geologic one, the boundaries should be geologically meaningful. In particular, we must strive to include the unconformity surfaces; this is often difficult, because unconformities are usually associated with reflection amplitudes which are highly variable (and sometimes zero). This point emphasizes an observation which has been made several times before: that correct selection of the horizons to be picked depends on the objective. When we are picking to make our initial time contour maps, we select those horizons which have good continuity over the prospect and those which are important geological markers. When we are picking for continuous interval-velocity studies, or for synthetic strength sections, or for purposes of correcting stacking velocities to migration velocities and depth-conversion velocities, we must select those horizons which bound geological units likely to have a definable rate-of-change. If we are fortunate, all these horizons may coincide; however, they may not, and the interpreter must therefore keep his objectives in focus while picking.

6. Reflections which disappear locally are always a major problem in this type of exercise. The interpreter needs to satisfy himself that the smoothing (whether directly under his control, or in the machine) is compatible with sensible geological explanations for the disappearance of the reflection. Thus the first implication from a major variation of reflection strength at a boundary between two interval-velocity units is that the contrast of interval velocity must change to be compatible with the strength variation (unless we have reason to suspect density variations); however, it is important to remember (particularly in shore-line situations) that the vertical gradient of velocity within one of the units can change — and the reflection strengths accordingly — without necessitating any change of seismic interval velocity as measured.

7. Areal smoothing of interval velocities is obviously illegitimate, in general, if the rate-of-change observed or allowed along the profile is incompatible with the line spacing.

8. The pattern of interval velocities which emerges from a prospect can be usefully checked, and often better understood, by assembling plots of interval velocity as a function of depth for each interval. This is sometimes an essential step for depth conversion in outcrop situations (though good geological thinking about the cause of velocity variation in rocks is necessary before velocity-depth curves should be extrapolated).

9. We recall that situations involving large rates-of-change of interval velocity produce a statics-type distortion of the stacking velocities at deeper levels (section 6.4). When this happens, and when it is not worth attempting specific correction for the effect, our smoothing at the deeper levels should properly be done in stacking velocity, and not in V^2t.

10. All of the foregoing material should be viewed in the context of each individual prospect as it arises. In particular, there is no intention to denigrate machine smoothing programs for regional and reconnaissance studies. However, for detail studies and for the estimation of reserves, there is no subsitute for the geologically-informed interpreter. And, in any case, the interpreter should not be lulled into false security when the processors talk of bi-cubic splines and least-squares fits; the programmers must do the best they can in the absence of specific geological information, but that is not always the best that can be done.

7.6.5 The derivation of depth-conversion velocities

We have seen in section 7.3 that the correction from stacking velocity to depth-conversion velocity requires the normal-incidence reflection time, the time gradients in two perpendicular directions, a good estimate of the stacking velocity, and an assurance that all the significant velocity breaks (or transitions) have been included.

Properly, the stacking velocities should be measured in the two directions; this is difficult to reconcile with any sort of smoothing applied before the correction process. The interpreter is once again required to balance one advantage against another; each case must be considered on its merits.

One solution, then, is to work solely along the profiles, and to use the time contour map for an estimate of the cross-dip at all conversion points. This method therefore assumes that the reflections can be picked without the help of time migration. The sequence might be this:

- Construction of the unmigrated time contour map for each picked horizon.

- Editing and smoothing of the stacking velocities, according to the principles set out in the previous sections.

- Estimation of the time gradient at right angles to the profile, from the time contour map.

- Division of each reflection into linear segments (preferably continuously), and association of each such segment with an in-profile time gradient, a cross-profile time gradient and a stacking velocity.

- Calculation of migrated position of each segment in depth, for each reflector in turn, using the three-dimensional version of the correction program (section 7.3).

- Examination of alignment of the migrated segments in three dimensions, and investigation of the velocities and velocity breaks if this is unsatisfactory.

- Posting and contouring of each segment centre in depth.

Another solution, also applicable to the case which allows picking without time migration, is effected on the contour map:

- Construction of the unmigrated time contour map for each picked horizon.

- Construction of lines of steepest gradient at suitable spacing, and division of these lines into suitable segments.

- Calculation of the time gradient at the centre of each segment.

- Smoothing and contouring of the stacking velocities, to yield a value at each segment centre.

- Calculation of migrated position of each segment in depth, for each reflector in turn, using the two-dimensional version of the correction program.

- Examination of the alignment of the migrated segments in two dimensions.

- Posting and contouring of each segment centre in depth.

Obviously, this method is readily modified to perform the migration at each grid node used in the contouring. This makes for an easy "hands-off" treatment of the data; the interpreter is then alert to the fact that it is difficult for him to monitor the geological reasonableness of the treatment in a particularly crucial zone. One interesting approach to this problem, concentrating on faults and on the criterion that the data must not offend the criterion of maximum convexity, is given by Kleyn (1976).

Another solution is required for the case where time migration of the sections is necessary before they can be picked. Since the process of time migration assumes no cross-dip, this is necessarily applied along the profiles:

- Identification of corresponding portions of the same reflection on migrated and unmigrated sections.

- Association of segments of these reflections on the unmigrated section with a time gradient and a stacking velocity.

- Calculation of the depth-conversion velocity applicable to the migrated segments, and association of this velocity with the correct horizontal offset along the profile.

- An intelligent combination of extrapolation and smoothing of these depth-conversion velocities, having regard to the geology evident on the migrated section.

- Application of the derived depth-conversion velocities to change the vertical scale of the time-migrated section to depth.

A final stage of all these approaches must be the reconciliation of the final depth map with the wells. The general matter of establishing the validity of check-shot data is discussed in section 8.1.1. We should note here, however, that wells are often drilled in places (structural high, fault zones, etc.) which in their nature involve local anomalies of interval velocity. Again the interpreter should try to understand the reason for failure to tie at particular wells, rather than to call for (as an example) a forced rotation of the velocity maps to yield a least-squares best fit to the wells. We should be able to do better than that.

But then, of course, we all grow tired, too.

PART 8

SPECIFIC INTERPRETATION PROBLEMS

8.1 THE INCORPORATION OF BOREHOLE AND OTHER DATA

8.1.1 The problem of discordance between stacking velocities and check-shot velocities

First, let us stress that there is nothing odd or unexpected about this discordance. And (provided the discordance is not due to something silly, like an inconsistency in the static correction treatment) there is nothing to be ashamed about either.

In the following material we assume that the check-shooting is done with an air-gun or similar source close to the wellhead, so that the old problems of oblique incidence are no longer with us. Then we see that one group of reasons for discordance are concerned with the geometry:

- The check-shooting represents an approximately vertical path; the stacking velocities represent an extensive triangle of ray paths. We remember from Part 7 the comment that wells are usually drilled in locations (structural highs, fault zones) where significant variations in velocity can be expected close to the hole.

- Even if the velocity gather is centred on the well, the derived stacking velocity requires several corrections before it represents an approximate average velocity. As set out in previous sections, it requires very careful attention to static corrections; further, it requires correction for refraction and dip at overlying and reflection levels (section 7.3). The last correction may yield an average vertical velocity which applies some distance from the centre point of the gather; a different gather may be required to obtain the vertical velocity at the well, and the gather may be different again for a different reflection.

- The stacking velocity is subject to significant errors in the presence of lateral changes of interval velocity, such as may be associated with the presence of hydrocarbons in the vicinity of the borehole (section 6.4).

- Also as discussed previously, the reflection alignment at early reflection times is not an exact hyperbola; the velocity analysis assumes it is, and seeks a best fit. The seriousness of this is easily tested by making analyses using the inner 60% and the outer 40% of the gather; the problem exists if the outer velocities are unacceptably higher than the inner velocities.

- The foregoing items suggest that the stacking velocities are subject to more geometric problems than the check-shot velocities. However, check-shot velocities are also subject to velocity lensing, blind spots, and the effects of dip and refraction generally; the consequences of wavefront healing are also different (section 3.3.1).

There are also the following non-geometric reasons for discordance.

The fact that check-shooting represents a one-way path obviously means that the effects of absorption, short-path multiples and scattering are less marked; we are making measurements on a pulse which has suffered less loss of the high frequencies, and its velocity must tend to be slightly higher for this reason (section 2.5.1).

8-2

The problem that our measurements are made on pulses of different shape is aggravated by the fact that the check-shot source is usually of different type, and even that the instruments may be of different type.

We have agreed in section 3.5.3 that our measurements of stacking velocity are made at or near to the pulse maximum, the time of which maximum includes the effects of absorption and short-path multiples on the way down and the way back, plus the effect of any interference which may be present at the reflecting level. The check-shot measurements, however, are invariably timed at a point earlier than that which would correspond to this maximum; we remember that the so-called "direct" arrival can have a saddle in its envelope (section 2.5.2) and no real first break, and under these conditions a pick is usually hazarded on some intermediate point of the waveform.

It is clear from the foregoing that we must estimate a T_o correction (along the lines of section 3.5.3 and 8.2), between the command-to-shoot and the part of the waveform we have decided to pick, for the stacking-velocity source and the check-shot source separately.

There seems no reason why knowledge of the outgoing pulse shape should not be used to force the check-shot pulse to be zero-phase (except for earth effects), just as we have agreed that we wish to do for the reflection data (sections 3.1.1.4; 3.2.8). The first break would be made even more unsatisfactory, of course, but the overall concordance between stacking and check-shot velocities should be improved.

(The problems of the optimum pick on check-shot data have been much aided by the advent of lock-in velocity-sensitive geophones; these are less afflicted by tube-wave interference and cable and casing breaks, which formerly complicated the picking even further.)

Finally, we must note that the velocities actually entering a conventional cdp velocity analysis are <u>horizontal</u> velocities (Herrenhefer and Ostrander, 1973). In the presence of anisotropy (which is known to be significant in some layered shales), the stacking velocity is caused to deviate from the check-shot velocity by this agency also.

Most of the above considerations favour the check-shot measurements over the cdp measurements. However, the stacking velocities, even at their best, can never be expected to be actually equal to the check-shot velocities unless proper attention is paid to the T_o corrections and to the place on the pulse which is selected for picking.

8.1.2 <u>The synthetic seismogram</u>

These may seem very late days to be talking about synthetic seismograms — how could there be anything still unsaid about them?

Well, there are a few things which certainly did not receive proper attention in the early literature (like sample interval, for example), and there are a few modern considerations also which should be stated.

1. First let us accept that the basic method of synthesizing all primaries and all multiples, and of adding the free-surface reflections above source and detectors, is now well established for plane waves, plane horizontal reflectors, and normal incidence.

2. These three restrictions constitute major departures from the practical case. However, the problems of constructing a synthetic seismogram for a full spread over arbitrarily dipping reflectors — while already

proven nor insuperable — nevertheless remain at least daunting.

3.   Therefore we are gratified to see that these problems may have been overtaken by events. In the early days of synthetic seismograms our single-fold field records were full of multiples, and so we were much concerned to generate synthetics which had all the multiples in the right place; then, we thought, we could identify the multiples on the synthetic, and strike them out of consideration on the real field records. But nowadays, in general, our real data take the form of stacked sections, each trace of which approximates a normal-incidence reflection signal with very small multiple content — substantially a record of the primaries only. Not quite that, because all the short-path multiples are still in it — but it is generally safe to say that all the long-period multiples are substantially removed by the stacking. And it is the long-period multiples which constitute the difficulty in the case of arbitrary dips. So we can defend the view that, for all but the most sophisticated purposes, our objective in making synthetics is achieved if we can compare a stacked trace against a synthetic trace made without long-period multiples. This is quite a different situation from that which obtained in the early days of synthetic seismograms.

4.   Perhaps the most surprising thing about past practice in the preparation of synthetic seismograms was the cheerful neglect of the basic principles of sampling theory. If the receivers in the logging tool are separated by say 2ft (60cm), then there is a notch in the system response at a spatial-wavelength value of 1ft (30cm), and this is the only (albeit imperfect) alias filter in the system. Therefore the resulting log must be sampled at a maximum interval of 1ft (30cm). If the log is sampled after conversion to time, this should really accommodate the highest velocity present; if that velocity is 20,000 ft/s (6000 m/s) then the sampling interval of two-way time should be 1/10ms.

5.   When the log is properly sampled in this manner, many or most of the old problems with synthetics disappear. There is no difficulty at all in balancing the calculated transmission loss against the observed transmission loss. The log can be subjected to an appropriate <u>digital</u> alias filter, and then resampled to a coarser interval; the transmission loss remains unchanged. The multiples have the correct amplitudes. Additionally, when the synthetic is convolved with an appropriate source pulse, it exhibits the correct change of frequency content with time. And, of course, the reflection times on the synthetic and on the observed data tie exactly, without any arbitrary fiddling with the time origin.

6.   For the usual comparison with <u>stacked</u> data, the synthetic is made without the long-period multiples; depending on the type of program in use, a modification may be required to include only the multiples up to a delay of say 100ms. Such a figure includes all the pulse-shaping action of multiple reflections, but omits the multiples likely to be attenuated on the stacked trace. With such an arrangement the comparison between stack and synthetic should be very good, even in the presence of a little dip.

7.   Of course the usual <u>practical</u> problems remain. Velocity logs always seem to start too deep and to finish too shallow; certainly it is important that wherever it is possible the velocity logging should start at the grass roots. Also, there remains a major problem with cycle-skipping in very-low-velocity formations; in these cases the velocity log becomes useless at a critically-important place, and it must be replaced by some sort of best guess based on other types of log. Obviously this is not very satisfactory; still it is essential that major log

excursions due to cycle-skipping should <u>not</u> enter the synthetic seismogram calculation.

8. Then there is the matter of <u>density</u>, which is the junior partner of velocity in the synthetic calculation.

The first approach, of course, is to use the density log (which must <u>also</u> be sampled correctly before use). Again we need to be very much on our guard against spurious values (particularly low values) associated with hole conditions and the limited penetration and accuracy of the density log. Further, it is wise to check the log calibration.

Better, perhaps, would be a borehole gravimeter.

Failing that, we may assume some simple general relation between velocity and density. For example, we could use the relation we cited earlier ($\rho = 0 \cdot 31 \, V^{\frac{1}{4}}$); it can be shown that the effect of using this is approximately the same as just increasing all the reflection coefficients by 25%. Or, better, we might use the knowledge of the geology in the well to point to a variation in this law appropriate to each rock type (for example, using our earlier Figures 2-58, 2-60, or an equivalent more specific to our own area). As always, we are particularly on guard against the anomalous densities of salt, coal, gypsum and anhydrite.

9. It often happens that a portion of the hole has no velocity log, but it does have some other logs. In the days when we were trying to synthesize all the multiples correctly, we were reluctant to accept these, because their "interfaces" do not necessarily correspond to seismic reflectors. Nowadays we are much less concerned about this. We accept, of course, that the primary reflections may not be correct in the zone where another log has been used. However, for the short-path multiples (which are the only multiples now relevant), the final effect depends mostly on the <u>statistics</u> of the reflection sequence, and not on individual <u>reflectors</u> (section 2.5.2). These statistics may be adequately represented by other logs.

Faust's proposal for modifying the resistivity log has been widely superseded by that of Rudman (1975). Rudman suggests the relation

$$\Delta T = A + BR^{-1/C},$$

where ΔT is the sonic travel time and R is the resistivity, and he shows how to estimate the constants A, B and C in practice. Salinity variations and high mud resistivity can necessitate adjustments in shale sections.

10. When a satisfactory match has been obtained between the synthetic and the section through the borehole, we come to the operation which is usually the most important one in the commercial sense — the attempt to understand the reservoir. We seek to modify the velocity log in the reservoir zone in a manner which represents geologically plausible behaviour of the reservoir zone away from the borehole and which simultaneously maintains the match with the seismic section away from the borehole. This is, of course, a trial-and-error exercise, guided by geological likelihoods and their expected physical expressions.

The general structural form of the reservoir is first estimated from the seismic sections. At the initial level, this may be checked by an appropriate modelling technique (section 6.1 or 6.2), using a first estimate of the reflection pulse shape. Then, having obtained an approximate shape for the reservoir and approximate values for most of the variables, we enter the highly detailed stage — modifying the thickness or velocity of the reservoir interval to yield a match to the section away from the borehole.

It is not generally necessary to generate the multiples in an exercise of this type; we are concerned with just a short interval of the synthetic, and unless the inter-reservoir multiples are major (which can be established on the check-shot data), our only concern whith multiples is with the short-path multiples which contribute to the pulse shape. So all we need, in general, is the velocity log, some idea of the reservoir configuration, and the effective pulse shape.

It is particularly important, in the present application, that our estimate of the pulse shape is a good one. We have additional tools available for this, by the time we have a synthetic; we have the usual methods of estimation from the reflection data (section 3.1.1), and estimation by recording of the source pulse and simulating the processing (section 8.2. but now we have the check-shot pulse recorded near the source and down the hole, and the additional facility of being able to compute the effect of the velocity-log stratification on the transmitted pulse (O'Doherty et al., 1971). Just as we have stressed so many times that reflection data should be self-consistent in respect of amplitudes, velocities, polarities and frequent content, so we now demand compatability between different methods of establishing the reflection pulse to be used on a detailed synthetic exercise. Thus the change of pulse shape between the check-shot source recording and the downhole recording should be due to short-path multiple effects and absorption and scattering effects; the former effects can be computed from the velocity log and auto-convolved to obtain the two-way version, and so the effects of absorption and scattering can be isolated. Auto-convolving these, and applying an operator to turn the check-shot source into the reflection source, should prove compatability between these two entirely independent approaches to effective pulse determination.

It is very clear that the check-shooting tool has been under-used in the past; there is far more utility in the data than the mere calibration of the velocity log (Galperin, 1974). The messages of the check-shooting should be thoroughly explored before detailed modelling of the reservoir synthetic is attempted.

11. The synthetic is also a valuable guide to the selection of field variables and processing techniques for subsequent surveys; it tells us (in conjunction with the tests described above) what source bandwidth is necessary for the resolution of a reservoir complex — and indeed whether the high-cut losses in the earth are of such magnitude that no practical source can be expected to resolve it.

8.1.3. The incorporation of sparker data

It often happens that, in order to see the near-surface expression of deep geology, it is desirable to add a sparker section to the shallow portion of a normal section (particularly if the latter has been recorded with a long offset to the first geophone array). For such purposes there is every incentive to keep the bandwidth provided by the sparker. However, where reflections are visible on both sections they do not occur at the same times (for reasons concerned with T_0 correction, pulse shapes, and dependence of velocity upon frequency); neither do they have the same amplitude relationships.

If it is desired to use sparker data to establish quantitative reflection coefficients on the normal data, it is necessary to narrow the sparker bandwidth to be the same as that of the normal section. Provided the processing of both surveys is done intelligently, and a few reflections are common to both, this is a legitimate method. The problem, usually, is to persuade the sparker operator to include frequencies down to 20 or 30Hz in his recording; this degrades the results as seen in the field, though good digital equipment should be at no risk in recording the whole signal for later filtering.

8.2 MISTIES BETWEEN SURVEYS

Misties between surveys are a perennial nuisance to the interpreter. Of course, they can be genuine; when two lines intersect, the reflection observed on the near traces should be identical, but in the presence of arbitrary dip the reflection observed on the far traces may have different reflection "points" on the reflector. After stacking, some mistie of shape or reflection time can be expected if the dip is steep.

More often, however, the interpreter is left feeling that the mistie is some sort of artifact. It helps, therefore, to set out the likely causes of such misties.

8.2.1. Survey and feathering problems

On land, the survey standards are easily specified (and fairly easily satisfied) to ensure that these problems do not arise. At sea, however, some of the popular navigation systems have accuracies which (although remarkable in a technological sense) can be specified only in a statistical manner — the chances are better than so-many percent that this shot-point is within so-many metres of the point mapped. Mistie errors are inevitable, and often the interpreter must ask for the original survey data to be reviewed to establish which line is most likely to be in error.

Streamer feathering can also introduce significant deviations of the common-depth-point from the line of profile. For detail work in areas of even mild water-currents, it is important that the angles between the streamer and the ship's axis and between the ship's axis and the line should be monitored frequently, and the appropriate correction incorporated in the maps.

If the current is massive (for example, tidal) the streamer remains straight when feathered, and the velocities remain unaffected. Across an estuary, however, the streamer may become curved, the distances may be shortened, and the velocities may appear high. This, depending on the spacing of velocity analyses and the interpolation between them, may cause stacking and velocity misties.

8.2.2 Polarity conventions

Many misties encountered in the past have been simple differences in display polarity, since there have been at least two widely-used conventions of polarity through the system. Previous versions of this course manual have sided with those who deemed it reasonable that a positive seismic signal should produce a positive deflection on the section; however, we now have an SEG convention to the contrary, and all parties should unite to adopt this as standard practice.

The SEG convention (Thigpen et al., 1975) is as follows:

A positive seismic signal (positive acoustic pressure in the water, or upwards motion of the geophone case) shall produce a negative number on tape, a negative deflection (downswing) on a monitor record, and a trough (white) on a section.

For work performed before adoption of this standard, the interpreter is at least entitled to ask what the polarity was.

8.2.3. Processing problems

1. The major cause of misties introduced in processing is static corrections. A fuller discussion of statics is postponed to section 8.3.2, but we should note here that the

effects can be troublesome and elusive. A first problem
can arise in that the model used for the field statics is
a ray model employing a very simple velocity picture; this
is likely to cause trouble, for example, at an intersection
between lines along and across a deep narrow valley. A more
general problem arises when an autostatics program is applied
as discussed in section 8.3.2, these programs may introduce
slowly-varying errors in the reflection times, and in the
nature of the basic programs there is no guarantee that the
introduced errors will not be different (or even in opposite
directions) at a particular line intersection. Advanced
versions of these programs guarantee some sort of optimized
fit at the line intersections, but even these do not guarantee
no misties.

2. Clearly, differences between the stacking velocities
used on two intersecting lines must produce a mistie of
reflection shape or time. This may be because of different
velocity picking or the erroneous picking of a diffraction
on one line, or different T_O correction, or different (or
geologically inappropriate) interpolation (section 3.5.1).
The mistie is likely to be more serious if the offset from
source to near geophone array is long; this is because the
stacked trace is made, in effect, by projecting back the
nmo hyperbola to the zero-distance line.

3. Because of nmo stretch on the far traces, misties are
often produced by simple differences in the mute pattern
applied on the two lines. This is a very common reason for
misties.

4. Some processing centres use zero-phase filters both
before and after deconvolution. Others hold that filters
applied before classical deconvolution (for example, digital
anti-alias filters) should be minimum-phase. This distinction
can produce a significant mistie of shape and time.

5.   Operators applied to bring the final reflected pulse to the zero-phase condition obviously introduce major (though desirable) changes of shape. The interpreter must ask what has been done about the T_o correction.

8.2.4. Other problems of reflection shape

The many ways in which differences of reflection shape can be caused are implicit in the illustration on p.8-14B. The figure represents successive stages in the life of the reflection pulse, <u>excluding</u> those effects which occur in the body of the earth. The source, in this case, is a small air-gun array from the published literature.

After the command to shoot, at nominal zero time, there is a delay during which the air-gun valve is accelerated. Then we observe the outgoing pulse as it would be recorded with a hydrophone very close to the array.

As this signal is transmitted downwards, it is followed by the free-surface or ghost reflection. In the present illustration the source is about 6m deep, so the ghost comes after 8ms; it is, of course, negative.

At this stage we imagine that the pulse is turned back by a perfect reflector, and enters the hydrophone array. It is followed by a new free-surface reflection from above the hydrophones; in the illustration the streamer depth is about 15m. This, then, is the signal received at the input to the amplifiers.

The anti-alias filter in the amplifiers has a marked effect, in the example shown; the illustration is for a standard DFS 3 equipment on an 8-62Hz (4ms sample) setting.

The next step illustrated represents the effect of a reasonably successful pulse-sharpening deconvolution. The last two are the effect of final zero-phase filtering — the first on a fairly broad-band filter likely to be used on the shallow reflections, and the second a lower and narrower filter suited to the deep reflections.

The illustrated changes of pulse shape and pulse time are those caused by the seismic system — not by the earth. Clearly they are major, even without any contribution from the earth.

The upper illustration on p.8-14D represents the same sequence for a different source — a gas-gun pulse, again from the published literature. The final reflection pulses at the bottom, as the interpreter would see them, are significantly different.

Also on p.8-14D we abstract from the last two illustrations the pulse as it is recorded on tape and as it appears after complete processing, and include the corresponding pulses for two other sources in common use. Significant changes are apparent.

Further, these illustrations assume that everything except the source — streamer depth, instruments, processing — all remain the same. In practice they may vary from one survey to another, even with the same source. It is not surprising that misties occur between different seismic vessels, or even between different surveys with the same vessel.

The illustrations drive home to us that where two recording or processing systems have different amplitude-frequency and phase-frequency responses, it is meaningless

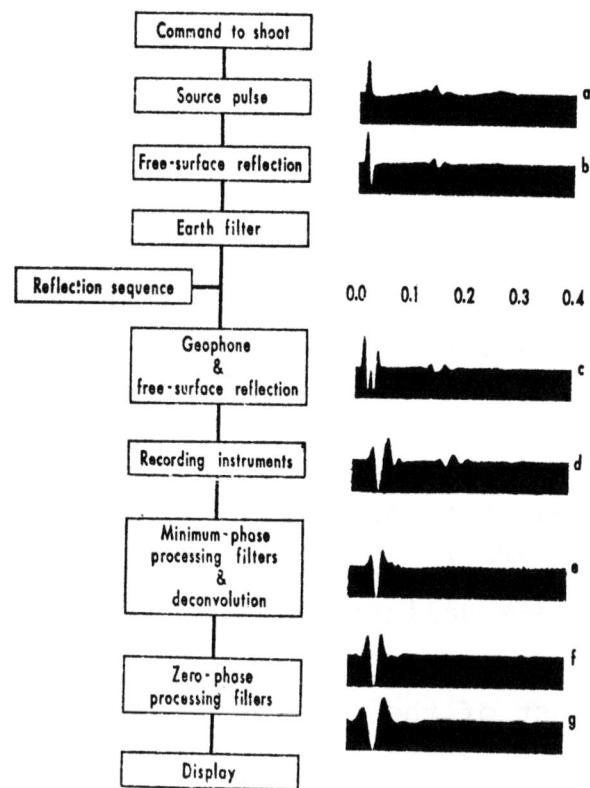

8-14B

to search for the "same" peak or the "same" trough — they do not exist.

Various solutions have been proposed for these mistie difficulties.

- The average pulse shape present on the section can be estimated from the data, in suitable short wide windows, by the techniques of section 3.1.1. It certainly helps the interpreter to have this shape determined, in that he is more likely to adjust his picking correctly in general. However, some vagueness inevitably remains in the timing and the polarities.

- When the average pulse shapes on the two intersecting sections have been estimated, it is theoretically possible to apply a deconvolution program to turn one into the other, or both into a suitable average. This is the cross-equalization solution.

- However, most of the advantage in the last proposal — certainly as regards the time misties on discrete reflections — is obtained more safely if the process is restricted to the phase spectrum. In this case the amplitude spectrum of the reflection pulse may be left unchanged, while an operator is applied to bring it to the symmetrical (zero-phase) shape. This is the phase-zeroizing process we have discussed in connection with polarity studies and resolution improvement; we now see that it is also helpful in minimizing time misties. If the amplitude spectra are left unchanged (rather than being brought to some chosen norm), the reflection pulses themselves are of different duration, but they align properly at the peak or trough representing the centre of symmetry.

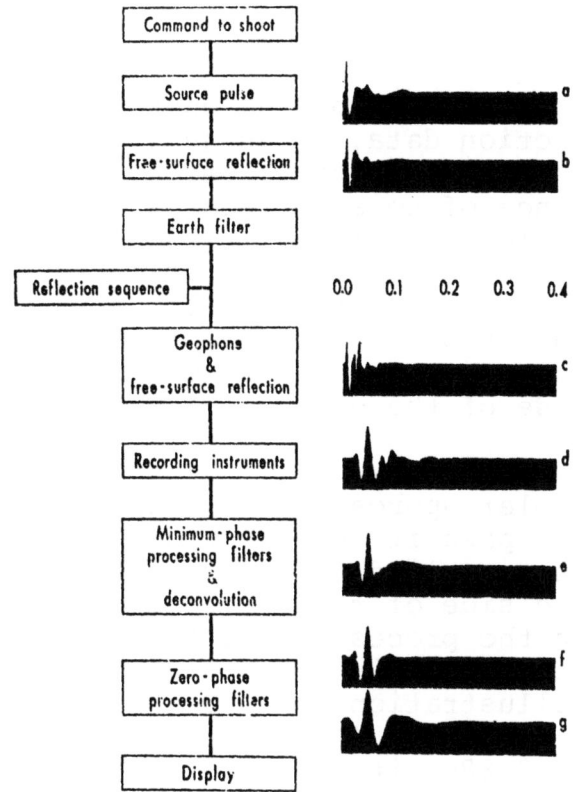

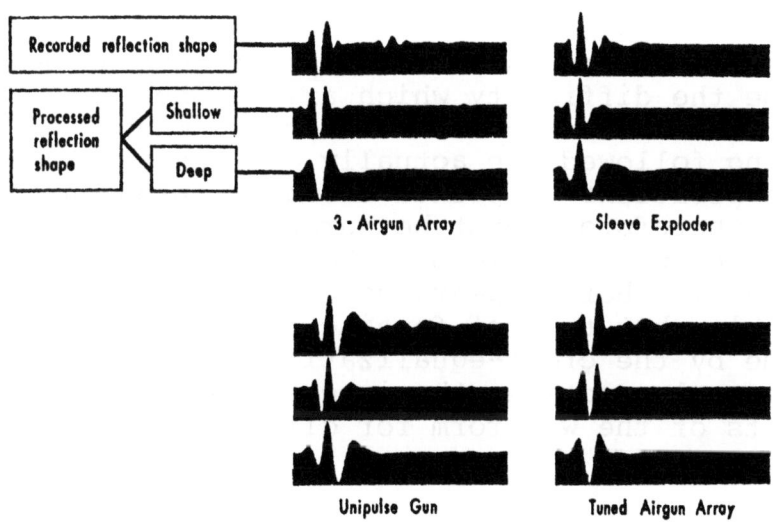

8-14D

- If the phase-zeroizing operator is derived from the reflection data, it is subject to the same errors as any other statistical deconvolution process in the presence of noise or poor window selection. Wherever possible, therefore, it should be derived from the recorded source pulse, passed through equivalent processing and perhaps some first estimate of the frequency-selective effect of the earth. Certainly it helps just to remove the different phase agencies outside of the earth.

- A simpler approach is to take the recorded source pulse, pass it through processing effectively the same as that applied to the section, and merely display it by the side of the conventionally-processed section. Since the processing is time-variant, the processed pulse may be displayed, for example, every second. The illustrations on p.8 - 15 show the results of this at 1s and 4s. In the case of the reflection at about 3.88s, it is reasonably clear from the 4s pulse that the reflection is positive, and that the necessary time correction for a pick on the major peak is about 47ms. If this is done on all survey lines of a prospect, the misties are removed. And then, of course, all the velocities come out correctly, and the multiples come at the right travel time.

- The methods of the last three paragraphs do not overcome the difficulty which arises when techniques of different bandwidth intersect and when the reflections being followed are actually complexes. In such cases the positions of picked troughs depend on the bandwidths, and therefore cannot be harmonized; not only do the reflections not tie but the time isopachs do not tie either. There appears no comprehensive solution to this problem but that of forcing the bandwidths to be the same by the cross-equalization method discussed above. In usual practice, the interpreter picks on different parts of the waveform for different techniques, but inevitably this leaves him with some remaining problems when the spacing of the reflector complex changes.

8.2.5. Diffraction problems

Misties are often observed at line intersections near small faults, and sometimes the resolution of these misties is very important in a reservoir sense. The cross-dip feature, discussed earlier, may be one contribution. Another may be that the reflections as picked are in fact interferences between reflections and diffractions; at a critical place the interfering diffraction may have an apparent velocity sufficiently low that the diffraction is attenuated by the arrays on the line obtuse to the fault, while it is not attenuated on the line acute to the fault.

In summary, it is clear that misties can be caused by many agencies. Some of these can be controlled or corrected; others are the consequences of our basic technology. The first step in the unravelling of a mistie problem is the recognition of the source of the trouble, under one or more of the headings discussed above.

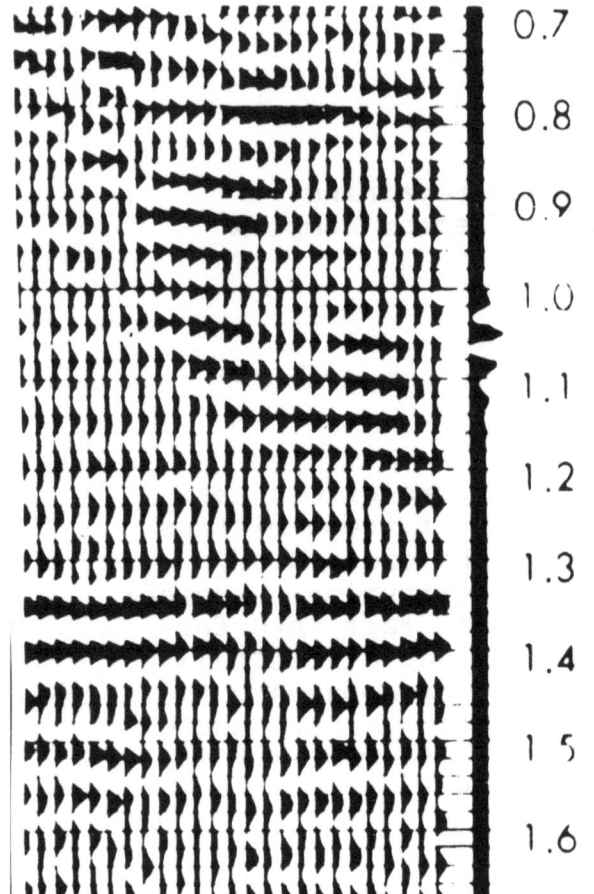
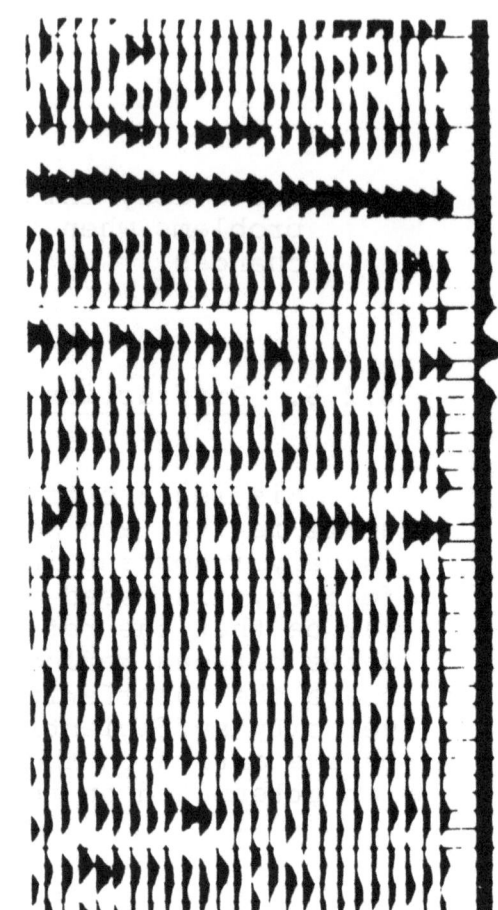

TAPE 36   8.3   THE SPECIAL PROBLEMS OF LAND DATA

8.3.1 Just differences

First a few points which are less problems than differences, and which we must note in going from marine work to land work.

1. We must remember that the geophones are now directionally sensitive. Charitably, we say that the geophones are vertically sensitive (despite the ones we know are lying on their side, or hanging in a bush, or stuck in a turnip). This means that an amplitude correction is required for the fact that the axis of the geophone is no longer aligned with the upcoming seismic wave. The correction is clearly the cosine of the angle of emergence; the angle of emergence is a function of the offset, the time and the velocity (and therefore of the nmo), and of dip. If a stacked section is made first, and the interpreter makes picks on the horizons of interest, then in principle the data can be reprocessed to provide correction for this emergence-angle effect.

2. Another difference is that on land our arrays are usually longer, which means that much more importance now attaches to the correction of the amplitudes and the frequency content of the shallow reflections. The old-timers among us remember a feeling of surprise in the early days of Vibroseis, when we realized that the dynamic range of a correlated Vibroseis signal was often less than 20 dB. In the days of dynamite shooting, of course, we were accustomed to dynamic ranges exceeding 60 or 80 dB. The reason for the greatly reduced dynamic range in the Vibroseis case was in part the limited bandwidth of the source, and in part the unusually great offsets which were then introduced, but also (and important in the present context) the effect of the unusually large array lengths.

3. Another difference, which is probably academic at this stage, offers intriguing processing possibilities in the future: the geophones in land work actually record the signal which is reflected back into the earth as the surface multiples. This, of course, is not so in the marine case, because of the submerged location of the hydrophones.

4. Another difference arises from a suggestion (which we may choose to doubt) that the base of the weathering is sometimes a stronger reflection than the surface. This is never so in the marine case. What is almost certain is that the weathering and the surface together, in land work, often represent a transitional reflector; this has the effect of imposing a greater-than-expected-frequency loss on the surface multiples.

So on to the problems.

8.3.2. <u>Statics</u> (ref. Duska, Taner)

1. The first operation in the application of static corrections is the calculation of the "field" statics. For some time after the introduction of autostatics programs, the importance of the field statics tended to be forgotten; there was a feeling that errors in the field statics were immaterial, because they would be removed by the autostatics program. Although this is almost true for static variations from trace to trace, it is emphatically not true (as we shall see in a moment) for slowly-varying statics errors. It is most important that the model we use for the computation of field statics is as appropriate as possible to the terrain. This requires good terrain descriptions to be logged on the observer's report (for example, "marshy ground between x and y, or "rocky outcrop here", or "loose fill in gully", or "I think this is the hard surface on which the dunes are moving"). It

also requires geological insight; in some situations (bare mountains and river fill) the weathering thickness is likely to be in anti-correlation with the elevation, whereas in others (moraines) it is likely to be correlated. A serious attempt must be made to match the field-statics model to the local weathering problem.

2. Autostatics programs operate by determining the static shift between pairs of traces, using the technique of cross-correlation. We should appreciate that this technique has some inescapable problems. First, there is a problem in selecting the window over which the cross-correlation should be computed. Clearly, as a protection against the removal of faults, the window should be long. There is also an advantage in long windows in that the cross-correlation process operates best with long windows; only if the waveform comprises more than a few cycles does the process of cross-correlation become better than the eye. However, if the nature of the autostatics program is such that it has to work on nmo-corrected data, then the nmo stretches the data, and the cross-correlation between a waveform and a version of itself stretched by one cycle is zero. Every processing centre has its own solution to this problem (multiple short gates, or whatever) but the solution tends to be hidden deep in the program, and normally interpreters may be unaware of the problem. This problem remains a major one whatever partial solution is used, since it is vitally concerned with the treatment of small faults (particularly vertical ones). The same problem of window length is also relevant to the well-known vicious circle between static corrections and dynamic

corrections — that is, that errors are likely in the stacking velocities when statics are present, and the statics are difficult to determine in the presence of errors in the velocity. The window-length problem arises because our ability to recognize nmo error in the presence of statics depends on having an nmo error which is substantially constant within the window; this forces us to adopt a short window, which is an undesirable course for the determination of statics.

3. Then there is another problem which is relevant to the risk of obscuring small faults, and that is the problem of lobe-jumping or cycle-skipping. A part of this difficulty arises because the cross-correlation function is unsatisfactory in some ways; in particular, it always has a narrower spectrum than either of its constituent signals, and therefore tends to be leggy. A leggy cross-correlation function makes it difficul to pick with assurance the correct peak on the cross-correlation function, and therefore lobe-jumping becomes a major problem. Some processing groups use the Cotas function (that is, the cosine transform of the amplitude spectrum) instead of the classical cross-correlation function, and this seems a very legitimate measure for minimizing the problem. However, the problem still remains to some degree, and somewhere down deep in everybody's program there is some subtle logic (but entirely seat-of-the-pants logic) intended to minimize the problem of lobe jumps.

Both of the last two problems — the window-length problem and the lobe-jump problem — are at least tackled in the programs available to the interpreter from his processin group, and he is able to make his own judgement on their efficiency. However, every interpreter must realize that there is one problem which cannot be solved with an autostatics program — and the problem is a very important one.

4. This serious and fundamental problem is that there is
no way of distinguishing, from reflection data alone, between
a thickening weathered layer and dip. Thus it is entirely
possible, in the present state of the art, to produce a
stacked section in which the static corrections are satisfactory
in the sense that they yield a perfect stack, but over which
there exists a steady dip at all reflecting levels which is
entirely due, in real fact, to statics. Similarly, of course,
autostatics programs can produce a perfect stack over a false
structure — a structure which is created solely by statics.
As is well known, this is the major problem with current
petroleum targets in the Middle East.

There are two classes of auto-statics programs, which
have different approaches to what can be done in this matter.

One says that it will take the average of all receiver
statics to be zero over a spread length, and the same for
the source statics. This, of course, is quite arbitrary,
and cannot be true in general. However, it is computationally
cheap and convenient. It has the effect of creating small
structures superimposed on the main structure, and having
a spatial wavelength related to the spread length (Hileman, 1968;
Blum, 1974).

The other approach does not make this assumption, but
replaces it by a requirement that the final section, after
autostatics, shall be a best fit, in the least-square sense,
to the original data to which field statics only have been
applied. This satisfactorily removes the spread-length effect.
However, as with the first class of autostatics programs, it
leaves the possibility that the structures remaining on the
section are created entirely by major weathering effects.
It seems there is no solution to this problem other than
independent measurements of the weathering thickness —
either by shallow refraction, or by drilled holes, or by
shallow reflection, or — if the data are good enough — by
auto-correlation measurements to differentiate the multiples
generated at the free surface from the multiples generated
at the base of the weathering. Ideally, we would wish to
have an uphole time (from below the weathering) for every
array; with some forms of shooting we actually obtain this,
and we can proceed to use either class of statics program
with an easy mind. Otherwise, the theory suggest that we
need some such independent weathering measurement twice per
cable length with the first class of program, or four times
per line with the second (or something in between with
hybrids).

It is most important that the interpreter should understand the limited objective of an autostatics program (which is to give a good stack) and should not confuse this with the separation of structure and weathering. He must be constantly watchful for indications of gross weathering changes (for example, changes in velocity and/or amplitude of the refracted and surface-wave arrivals on the monitor records), and quick in demanding an appropriate field technique whenever the problem arises.

5. This appreciation of the inherent problem of statics is very useful when we come to the next major difficulty in land work — how are we to correct the <u>amplitudes</u>? Clearly the difficulty is a major one, because we know that (whether the source is dynamite in drilled holes, or vibrators, or other surface sources) there must be major changes in the source output from place to place along a line. And, of course, there must be major changes in the coupling of the geophones to the ground from place to place along a line. The problem is therefore much compounded, relative to the marine case. It is further compounded by our long-held belief that on land we are not able to obtain any trustworthy recording of the outgoing signal. (The signal obtained from an uphole geophone for example, depends not only on the output from the explosive but also on the acoustic impedance of the material on which the geophone is planted and the manner of gradation of acoustic impedance up the hole. But perhaps the time has come to look at this problem more carefully; certainly in Vibroseis, by separate measurements of force and motion in the vibrator, we should be able to obtain a fair assessment of the signal actually sent into the earth.)

If we do a simple base-level on a record-by-record basis (the equivalent of a source static in the autostatics case) we increase the noise on the weak shots and we increase the brightness of a shallowing section. If we base-level on a trace-by-trace basis (the equivalent of source and receiver statics in the autostatics case) we likewise increase the noise on the weak traces, and run the risk of making the amplitudes on the far traces much too high because of

inadequate reflection section in the window; we still
retain, of course, the problem of brightening a shallowing
section.

If we are prepared to do much more computation, we
can perform the equivalent of the second class of autostatics
programs (which, we remember, accomplishes a least-squares
best-fit of the final section to the original section).
This removes all the local variations of source strength and
receiver coupling without producing an artificial brightening
of the areas where the section is shallow.

Basically, we have a great mass of traces, whose
amplitudes represent many effects, and we search to separate
the different effects and to establish their individual
relation to their causes. So: we observe amplitudes on
traces, we ascribe these amplitudes to effects at the source,
effects at the receiver (both including the near-surface),
effects associated with the offset distance, and effects
due to the deep geology. This yields an enormous matrix
solution for the whole line, under the constraint that
the effect of the near-surface , at any one surface point,
is assumed to be the same for the downgoing and upcoming
waves.

This is therefore clearly better than the base-levelling
approach. However, just as the fanciest autostatics program
cannot by itself distinguish between a gradually thickening
weathering and a dipping section, so the fanciest true-amplitude
program cannot distinguish between a gradually more efficient
source coupling and a gradually brightening sequence of
reflectors. (The interpreter, watchful as ever, will be
asking whether it is geologically reasonable for all reflectors
to brighten slowly across the section.)

In the autostatics case, we remember, the only solution
was some sort of independent measurement of the gross weathering
change along the line (actually, four independent measurements
as a minimum on each line). In the amplitude case the same

argument applies, except that (because there is no parabolic term representing residual nmo) the number of independent measurements is reduced to three.

In the autostatics case, the independent measurements are made by refraction or uphole shooting. In the amplitude case the equivalent problem (if we cannot measure the outgoing signal) is to find the actual reflection coefficient of a reflector — any reflector — at three places on the line, and then to force the amplitudes on the section to fit these reflection coefficients. For poor quality data, there appears to be no way of doing this, and for the moment we must admit ourselves defeated. For good quality data the reflection coefficients of one or more very good reflections may be computed from their multiple reflections on the raw data, or from appropriately-processed auto-correlation functions (as we did in Section 4.1.3); then the true reflection amplitudes are thereby established, free of the disturbing effects of the source and the near-surface.

When we are able to do this reflection-coefficient calculation at selected locations, then, obviously, we also obtain the bright-spot calibration which we sought in the marine case by using the sea-floor reflection. (Actually it is not quite as safe, for while the sea surface clearly has an abrupt reflection coefficient very close to - 1, the free-surface reflection in the land case may be transitional.

And a final caution in this unpleasantly cautionary section: the available statics and amplitude-correction programs — although they represent triumphs of technology — still view the effect of the near-surface as a simple _time_ shift and a simple _amplitude_ loss. In fact, of course, the effect of the near-surface is _frequency-selective_, and hence much more complicated than this.

Perhaps one day we shall have _real_ corrections for the near-surface — rather like decon operators.

Additional references:

Wiggins et al., 1976
Booker et al., 1976

8.4 ENHANCING THE SECTION FOR PARTICULAR OBJECTIVES

This course is directed fairly specifically to the interests of the interpreter. However, almost everything in the course drives home the message that there is no clean-cut point at which the interpreter first becomes involved. If it was ever true that the interpreter could take a seismic cross-section at its face value, and start his interpretation there, it is certainly true no longer.

So it seems appropriate to speed a little time on what the interpreter should know about improving the seismic data before they ever reach him. Then he is equipped to commission enhancement work for his own objectives, without necessarily having to explain to an outsider what his objectives are.

Of course it would be convenient to concentrate on the processing, which is readily adapted for specific objectives. However, we must make a few comments also on the field work; this is partly because it is so germane to the final interpretability and validity of the sections, partly because some interpreters have to contain a whole geophysical department in their own persons, and partly because it is very clear that in any organization the interpreter of the reconnaissance survey should be a party to the planning of the detailed survey.

8.4.1 Signal-to-noise ratio, and resolution

So we start with the sort of land section that makes one recoil in horror: "How the blazes can anyone expect me to interpret that?" This is the problem that we rather vaguely call signal-to-noise.

Of course if a section has been badly stacked, or had the statics put in backwards, or has been run through inadequate programs, then obviously it must appear to have poor signal-to-noise. However, in general, the most important likely cause of poor signal-to-noise is in the earth or the field (or a combination of both) rather than in the processing.

To find that cause, it seems that two tests are of particular value — one old one, and one new one.

The old one is also a very simple one, and yet one that is not much used. It is the simple test of repeating a record, with everything identical (or as near identical as the earth allows). Then we look at the wiggles very carefully. Ordinarily the wiggles are identical early in the record (where the record is dominated by source-generated signal) and different late in the record (where the record is dominated by ambient noise). Then we say that if the stacked section is going to be good enough for positive interpretation, the individual field records must have some reflections visible on them (it just is not true that processing is capable of producing something from nothing). So if the wiggles are the same — say in the middle of the zone of interest — but the reflections are very poor or altogether missing, then the problem is one of source-generated noise (in particular, of surface waves). In this case the approach is to change the field technique to provide better attenuation of radial or reflected or scattered surface waves. But if the wiggles on the two records are not the same, in the middle of the zone of interest, but the reflections are very poor or altogether missing, then the problem is one of ambient noise. In this case, of course, the approach is to provide a quieter spread, or more useful source energy.

In many or most cases of what would be called poor signal-to-noise on land records, the wiggles on the two records (shot under identical conditions in the same place) are identical to times well beyond the zone of interest; the problem is one of source-generated noise.

The second test for the cause of poor signal-to-noise — the new one — is just to look at the amplitudes. Let us say that we go through a normal processing sequence — geometrical divergence corrected for refraction (using the nearest known velocities), then time corrections and stack, and finally a filter low enough to exclude any frequencies that might be affected by absorption or short-path multiples (say 10 - 30 or 10 - 35 Hz) — and then display the amplitudes without any further manipulation. If there is a very marked decay of amplitudes, we have to accept that the acoustic contrasts at depth are just very weak, and the reason for the poor signal-to-noise is entirely geological. With that established, the geophysicist may choose to mount a proportionately higher field effort, or he may decide to abandon the area.

So where signal-to-noise problems are concerned, the real answer almost always lies in a careful matching of the field technique to the subsurface, near-surface and surface geology.

This point is emphasized by a clear rule which follows from the energy principle: that once the field work is done, with a certain source energy and a certain noise power and a certain bandwidth, then subsequent processing can improve the signal-to-noise only by trading something else to get it. What it trades is usually resolution — either vertical resolution (by frequency filtering) or horizontal resolution (by velocity filtering). There is no escape from this.

There is another class of signal-to-noise problems which we should mention — one which can still be present when the section as a whole is good enough to make one gasp. This rises when the interpreter is vitally concerned to improve his vertical resolution — for example, in the Rotliegendes of the southern North Sea, or the Mesozoic sands of the northern North Sea, or in reef and stratigraphic plays all over the world — and therefore wishes to trade signal-to-noise in order to improve resolution (or, in other words, to deconvolve very strongly). Then the interpreter may find he has a signal-to-noise of 100:1 on every other reflector in the section, but not enough to trade for vertical resolution at the target level. Again, the first answer lies in the field: in paying what it takes to get more source energy in the right bandwidth, or what it takes to get quiet recordings, and in being sure to record many samples of the desired data. Then in the processing the techniques are well known: apply a deterministic deconvolution for the source and the free-surface reflections (derived, of course, from the source pulse); stack; use the stack as a reference against which the individual traces are correlated; select only the traces whose addition improves the signal-to-noise ratio; restack; velocity filter; apply statistical spiking deconvolution; velocity filter again; deconvolve again, and so on until the limit is reached — when all the signal-to-noise is gone, and the horizontal resolution is degraded to the point of jeopardizing the faults. In other words, we are careful to optimize the signal-to-noise ratio <u>before</u> each statistical deconvolution. And, of course, we <u>obtain</u> the necessary many samples of the pulse shape <u>not</u> by using long thin auto-correlation windows but by using short fat ones.

So resolution (which is a major concern of section enhancement techniques) becomes intertwined with our discussion of signal-to-noise. But before we leave signal-to-noise, it is worth making a couple of points about display. Both emphasize the observation that while the final geological judgements continue to be made by a human interpreter (as they should be), so the display of the results should be that best suited to the eye of the interpreter. Thus virtually every attempt to devise an automatic selection of final filters, based on some machine-compatible criterion of signal-to-noise ratio, has failed; most of us accept that the best thing to do is to run a suite of filter tests, and to select those filters which optimize the <u>visual</u> clarity of the geology. By the same token, we continually see wisdom in providing <u>two</u> separate sections for the interpreter: one at approximately natural scale, on which the proportions of the features are about correct, and a second at perhaps 1/10 of the horizontal scale, on which the geological continuity is made most apparent This simple and inexpensive artifice is usually worth more than all the signal-to-noise enhancement techniques known to man. What is more, the squashed display brings a major increase of clarity to the faults.

We should remember also (from section 4.3.3) another important aspect of display: that on one section at least the interpreter must be able to see the whole waveform. Displays where the waveform is clipped on one side and biased off on the other may be fine for coarse structural studies, or for statics problems, but they are quite unsuitable for detailed studies of subtleties associated with hydrocarbon accumulations. Which type of display the interpreter chooses is, and should properly be, an individual affair decided between him and his eyes; some interpreters are prepared to work with the two section (positive and negative) required by va-wiggle, others prefer an unclipped variable-area with no bias, and still others (taking advantage of the industry's new ability to make and reproduce high-quality variable-density) feel that variable-density best simulates the realities of the pulse in the earth. But whatever is used, it is very important that the whole waveform be displayed faithfully; there is no reason at all to suppose that subtleties of waveform will not appear at or near a peak or a trough.

Further, since it is subtleties with which we are concerned here, we should note the importance of keeping the bandwidth open. Of course everyone knows that we lose resolution if we cut the high frequencies, but fewer take note of the fact that we lose stand-out (amplitude contrast) if we cut the low frequencies.

Often the geological character of a reflection complex is carried by the low frequencies, and very obvious character correlations across a disturbed zone can be destroyed if the low frequencies are cut. Further, the preservation of the low frequencies is often important in making reefs stand out from the section as anomalous.

We note, however, that the preservation of the low frequencies increases the problem of display without clipping; it must then be possible to see the high frequencies riding on the low frequencies. This matter of display is very important in the application of modern interpretation techniques.

8.4.2 Multiples, reverberation and ringing

Again we should not occupy time dwelling on things which are standard practice, but there are a few observations which should probably be included as guides for the occasions when the interpreter has to specify a new survey, or decide on a reprocessing sequence.

As is well known, there are multiples and there are water reverberations. And water reverberations come in three sizes: shallow (which we call ringing), medium (which we call water reverberations), and deep (which we call deep-water multiples) Deep-water multiples are troublesome even as simple trapped systems in the water, particularly if the sea floor is acoustically hard (a major problem offshore Eastern Canada, and in some parts of the Mediterranean). All the others are most troublesome in the form of reflected reverberations (that is, reverberant trains following each primary reflection).

The two widely accepted methods for the removal of multiples and reverberations are deconvolution and stacking. If the delay and velocity of the reverberant system are such that the multiples have significant difference of nmo from the primaries coming in at the same time, significant benefit is obtained from stacking; otherwise deconvolution is the best hope.

Every processing centre has one or more proprietary techniques to enhance the benefit obtainable from stacking; these techniques may be directed either at a particular multiple train of given multiple velocity function or at everything except the primaries (as defined by a primary velocity function). These programs are not much used for reconnaissance work, where one dare not usually risk the implicit assumption that the primary stacking velocities are known very accurately. However, for detail surveys (when, after a full velocity analysis using all the ray paths, the velocities are extremely well known) such techniques represent an entirely legitimate way of suppressing the multiples — legitimate, that is, in the sense of leaving the primary pulses intact in waveform, and to a lesser extent in amplitude. As noted in section 7.2, these techniques are particularly desirable as a prelude to migration, since it is important that multiples are not migrated.

For simple stacking to do its best in suppressing multiples, it is desirable to decrease the weight of the inside traces of the gather, relative to the outside traces; this is an obvious consequence of the hyperbolic nature of normal move-out. Decreasing the weight of any traces (or, in the limit, just omitting them) decreases the signal-to-noise benefit of the stack. Therefore a defensible course, where multiples are a major problem, is to perform the decreased weighting in the early and middle parts of the record (where the signal-to-noise is tolerably good) and gradually to remove the weighting toward the end of the record (where signal-to-noise may be more important than multiples). The time-variant effect on primary amplitudes must be compensated, of course.

As is well known, an N fold stack with N channels gives less multiple suppression at early times than a stack of fold N/2; as far as shallow multiples are concerned, a 24-fold stack on 48 channels is better than a 48-fold stack. But a buried river-channel in the near-surface is clearer on the 48-fold.

Which takes us to the water reverberations, and to the need to extend our conclusions of section 3.2.8.3. Each class of water reverberation has its own problem, but its own relieving feature.

In the case of ringing, for example, classical deconvolution is very satisfactory. It is no hardship to provide an operator of length perhaps 40 ms more than double the reverberation time, and the results are almost always good. However, we remember the problem — that in the case of ringing the reverberation time is less than a pulse length, and then it is not possible to remove the reverberations without jeopardizing the pulse shape and the amplitude of the primary. Of course there are reasonable things to do, which minimize our apprehension, but the fundamental risk remains. The shallower the water, obviously, the worse the risk. In fact, one can deconvolve the primary pulse to a surprising extent before serious effects on the amplitudes are seen in the gross, over a whole section; however, in bright-spot work it often happens that we wish to place firm reliance on a few traces in quite a small zone, and for this type of work we have already expressed our concern about shallow-water deverberation.

The medium range of water reverberations extends from reverberation times just longer than a pulse length up to those of 300 or perhaps 400 ms (that is, in water depths from 30 m to perhaps 300 m). In this range there is no problem with providing the dereverberation without any risk of affecting the primary amplitude or shape. However, three new problems arise:

- The cost of providing an operator more than twice as long as the reverberation time.

- The presence of _two_ reverberating systems over the early part of the _record_ (the trapped and the reflected-reverberation systems) having _different_ reverberation statistics. Fortunately, as _it happens_, the operator computed for the reflected-reverberation system works well on the trapped system with the exception of the first bounce, which is preserved at its original amplitude but reversed in polarity.

- The problem of the length of the autocorrelation window. As is well known, the fundamental requirements for the autocorrelation window are that it shall be long enough to contain several samples of the reverberation period, and that it shall not be _dominated_ by one or two events; if it is so dominated, then _it is_ inevitable that the

dereverberation must attack the primaries. If the
risk of domination arises because the window is too
short, then some means must be found to lengthen or
broaden the window to remove the domination. If the
domination arises because there is one large reflection
coefficient followed by a succession of weak but
important ones, then it is permissible and desirable
to apply an exponential expansion to the trace before
deconvolution. Such exponential expansion artificially
balances the amplitudes within the autocorrelation
window without distorting the ratio of primary to
reverberations on overlapping reverberatory trains.
After deconvolution the exponential expansion should
be removed; we remember (section 2.5.1) that exponential
expansions do not compensate absorption.

All of this means that the interpreter concerned to
know that all his primaries are present — and in particular
that all of their amplitudes are correct — has no alternative
but to involve himself with the deconvolution, and to make sure
that it is appropriate to the problem.

The third range of water reverberations — the deep
range — takes over at those depths for which nmo considerations
mean that the reverberation period is no long approximately
constant. This condition defeats classical deconvolution.
Further, it may be aggravated by the increase of the sea-
floor reflection coefficient which occurs near the critical
angle; this effect changes the amplitude relationships for
successive reverberations in a way which again defeats classical
deconvolution.

Presumably all processing groups currently have their
own proprietary program to combat this problem, and presumably
all of them are by now more or less successful. However, to
the interpreter concerned with details of amplitudes and
subtleties of waveform, the counsel must again be caution.

If the interpreter is trying to make a careful study of a
particular hydrocarbon accumulation, he might be better
advised to go back to the original field records, to mute
them very carefully to exclude all the trapped reverberations,
to restack them, and to hope that there are sufficient data
peeping through these windows to resolve the feature.

This problematical inter-relation between processing
and interpretation would allow discussion for ever. If
there is one point which we should single out for stress,
it is the view of interleaved or iterative processing and
interpretation. The stack of the reconnaissance survey is
just to identify regions which may be discounted. Then
comes the detailed study of the possible areas of interest.
For these the original stack is only the start. From then
on the processing does not have to consider all possibilities
at all times, for the interpreter can immediately make some
judgements, if gross ones, which allow the processing to
be refined. When this is done he can narrow his judgements
still further, and the processing-interpretation team must
then start looking at individual wiggles. Finally all the
validation checks must be made, to check that the final
interpretation is such that it would produce the original
field records.

The geophysicist should shiver when he sees $5 M wells
being drilled on the basis of what he know is really semi-
detail seismic work.

8.5 DELINEATION WITH FEWER WELLS

The basic thesis here is that as soon as a hydrocarbon trap has been located by the discovery well, the range of possibilities for detailed seismic interpretation increases dramatically. So we are quite out of sympathy with those companies who hold that, as soon as a discovery is made, the engineers take over and the geophysicists go and do something else. On the contrary, it is important that the interpreter should go back to the seismic data at that time, and review the data very carefully with the benefit of the new knowledge obtained from the well.

1. As soon as the producing formation is known, its extent is sometimes perfectly clear on the sections. One feels almost apologetic to mention this, but there have been cases where dry holes were drilled because no one had gone back to look at the seismic sections -- where the producing sand was as plain as a pikestaff.

2. Sometimes the identification of the producing formation at least establishes what the seismic problem is. Perhaps the delineation problem then emerges as the clarification of minor faults. In that case the whole seismic toolbox can be applied to this particular problem — with the resolution requirements (and hence the objective bandwidth) established from the geologic logs, and the migration velocities enhanced by knowledge of the check-shot data.

3. Then comes the detailed study of the seismic data at and near the production zone, in the light of the new knowledge from the logs. For delineation purposes, it is desirable to have a dozen lines radially through the location, with a square to tie the ends. Naturally, in the geophysicist's view these data should be shot and analyzed _before_ the well is drilled; however, after is better than not at all. For this type of shooting, usually severe QC constraints are placed on the contractor (for which he is paid, of course!); in particular, no variations of source pulse or streamer depth are allowed.

So to the detailed studies. The objective of these, of course, is the increased understanding of the seismic response of the production zone, and then the tracing of this zone, seismically, away from the discovery well. We remember from section 8.1.2.10 that the first things are the proper calibration of the velocity log and the density log for instrumental and operational bad-spots, and the construction of the synthetic seismogram. Provided there are no steep dips, the final synthetic seismogram must agree well with the seismic section, with no arbitray delays or empirical fiddling. If it is not so, something is not being done correctly, and it is unsafe to base any delineation conclusions on the synthetic seismogram.

Then we repeat the synthetic seismogram with likely variations of thickness, porosity and hydrocarbon saturation in the production zone, and study these in great detail against the variations evident on the radial lines through the well. This requires great patience, but is a most instructive exercise (instructive, usually, in reminding us what a blunt instrument is a seismic wavelength). As the picture becomes clear, the synthetics representing the postulated changes are assembled into a synthetic section across the well, for direct comparison with the actual field section. We need to be watchful, of course, for disturbing circumstances <u>above</u> the target zone on the field section, which might change the apparent response at the target zone and invalidate the conclusion from the synthetic; we must also be prepared to go back to the individual field records before stack, for the finest detail.

Then we use the check-shot data, as discussed in section 3.3.3, to establish the actual reflection coefficients obtained at the interfaces in the production zone. From these we calibrate the field sections in terms of actual reflection coefficient at the target level, as a refinement of what we have previously been able to do from the sea-floor reflection or some other shallow reflection.

4. There are a few things we should do to the comparison field sections. In any case, of course, we will be making these both without decon and with extreme spiking decon. We should also make a migrated version with the benefit of the velocity validation given by the check-shot data; by this time the structural picture must be fairly clear, and we select for this detailed migration only those of the radial lines which are seen to have no significant cross-dip. Then we should try a cycle of severe horizontal velocity filtering and severe spiking deconvolution in an attempt to isolate the fluid contact or contacts within the production zone. It is also very important, of course, to be sure that both the stacking and migration velocities are appropriate to the fluid contact.

536
537
538
O.PROJ.

539(B)   Incidentally, if the depth of the production and the length of the spread allow it, we should search for this fluid-contact reflection <u>at wide angle</u>. This will probably mean on the outside traces. Since the wide-angle brightening occurs only on <u>positive</u> reflectors, we have a means of emphasizing the fluid-contact reflection relative to the probably-negative reflectors above it.

5. Finally we turn to the delineation possibilities available if a seismic crew or vessel can be utilized when the check-shooting is done.

The concept is that, for each of several highly critical positions of the well geophone relative to the hydrocarbon accumulation, the seismic source pops its way along a radial line or lines through the well. The well-geophone output is recorded separately for each pop, and these recordings are subsequently displayed in sectional form.

It is straightforward to configure this technique for particular delineation problems, and the details are left (as the real text-books say) " as an exercise for the reader". However, we can readily see the behaviour of the direct arrival for the simple cases sketched on p.8-40.

The first arrivals basically form a hyperbola, and we are concerned to study the deviations of these first arrivals from a regular hyperbola -- both in time and in amplitude. Let us consider first the situation (top left) when the pops are recorded into the well-geophone at the position A, and when the stratification above A is generally horizontal. Then if we arrange the geophone signals in sectional form (top right), we obtain the classical nmo hyperbola corresponding to the rms velocity down to point A. Over the hyperbola, the changes in amplitude and form of the first arrivals are progressive and minor, being due primarily to the slight changes in the short-path multiple reflections as the angle of incidence increases and the spacing between beds therefore appears to increase also.

Now let us move the geophone down to B (lower left), and repeat the operation. B represents a position in the hole significantly below a hydrocarbon accumulation in a classical domal trap. Let us say that the velocities within the hydrocarbon-saturated zone are lower than those above and below (as they certainly will be, for example, if the reservoir is a sand and the hydrocarbons are gas). Then we have a simple plano-convex lens formed by the zone of hydrocarbon saturation.

As is easily demonstrated, in the optical analogy there is a range of positions on the surface from which the well geophone cannot be seen. If we construct the assembly of traces which we expect to get in the seismic case (but using only geometric ray paths) we obtain the basic hyperbola representing the situation in the absence of the hydrocarbons (with its sides distorted to some extent by the dip of the dome flanks), modified by the velocity pull-down contributed by the hydrocarbon-saturated zone. We can see also the brightening produced by the "magnification" of the lens.

Between the alignments there are gaps, representing the
blind zones in the optical analogy. The outer limits of
these gaps define the edge of the hydrocarbons. In fact,
as we well know, the optical case is an oversimplification,
and we must bring in diffractions. These, as we would
expect, make the edges of the gaps (particularly the
important outer edges) less abrupt.

Incidentally, we note that the principle of reciprocity
allows us to think of the source as being at the geophone
location, if it aids the thinking in any way. The reason
for the gap at the surface, and the nature of the diffractions,
are perhaps a little easier to see when the source is visualized
as being down the hole.

The amplitude changes and the velocity pull-down can be
seen more clearly if nmo is applied (for one-way time), on the
basis of the rms velocity appropriate to the depth B away from
the hydrocarbon accumulation. Then the final display appears
as at the bottom right of p.8-40.

We can make such a display for each of the radial lines
through the well. The wanted edges are then located in space
by a simple graphical construction, tracing the rays which
emanate from the geophone and arrive at the surface positions
to which the edges are ascribed. This is, of course, the
basic operation performed by an nmo correction program.

The method requires, obviously, that the hole should
be drilled beyond the production zone, by an amount which
depends on the distance from the hole to the edge of the
hydrocarbon-saturated zone. However, it often happens that
such continuation is made on the discovery well or on one
of the early delineation wells, so the method is realistic
in these cases.

The example of a simple dome, with velocities in the saturated zone being less than in the surroundings, is scarcely the problem in practice, for in the presence of such simplicity we do not need to know much more than the depth of the fluid contact. However, virtually any type of limit to a reservoir — whether due to folding, faulting or facies change — produces some blind zones and some brightened zones, by reason of the velocity lenses involved.

If the hole is deep enough to allow several geophone positions below the hydrocarbon zone we can select from the well geophone records the traces showing blind zones, and we can juxtapose them and shift them in time so that groups of them represent different samples of travel path just within and just outside the hydrocarbon limits.

Having the geophone down the hole, of course, increases very considerably the resolution available, compared to reflection measurements at the surface. Down the hole the signal-to-noise is good enough to allow us to work with single traces, and stacking is not as important. Further, the picture is not complicated by negative interfaces, as is the reflection picture; we remember that the direct transmitted signal down the hole is always erect. Again, diffractions (while still present, of course) are less of a problem.

However, when we have established by these means which are the ray paths which do go through the edge of the hydrocarbon accumulation, we naturally return to the individual reflection traces from the original detail survey, to check whether the blind-zone effect can be seen on them.

These radial lines shot into a down-hole geophone also allow us to look carefully at reflections off the producing zone at oblique incidence. With reflection records only we would not dare to draw many conclusions about Poisson's ratio from ray paths at oblique incidence, but with a geophone down the hole and the opportunity to place it correctly, we

Surface
A
Borehole
x
t
Nmo hyperbola

Surface
B
Before nmo correction

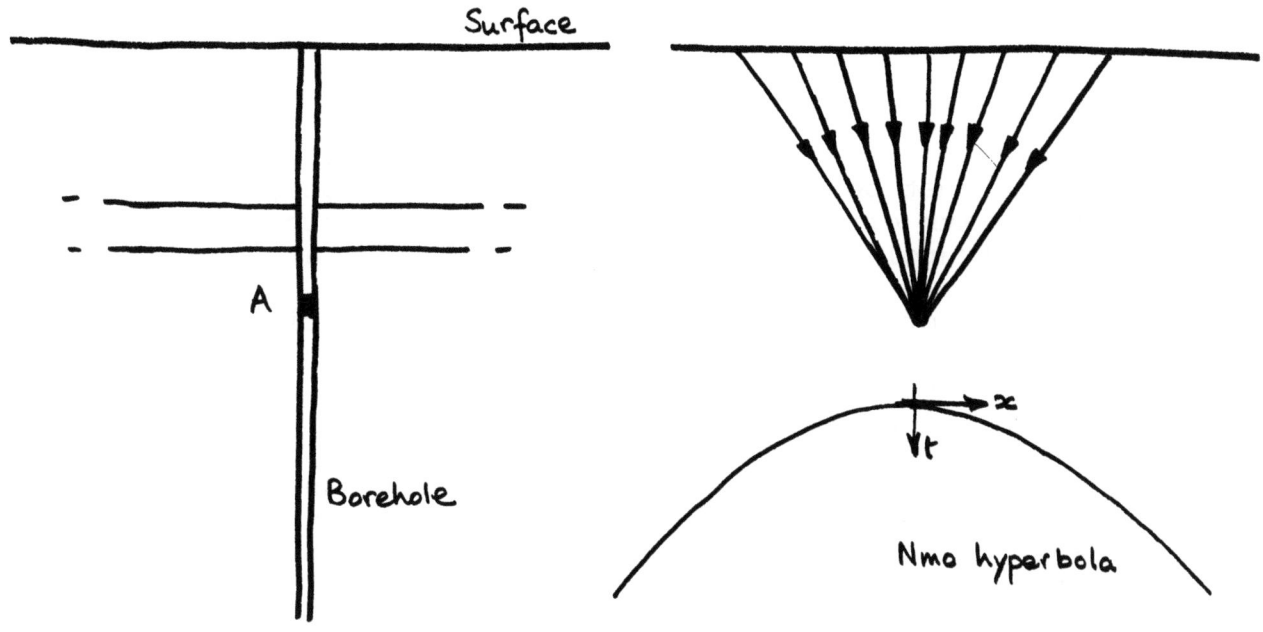

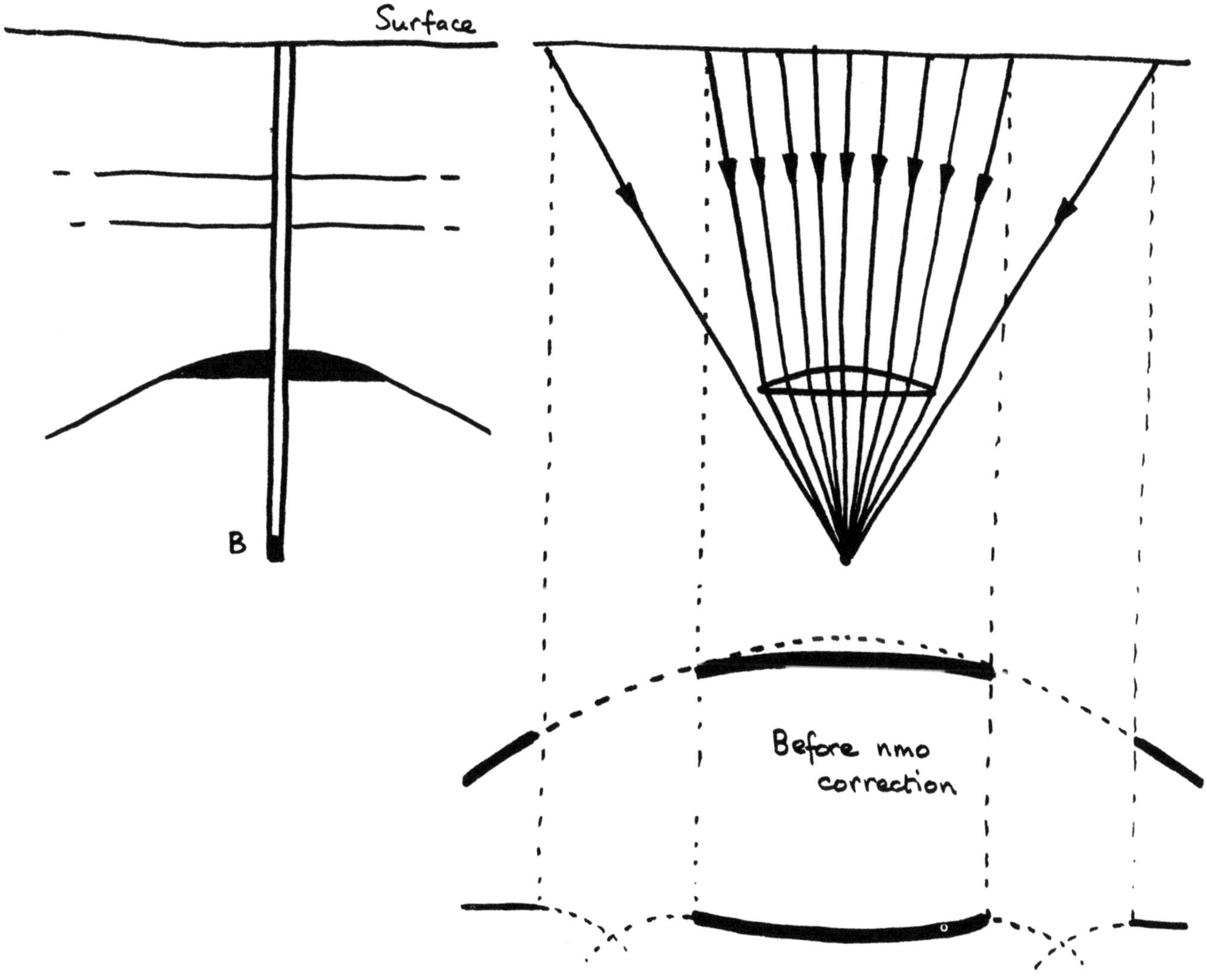

might just be able to make such conclusions, to associate them with effects observed on the reflection records, and to extrapolate them accordingly over the area.

Before we leave the topic of delineation with fewer wells, we should refer to the work of Bois. He considers what is perhaps a once-in-a-lifetime case, but one with which interpreters concerned with delineation should know. In his work he actually shot between wells, using downhole geophones as detectors and perforating guns as sources. Considerable clarification of a complicated fault pattern was obtained by studying the details of the refracted paths between the wells.

Additional reference:
Wuenschel, 1976

8.6 THE PREDICTION OF DRILLING HAZARDS

The hazards which the seismic method now appears to be able to define in advance are:

- hydrocarbon seeps at the sea floor
- buried river channels
- gas pockets
- unconsolidated shales.

We consider these in turn.

8.6.1 Hydrocarbon seeps at the sea floor

The presence of gas seeps through the sea floor (particularly if aggravated by the drilling process) is a hazard to jack-up rigs because of the risk of movement of the sea floor, and to buoyant rigs because of the risk of reduced buoyancy of the water.

In section 5.1.4 we have looked at some of the indications of sea-floor gas seeps on conventional echo-sounders, on sparkers, and on side-scan sonar. Obviously these indications can often be usefully supplemented by sniffers, or the sniffer survey can be usefully be pin-pointed with the benefit of such indications.

But our concern here is with the gross early-warning of seeps provided by the routine seismic work. Of course such work is not ideal for the detection of seeps — the wavelengths are too long. Yet in a way the seismic results are more significant than the sparker results, because (although the zone of seepage needs to be fairly large for detection by seismic), the seismic section often reveals the deep mechanism of the seepage.

The course slides include many seismic indications of extensive gas seeps associated with known oil and gas fields.

Quite a fine degree of detail is obtainable with conventional seismic, provided that it includes detectors close enough to the source to record the sea-floor reflection. Major seep lineaments can be mapped, sometimes yielding an impressively coherent picture.

If the zone affected by gas is small or the water is deep, wavefront healing makes the weakening of the sea-floor reflection less apparent. However, the point remains sound that the conventional seismic survey can at least furnish a warning that seep conditions exist, and that more detailed investigations are warranted before the final location of a rig is decided.

8.6.2 Buried river channels

These represent a risk to the geophysicist (because of the possibility of creating a structural turn-over where there is none), and a physical risk to the safety of jack-up rigs.

The section on p.8-44 illustrates a very large channel which gives no expression at all in the relief of the sea floor. It is about 2·5 km in width, and 500 m deep. The pull-up underneath it is primarily caused by the high velocities of the glacial debris in the channel. Dry holes have been drilled on such pull-ups — because the seismic work had been done with a long offset to the near geophones, so that the section did not really start until about 0·5s.

Of course there is a continuum of size from the mammoth channels down to those whose presence can only just be detected, in the sense of reflection alignments, under the sea-floor reflection. The small ones may not be significant as pull-up or pull-down agencies, but remain hazardous for the driller. At small scale, the presence of shallow buried channels may show a weakness in the sea-floor reflection — a weakness which should be traceable from line to line (if the lines are sufficiently closely spaced) to confirm the conclusions of a

8-44

buried river system. In this we have an analogy with the mapping of seep lineaments; in fact the judgement as to whether the cause is a seep system or a channel system may by made on the basis of the configuration evident on the map.

Again, we must not suggest that the routine seismic data are usually sufficient in themselves. Obviously the correct approach is that the routine seismic data should be scanned by the interpreting geophysicists (as a matter of normal concern) for buried river channels, and that any indications of these in the prospective drilling area should then be detailed by one of the shallow profiling methods.

A buried river channel such as the one illustrated is a perfect example of a case where it is necessary to switch from the type of statics program which averages over a spread length to the type which does a least-square fit over a line. The type which averages over a spread length would quite fail to correct the deep structure for the buried river channel, and would impose additional undulations on the deep structure each side of the edges and bottom of the channel. Further, the appearance of the deep structure (and its amplitude indications) would be jeopardized seriously if one velocity analysis had its gather down one flank of the valley, the next had its gather up the other flank of the valley, and normal interpolation routines were used for the stacking corrections in between. Indeed, the stacked section derived from virtually any disposition of spaced-apart velocity analyses must be in error because of the channel. This highlights the importance of knowing from the seismic data, at an early stage of the processing, that the buried river channel is there, so that appropriate correction measures can be taken.

8.6.3 Gas pockets

As we have seen, there is a systematic tendency (in some petroleum provinces at least) for more-than-usual minor accumulations of gas to be found at shallow depths over hydrocarbon accumulations. Therefore, in such provinces, the driller must be doubly conscious of the risk of blow-outs. The geophysicist's problem is to determine whether the hole can be sited to avoid such gas pockets, and if not to warn the driller of the depth at which he should expect gas.

Basically this is a simple bright-spot exercise. If gas pockets are likely at depths beyond 0·6 or 0·7 seconds we should use full-scale seismic techniques on a rather close grid. If the problem is confined to the first ½ second or so, then it is cheaper , and adequate, to use a big sparker. In both cases, it is of extreme importance to apply proper corrections to the amplitudes, and to ensure that the gas zones are displayed in a correct amplitude relationship to the sea floor. As we discussed in Part 4, the interpreter should not just assume that his processing centre is doing it correctly — he should ask for details.

One of the early wells on a major North Sea oilfield struck a gas pocket when drilling with sea water. The gas blew out, pipe was thrown in the air, and the rig was covered with sand. Altogether a harrowing experience. So the company involved ran a sparker survey before siting further wells, and p. 8-48 shows one line from that survey (Lucas, 1975).

Having defined the gas pockets, the geophysicist must decide whether it is practicable to drill between them. Is it safe, for example, to drill a hole 50 m from the point at which the bright-spot amplitude has fallen by say 12dB? In one example where such a thing was done (actually at 100 m), no trouble was experienced; however, the risk would obviously depend critically on the characteristics of the reservoir in terms of seal, permeability, structure, etc.

If, however, the approach has to be to drill through such a bright-spot, then at least the geophysicist can tell the driller where to expect trouble. He should be able to do this to within a few metres, if he takes care with the correction of zero time and if he does not have buried-river-channel problems to create statics. At the worst, under reasonable conditions, he should narrow the range to ± 10 m; after all, in the nature of things, we always obtain very clear velocity indications on a bright-spot. Drillers are now looking at geophysicists with an altogether new respect.

TAPE 37   8.6.4 Unconsolidated shale   (ref. Pennebaker, Reynolds, Herrin

The drilling hazard which may arise in the drilling of an unconsolidated shale is twofold: the risk of blow-out from pockets of very high pressure gas, and the risk of the shale flowing into the hole and squeezing the drill stem.

549
The first of these — the high-pressure gas — is tackled, or at least corroborated, by looking for small bright spots within the shale (so small, incidentally, that they may appear just as diffraction patterns with a bright apex). One of the earlier slides (311) showed a large shale mass in which small local inhomogeneities were present, these might well be such local gas pockets. In general, of course, this is primary gas — gas formed from organic material trapped in the shale at the time of its deposition, and blocked by the impermeability at the shale boundaries from migrating into a reservoir trap. Tiny pockets of this gas may accumulate into larger (and dangerous) pockets — as a result of the permeable path represented by the borehole, and the general disturbance — during the time that the drill stem is withdrawn to change bits ("trip gas").

The second risk — that of the shale flowing into the hole and squeezing the drill stem — is tackled by attempting to identify the abnormal pressure conditions which make such flowage possible. We look, therefore, for seismic measurements which reveal abnormal pressure conditions, and of course the seismic measurement most helpful in this regard is velocity We are therefore very much concerned with the discussion of the effect of overpressure on shale velocities (section 2.3.4).

In an overpressured shale, we remember, the water is contributing to the support of the overburden (instead of the usual situation where the overburden is supported by the rock matrix) This means that the forces holding the particles of shale in contact are small, and therefore that movement between the particles is much more easily provoked — the shale is likely to flow or slide into the borehole. By the same token, we have a change from the normal situation where the seismic wave is carried by the rock matrix (with minor modification by the contained water) to

Color plate 8-48 follows page number 349.

a new situation where the elastic characteristics of the water are a major determining factor in the propagation of the wave.

From section 2.3.4 we know that the interval velocity in an overpressured shale shows much less increase with depth than would be the case with any normally-pressured rock. This is, of course, because it is not experiencing the usual increase of compaction with depth of burial.

So the key to the identification of a hazardous overpressured shale is the recognition of a velocity distribution indicating a deviation from the usual increase of compaction with depth of burial. The technique can work only in sedimentary basins in which the observed change of velocity with depth is primarily the effect of compaction, in a column of substantially constant shale or sand-shale lithology.

Beyond this introductory stage, we are projected into the controversy between the log-loggers and the semi-loggers. We shall be able to see the most glaring overpressured zones without resolving this controversy, but we shall be limited in our efforts to see the less glaring ones, and in particular to predict the necessary mud weights successfully, unless we can resolve it for our own area.

The basic approach is set out by Hottman and Johnson, 1965; Herring, 1973; Pennebaker, 1969. It is usual to work in log-analyst's terms, so we invert our normal velocity thinking into terms of sonic travel-time (plotted with slow values to the right).

The first step is the establishment of a "normal" compaction curve for the shale or sand-shale sequence. This would ordinarily be done from the sonic logs of wells in the area, as described in Part 2; reference to "normal" velocity analyses, such as that on p.8-52, may also help.

Then we study the sonic behaviour of overpressured zones encountered in the area, and relate this to the mud weight used in the drilling. We are then ready to tackle the task of predicting abnormal pressure in future wells.

8-49

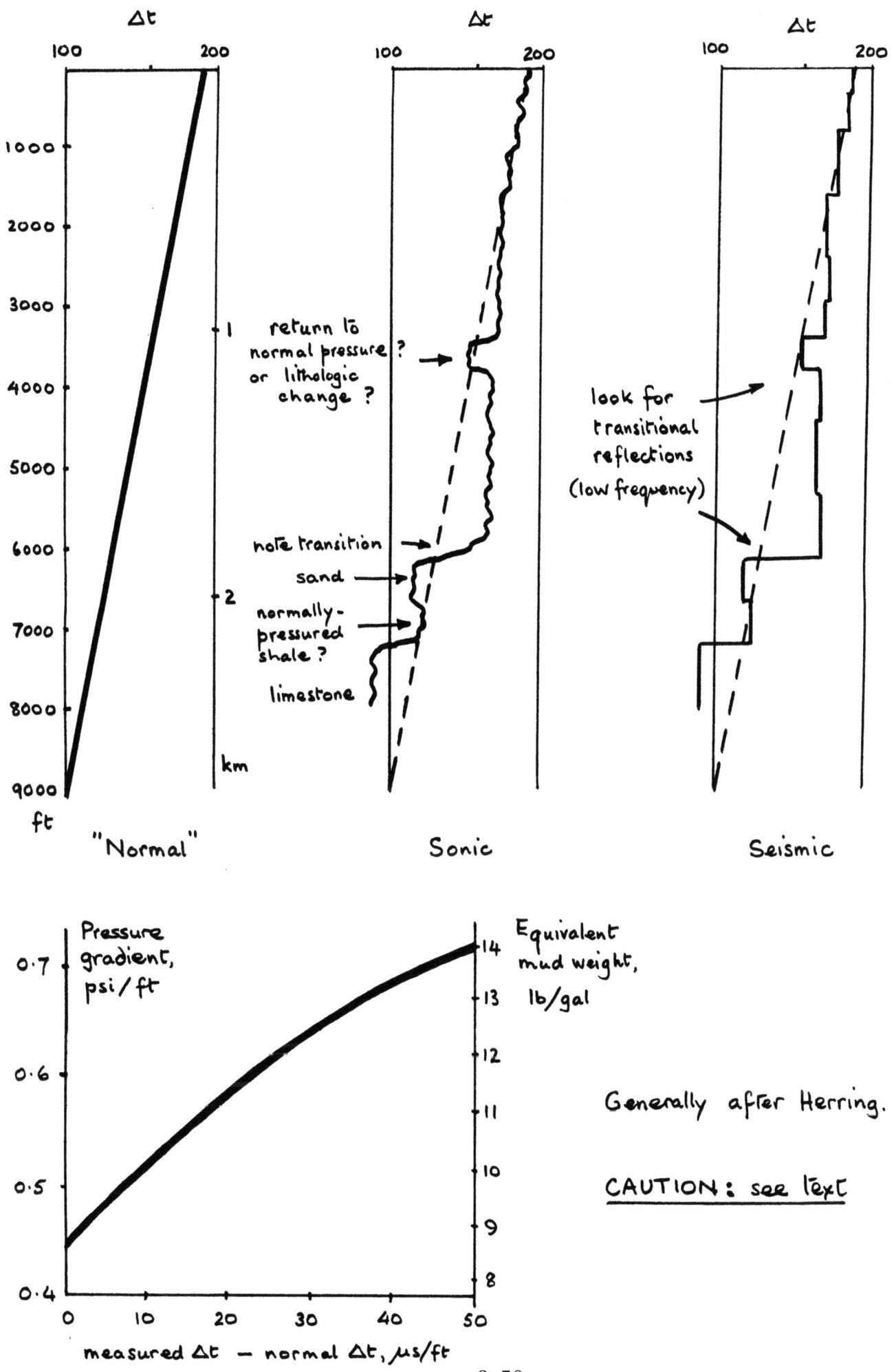

8-50

554   This we do by computing interval velocity from the velocity analyses of the detail seismic survey, for several or many intervals in the shale or sand-shale sequence. Then we analyze these for indications of interval velocities markedly lower than expected, and, after validating these indications on as many velocity locations as possible, we convert the "distinction" of interval velocity (or its reciprocal the sonic time) into a predicted mud weight.

The illustrations on p.8-50 paraphrase those of Herring, who is a semi-logger. The velocity analysis on p.8-53 illustrates the general appearance associated with a very marked and very extensive over-pressured zone; the interval velocity remains at or below 2100m/s (7000 ft/s) from 2·2s to 3·2s. The velocity analysis on p.8-54 shows a major degree of overpressure from 1·8 to 2·8s.

The interpreter should be alert to the weaknesses of the approach. First and foremost is the difficulty of defining "normality" — the selection of the sonic intervals judged to represent normally-pressured conditions. Second, if he accepts the semi-logarithmic plot, he must know that it cannot be true at great depths (p.2-35). This is most germane in the matter of the relation between the sonic time and the mud weight; what we have loosely called a
555 "distinction" above is a _difference_ according to Herring
556 and a _ratio_ according to Pennebaker. One "distinction"
557 could _lead_ to mud weights of 5lb/gal higher than the other, which is clearly unacceptable as a prediction.

Therefore we need to address the controversy between semi-log and log-log more carefully. We have agreed in Part 2 that it does not matter too much for purposes of bright-spot calibration, but it clearly does matter for purposes of drilling pressure prediction.

The resolution of the controversy probably lies in the observation that the very shallow zones of all rapidly-deposited shale sequences are _ordinarily_ over-pressured to some degree. Therefore the velocities observed in the first thousand meters are almost always too low. Nevertheless, because these values are the ones generally observed, the

8-51

8-52

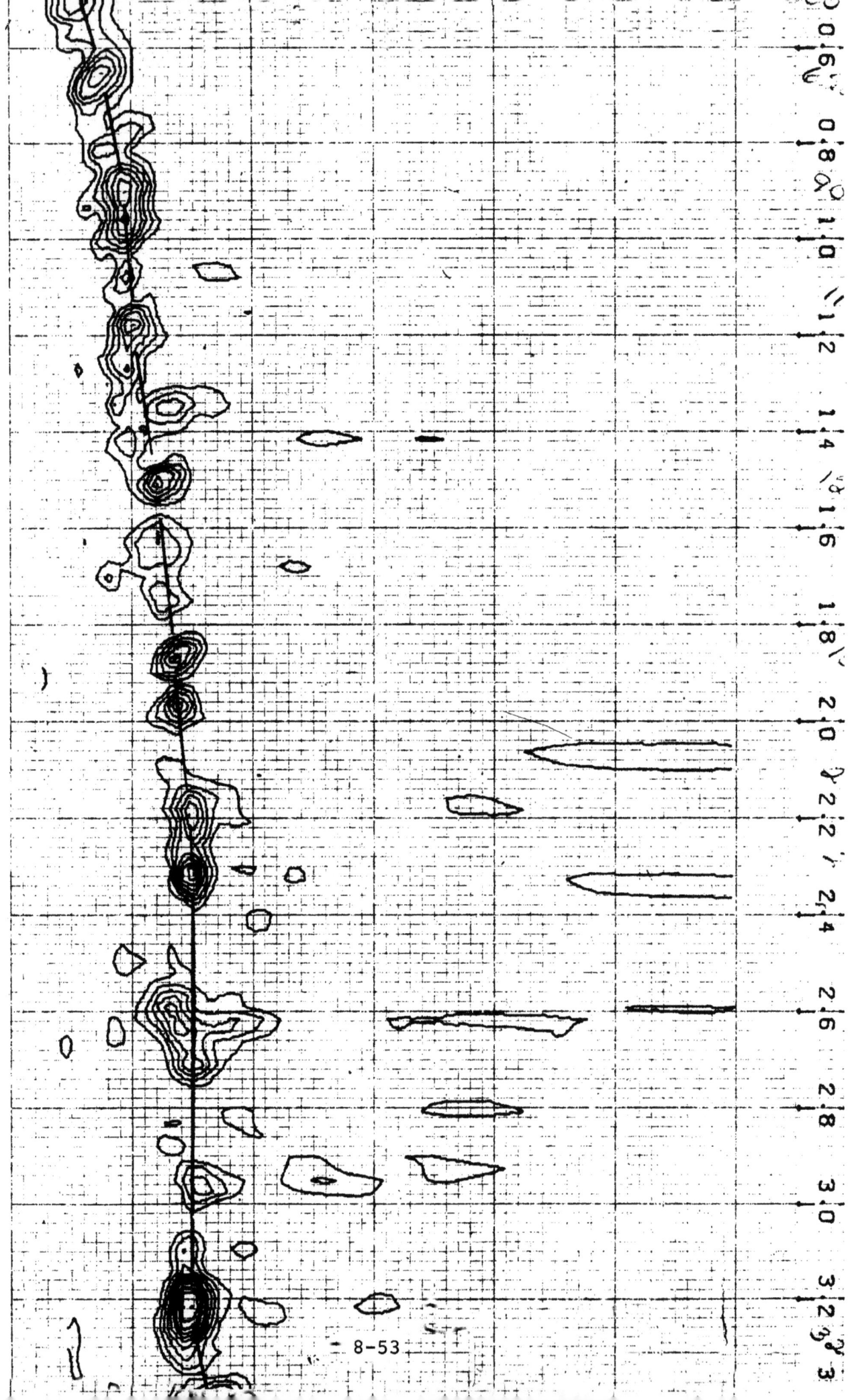

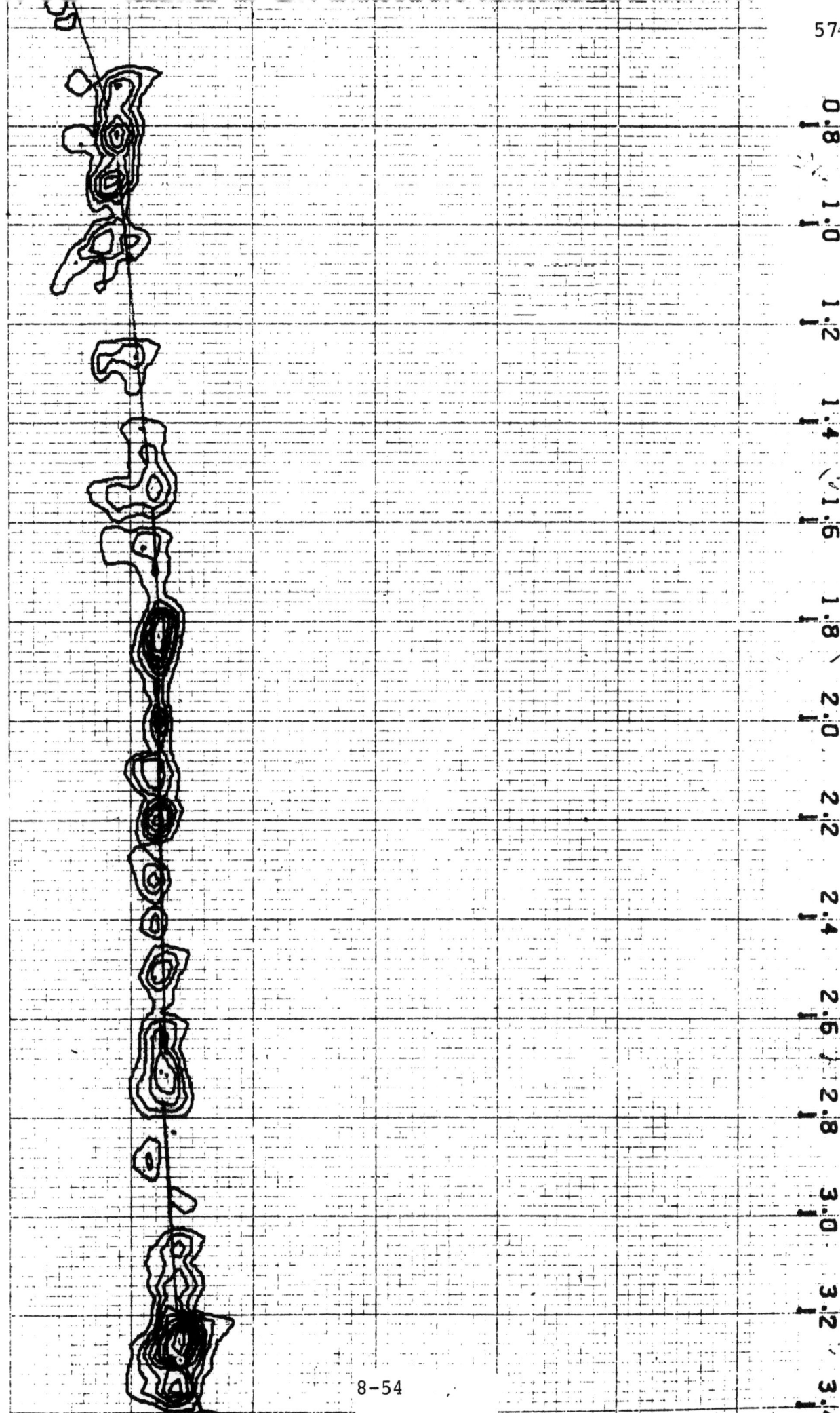

analyst calls them "normal".

If this is indeed the truth of the matter, we must incline toward the log-loggers. In expecting the curve of sonic time against depth to be linear on log-log paper, they are saying that it should be convex toward 10 o'clock on semi-log paper, and this is what we would obtain if we corrected the shallow values for some degree of overpressure.

Even accepting the semi-logger's data, we might come to a similar conclusion by drawing our "normal" curve toward the high-velocity side of the data points instead of through the middle; this would be the correct course if we are satisfied that the low values are low because of overpressure.

In his own area, then, the interpreter analyses the sonic and drilling data for the best inter-relation between interval velocity and depth in the normally-pressured state, and checks carefully the consequences to his curve if there is slight overpressure at shallow depth. Then he establishes his empirical relation between mud weight and the appropriate function of sonic time, and proceeds happily to predict mud weights for future wells.

In all of this, of course, the interpreter is confined to shale sequences with no more than minor fractions of sand.

L(B)

8.7 THE APPLICATION OF SEISMIC WORK AFTER A FIELD IS IN PRODUCTION

560

This section is based — perhaps whimsically — on the published example of a gas-induced bright spot whose dimensions decreased as the field became depleted, and increased again as it was used for gas storage.

561(B)

The producible reserves of an oil or gas field are only a proportion of the total hydrocarbons in place. In making an assessment of reserves, therefore, we need to estimate this proportion. The estimation is traditionally done using engineering data obtained from the wells (particularly the curve showing decline of reservoir pressure against cumulative production, and the changing gas-oil ratio). However, these data may advantageously be supplemented by data obtained from reshooting the field seismically at periodic intervals. Needless to say, the seismic technique used should preferably be the same for the periodic reshooting as for the original shooting (or perhaps each reshooting should be done both with the original technique and also with the latest technique of the day).

The proportion of hydrocarbons in place which may actually be produced can vary from 20% to 90%, so that major importance attaches to making a good estimate. The recovery factor is a function of the type of drive existing in the reservoir, and so we are concerned to establish what help the seismic results can give to the identification of the type of drive. Basically, we are working with the same variables we use for the original identification of the hydrocarbons: the strength and polarity of the reflection from the top and/or bottom of the reservoir, the velocity within the reservoir, and, of course, the reflection from the fluid contact if we can see it. With these, we do the best we can to differentiate between the types of drive, according to the following scheme.

8.7.1 Gas reservoir

The two types of drive are the simple gas-expansion drive, and water drive. Gas-expansion drive is due to the totally confined (and therefore declining) gas pressure; it involves no change in the dimensions of the hydrocarbon-saturated part of the reservoir. Therefore it is established if the seismic reshooting yields identical results (and is confirmed by a straight-line decay of well pressure with time). It is also worth noting that sometimes the original seismic data can even predict the conditions for gas-expansion drive, because they (in combination with the known geology of the area may show the likelihood that the gas reservoir is contained entirely within rocks of very low permeability, so that no water drive would be possible. An example is shown on p.8-58. So in either of these events — no change in the seismically indicated dimensions of the gas accumulation, or impermeable surroundings — the interpreter can assume gas-expansion drive, and can therefore hope for 80 - 90% recovery. If the bright-spot dimensions reduce with time, however, he must accept water drive, and his estimate of recoverable reserves must be downgraded to 70 - 80%.

8.7.2 Oil reservoir

This is more complicated, because there are four possible types of drive.

1. A simple water drive, which ordinarily yields a recovery of 40 - 60%, is probably best recognized from the reservoir pressure information. Structurally, the seismic picture is valuable, since this must be compatible with the large aquifer characteristics required for a water drive. The contribution o

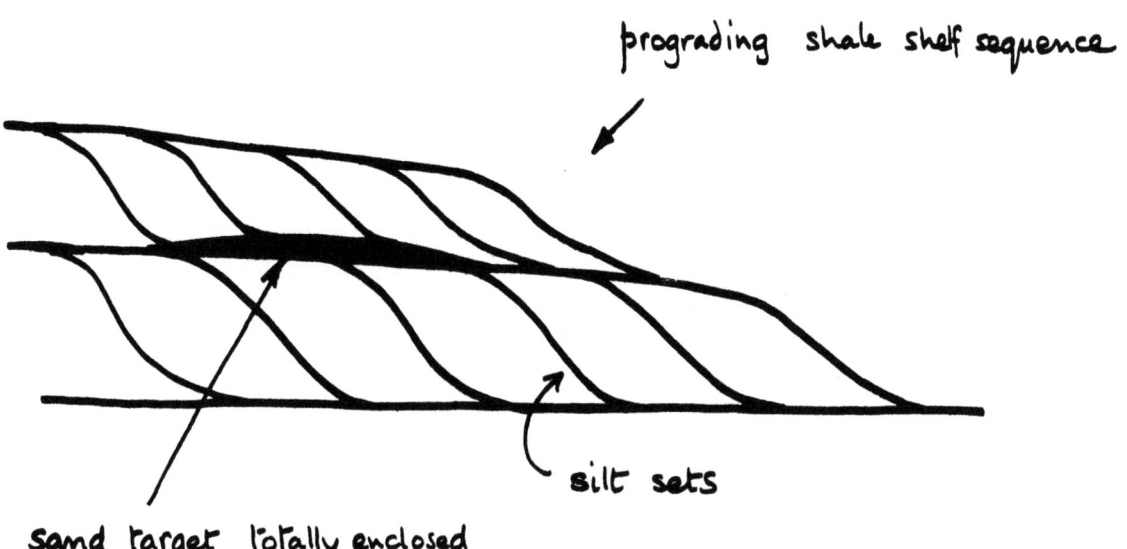

the auxiliary seismic variables, however, is likely to be
small. If seismic work were to help, it would be by
demonstration of the encroaching water — by the small
change of reflection strength associated with the replacement
of oil-saturation by water-saturation. But positive
identification of such a change, as we know, is fairly
unlikely. There would also be a small change (typically
a few percent) in the seismic velocity within the reservoir,
but in the early years of production (when the information
on reserves is most required — or most in doubt) this could
be detected only at the edges of the reservoir, where the
interval is thin and the velocity therefore uncertain.
Perhaps it is best to say, at this stage of the art, that
seismic work cannot help in positively identifying a simple
water drive in an oil reservoir. However, as we shall see,
it can often eliminate the other possibilities, and so
indicate water drive in a negative manner. Notice this is
the reverse of the case of the gas reservoir.

The three other types of drive produce a change much
more likely to be observed seismically.

2. The presence of a gas cap before the field is produced
normally appears, of course, as an anomaly of reflection
strength, and this may occur in combination with a polarity
change (according to the local circumstances of depth and
lithologic environment).

The first requirement for a gas-cap drive (which typically
yields recovery factors of 25-35%) is that the gas-cap shall
be volumetrically sufficient to maintain the drive in the oil
zone. In a simple reservoir, this can be assessed from the
original seismic data. Then the actual existence of gas-cap
drive may be demonstrated from the seismic reshooting if the
change in the volume of the gas cap is exactly that which

would be expected from the known cumulative production of oil, having regard to the known reservoir pressures.

3. Gravity-drainage drive (which typically yields recovery factors of 80%, and which is therefore a highly desirable condition) is suggested if the seismic reshooting indicates a change in gas-cap volume which is greater than that which would be expected from the cumulative production of oil. As the simplest example of this, gravity-drainage drive is indicated if the original seismic shooting shows no reflection-strength anomaly interpretable as a gas cap, while the reshooting shows a clear gas cap. This positively establishes that the permeability and other reservoir conditions are such that the gas released from solution by the declining reservoir pressure is able to move upwards through the reservoir, under gravity.

4. The existence of dissolved-gas or solution drive (which may yield recovery factors of only 15-20% of the oil in place) is detectable seismically, at least in principle, by velocity measurement. Physically what happens with solution drive is that the gas released from solution by the production of oil remains within the pores and interstics of the rock, and is not able to migrate upwards as it does when the reservoir conditions allow gravity-drainage drive. As we know from our earlier studies, the presence of just a tiny bubble of free gas within each pore may materially alter the elasticity of the material if the reservoir is at fairly shallow depth (though it does not at first make much change to the density). Therefore the seismic indications of solution drive are a significant reduction in the interval velocity within the reservoir, associated with a reflection-strength anomaly whose degree is commensurate with the velocity change acting almost alone. Both the velocity change and the concomitant reflection-strength anomaly extend over the entire oil-field; they are thus distinguishable from the gas-cap

situations not only by the distinction of density but also by their presence within the zone known to be currently productive of oil.

Of course, we should not delude ourselves about the degree of precision of these seismic indicators. When it comes to the identification of a drive mechanism, the seismic method is a fairly blunt instrument. Therefore the seismic indications should be viewed only as additional or supplementary input to the resolution of the drive problem; the primary input must continue to be the pressure decline curves, the gas-oil ratios, the cores, the logs, the traditional seismic volumetric determinations, and the changes of the pressure and gas-oil-ratio conditions with structural height. However, it is worth noting that several of these primary inputs represent measurements very local to a borehole, while the seismic observations represent gross measurements of average conditions over considerable extents of the reservoir. Each, therefore, has its value.

In particular, if the performance of a field proves to be different from the original estimates based on volumetric calculations, then the original seismic work should be repeated and the answer sought by means of these detailed seismic analyses. In addition to the conclusions concerning the type of drive, the seismic work may show that the performance of the field is being affected -- for better or worse -- by the presence of faults or impermeable shale breaks which were not previously recognized. These may produce, for example, reflection-strength anomalies over just part of the reservoir, and show that the total recovery could be materially increased by additional drilling into a hitherto-unsuspected extension or separated sub-division of the reservoir.

8.7.3 Summary

Gas reservoir: Water drive (70-80%) if connected aquifer volume sufficient and areal extent of bright spot decreases with time.

                    Gas-expansion drive (80-90%) if the
                    extent of the bright spot is unchanged
                    with time.

Oil reservoir: Gas-cap drive (25-35%) if the bright
                    spot anomaly indicates a gas cap
                    volumetrically sufficient to maintain
                    the drive and the areal extent of the
                    bright spot increases with time to
                    a degree which volumetrically keeps
                    pace with cumulative oil production.

                    Gravity-drainage drive (70-80%) if a
                    bright spot sufficient to indicate
                    the formation of a previously non-
                    existent gas cap appears during
                    production, or the areal extent of an
                    existing gas-cap bright spot increases
                    faster than would volumetrically keep
                    pace with cumulative oil production.

                    Solution drive (15-20%) if the seismic
                    velocities over the entire extent of
                    the oil production decrease markedly
                    with time.

                    Water drive (40-60%) if the connected
                    aquifer volume is sufficient and there
                    is little or no detectable change in
                    the seismic appearance of the field
                    (either in velocity, reflection strength,
                    or other auxiliary variable).

8-62

PART 9

ASPECTS RELATED TO

EXPLORATION MANAGEMENT

9.1 METHODS OF CONVEYING THE ESSENCE OF AN INTERPRETATION TO MANAGEMENT, WHILE RETAINING POINTERS TO SIGNIFICANT UNCERTAINTIES

9.1.1 The contour map

The basic rules of contouring are summarized in the pre-course notes (section 1.2.1). In the present section we assume that all participants are already well skilled at contouring; our concern is with matters beyond basic technique.

If we ever get a chance to finish an interpretation, it emerges as a set of fair-drawn contour maps and isopachs, mostly in full lines, with the faults appearing as thick black lines redolent with authority. Here and there, perhaps, is a dashed line suggesting an element of doubt — usually on the edge of the map where it matters little. But generally, the maps appear definitive and convey an impression of finality. Between ourselves, however, we know that this is nonsense. We have guessed a continuity here, accepted a leg jump there, phantomed here, put in a questionable fault there, failed to resolve a mistie here, not bothered to include the feathering angle there, and finally accepted one contour solution (whether by hand or by machine) without appraising the alternative solutions.

And, of course, we cannot be blamed for that (or at least for most of it) because that is the inevitable consequence of being forced to present the final results in the form of contour maps. If something is to be contoured it must, sooner or later, have a single accepted value; we might be able to accommodate a single zone of doubt, but it is virtually impossible to accommodate multiple zones of doubt on contour maps (and even less so on isopachs).

Yet, while not accepting any blame for what we have to do to get a contour map, we surely do not hide from ourselves that these uncertainties could be really important. They are by no means always trivial. To establish that, we have only to do the exercise of adapting an old contour map to some additional shooting — invariably finding that at least the details of the contours (and sometimes their essential form) require changes to accommodate the new information. Not only that, but we usually realize (deep inside, if not publicly) that we really ought to go back and rework some of the original picking.

So, some part of the problem lies in the nature of contour maps, and the difficulty of incorporating in them some quantitative indication of the effect of an alternative picking judgement. We note that this difficulty — of accommodating an alternative picking judgement — is quite different in type from the error accommodation which may be provided for potential-field data in an automatic contouring program. Mathematically sophisticated programs exist which make it easy to insert plausible limits on the spatial frequencies of the data, or the order of the polynominal to which they are fitted, and to obtain computed variances distinguishing zones of reliable and less-reliable data (Wren, 1973). However, each data point in a potential-field map — though it may have significant error — has only one original value. We can contrast this with the seismic situation where we are following a reflector with perfect confidence and virtually no error, and then we go into a disturbed zone where the reflector could be almost anywhere, and then we emerge into a region where on broad geological correlation we can be sure that our pick should be on one of two legs (but we cannot say which), and then we have another disturbed zone, and finally back into unquestionable clarity. It is clearly quite wrong to apply statistical-type error measures to all parts of such a case.

As happens so often in seismic technology, we are at risk from our friends the mathematicians and the computer programmers, who make our decisions for us — sometimes without understanding the intricacies of the problem, and almost always without telling us the arbitrary component in what they have done.

Let us spend a moment reviewing the differences between contouring as they see it — as an exercise in mathematical morphology — and contouring as the geophysical interpreter sees it.

The basic assumptions in the mathematical approach are that the variable to be contoured is single-valued, continuous, and isotropic.

The single-valued assumption is good for gravity data, magnetic data, elevation data and fathometer data; it is also good for seismic horizon data provided that there are no overthrusts or reverse faults. So there are no real problems in this assumption.

The continuous assumption is also good for the above applications, except where faults exist. Faults, of course, represent a major problem for any machine-contouring program; such programs normally accommodate faults by ignoring the data points on the far side of the fault, and thus fail to honour the tear behaviour of the region near the fault in the way that a geologist would do.

But it is in the third assumption — that the variable contoured is isotropic — that the machine programs are most in error. For the earth is clearly not isotropic — the whole purpose of our endeavour is to explore the techtonic consequences of enormous horizontal forces and local variations in the vertical force, and these consequences inevitably include structural grain, piercement, flow and fracture adjustment, and other dominant violations of the isotropic assumption. Which is just another way of saying

that the machine programs (or at least the widely-available
ones) have not yet been taught sufficient geology to yield
a proper representation of the earch. This remains a good
topic for academic research.

We, however, have to do what we can with what we have,
accepting that machine-contoured maps can show neither a
geological nor a management bias. The point is illustrated
by the simple example of a square NS-EW grid (p.9-6).

Confronted with the basic data posted on the observed
lines, any contouring program used in a normal manner
yields two separated circular highs. If it is known,
however, that the structural grain of the area runs NW-SE,
then the geological bias suggests that there is a single
high with its axis in this direction. (Further, if company
policy is "Always contour optimistically" — and in
particular if it is known that management is seeking a
farm-out — then it could happen that some of the flank
contours somehow become bowed-out, and that an even higher
high appears in the centre.)

Because of the accepted need for the geological bias,
everyone pays lip service to the concept of machine
contouring as a first step only, followed by a human
inspection and adjustment. This is fine as a philosophy,
but perhaps less than perfect in practice. Of course the
machine-generated map will highlight a bad data point;
the bad point may then be removed and the map re-run. But,
bad data points apart, the problem is in distinguishing,
on the machine contour map, between those areas in which the
program has done well and those where it requires a
correction to represent geological bias.

To this end it is most important that the interpreter
should know what the program is doing (Walters, 1969). First
he must realize that, while it may be true that — as his
suppliers or his research department tell him — his program

is "better" than others, it is also true that all contouring algorithms are arbitrary. The arbitrary nature expresses itself in several features (which may represent options which the interpreter himself has to select), and which typically include these items:

1. One of the initial steps in a contouring program is to change from the original distribution of data points to a new set of computed values at the intersections or nodes of a square grid; the dimensions of this grid must first be decided.

2. The value assigned to any node is computed by searching for and selectively using the original data points in the region of the node. The simplest type of program uses some minimum number of data points nearest to the node; this number must be decided. The objection to this type of program is that it may find all its points in a narrow wedge in one direction; this is particularly likely in seismic work. So more refined programs require some minimum number of data points in each quadrant, and still more refined programs require some minimum number (usually one or two) in each octant. These directional constraints improve the chances that the contouring will detect a grain in the data, but do so only at the expense of taking in data from farther away (and therefore of introducing a major degree of smoothing in the map).

3. When the data points to be used in any node computations are established, they are weighted according to their distance from the node. Some programs use weighting inversely proportional to the distance, others to the square of the distance, some to as much as the sixth power of the distance; this arbitrary weight is another item to be decided.

4. If the calculation of the node value is done by simple averaging of the selected neighbouring data-points, then its value cannot lie outside the range of those points. Such a

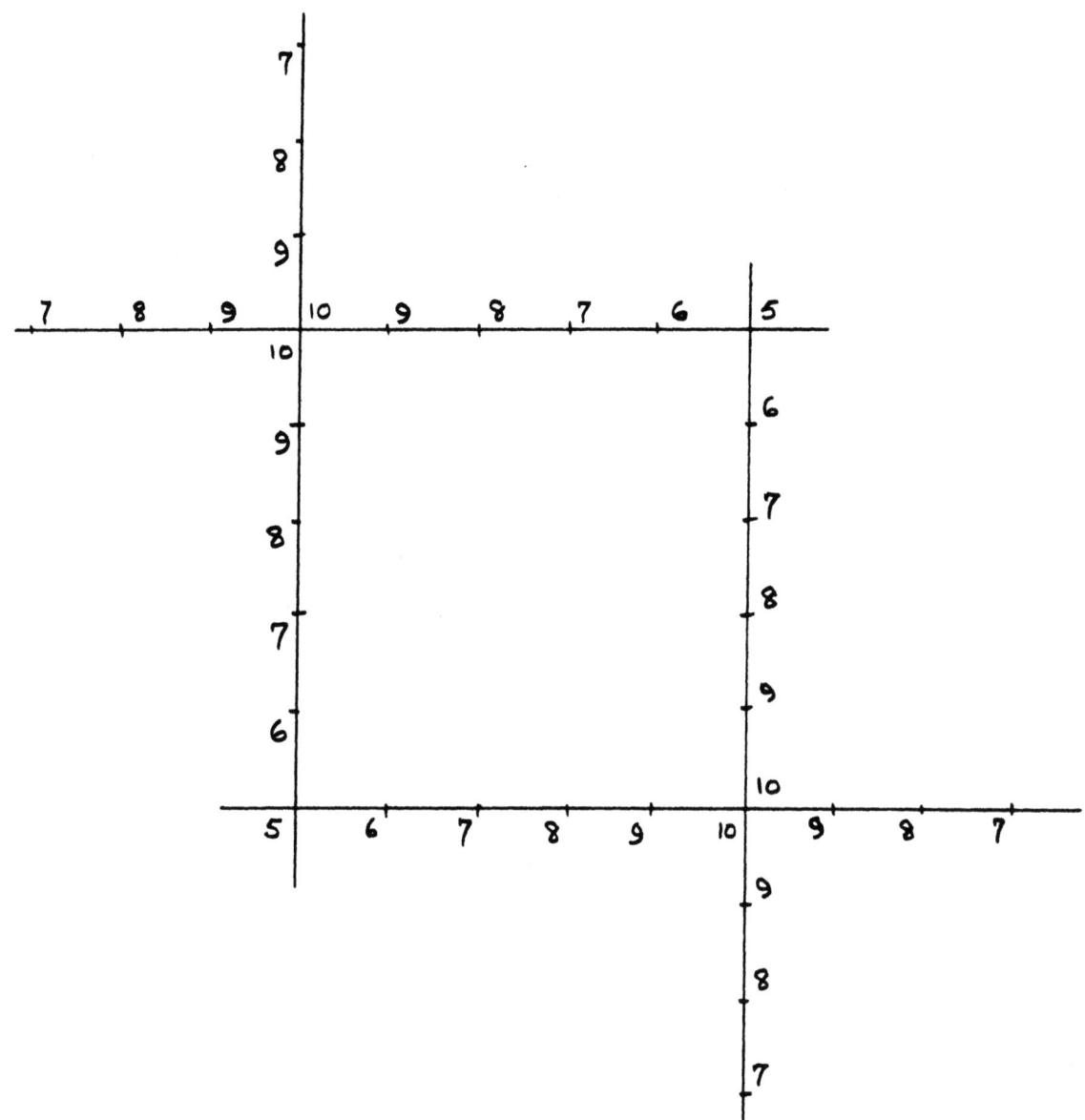

calculation, therefore, cannot properly indicate a contour high or a contour low (which must lie outside the range of the neighbouring data points). Therefore some programs measure the dip between neighbouring data points (in pairs or triplets) and project this back to find a dip-projected value at the node point; these dip-projected values are then weighted and averaged. In such programs the manner of using the data points in the dip measurement must be decided, as well as the distance-weighting relation.

5. Finally we must decide the algorithm used for actually drawing the contours, using the gridded values.

So the interpreter must know (or must himself define) the decisions called for within the contouring program. If he defines them himself he does so having regard to these facts:

- Where the data are dense the final map depends little on the algorithm; the simplest (which is also the cheapest) produces the least smoothing, and hence honours the individual data points most accurately.

- Where the data are sparse the final map depends critically on the number of data points used for each node value, with the greatest number of data points and the least severe weighting producing the greatest smoothing (and hence the least honouring of individual data points).

- The use of dip projection produces highs and lows where there should be highs and lows, but also produces highs and lows where there are no data.

- Once the node values have been computed the errors introduced by the contouring program are defined, and these are not reduced by calling for a smaller contour interval in the final contouring operation.

- For randomly distributed data, or for gravity data on land, quadrant or octant searches should not be used unless a considerable degree of smoothing can be tolerated. For data obtained on a network of lines, such as seismic data or aeromagnetic data, it is virtually mandatory to use at least a quadrant and preferably an octant search; in this case there is no choice but to accept the smoothing.

- If faults are inserted when contouring seismic data with a dip-projection program, the contours close to the faults are almost certainly in error. If they are not inserted, they become grossly smoothed.

Therefore the interpreter, accepting that the contouring program must be in part arbitrary, and that it must involve errors, is concerned to know the effect of the arbitrary decisions and the probably nature of the errors. It helps that he should have before him several examples of what his program does to particular types of data — what it does if the data are dense, what it does if the data are sparse, what it does toward the edge of the map, what it does to highs and lows, what it does near faults, what it does to the NW-SE example we considered earlier, and what it does when (having regard to the spatial frequencies evident along the seismic lines) the distance between lines is so great that no contouring could possibly be honest.

Thereafter, when the interpreter has inspected a production contour map and inserted, as best he can, the bias represented by his knowledge of the local geology, there is merit in reading back the final map into the machine, and calling for a plot of the time section along each line, using the values actually contoured. Then the interpreter may check, by overlaying, that every pick on every trace remains a geologically plausible one.

TAPE 38   So we retain some anxiety about contouring practices. If the contouring is done by machine, then sophisticated programs are required to accommodate the particular nature of seismic data, and these programs of necessity must mask or obscure subtleties which may be geologically significant. Further, there is a risk that the interpreter will be too far from the decisions made deep in the program, and that — just for reasons of indolence or overwork — he will not check the significance of those decisions by converting the final data back into an overlay time section.

If on the other hand the contouring is done by hand, then there is a risk that preconception may be _too_ strong, and that the treatment of alternatives will not be consistent.

In the industry generally, probably about 75% of the work submitted to machine contouring is accepted, on an area basis. That is not to say that it is right or wrong, of course — just that it is accepted. However, variables whose interpretation is highly dependent on geological insight tend to be exclusively hand-contoured; representative variables here are interval velocity, rms velocity, and permafrost surface corrections.

The problem of contouring velocities is, of course, an extreme one and an extremely important one (section 7.6). The validity of a migration and the accuracy of a depth conversion both depend critically on a sound method for interpolating and smoothing velocities, and the interpreter must stay very close to these velocity operations whether they are done by machine or by hand (ref. Duval).

In one sense, the velocity problem is more tractable than the contouring of reflection times: although the individual velocity measurements may have a much greater scatter than the reflection times, at least they are single-valued — the leg-jumping problem has only minor significance. This has led to an interesting application of the Krige method to velocity contouring (Haas and Viallix, 1974; Delfiner, 1976).

The concept is to derive the weighting factors from the measured spatial continuity of the variable, the spatial continuity being calculated by consideration of all possible pairs of data values. In practice, of course, this high-sounding generality has to be curtailed for reasons of cost. But, apart from the cost factor, the interpreter does not have to decide what shall be the degree of smoothing; the trends or regionals are measured automatically, and used to define the smoothing.

The same method can be used for interpolation of the time values, without smoothing; thus the time gradients are available at every point, and automatic contour migration is readily accomplished (Kleyn, 1976). Again, it all sounds fine; we are still left with some apprehension about the decisions which have been made somewhere deep in a program, and about how we check the geological reasonableness of what has been done.

However, after all those worries about contour maps, the arguments for making them — that we must have some way of estimating what happens between lines, and that exploration management (and indeed we ourselves) must have some means for distilling the essence of a survey into some readily-assimilable form — these arguments remains sound.

In a moment we shall question whether the contour map is the only way of making such a distillation. However, before we go to that, let us ask what we should do about the auxiliary variables.

All that our contour map has done for us, even if it is perfect, is to distill the essence of the structure. And if there is one thing that distinguishes the new seismic interpreter from the old-style seismic interpreter, it is his insistence that every useful property of the seismic signal — in addition to just reflection time — must now be extracted and displayed.

Consider reflection strength, for example. We have learned:

- that the variations of reflection strength along a reflector are much larger than we used to think,

- that these variations are sometimes immensely significant, and

- that even the trace-to-trace <u>details</u> are significant, and must be preserved.

So a very important problem emerges — how to transfer to the structural contour map the significant variations in reflection strength.

The first approach, clearly, is to transfer to the contour map the reflection strength information evident on the profiles at the contoured horizon, and then to contour this strength information superimposed on the structural information. For example, if we have the reflection strength displayed on the profile sections in colour, then we can manually transfer at least the gross variations of colour to the contour map, by colouring the lines accordingly. Then, ordinarily, we see new <u>areal</u> significance in the colour variations. Often this areal view clinches an indication of hydrocarbons, beyond doubt.

Though this simple approach is a very useful forward step, it clearly requires developing further. The problem is that the <u>gross</u> variations are not enough; frequently the <u>detailed</u> variations are significant also (particularly of small faults and fractures, which are highly germane to the prospects of hydrocarbon accumulation). The truth of the small-scale faulting at cap-rock level is much more likely to be revealed by these detailed strength variations than by the interpreter's basic picking of the horizon. But these variations, obviously, cannot actually be contoured themselves

unless the spacing of the seismic lines is sufficiently close to allow this — which is unlikely if the line spacing has been selected on the usual considerations concerned with the dimensions of the structure. So we seek some way of allowing the display of these strength variations to <u>ride on</u> the basic structural display, each adding new <u>significance</u> and interpretability to the other.
We seek, in fact, a <u>four</u>-dimensional display.

9.1.2 The fence diagram

At this stage let us leave contour maps hanging in mid-air, so to speak, and consider the other established method of representing data from a network of lines. In particular, let us consider the old fence diagram.

Probably every interpreter has made one fence diagram at some time, when a particularly thorny interpretation problem demanded every insight which could be obtained into the 3-dimensional configuration of a feature. In so doing, he will have noticed:

- that a fence digram, since it is built of sections, retains all the detailed information on these sections,

- that a major 3-dimensional clarification is obtained by being able to look into the model from different angles,

- that the view which is of particular importance is always obscured by an irrelevant piece of section on another line, and

- that all that fiddle-faddle with sticky tape and glue is so much of a bother that he will never do it again.

9.1.3 The sculpted fence diagram

The last disadvantage is just an engineering problem — it can be overcome by having made some prefabricated click-together parts like a Lego set. The other one — the difficulty of looking down into the fence diagram — suggests a solution which perhaps contains within it the best features of a fence diagram <u>and</u> of a contour map. It suggests, in fact, a <u>sculpted fence diagram</u> (Meek, 1971).

1. Let us suppose that we have picked a particular horizon on a grid of seismic lines, and that we estimate that in the regions of doubt our picking could not be in error by more than 50 ms. So we construct a fence diagram of sections in which everything more than 50 ms above the contoured horizon is stripped off.

2. Perhaps we start with just the dip lines (p.9-14, top). Then we may find that, by looking into the model from different aspects, we can already see how to resolve the picking doubt. Perhaps we have to add the strike lines before we can be sure. It is, of course, easy to run round loops checking the line ties and the consistency of the fault-picking on such a display. When we are sure, then we may choose to trim off the extra 50 ms, and so give ourselves a model on which the three-dimensional surface of the picked horizon can be clearly visualized from the line data (p.9-14, bottom). Alternatively we may choose to leave the extra 50 ms on, to retain the indications of other horizons which may be onlapping the picked horizon and thereby producing interference patterns.

3. In either case, the cut edges define for us the surface of the reflector we have picked, in a way which contains all the valid information which may be displayed on a contour map — but without the arbitrary decisions present in the contour map.

If the timing lines are marked on the cut edge of the sections, these, in effect, suggest contour lines. What is more, all the reflection-strength information and all the subtleties of pulse shape and polarity change are preserved

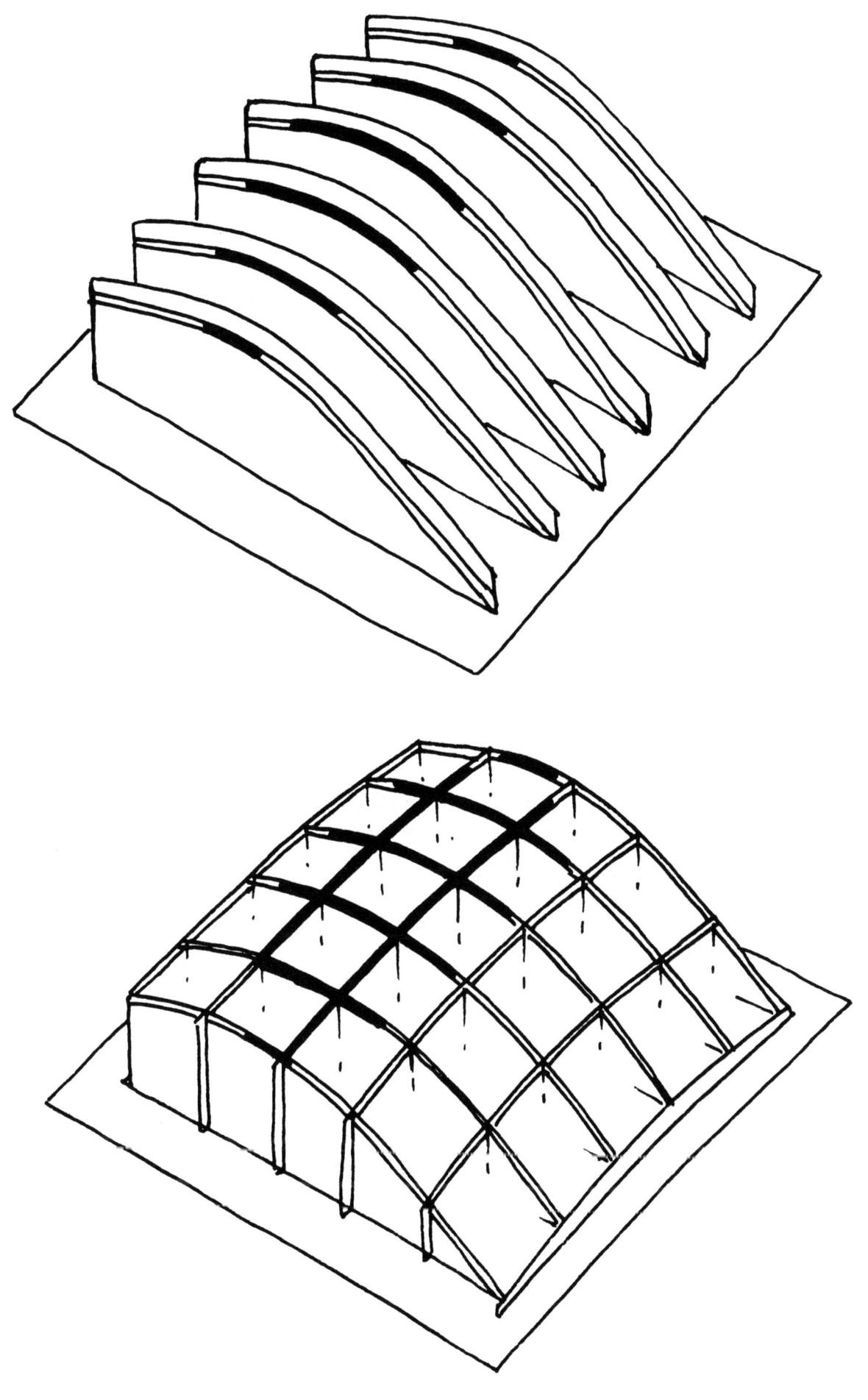

9-14

also. For example, let us suppose that in the figure the blacked-in trough represents a certain high level of reflection strength; then the sculpted fence diagram shows the areal extent of this high reflection strength at a glance.

4.　But far more than this, it also preserves all the minor variations of reflection strength, so that all the highly-significant indications of minor faults, fractured zones, gas pockets and fluid contacts are retained <u>in their original form</u> on the display. There is no way, obviously, in which all these subtleties could be transferred by the interpreter from the sections to a conventional contour map.

5.　In the case of reconnaissance grids, one of the immediate benefits of the display is often a realization that the data in some parts of the grid area cannot be honestly contoured. Sometimes a gross geological grain can be established by rotating the aspect, and these allow a bold interpreter to do a tentative contouring by hand; in such a case, of course, a machine contour map would need to be excessively smoothed.

　Also evident, usually, is the remarkable degree of areal consistency which can be expected of certain types of reflection-strength anomaly (such as the feather-edge of a disappearing layer at an unconformity surface).

6.　As far as structural traps are concerned, it is obviously a major advantage not only to have the structural highs depicted as real three-dimensional highs, but to have visible below the high the possible migration paths by which hydrocarbons might have been transported from the source rocks into the trap.

7.　So these models have major value for the petroleum geologist, and justify themselves on that benefit alone. They also have great value in giving management a real feeling for the sort of anomaly that it is proposed to drill, while also making it clear what are the regions and extent of the doubtful features of the interpretation. In removing the partial magic between the seismic sections and the contour map, these displays allow management to be a party to the decision-making and risk-taking performed by the seismic interpreter, but without the intense and detailed labour which that would have required with conventional contour maps.

Cases have occurred of geophysicists who, being themselves highly skilled in the art of contour interpretation, have agreed to use this type of display solely for the reason just given — as a sop to simple-management. But, in the event, they have found that the benefit to the geophysicist is greater than to anybody else — primarily because of the fact that a sculpted fence diagram adds so much more of the reason for things being as they are.

8. This is particularly true of the auxiliary variables, and even more so of reflection strength. Let us suppose that we are in an area where stratigraphic traps are of economic importance. Then we are searching for features which have a certain natural pattern: meandering river channels, delta distributaries, sand bars, beaches, and so on. Then at every unconformity level at which such features might exist, we strip off everything above the unconformity, and look for small (possibly very small) local anomalies of reflection strength. We squint along the top of the model in all directions, looking for the strongly sinous alignments of reflection strength which would indicate a river channel, or the weakly sinuous alignments which would indicate a beach, or the finger-type alignments which would indicate a delta system, or the linear alignments parallel to a coast and fringing a delta which would indicate sand bars. Notice that we are not squinting to see structure, but to see alignments of anomalous reflection strength which could indicate structurally-insignificant but highly-explorable sand bodies. In this the geophysicist has a significant new tool in his old struggle to evolve a way to find stratigraphic traps (Hun and Boisse, 1976).

Even the minor variations within the aligned anomaly may be significant. Thus the geophysicist will be looking for such minor variations representing thicker channel sands on the outside of bends in a river channel, and similar indications which, even on a shelf sand, will allow a guess as to the direction of transport.

9. We can, of course, photograph our model, from any chosen viewpoint. Then what we have is a two-dimensional representation of a three-dimensional model. If this is sufficient, we could save ourselves some handicraft work by constructing the two-dimensional representation directly within the computer. And, since we gain nothing by the perspective, we might as well keep our scales isometric.

566
567

The figure opposite shows such a computer-generated isometric display, for a portion of a known gas field. The display represents the view looking east; the next figure shows the view looking west. These two figures are stripped off to the top of a known gas-bearing sand. In this case the brightening of the gas-sand complex over its high can be seen in the black-and white display; it is, of course, much clearer and better quantified in the colour display.

568

569L
570L
571

The only input necessary to construct these isometric displays, obviously, is the interpreter's picked section, the final output tapes, and the map. The first benefit is an immediate visual proof that the picking has been correct, and an indication of any misties inherent in the map or the sections. The second benefit is the preservation of the important structural indications under the feature of interest. The third is a release from some of the risks and worries of contouring (or at least a proof that they have been properly handled). The fourth is the preservation of the subtleties of waveform and reflection strength associated with the target. And the fifth is for management — an immediate demonstration of the scale of the feature, and the certainties and the doubts in its identification.

572
573L
574

In the gas field illustrated there are two prospective zones: the simple well-behaved structural trap in the sandston at the contoured level, and a deeper problematical zone beneath the salt and a limestone layer. To study the deeper zone, we strip off the section to the top of the salt (pp.9-21 and 9-22)

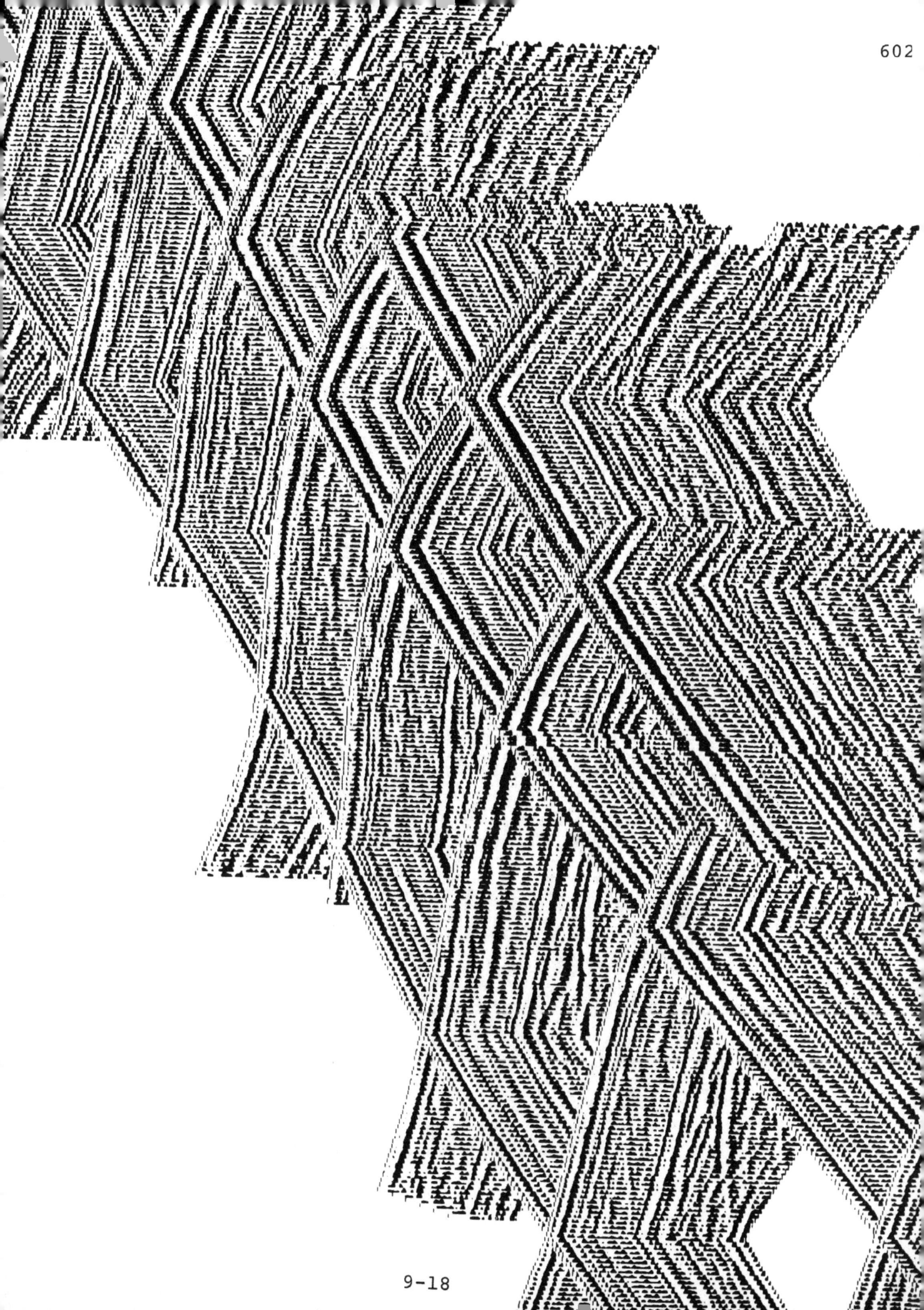

In this case the black-and-white display gives us the
structural picture reasonably clearly, but quite misses
the subtleties of reflection strength. These subtleties,
as discussed in section 4.6, are the key to avoiding the
tight zones in the producing sand.

The illustration on p.9-23 shows the ultimate validation
of a machine contour map — its superposition on the original
data from which it was derived.

Our next discussion is about cost-effectiveness in
exploration. Perhaps we could anticipate that discussion
on this one single point — that of conveying the essence
of an interpretation to management. This is a critically
important part of the exploration sequence, and, whatever
solution we adopt, its cost could scarcely amount to 1%
of the total geophysical cost. Everything tells us to
pay more attention to this matter than we have generally
done in the past.

And surely we cannot agree with those few geophysicists
who say that they do not want to convey anything other than
a drilling recommendation to management, who are scared stiff
that management will start poking its nose into the details
of the geophysics. A comprehensible integrated display
will first assist the geophysicist to be sure of his own
conclusion, then permit him to make a clearer exposition
of the prospect (including its risks) to management, and
finally exact more comprehending and more committed support
from management. We are all hesitant to back what we
understand only imperfectly.

These comments seem particularly apt in the case of
the rush job — where the drilling department needs a
location by (as always) Monday morning. If we know the
picking and contouring is going to be troublesome, we
would be better advised to base the location on one of
these diagrams (which require no final picking judgements),
than on contouring which we know we shall have to revise
after the rush is over.

9-22

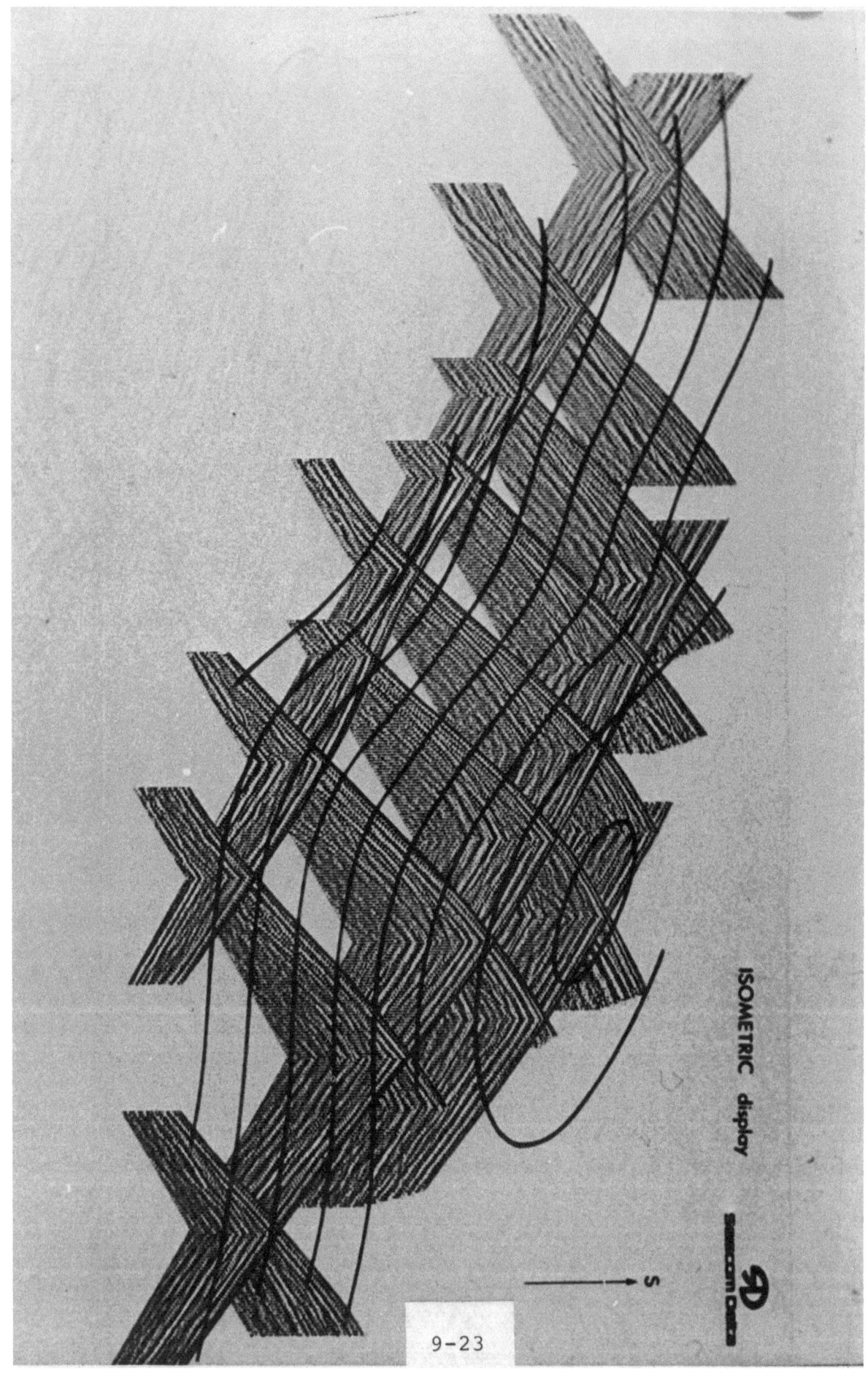

TAPE 39   9.2   COST-EFFECTIVE EXPLORATION

9.2.1   Cost-effectiveness in the total exploration scheme

First, let us consider a few observations about exploration in the broad sense — all the way from the first aeromagnetic survey to the siting of the first development well. Thus we are concerned with a chain including an aeromagnetic and/or gravity survey, a reconnaissance seismic survey, a detail seismic survey, a wildcat, a set of borehole logs and cores, seismic delineation studies, delineation wells, more borehole logs, and finally the decision to develop.

Then by cost-effective exploration in the broad sense we mean the minimization of exploration costs which do not result in development, and an optimum balance between cost and certitude along the those exploration chains which do result in development.

It is self-evident to the point of being trivial to say that the optimization of any chain as complex as the exploration chain must depend on the relative costs and effectiveness of the links in the chain. Yet it often happens that, largely due to the rigid departmentalized structure of some oil companies, there is in practice surprisingly little flexibility in this chain, or interdepartmental attempt to optimize it for the circumstances.

Perhaps we geophysicists can help the situation by being perfectly honest about our inability to solve certain problems economically, while making vocal nuisances of ourselves

when we see problems that we can solve being tackled by
methods costing a hundred times as much.

   For example, we will cheerfully allow that circumstances
exist where the production zones are relatively shallow,
where oilwell drilling is rapid and predictable, where rough
topography makes seismic access difficult, where static
corrections are extreme, where reflection coefficients are
small, and where seismic results are plagued with scattered
energy and all forms of noise. Then undoubtedly the majority
of the exploration effort must be in the drilling of holes,
and the humble geophysicist will site his seismic work with
great care to get the maximum benefit from what he accepts
must be relatively few profiles.

   On the other hand, we will shout from the housetops
that an error is about to be made if a 5-million dollar
wildcat is proposed, in the offshore, on the basis of a
traditional grid of traditional seismic work. We will
remonstrate that, once the feature to be explored is determined
the full might of the most advanced seismic techniques could
be directed against that feature for less than the cost of
one day of the drilling rig. And that then the well might
be drilled, or it might not, or it might be drilled a little
to the side. We will insist with the full earnestness of
conviction that it is crazy to persevere with an old-style
balance between geophysics and the drill, in the presence
of such improvements in the seismic method and such increases
in the cost of the drill.

9.2.2 Cost-effectiveness in seismic exploration specifically

But let us turn from the grand picture, and address specifically the problem of getting the most for our seismic expenditure.

1. The first principle in planning a seismic survey is to use all available prior knowledge. First, of course, we evaluate all the previous geology and geophysics to make sure that we position only a necessary minimum of line on basement. Second, we use all previous seismic indications to help us position the lines (and in particular the line intersections) for good seismic quality. Third, we lay out a pattern of lines which, as far as we can judge, cross all the major faults at right angles and traverse all dips on the line of steepest gradient. The correct positioning of the lines is certainly the most important single step in securing the maximum geological certitude from a given extent of profile. For example, the correct network of lines for a roughly circular structure is emphatically not a rectangular grid, but a system of radial lines linked by suitable strike lines down the flanks.

2. Now to the second principle for cost-effective seismic surveys — design the surveys to tackle the problem. The geophysicist should not just accept off-the-peg technology because it is described as state-of-the-art. He should think about it, defining the problem in his own mind and then adjusting the survey specifications to optimize the chances of solving that problem. Too often, the survey specifications and the quality-control requirements sent to the contractor are just a xerox copy of what was done last time — somewhere else, for a different problem.

So, in the very briefest outline, let us identify some of the considerations which should relate the survey specification to the survey problem.

The spread

The following notes are based on end-on spreads, and are subject to obvious modifications for split or cross spreads. It is important to note that the variables are decided <u>in the given order</u>; for example, in practice we often feel obliged to start with a defined number of channels — but that is not really correct, and it can waste money.

1. The distance from the source to the far array is defined by the depth of the target (being about equal to that depth). This often approximates to the other desirable situation — that full fold of stack becomes effective at target time.

2. The offset distance (source to near array) is defined by the depth of the shallowest reflector of interest, again being approximately equal. For marine work, this depth is that of the sea floor; adherence to this tenet is very important for bright-spot calibration.

3. The number of groups (or channels) is defined by:

- the horizontal resolution necessary to establish important features of the geology;

- the reflected bandwidth;

- the wavefront curvature, for interpretation without migratio this is a function of the migration velocity, which is of course time-variant;

- the alias problem (section 7.2) in migration.

There is no point (just a great waste of money) in providing more channels than these considerations require.

4. The maximum geophone-array length, for a linear array, is defined by the step-out time t between the two farthest groups and by the highest frequency of interest; for 4 ms recording (62 Hz cut-off) a suitable value is $16x/t$ and for 2 ms recording $10x/t$, where x is the group interval. As a simple general rule, this dimension yields a good blend of noise attenuation, surface-wave attenuation and plant sampling.

5. The maximum useful number of geophones per group is usually defined by the correlation distance of the noise; it is the array length divided by this distance, plus one.

6. The streamer depth is defined by the relative importance of ambient noise and reflection bandwidth; in general, shallow targets requiring good vertical resolution imply shallow streamers, and deep targets requiring great penetration imply deep streamers.

7. The maximum feathering angle is defined as a function of the maximum expected cross-dip, the length of the spread, and the highest frequency of interest; for many practical cases the product of feathering angle and cross-dip should not exceed 100.

The source

It is most important that the interpreter first define the bandwidth necessary to tackle the exploration problem. Then the source bandwidth should be matched to this — neither more nor less. Less bandwidth means that the problem will not be solved; more means that energy is wasted which could have been employed usefully (or whose cost could have been saved). Then the energy radiated within that bandwidth must be sufficient to overcome noise level, at the target depth, after summing and stacking.

If it is likely that the data are to be transformed into acoustic impedance logs ("pseudo-velocity logs" or "seislogs") the source bandwidth should extend to lower frequencies (and preferably high frequencies also) than are usual.

The instruments

The filter settings, of course, must preserve the desired bandwidth. It is most important, for the practitioner of the new seismic interpretation, that there be no instrumental clipping or overloading; this is particularly true for the sea-floor reflection (where it can happen, with any but the most modern instruments).

The processing

The rule must be to abstract the most from the seismic data, once recorded. The interpreter defining the processing must be prepared to look at many intermediate plots, to satisfy himself of the validity of the final data. He should be watchfu that the money is spent where the problem is, and not wasted elsewhere.

Because digital anti-alias filters are more effective than the anti alias filters in the field, he can save considerable money by making a very careful match between the processing bandwidth and the recording bandwidth. The clearest illustration of this is in Vibroseis (how many hundreds of thousands of dollars have been wasted by processing 10-40 Hz sweeps, after correlation, at 4 ms sample interval?); however, the point is valid in all processing.

Often he can save money also by checking very carefully whether there is benefit or harm (there can be either) in deconvolution before stack and in vertical summing.

It is most important that he should not allow costly and sophisticated processing to be applied to data on which the first stages of processing (particularly the amplitude processing) have been done cheaply or poorly.

In local areas of great interest, when the target has been defined, the interpreter must be prepared to iterate several times round the processing-interpretation loop.

Finally, it is most important that seismic and borehole data should be related whenever possible.

So we put the money where the problem is, and are on our guard against the fiction that more money ensures a more appropriate result.

81
82
83(B)
 L(B)

9.3 QUESTIONS THAT MANAGEMENT SHOULD PROPERLY ASK

9.3.1 General questions

1. Have you geophysicists fully adjusted to the requirement that you should be hydrocarbon-finders — not just structure-finders?

2. In each area independently (according to the different merits of seismic work and the drill), is our balance of expenditure optimum?

9.3.2 Questions before a seismic survey is approved

1. Have you carefully defined the problem (the hydrocarbon problem, not necessarily the seismic problem) with the geologists and/or the reservoir engineers, before you decided how the survey would be done?

2. Is there a proper integration of the specifications for the field work, the processing and the interpretation? Or are these function being handled by different people without enough interrelation?

3. Are we setting survey specifications either less than we want, or more costly than would otherwise be, in order that we shall be able to trade the data? Have you thought out whether this is the correct course?

9.3.3 Questions before a well is drilled

1. Before we drill, have you exhausted what can be done economically with the geophysics? Are you satisfied that the drilling recommendation is based on sound data? Have you made an honest effort to explain all anomalous results near the feature, and to interpret their hydrocarbon significance?

2. Have you done everything that can be done seismically to identify fluid contacts?

3. Have you talked over the contouring and other uncertainties with the geologists?

4. What is the tolerance on the depth prognosis?

5. Have you notified the drilling department of any drilling hazards evident on the seismic data?

6. Are there any doubts on the positioning?

7. Can you show me what the feature looks like, in a way which will be meaningful to me?

9.3.4 Questions after a discovery

1. What is the tolerance on the volumetric estimate of reserves?

2. If we put the rig at your disposal for a day, your work and the loss of rig time may cost us $60,000. Are the chances better than 1 in 50 that it will save us a $3 million delineation well?

3. You fellows seems to be much more important to us nowadays — care for a salary increase?

REFERENCES

PART 2

Averbuch and Trapeznikova, Izvestiya, Sept. 1972, 616.

Balachandran, GE 39-1-73.

Baldwin, "Ways of Deciphering Compacted Sediments", J.Sediment. Petrol., 41-1-293.

Biot, JASA 28, 168.

Blum, CSEG April 1974.

Bortfeld, R., GP 10-4-517.

Brandt, Jour. Appl. Mech. (ASME) 22, 479.

Burcik, E., "Properties of Petroleum Reservoir Fluids," IHRDC.

Chapman, "Petroleum Geology", publ. Elsevier, 1973.

Crowe and Alhilali, "Amplitudes of Seismic Events and their Dependence on the Absorption-Dispersion Pairs of the Media" SEG Dallas, 1974. Also AAPG Dallas, 1975.

Dapples, Bull. AAPG 56.1.3.

Davis, Journal CSEG, December 1972, 1.

Dix, C.H., GE 19-4-722.

Domenico, N., "Effect of Water Saturation on Seismic Reflectivity of Sand Reservoirs Encased in Shale", SEG 1973, GE 39-6.

Duska, GE 28-6-925 (and references).

Faust, GE 16-2-192, 18-2-271.

Garner, French and Matzuk, SEG 1973.

Gardner, Gardner and Gregory, "Formation Velocity and Density: the Diagnostic Basics of Stratigraphic Traps". SEG 1968, GE 39-6.

Geerstma, Trans AIME 210, 331.

Hagedoorn, J.G., GP 2-2.

Herring, "Estimating Abnormal Pressures from Log Data in the North Sea". 2nd annual European meeting of SIPM of AIME, London, April 1973.

Hileman, GP 16-3-326.

Hilterman, GE 40-3.

Hornabrook, 1974; see part 7 references.

Kaufman, GE 18-2-289.

Kelly et al., GE 41-1-2.

Kinsler and Frey, "Fundamentals of Acoustics", publ. Wiley, 2nd ed.

Korvin, "Wave propagation in a medium with random inhomogeneities", Geophys. Trans. 21-1-4, Eötvös Institute, Hungary.

Kuster and Toksoz, GE 39-5-587.

Matthews, "Well Logs: A Drilling Tool", publ. Oil and Gas Journal, 1972.

Miles, GE 25-3-642.

Newman, GE 38-3-481.

Nikolayev and Averyanov, Izvestiya, May 1973, 330.

O'Brien, P., GP 19-1-1.

O'Doherty and Anstey, GP 19-3-430.

Pickett, GE 25-1-250.

Pirson, "Geologic Well Log Analysis", publ. Gulf Publ. Co. 1970.

Poley, P. GP 12-4-397.

Rieke and Chilingarian, "Compaction of Argillaceous Sediments", publ. Elsevier, 1974.

Sakhumuradova, Izvestiya, July 1973, 469.

Sarmiento, Bull. AAPG 45-5-633.

Schoenberger and Levin, GE 39-3-278.

Shah, GE 38-4-643.

Smirnova, Izvestiya, June 1973, 403.

Spencer, GE 25-3-625.

Strick, GE 35-3-387.

Taner, Koehler and Alhilali, GE 39-4-441.

Tegland, E., 1974, see part 7 refs.

Trorey, GE 25-5-762.

White, GE 25-3-613.

Widess, GE 38-6-1176.

PART 3

Al-Chalabi, GP 21-4-783.

Balch, GE 36-6-1074 (and references).

Beitzel and Davis, "A Computer-Oriented Velocity Analysis Interpretation Technique". SEG 1973.

Carroll, "Signature Deconvolution", CSEG Bull, 1973.

Cochran, GE 38-6-1042.

Dedman, Lindsey and Schram, "Stratigraphic Modeling", Geoquest 1975.

Dunkin and Levin, GE 38-4-635.

Gupta, GE 30-1-122 (and refs).

Hales and Edwards, GP 3-1-65.

Kennett and Ireson, "Well Geophone Signals as an Aid to Hydrocarbon Indication", SEG 1973.

Larner, Mateker and Wu, GSH Symposium Sept. 1973.

Michon and Muniz, 1975, "Broadening of the Spectrum", CGG publication.

Michon and Tariel, "Wide-line Profiling", EAEG 1972.

Morgan, GE 35-3-447.

Mossman and Schoellhorn, "Causes of Reflection Amplitude Variances", Dallas Geophysical Society Symposium, Dec. 1973.

Newman, GE 38-3-481.

O'Brien, GP 19-1-1.

Prescott and Scanlan, "The Effects of Weathering on Derived Velocities", SEG 1971.

Schoenberger, GE 39.

Shah, GE 38-4-643.

Sheriff, GP 23-1-125.

Sheriff and Taner, 1977, "Seismic Stratigraphy", AAPG publication.

Taner, GE 39-4-441.

Walton, GE 36-6.

White and O'Brien, GP 22-4-627.

PART 4

All papers presented in the Symposium of the Houston Geophysical Society "Lithology and Direct Hydrocarbon Detection by Geophysical Methods", Oct. 1973.

Gierasimow, "Effect of oil and Natural Gas reservoirs on the Character of Seismic Records", Nafta (Katorice) 28-4-154, 1972.

Paturet, GP 19-1-27.

PART 5

Albright, Offshore, May 1973.

Chombart, GE 35-4-779.

Dobrin and Sheriff, AAPG Continuing Education Course on "Principles of Seismic Stratigraphy", Dallas 1975.

Harper and Shaw, "Cretaceous-Tertiary Carbonate Reservoirs in the North Sea", ONS Stavanger, 1974.

Jankowsky, GP 19-1-103.

Norris, "Geologic Features of Continental Margins", SEG Dallas, 1975.

Sangree and Widmier, "Interpretation of Depositional Facies from Seismic Data", Houston Geophysical Society Symposium, Dec. 1974.

Sieck, O&GJ, 16 July 1973.

Stuart, "Sedimentologic and Stratigraphic Interpretation of Seismic Data in the Gulk of Mexico", AAPG Dallas, 1975.

Todd and Mitchum, "Seismic Stratigraphic Indentification of Eustatic Cycles", AAPG Dallas, 1975.

Vail, Michum, Sangree and Thompson, "Stratigraphic Framework and Eustatic Cycles from Seismic Stratigraphic Analysis", AAPG, Dallas 1975.

Zaaza and Visher, "Worldwide Distribution of Giant Oil and Gas Fields as Controlled by Stratigraphic Sequences", AAPG Dallas, 1975.

PART 6

Alford, Kelly and Boore, 1974, GE 39-5-834.

Barry and Shugart, "Zero-Phase Seismic Sections", SEG 1975.

Dedman and Lindsey, "Structural Modeling", Geoquest International Ltd., 1975.

Dedman, Lindsey and Schram, "Stratigraphic Modeling", Geoquest International Ltd., 1975.

Hodgson and Neidell, "Modeling Methods in Direct Hydrocarbon Detection", Geophysical Society of Houston Symposium, Sept., 1973.

Larner, 1974, "An Overview of Continuous Velocity Analysis", Geophysical Society of Houston Symposium, Dec. 1974.

Levin, 1971, GE 36-3-510.

Levin, GE 38-4-771.

Lucas, Al-Chalabi and Shaw, "The Calculation of Laterally Varying Time Delays from Stacking Velocity Anomalies", SEG 1975.

May and Hron, "Synthetic Seismic Sections of Typical Petroleum Traps", SEG 1975.

Patch, 1973, "Modeling for Lithology", Geophysical Society of Houston Symposium, Sept. 1973.

Pollet, "Simple Velocity Modeling and the continuous Velocity Section", Geophysical Society of Houston Symposium, Dec. 1974.

Shah, GE 38-3-600.

Taner, Cook and Neidell, 1970, GE 35-4-551.

Wilson and Embree, "Interactive Modeling of Direct Hydrocarbon Indicators", SEG 1975.

PART 7

Claerbout, J.F. and Doherty, S.M., GE 37-5-741.

French, W., GE 39-3-265.

Gardner, French and Matzuk, "Elements of Migration Velocity Analysis", SEG 1973.

Haas, A., and Vialliz, J-R., 1974, "Krigeage Applied to Geophysics: The Answer to the Problem of Estimates and Contouring", EAEG Madrid, 1974.

Hagedoorn, J.G., 1954, GP 2-2.

Hornabrook, J., "Seismic Re-Interpretation Clarifies North Sea Structure", Petroleum International, March and April 1974.

John, P.B., 1975, The Radio and Electronic Engineer, May 1975.

Larner, K., 1976, "Wave Equation Migration: Two Approaches", SEG 1976 (also Western Geophysical Company).

Loewenthal, Lu, Roberson and Sherwood, "The Wave Equation Applied to Migration and Water-Bottom Multiples", SEG Dallas 1974.

Michon, D., "Wide-line Profiling", EAEG 1972.

Michon, D. and Tariel, P., 1975, "Amplitude Preservation during Migration", EAEG Bergen, 1975.

Newman, P., 1975, "Amplitude and Phase Properties of a Digital Migration Process", EAEG Bergen, 1975.

Peterson, R., "Through the Kaleidoscope — a Dooddlebuger in Wonderland", UGC, 1974.

Sattlegger, Dohr and Stiller, GE 40-1-1; also EAEG 1975.

Tegland, E., "Sand-Shale Ratio Determination from Seismic Interval Velocity", SEG-AAPG Regional Meeting Dallas, 1970.

PART 8

Albright, J., Offshore, May 1973.

Bois, P., GP 19-1-43.

Booker et al., 1976, GE 41-5-939.

Cassano, R., GE 38-6-1053.

Galperin, 1974, "Vertical Seismic Profiling", translation published by SEG.

Herkenhoff and Ostrander, "Effect of Elliptical Velocity Anistrop y on Surface Velocity Measurements", SEG 1973.

Herring, see Part 2 refs.

Thigpen, Dalby and Landrum, 1975, GE 40-0-694.

Paturet, GP 19-1-27.

Pennebaker, "The Use of Geophysics in Abnormal Pressure Applications", SEG 1969.

Reynolds, O and GJ, 11 March 1974, p.113.

Robinson, GE 35-3-436.

Rudman, 1975, AAPG Bull July 1975; also AAPG Dallas April 1975.

O'Doherty and Anstey, 1971, GP 19-1-430.

Taner, Koehler and Alhilali, GE 39-4-441.

Wiggins et al., 1976, GE 41-5-922.

Woeber and Penhollow, GE 40-3-388.

PART 9

Delfiner, P., 1976, "Geostatistical estimation of hydrocarbon", Ecole des Mines de Paris, 1976.

Duval, Fourmann, Haas and Viallix, "Determination of Coherent Three-Dimensional Velocity Distributions", CGG, 1972.

Haas and Viallix, "Krigeage Applied to Geophysics" EAEG 1974.

Meek, GE 38-2-295.

Walters, Bull. AAPG 53-11-2324.

Wren, E., Journal CSEG, 1973, P.39.

Wren, E., GP-23-1-1.

MIX
Papier aus verantwortungsvollen Quellen
Paper from responsible sources
FSC® C105338

If you have any concerns about our products,
you can contact us on
ProductSafety@springernature.com

In case Publisher is established outside the EU,
the EU authorized representative is:
**Springer Nature Customer Service Center GmbH
Europaplatz 3, 69115 Heidelberg, Germany**

Printed by Libri Plureos GmbH
in Hamburg, Germany